D0084602

THE INTERNATIONAL SERIES OF MONOGRAPHS ON PHYSICS

INTERNATIONAL SERIES OF MONOGRAPHS ON PHYSICS

110. N.B. Kopnin: *Theory of nonequilibrium superconductivity*
109. A. Aharoni: *Introduction to the theory of ferromagnetism 2/e*
108. R. Dobbs: *Helium three*
107. R. Wigmans: *Calorimetry*
106. J. Kubler: *Theory of itinerant electron magnetism*
105. Y. Kuramoto, Y. Kitaoka: *Dynamics of heavy electrons*
104. D. Bardin, G. Passarino: *The Standard Model in the making*
103. G.C. Branco, L. Lavoura, J.P. Silva: *CP Violation*
102. T.C. Choy: *Effective medium theory*
101. H. Araki: *Mathematical theory of quantum fields*
100. L. M. Pismen: *Vortices in nonlinear fields*
99. L. Mestel: *Stellar magnetism*
98. K. H. Bennemann: *Nonlinear optics in metals*
97. D. Salzmann: *Atomic physics in hot plasmas*
96. M. Brambilla: *Kinetic theory of plasma waves*
95. M. Wakatani: *Stellarator and heliotron devices*
94. S. Chikazumi: *Physics of ferromagnetism*
92. J. Zinn-Justin: *Quantum field theory and critical phenomena*
91. R. A. Bertlmann: *Anomalies in quantum field theory*
90. P. K. Gosh: *Ion traps*
89. E. Simánek: *Inhomogeneous superconductors*
88. S. L. Adler: *Quaternionic quantum mechanics and quantum fields*
87. P. S. Joshi: *Global aspects in gravitation and cosmology*
86. E. R. Pike, S. Sarkar: *The quantum theory of radiation*
84. V. Z. Kresin, H. Morawitz, S. A. Wolf: *Mechanics of conventional and high T_c super-conductivity*
83. P. G. de Gennes, J. Prost: *The physics of liquid crystals*
82. B. H. Bransden, M. R. C. McDowell: *Charge exchange and the theory of ion-atom collision*
81. J. Jensen, A. R. Mackintosh: *Rare earth magnetism*
80. R. Gastmans, T. T. Wu: *The ubiquitous photon*
79. P. Luchini, H. Motz: *Undulators and free-electron lasers*
78. P. Weinberger: *Electron scattering theory*
76. H. Aoki, H. Kamimura: *The physics of interacting electrons in disordered systems*
75. J. D. Lawson: *The physics of charged particle beams*
73. M. Doi, S. F. Edwards: *The theory of polymer dynamics*
71. E. L. Wolf: *Principles of electron tunneling spectroscopy*
70. H. K. Henisch: *Semiconductor contacts*
69. S. Chandrasekhar: *The mathematical theory of black holes*
68. G. R. Satchler: *Direct nuclear reactions*
51. C. Moller: *The theory of relativity*
46. H. E. Stanley: *Introduction to phase transitions and critical phenomena*
32. A. Abragam: *Principles of nuclear magnetism*
27. P. A. M. Dirac: *Principles of quantum mechanics*
23. R. E. Peierls: *Quantum theory of solids*

Theory of Nonequilibrium Superconductivity

NIKOLAI B. KOPNIN

Low Temperature Laboratory,
Helsinki University of Technology, Finland
and
L.D. Landau Institute for Theoretical Physics,
Moscow, Russia

CLARENDON PRESS • OXFORD

2001

OXFORD
UNIVERSITY PRESS

Great Clarendon Street, Oxford OX2 6DP

Oxford University Press is a department of the University of Oxford.
It furthers the University's objective of excellence in research, scholarship,
and education by publishing worldwide in

Oxford New York

Athens Auckland Bangkok Bogotá Buenos Aires Cape Town
Chennai Dar es Salaam Delhi Florence Hong Kong Istanbul Karachi
Kolkata Kuala Lumpur Madrid Melbourne Mexico City Mumbai Nairobi
Paris São Paulo Shanghai Singapore Taipei Tokyo Toronto Warsaw
with associated companies in Berlin Ibadan

Oxford is a registered trade mark of Oxford University Press
in the UK and in certain other countries

Published in the United States
by Oxford University Press Inc., New York

First published 2001

A catalogue record for this title is available from the British Library.

Library of Congress Cataloging in Publication Data

Kopnin, N. B.
Theory of nonequilibrium superconductivity / Nikolai B. Kopnin.
(International series of monographs on physics; 110)
Includes bibliographical references and index.
1. Superconductivity. 2. Superconductors–Effect of radiation on. I. Title II.
International series of monographs on physics (Oxford, England); 110.
QC611.92.K67 2001 537.6′23–dc21 2001018508

ISBN 0 19 850788 7 (acid-free paper)

Typeset by the author in LaTeX
Printed in Great Britain
on acid-free paper by
T. J. International Ltd,
Padstow, Cornwall

PREFACE

After more than four decades of existence, the microscopic theory of superconductivity due to Bardeen, Cooper, and Schrieffer (Bardeen *et al.* 1957) (BCS) has established itself as one of the most beautiful theories in condensed matter physics. Based on quite simple principles, it gives surprisingly good description of many properties of superconductors. Before the discovery of high temperature superconductors, the BCS theory with a phonon mediated electron coupling could be regarded as an almost perfect piece of art. The high temperature superconductivity appeared to be a challenge for the microscopic theory. Apparently, its complicated nature goes, strictly speaking, beyond the BCS model, and many attempts have been undertaken to build a microscopic picture of this mysterious phenomenon. However, no reliable and commonly accepted theory has emerged. Instead, the last decade of intensive studies in the high temperature superconductivity revealed yet another important advantage of the classical BCS theory: Sometimes it works reasonably well even when it is not expected do so! It still remains the best available theory to describe the new superconductors. It also appears that the BCS model originally designed to describe the *s*-wave pairing in conventional (low temperature) superconductors, can easily be generalized to deal with unconventional superfluids and superconductors. The first successful application was to the *p*-wave superfluidity in ^{3}He Fermi liquid. Here it provides the theory of a very exciting state with much more complicated and rich structure of the superfluid order parameter (Leggett 1975, Wölfle and Vollhardt 1990, Volovik 1992). Heavy-Fermionic and high-temperature *d*-wave superconductors are examples where the BCS model can also be used with a great chance for a success (Mineev and Samokhin 1999). This is why the interest in the BCS theory remains alive despite its quite respectable age.

The BCS model grew into a highly powerful theory of superconductivity also because of its formulation in terms of the Green functions by Gor'kov (1958). The Green function technique constitutes a complete tool for solving almost any problem within the BCS theory. A very important and useful improvement in the Green function theory of superconductivity has been provided by the so-called quasiclassical method initiated by Eilenberger (1968). The method operates with the Green functions integrated over the energy near the Fermi surface of the normal state. This method is based on the fact that the energies involved in the superconducting phenomena are normally much smaller than the Fermi energy which is the scale of variations of the Green function in the normal state. In the quasiclassical scheme, the calculations are reduced to a more or less automatic action provided the problem is adequate for the model. It is the quasiclassical version of the Green function formalism which is now most frequently used for calculations of various properties of superconductors.

Despite the quasiclassical methods being widely used for practical purposes, their description is hard to find in the textbook literature. The review by Serene and Rainer (1983) gives an introduction to the quasiclassical approach. The book by Svidzinskii (1982) deals with the quasiclassical theory of static properties of weak links. Applications of quasiclassical methods to various problems in nonstationary superconductivity can only be found in original papers and specialized reviews such as, for example, a review by Larkin and Ovchinnikov (1986). The aim of the present book is to provide a basic knowledge of the microscopic theory of nonstationary superconductivity including the most advanced quasiclassical methods. We assume that the reader is familiar with the main ideas of the BCS theory of superconductivity. There are many books on the principles of the BCS theory, and we do not intend to give a comprehensive list of them here. The personal preference of the author is with the book by de Gennes (1966) which contains all the fundamentals which we would need to proceed with the more recent microscopic theory.

We try to describe the quasiclassical method in such a way that a newcomer to the field could learn it in a short time. First, we discuss stationary, time-independent properties. We shall learn about the Green function, how to find it for a particular superconducting state, and how to calculate the superconducting properties once the Green function is known. The nonstationary theory of superconductivity requires more efforts, which necessarily use principles developed for stationary problems. In the theory of nonstationary phenomena, the basic concept is the distribution function of excitations. The specifics of superconductors is that the interaction of a superconductor with an external electromagnetic field normally causes a considerable distortion of the quasiparticle distribution on the energy scale relevant for the superconducting parameters. As a result, the behavior of nonequilibrium excitations becomes one of the most important factors which govern the response of the superconductor to applied fields. This is why the problem of formulating the correct description of the quasiparticle distribution is of the major concern throughout the book.

We consider several applications of both stationary and nonstationary quasiclassical theory. For example, we derive and analyze the time-dependent Ginzburg–Landau theory which is most frequently used to describe the dynamics of superconductors. We demonstrate how powerful this theory is within certain limits and, at the same time, we emphasize that it is far from being a complete story about the kinetics of superconductors.

The great deal of attention is paid to the dynamics of vortices in the mixed state of type II superconductors. There are two good reasons for the choice. First, it is well established now that the dynamics of vortices controls almost all the magnetic properties of type II superconductors, especially those of high-temperature superconductors. It also determines hydrodynamics of superfluids. Moreover, dynamics of superfluid vortices is now believed to have a close relation to other fields of physics such as high energy physics and cosmology (Achucarro and Vachaspati 2000, Shellard and Vilenkin 1994). Second, the vortex dynamics has been the major interest of the author's research during many years, and it is hard to resist the temptation to say something about this fascinating subject. The reader is assumed to be familiar with the basic properties of vortices at least within the framework of the usual Ginzburg–Landau theory. Here we concentrate on the motion of vortices under the action of a current passing through a superconductor in the so-called flux flow regime. We shall see that, in presence of vortices, the superconductor is no longer "superconducting" in a practical sense: it offers a resistivity to the current! This is why the vortex dynamics plays an important role in the physics of superconductors.

Of course, there are very many interesting and important phenomena left beyond the scope of this book. For example, we do not consider Josephson junctions and weak links; some aspects of this problem can be found in review by Aslamazov and Volkov (1986), and in books by Likharev (1986) and Tinkham (1996). We do not discuss propagation of sound through superconductors (see, for example, Bulyzhenkov and Ivlev 1976 a, 1976 b). Neither we consider thermoelectric phenomena, etc. Each of these topics deserves a book of its own. In this context, we mention the book by Geilikman and Kresin (1974) which treats acoustic, thermoelectric, and some other effects with the standard Boltzmann kinetic equation. Nevertheless, the quasiclassical scheme is not included there. We hope, therefore, that our presentation provides a basis for description of these and many other properties of superconductors in a coherent way using the simple and more advanced quasiclassical method.

The choice of the contents and of the presentation throughout the book is to a larger extent affected by the research in the theory of superconductivity which has been conducted at the Landau Institute for Theoretical Physics in Moscow. I had the privilege to work together with many brilliant scientists during the time when the Landau Institute was blooming with great scientific discoveries in a unique unforgettable creative atmosphere. I have benefited a lot from the Landau Institute seminars and discussions with A. Abrikosov, S. Brazovskii, I. Dzialoshinskii, G. Eliashberg, M. Feigel'man, L. Gor'kov, S. Iordanskii, B. Ivlev, I. Kats, I. Khalatnikov, D. Khmelnitskii, A. Larkin, V. Mineev, Yu. Ovchinnikov, V. Pokrovskii, G. Volovik and many, many more. It is hard to over-estimate their influence and the effect of great ideas that are generously shared by them with everybody who is interested in physics. My special thanks are to D. Khmelnitskii who read the manuscript and made valuable remarks helping to improve the presentation.

The book is partially based on the lecture courses given at the Université Paris-Sud, Orsay, France, and at the Low Temperature Laboratory, Helsinki University of Technology, Finland. It is intended for graduate and post-graduate students, and for researchers who work in the condensed matter theory.

Moscow and Espoo
1997 – 2000

N.B.K.

CONTENTS

III NONEQUILIBRIUM SUPERCONDUCTIVITY

Part I

Green functions in the BCS theory

1
INTRODUCTION

We give a brief outline of general ideas of the theory of superconductivity and introduce the basic quantities that characterize the superconducting state. We discuss the Ginzburg–Landau theory which provides the simplest description of stationary superconductors and consider its application to the vortex state. The microscopic Bogoliubov–de Gennes theory is introduced together with the concept of quasiclassical approximation. We also formulate the typical problems of non-stationary theory and consider its simplest methods such as the kinetic equation approach and the time-dependent Ginzburg–Landau model.

1.1 Superconducting variables

The most important characteristic of a superconductor is the superconducting *order parameter* Δ. The order parameter is proportional to the wave function of "superconducting electrons" which form a "Bose condensate". It is normalized in such a way that its modulus is equal to the energy gap in the electronic spectrum which usually appears after a transition into the superconducting state. The order parameter is a complex function $\Delta = |\Delta|e^{i\chi}$ where the phase χ is the same for all condensate particles if there is no current. In presence of a supercurrent, the phase χ acquires a spatial dependence slowly varying from one point in the superconductor to another. The existence of a coherent phase of the wave function for all superconducting particles is the very essence of superconductivity.

Another important characteristic is the energy spectrum of single-particle excitations in a superconducting system. For a homogeneous clean superconductor in absence of currents and magnetic field the energy spectrum has a gap

$$\epsilon_{\mathbf{p}} = \sqrt{\xi_{\mathbf{p}}^2 + |\Delta|^2} \tag{1.1}$$

such that all excitations have energies above $|\Delta|$. In eqn (1.1),

$$\xi_{\mathbf{p}} = E_n(\mathbf{p}) - E_F \tag{1.2}$$

while $E_n(\mathbf{p})$ is the electronic spectrum in the normal state, and E_F is the Fermi energy. The energy is counted from E_F because, as in any Fermi liquid, relevant excitations are concentrated near the Fermi level. The normal spectrum $E_n(\mathbf{p})$ of the metal can have a complicated dependence on the momentum $\mathbf{p}$ reflecting the band structure of the metal. In many cases, if we are not interested in the particular band-structure effects, it is sufficient to consider a simple parabolic spectrum $E_n = \mathbf{p}^2/2m$.

The order parameter determines both thermodynamic and transport properties of a superconductor. To learn more of the order parameter as well as of other important superconducting quantities we start with the Ginzburg–Landau theory for a time-independent superconducting state.

1.1.1 *Ginzburg–Landau theory*

The Ginzburg–Landau (Ginzburg and Landau 1950) theory is a generalization of the Landau theory of second-order phase transitions (Landau and Lifshitz 1959 b). Consider the free energy of a superconductor. Assume that we can expand it in terms of Δ and its gradients:

$$\mathcal{F}_{sn} = \int \left[\alpha|\Delta|^2 + \frac{\beta}{2}|\Delta|^4 + \gamma|(-i\hbar\nabla - \frac{2e}{c}\mathbf{A})\Delta|^2 \right] dV. \tag{1.3}$$

The free energy expression is supplemented with the Maxwell equation for the magnetic field

$$\operatorname{curl}\mathbf{H} = \frac{4\pi}{c}\mathbf{j} \tag{1.4}$$

where

$$\mathbf{H} = \operatorname{curl}\mathbf{A}. \tag{1.5}$$

The average of $\mathbf{H}$ gives the magnetic induction $\mathbf{B}$.

The gradient term in eqn (1.3) is the momentum operator in presence of the magnetic field

$$\mathbf{P}_s = -i\hbar\nabla - \frac{2e}{c}\mathbf{A}. \tag{1.6}$$

It refers to a Cooper pair having the charge $2e$ (e is the electronic charge). Equation (1.6) implies a gauge invariance: the free energy of the system and the magnetic field do not change if one makes a simultaneous transformation

$$\chi \to \chi + f(\mathbf{r}) \; , \; \mathbf{A} \to \mathbf{A} + \frac{\hbar c}{2e}\nabla f(\mathbf{r}) \tag{1.7}$$

where $f(\mathbf{r})$ is an arbitrary function of coordinates.

At the transition temperature $T = T_c$, the coefficient α changes its sign and becomes negative for $T < T_c$, while β and γ remain constant. Microscopic theory (Gor'kov 1959 a, 1959 c) gives

$$\alpha = -\nu(0)\frac{T_c - T}{T} \; , \; \beta = \frac{7\zeta(3)\nu(0)}{8\pi^2 T_c^2} \tag{1.8}$$

where $\nu(0)$ is the density of states at the Fermi level, and $\zeta(3) \approx 1.202$. We use the units with the Boltzmann constant $k_B = 1$. We will derive eqn (1.8) later in Section 6.1.2. Equation (1.8) demonstrates that the free energy expansion in eqn (1.3) goes in powers of Δ/T_c which suggests that Δ should be much smaller than T_c.

The coefficient γ depends on purity of the sample. The purity is characterized by the parameter $T_c\tau/\hbar$, where τ is the electronic mean free time due to the scattering by impurities. Superconductors are called clean when this parameter is large, and they are dirty in the opposite case. We discuss the microscopic justification of the Ginzburg–Landau theory later. Here we just write down the expression for γ in two limiting cases. For dirty superconductors, it is

$$\gamma = \frac{\pi\nu(0)D}{8\hbar T_c} \tag{1.9}$$

where D is the diffusion coefficient. In the clean case

$$\gamma = \frac{\nu(0)\,7\zeta(3)\,v_F^2}{48\pi^2 T_c^2}. \tag{1.10}$$

Here v_F is the velocity of electrons at the Fermi surface.

The total free energy is

$$\mathcal{F}_{\rm tot} = \mathcal{F}_{sn} + \int \frac{H^2}{8\pi} dV. \tag{1.11}$$

Variation of $\mathcal{F}_{\rm tot}$ with respect to Δ, Δ^* and $\mathbf{A}$ gives

$$\begin{aligned}\delta\mathcal{F}_{\rm tot} = \int \Bigg(& \left[\alpha\Delta + \beta|\Delta|^2\Delta + \gamma\left(-i\hbar\nabla - \frac{2e}{c}\mathbf{A}\right)^2\Delta\right]\delta\Delta^* + \text{c.c.} \\ & + \left[\frac{\text{curl curl}\mathbf{A}}{4\pi} - \frac{2e\gamma}{c}\left[\Delta^*\left(-i\hbar\nabla - \frac{2e}{c}\mathbf{A}\right)\Delta + \text{c.c.}\right]\right]\delta\mathbf{A} \\ & + \text{div}\left[\hbar\gamma\delta\Delta^*(\hbar\nabla - \frac{2ie}{c}\mathbf{A})\Delta + \text{c.c.} + \frac{1}{4\pi}\delta\mathbf{A}\times\text{curl}\mathbf{A}\right]\Bigg) dV.\end{aligned}$$

The requirement of extremum of the free energy leads to the GL equations. Vanishing of $\delta\mathcal{F}_{\rm tot}/\delta\Delta^*$ results in

$$\alpha\Delta + \beta|\Delta|^2\Delta + \gamma\left(-i\hbar\nabla - \frac{2e}{c}\mathbf{A}\right)^2\Delta = 0. \tag{1.12}$$

Condition $\delta\mathcal{F}_{\rm tot}/\delta\mathbf{A} = 0$ gives the expression for supercurrent

$$\mathbf{j}_s = 2e\gamma\left[\Delta^*\left(-i\hbar\nabla - \frac{2e}{c}\mathbf{A}\right)\Delta + \Delta\left(i\hbar\nabla - \frac{2e}{c}\mathbf{A}\right)\Delta^*\right]. \tag{1.13}$$

1.1.1.1 *Discussion of the GL equations* Consider first the equation (1.12) for the order parameter. In a homogeneous case without a current and a magnetic field the order parameter is

$$\Delta \equiv \Delta_\infty = \sqrt{|\alpha|/\beta} = \left(\frac{8\pi^2}{7\zeta(3)}\right)^{\frac{1}{2}} T_c\,(1 - T/T_c)^{\frac{1}{2}}. \tag{1.14}$$

We denote it by Δ_∞ for the reason which will be clear later. The ratio Δ_∞/T_c should be small. This implies that the GL theory works for temperatures close to T_c, i.e., when $1 - T/T_c \ll 1$. Simultaneously we see that $\Delta \sim T_c$ when temperatures are well below T_c.

Equation (1.12) defines the length

$$\xi(T) = \sqrt{\gamma\hbar^2/|\alpha|} \propto (1 - T/T_c)^{-\frac{1}{2}} \tag{1.15}$$

which is a characteristic scale of variations of the order parameter. It is called the *coherence length.* The notation ξ for the coherence length is to be distinguished from $\xi_{\mathbf{p}}$ that denotes the energy variable in eqn (1.2). We keep both these notations because they are commonly accepted in the literature.

In the clean case $T_c\tau/\hbar \gg 1$ we have from eqn (1.10)

$$\xi(T) = \sqrt{7\zeta(3)/12}\,\xi_0(1 - T/T_c)^{-\frac{1}{2}} \tag{1.16}$$

where

$$\xi_0 = \frac{\hbar v_F}{2\pi T_c} \tag{1.17}$$

is the "zero-temperature" coherence length. In the dirty case eqn (1.9) gives

$$\xi(T) = \left(\pi/2\sqrt{3}\right)\sqrt{\xi_0\ell}\,(1 - T/T_c)^{-\frac{1}{2}} \tag{1.18}$$

where $\ell = v_F\tau$ is the electron mean free path. The impurity parameter can be expressed through the ratio of ξ_0 and ℓ:

$$T_c\tau/\hbar = \ell/2\pi\xi_0 \tag{1.19}$$

so that a dirty limit corresponds to $\ell \ll \xi_0$ while a clean limit is for $\ell \gg \xi_0$.

Consider now expression (1.13) for the current. Using the definition of the momentum operator eqn (1.6) we introduce the superconducting velocity

$$2m\mathbf{v}_s = \mathbf{P}_s \tag{1.20}$$

for a Cooper pair with the mass $2m$. We use here the bare mass m of an electron. However, the bare electronic mass is not a good quantity for a metal where the normal-state electronic spectrum may have a complicated form. In this case also the "superconducting velocity" is not a real velocity of Cooper pairs. Bearing this in mind we write the supercurrent as

$$\mathbf{j} = -\frac{e^2 N_s}{mc}\left(\mathbf{A} - \frac{\hbar c}{2e}\nabla\chi\right) = N_s e\mathbf{v}_s \tag{1.21}$$

where

$$\mathbf{v}_s = \frac{\mathbf{P}_s}{2m} = \frac{\hbar}{2m}\left(\nabla\chi - \frac{2e}{\hbar c}\mathbf{A}\right) \tag{1.22}$$

and

$$N_s = 8m\gamma|\Delta|^2 \tag{1.23}$$

is the density of "superconducting electrons".

The Maxwell equation (1.4) combined with eqn (1.21) gives

$$\text{curl curl}\,\mathbf{A} = -\frac{4\pi N_s e^2}{mc^2}\left(\mathbf{A} - \frac{\hbar c}{2e}\nabla\chi\right). \tag{1.24}$$

Equation (1.24) defines the characteristic length

$$\lambda_L = \left(\frac{mc^2}{4\pi N_s e^2}\right)^{\frac{1}{2}}. \tag{1.25}$$

In a homogeneous case

$$\lambda_L = \left(\frac{c^2\beta}{32\pi e^2\gamma|\alpha|}\right)^{\frac{1}{2}} \propto (1 - T/T_c)^{-\frac{1}{2}}.$$

This is called the London penetration length. It determines the characteristic scale of variations of the magnetic field. The supercurrent can be written as

$$\mathbf{j} = \frac{c^2}{16\pi e\lambda_L^2}\left[\Delta^*\left(-i\hbar\nabla - \frac{2e}{c}\mathbf{A}\right)\Delta + \Delta\left(i\hbar\nabla - \frac{2e}{c}\mathbf{A}\right)\Delta^*\right]\Delta_\infty^{-2}. \tag{1.26}$$

Sometimes, it can be convenient to use the normalization of the order parameter such that it has the form of the wave function Ψ of superconducting electrons. The free energy becomes

$$\mathcal{F}_{sn} = \int\left[a\,|\Psi|^2 + \frac{b}{2}\,|\Psi|^4 + \frac{1}{2m}\left|\left(-i\hbar\nabla - \frac{2e}{c}\mathbf{A}\right)\Psi\right|^2\right]dV. \tag{1.27}$$

The constants a and b satisfy

$$\frac{|a|}{b} = \frac{mc^2}{16\pi e^2\lambda_L^2}$$

and determine the new order parameter magnitude $|\Psi_{GL}|^2 = |a|/b$ which is

$$|\Psi_{GL}|^2 = 2m\gamma\Delta_\infty^2$$

in terms of the previous definition of Δ. The coherence length is now expressed through the electronic mass

$$\xi^2 = \frac{\hbar^2}{2m|a|}.$$

1.1.1.2 *The Ginzburg–Landau parameter. Type I and type II superconductors* The ratio of the two characteristic lengths is called the Ginzburg–Landau parameter

$$\kappa = \frac{\lambda_L(T)}{\xi(T)} = \left(\frac{\beta c^2}{32\pi\hbar^2 e^2 \gamma^2}\right)^{\frac{1}{2}} . \tag{1.28}$$

It is independent of T near T_c and is determined by the material characteristics. For clean superconductors with γ from eqn (1.10) it is

$$\kappa = \left[\frac{9\pi^4}{14\zeta(3)}\right]^{\frac{1}{2}} \left[\left(\frac{a_0 p_F}{2\pi\hbar}\right)\left(\frac{e^2/a_0}{E_F}\right)\right]^{\frac{1}{2}} \frac{\hbar c}{e^2}\frac{T_c}{E_F}. \tag{1.29}$$

Here a_0 is the interatomic distance. Usually, it is of the order of $1/2\pi\hbar p_F$. The ratio of the Coulomb interaction energy of conducting electrons e^2/a_0 to the Fermi energy is of the order of unity for good metals, but it may become larger for systems with strong correlations between electrons. The last factor in eqn (1.29) is usually small: T_c/E_F is of the order of 10^{-3} for conventional superconductors, but it is of the order of $10^{-1} \div 10^{-2}$ for high temperature superconductors with $T_c \sim 100K$ and $E_F \sim 1000K$. The fine structure constant $e^2/\hbar c = 1/137$.

We see that the Ginzburg–Landau parameter is normally small $\kappa \ll 1$ for conventional clean superconductors, though, in some cases it may be of the order of 1. On the contrary, for high temperature superconductors, which have a tendency to be strongly correlated systems with a not very low ratio of T_c/E_F, the parameter κ is usually very large. The Ginzburg–Landau parameter increases for dirty superconductors. According to eqn (1.9) it becomes

$$\kappa_{dirty} \sim \kappa_{clean}(\hbar/T_c\tau). \tag{1.30}$$

Therefore, dirty alloys normally have a large κ.

The value of the Ginzburg–Landau parameter divides all superconductors into two classes: type I and type II superconductors. Those with $\kappa < 1/\sqrt{2}$ belong to the type I, while those with $\kappa > 1/\sqrt{2}$ are the type II superconductors.

1.1.2 *Example: Vortices in type II superconductors*

Vortices are very important topological objects in superconductors and superfluids. As already mentioned in the Preface, vortices determine the most fundamental properties of superconductors and their responses to external d.c. and a.c. electromagnetic fields. They play a crucial role also in hydrodynamics of superfluids. We will study the dynamics of vortices in detail later in this book. Here we summarize the main features of vortices which can be deduced from the Ginzburg–Landau model. A more detailed Ginzburg–Landau theory of the vortex state in superconductors can be found, for example, in the book by Saint-James *et al.* (1969).

1.1.2.1 *Transition from normal into the superconducting state* The transition from normal into the superconducting state in a magnetic field is of the second order in type II superconductors. Let us find the magnetic field when a nonzero order parameter first appears (Abrikosov 1957). Close to the transition point the order parameter is small, $\Delta \ll \Delta_\infty$. We linearize the GL equations in a small Δ:

$$\xi^2 \left(\nabla - \frac{2ie}{\hbar c}\mathbf{A}\right)^2 \Delta + \Delta = 0. \tag{1.31}$$

Let the magnetic field be along the z-axis. The vector potential can be taken in the Landau gauge $\mathbf{A} = (0, Hx, 0)$. The order parameter depends on x and y. Now we have

$$\frac{\partial^2 \Delta}{\partial x^2} + \left(\frac{\partial}{\partial y} - \frac{2ieHx}{\hbar c}\right)^2 \Delta + \xi^{-2}\Delta = 0. \tag{1.32}$$

This is the Schrödinger equation for a charge in a magnetic field. We put

$$\Delta = e^{iky} Y(x)$$

and obtain the oscillator equation

$$\frac{\partial^2 Y}{\partial x^2} - \left(k - \frac{2eHx}{\hbar c}\right)^2 Y + \xi^{-2} Y = 0 \tag{1.33}$$

where the oscillator frequency ω_0 and energy E are

$$\omega_0 = \frac{2|e|H}{mc}, \quad E = \frac{\hbar^2}{2m\xi^2},$$

respectively. The energy spectrum $E = \omega_0(n + 1/2)$ gives

$$\frac{\hbar}{2m\xi^2} = \frac{2|e|H}{mc}\left(n + \frac{1}{2}\right).$$

The highest magnetic field $H = H_{c2}$ is for $n = 0$:

$$H_{c2} = \frac{\hbar c}{2\,|e|\,\xi^2} = \frac{\Phi_0}{2\pi\xi^2} \tag{1.34}$$

where

$$\Phi_0 = \frac{\pi \hbar c}{|e|}$$

is the magnetic flux quantum (see below). H_{c2} is the upper critical magnetic field below which the transition into superconducting state occurs. It is proportional to $1 - T/T_c$ near the critical temperature. Comparing it with the thermodynamic critical field (de Gennes 1966) H_c we observe that $H_{c2} = \sqrt{2}\kappa H_c$. For type II superconductors with $\kappa > 1/\sqrt{2}$, the upper critical field $H_{c2} > H_c$.

The solution for the lowest energy level is a Gaussian function

$$Y = C \exp\left[-\frac{1}{2\xi^2}\left(x - \frac{\hbar ck}{2eH_{c2}}\right)^2\right]. \tag{1.35}$$

The function in eqn (1.35) is centered at $x = \hbar ck/2eH$. Actually, the full solution is a linear combination of these functions with different k. One can construct a periodic solution in the form

$$\Delta = \sum_n C_n e^{iqny} \exp\left[-\frac{1}{2\xi^2}\left(x - \frac{\hbar cqn}{2eH_{c2}}\right)^2\right]. \tag{1.36}$$

This is periodic in y with the period $y_0 = 2\pi/q$. It will be periodic in x, as well, if the coefficients C_n satisfy periodicity condition $C_{n+p} = C_n$. Then,

$$\Delta(x + \frac{p\hbar cq}{2eH_{c2}},\, y) = e^{ipqy}\Delta(x,\, y).$$

The simplest case is realized when all the coefficients C_n are equal. The period in x is then $x_0 = \hbar cq/2eH_{c2}$. The solution eqn (1.36) forms a rectangular lattice with the area $S_0 = x_0 y_0 = \Phi_0/H_{c2} = 2\pi\xi^2$ which corresponds to exactly one flux quantum per unit cell. If q is chosen such that $x_0 = y_0$, we obtain a square lattice.

The $|\Delta|$–pattern has zeroes at the points $x = x_0/2 + x_0 n$, $y = y_0/2 + y_0 m$, surrounded by supercurrent flow lines. Indeed, the supercurrent is

$$j_x = -\frac{\hbar c^2}{16\pi\lambda_L^2 e\Delta_\infty^2}\left(i\Delta^*\frac{\partial\Delta}{\partial x} - i\Delta\frac{\partial\Delta^*}{\partial x}\right),$$

$$j_y = -\frac{\hbar c^2}{16\pi\lambda_L^2 e\Delta_\infty^2}\left[\Delta^*\left(i\frac{\partial\Delta}{\partial y} + \frac{2eH_{c2}x}{\hbar c}\Delta\right) - \Delta\left(i\frac{\partial\Delta^*}{\partial y} - \frac{2eH_{c2}x}{\hbar c}\Delta\right)\right].$$

To transform this further we use the identity which holds for the function of the type of eqn (1.36):

$$\frac{\partial\Delta}{\partial x} = \left(-i\frac{\partial}{\partial y} - \frac{2eH_{c2}x}{\hbar c}\right)\Delta. \tag{1.37}$$

With help of eqn (1.37) we get

$$j_x = -\frac{\hbar c^2}{16\pi\lambda_L^2 e\Delta_\infty^2}\frac{\partial|\Delta|^2}{\partial y}, \tag{1.38}$$

$$j_y = \frac{\hbar c^2}{16\pi\lambda_L^2 e\Delta_\infty^2}\frac{\partial|\Delta|^2}{\partial x}. \tag{1.39}$$

These expressions suggest that $|\Delta(x, y)|^2$ is a stream function, i.e., that the current flows along the lines of constant $|\Delta|$. If we place the node of $|\Delta|$ in the center of the Bravais unit cell, the current along the boundary of a unit cell is zero: due to periodicity, the lines of constant $|\Delta|$ are perpendicular to the boundary. We now calculate the contour integral along the unit cell boundary

$$\oint \left(\mathbf{A} - \frac{\hbar c}{2e} \nabla \chi \right) d\mathbf{r} = 0. \tag{1.40}$$

It vanishes because $\mathbf{A} - (\hbar c/2e)\nabla\chi = 0$ at the boundary, as we have just proven. We obtain

$$\Phi_0^{-1} \int_{S_0} \mathbf{H} \cdot d\mathbf{S} = (2\pi)^{-1} \delta\chi$$

where the integral is over the unit cell and $\delta\chi$ is the variation of the order parameter phase along this contour. It is $\delta\chi = 2\pi$ since the flux through the unit cell is equal to one flux quantum. Therefore, the phase of the order parameter acquires the increment of 2π after encircling the point where the order parameter is zero. The phase variation by 2π is also necessary for single-valuedness of the order parameter. Here we come to a vortex: A quantized vortex is a linear (in three dimensions) topological object which is characterized by a quantized circulation of the order parameter phase around this line. In principle, vortices with a phase circulation of an integer multiple of 2π are also possible. The axis of vortex circulation $\hat{\mathbf{z}}$ is parallel to $\mathbf{H}$ if the charge of carriers is positive and antiparallel to it in the opposite case: $\hat{\mathbf{z}} = \operatorname{sign}(e)\hat{\mathbf{H}}$. For electrons, $\hat{\mathbf{z}}$ is antiparallel to $\mathbf{H}$. We see that transition into a superconducting state in a magnetic field below H_{c2} gives rise to formation of vortices. Vortices in superconductors were theoretically predicted by Abrikosov (1957).

Let us consider an applied magnetic field H slightly below H_{c2} such that $H_{c2} - H \ll H_{c2}$. The solution of the GL equation is the function eqn (1.36) plus a small correction Δ_1. This correction is caused by (i) nonlinear term in the GL equation, (ii) local variations in $\mathbf{A}$ due to the supercurrent eqns (1.38), (1.39), and (iii) deviation of H from H_{c2}.

Using the Maxwell equation

$$j_x = \frac{c}{4\pi}\frac{\partial H_z}{\partial y}, \; j_y = -\frac{c}{4\pi}\frac{\partial H_z}{\partial x}$$

we obtain variations in the local field

$$\delta H_z = -\frac{\hbar c |\Delta|^2}{4\lambda_L^2 e \Delta_\infty^2}. \tag{1.41}$$

Therefore, the vector potential becomes $\mathbf{A} = \mathbf{A}_0 + \mathbf{A}_1$ where

$$\mathbf{A}_1 = (0, \, (H - H_{c2})x, \, 0) + \delta\mathbf{A}$$

where $\delta\mathbf{A}$ is due to the magnetic field δH_z.

Since the non disturbed function of eqn (1.36) satisfies the linearized GL equation with $\mathbf{A} = \mathbf{A}_0$, we obtain for the correction Δ_1

$$\left(\nabla - \frac{2ie}{\hbar c}\mathbf{A} \right)^2 \Delta_1 + \xi^{-2}\Delta_1$$

$$= \frac{2ie}{\hbar c}\left(\nabla - \frac{2ie}{\hbar c}\mathbf{A}\right)(\mathbf{A}_1\Delta) + \frac{2ie}{\hbar c}\mathbf{A}_1\left(\nabla - \frac{2ie}{\hbar c}\mathbf{A}\right)\Delta + \xi^{-2}|\Delta|^2\Delta/\Delta_\infty^2.$$

Now we apply the orthogonality condition: we multiply this equation by Δ from eqn (1.36) and integrate it over $dx\, dy$. After integration by parts using eqn (1.37), we obtain

$$\frac{2e}{\hbar c}\int |\Delta|^2(H - H_{c2} + \delta H)\, dx\, dy + \Delta^{-2}\xi^{-2}\int |\Delta|^4\, dx\, dy = 0.$$

We introduce the average over the area occupied by the vortex array

$$\langle|\Delta|^2\rangle_L = S_0^{-1}\int_{S_0} |\Delta|^2\, dx\, dy$$

and obtain using eqn (1.41)

$$\Delta_\infty^2 \langle|\Delta|^2\rangle_L \left(1 - \frac{H}{H_{c2}}\right) = \left(1 - \frac{1}{2\kappa^2}\right)\langle|\Delta|^4\rangle_L.$$

It is convenient to introduce the ratio

$$\beta_A = \frac{\langle|\Delta|^4\rangle_L}{\langle|\Delta|^2\rangle_L^2} \geq 1 \tag{1.42}$$

which is called the Abrikosov parameter. The value β_A is determined by the structure of the vortex lattice. Now we get

$$\langle|\Delta|^2\rangle_L = \frac{2\kappa^2\Delta_\infty^2}{(2\kappa^2 - 1)\beta_A}\left(1 - \frac{H}{H_{c2}}\right). \tag{1.43}$$

Equation (1.43) shows that $|\Delta|^2$ has a small magnitude proportional to $1 - H/H_{c2}$ if $\kappa > 1/\sqrt{2}$. In the opposite case, the order parameter cannot be small for H close to H_{c2}: it jumps to a finite value making the transition of the first order. This is exactly the condition which separates type I and type II superconductors: A superconductor can accommodate vortices and allow a magnetic field below H_{c2} to penetrate into it if $\kappa > 1/\sqrt{2}$. On the contrary, the order parameter is always finite and no magnetic field can exist in the superconductor if $\kappa < 1/\sqrt{2}$.

Using eqn (1.41) we can find the magnetization of the superconductor

$$M_z = \frac{B - H}{4\pi} = \frac{\langle\delta H_z\rangle}{4\pi} = -\frac{H_{c2} - H}{4\pi\beta_A(2\kappa^2 - 1)}$$

where B is the magnetic induction. One can also calculate the total free energy density. It becomes

$$F = \frac{B^2}{8\pi} - \frac{(H_{c2} - B)^2}{8\pi[1 + (2\kappa^2 - 1)\beta_A]}. \tag{1.44}$$

It decreases with decreasing β_A. For a square lattice $\beta_A = 1.18$ while for a hexagonal lattice $\beta_A = 1.16$. A hexagonal lattice corresponds to the coefficients

$$C_{n+2} = C_n,\ C_1 = iC_0.$$

It is the stable configuration in an isotropic environment. On the other hand, a square lattice is unstable: it corresponds to an extremum rather than to a minimum of the vortex free energy (Saint-James *et al.* 1969). It can be stable, however, if the interaction with the corresponding underlying crystalline structure is strong enough.

1.1.2.2 *Single vortex* The previous case corresponds to the situation where vortices are closely packed together: the distance between them is of the order of the coherence length. We now consider the situation when vortices can be treated separately. A single vortex has the order parameter phase which changes by 2π after encircling its axis which we choose to be the z-axis. We take $\chi = \phi$ where ϕ is the azimuthal angle in the cylindrical frame $(\rho,\ \phi,\ z)$. We thus assume a cylindrical symmetry of the vortex and look for a solution in the form

$$\Delta = \Delta_\infty f(\rho) e^{i\phi}.$$

The vector potential has only a ϕ-component: $\mathbf{A} = (0,\ A_\phi,\ 0)$. We have for $f\ (\rho)$

$$\xi^2 \left(\frac{\partial^2}{\partial \rho^2} + \frac{1}{\rho}\frac{\partial}{\partial \rho} - \frac{4e^2 Q^2}{\hbar^2 c^2} \right) f + f - f^3 = 0. \tag{1.45}$$

Here we introduce the gauge-invariant vector potential

$$\mathbf{Q} = \mathbf{A} - \frac{\hbar c}{2e} \nabla \chi. \tag{1.46}$$

In our case $\mathbf{Q}$ has only an azimuthal component $A_\phi - \hbar c/2e\rho$.

Equation (1.24) becomes

$$\lambda_L^2 \text{curl curl}\, \mathbf{A} + f^2 \mathbf{Q} = 0.$$

For $\rho \neq 0$ it is

$$\frac{\partial^2 Q}{\partial \rho^2} + \frac{1}{\rho}\frac{\partial Q}{\partial \rho} - \frac{Q}{\rho^2} - \frac{f^2 Q}{\lambda_L^2} = 0. \tag{1.47}$$

This equation can be solved in the limit $\kappa \gg 1$ where $\lambda_L \gg \xi$ such that one can put $f = 1$ for $\rho \gg \xi$. Equation (1.47) gives

$$Q = -\frac{\hbar c}{2e\lambda} K_1(\rho/\lambda_L).$$

Here $K_1(z)$ is the first-order Bessel function of an imaginary argument. For $z \ll 1$ the function $K_1(z) = 1/z$, and it decreases exponentially for large z:

$$K_1(z) = \left(\frac{2}{\pi z} \right)^{\frac{1}{2}} \exp(-z).$$

The constant at Q is chosen in such a way that $A_\phi = Q + (\hbar c/2e\rho)$ is not diverging for $\rho \ll \lambda$. The magnetic field is

$$H_z = \text{curl}_z \mathbf{Q} = \frac{1}{\rho}\frac{\partial(\rho Q)}{\partial \rho} = \frac{\hbar c}{2e\lambda_L^2} K_0(\rho/\lambda_L)$$

where $K_0(z)$ is the zero-order Bessel function of imaginary argument. For $z \ll 1$ it is $K_0(z) = -\ln z$, and it decreases exponentially for large z in the same

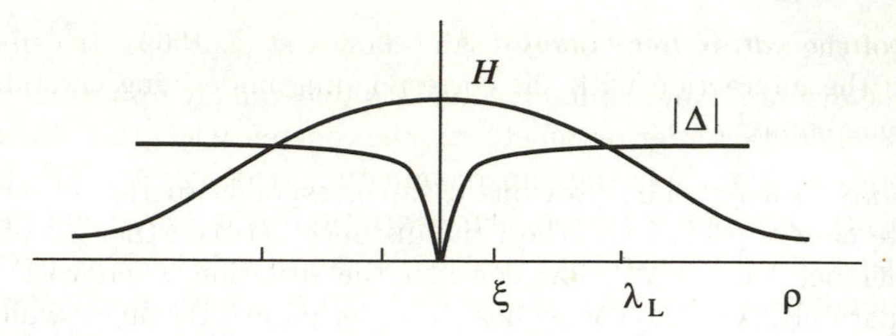

FIG. 1.1. The order parameter and the magnetic field near the vortex. The vortex core, i.e., the region where Δ is suppressed has the size of order of the coherence length ξ; the magnetic field and currents decay on the scale on order λ_L.

way as K_1. Therefore, the magnetic field produced by a single vortex decays exponentially for $\rho \gg \lambda$ and it is

$$H_z \approx \frac{\Phi_0}{2\pi\lambda_L^2} \ln \frac{\lambda_L}{\rho}$$

for $\rho \ll \lambda$. For small ρ the logarithm is cut off at $\rho \sim \xi$ where f starts to decrease.

Equation (1.45) for the order parameter magnitude f takes a simple form in the region $\rho \ll \lambda_L$ where $Q = -\hbar c/2e\rho$. It is

$$\xi^2 \left(\frac{\partial^2}{\partial \rho^2} + \frac{1}{\rho}\frac{\partial}{\partial \rho} - \frac{1}{\rho^2} \right) f + f - f^3 = 0. \tag{1.48}$$

The function f decreases as $f \propto \rho$ for $\rho \to 0$. At large distances $\rho \gg \xi$, the function $f \to 1$; it remains $f \approx 1$ also for longer distances of the order of λ_L where Q decays due to the screening currents.

The region near the vortex axis with the size of the order of ξ where the order parameter is decreased from its value in the bulk is called the vortex core; the order parameter magnitude $|\Delta|$ vanishes at the vortex axis. The vortex core is surrounded by supercurrents which, together with the magnetic field, decay away from the vortex core at distances of the order of λ_L. The schematic behavior of $|\Delta|$ and H_z is shown in Fig. 1.1.

The cores of neighboring vortices do not overlap when the distance between vortices is larger than ξ. The intervortex distance can be found from the condition that each vortex unit cell carries one magnetic flux quantum. If we replace a Bravais unit cell by a circle of a radius r_0, we have

$$\pi r_0^2 B = \Phi_0$$

which gives

$$r_0/\xi \sim \sqrt{H_{c2}/H}.$$

Therefore, vortices can be considered separately if $H \ll H_{c2}$.

1.1.3 *Bogoliubov–de Gennes equations*

The simple Ginzburg–Landau description introduces many important characteristics. We see here the order parameter Δ, the coherence length ξ, the magnetic-field penetration depth λ_L, the superconducting velocity, etc. The Ginzburg–Landau theory provides a reasonable description of a superconductor in the vicinity of the critical temperature. To learn about properties of superconductors at lower temperatures, however, one needs to go to a more microscopic level of description. A comparatively simple approach is provided by the Bogoliubov–de Gennes theory (Bogoliubov *et al.* 1958, see also de Gennes 1966). Unfortunately, it has a manageable form for clean superconductors only. We briefly summarize here the main ideas of the Bogoliubov–de Gennes method. We derive the Bogoliubov–de Gennes equations later in Section 3.2.

The single-particle wave function has two components: the particle-like function u and the hole-like function v which satisfy the so-called Bogoliubov–de Gennes equations

$$\left[\frac{1}{2m}\left(-i\hbar\nabla-\frac{e}{c}\mathbf{A}\right)^2-E_F\right]u+\Delta v=\epsilon_K u,$$
$$\left[\frac{1}{2m}\left(-i\hbar\nabla+\frac{e}{c}\mathbf{A}\right)^2-E_F\right]v-\Delta^* u=-\epsilon_K v. \quad (1.49)$$

K denotes the set of quantum numbers. Here we encounter the excitation spectrum ϵ_K as an important characteristics of superconductors. For simplicity, we consider a parabolic spectrum of normal electrons. The wave function is normalized such that

$$\int (u_K u_{K'}^* + v_K v_{K'}^*)\, dV = \delta_{K,K'}. \quad (1.50)$$

The order parameter itself is determined self-consistently from the BCS equation in the form (Bogoliubov *et al.* 1958)

$$\Delta_K = -\sum_{K'} U_{K,K'}\,(1-2n_{K'})\,u_{K'}v_{K'}^*. \quad (1.51)$$

Here $U_{K,K'}$ is an attractive pairing interaction. The sum is over the states of the system and ϵ_K is the energy spectrum which is found from eqn (1.49). A new quantity appears in eqn (1.51), namely the distribution function n_K of excitations. In equilibrium, it is the Fermi function

$$n_K = \frac{1}{e^{\epsilon_K/T}+1}.$$

We shall see later that the distribution function of excitations is of a crucial importance for nonequilibrium processes in superconductors.

If the magnitude of the order parameter is constant in space, we can look for a solution in the form

$$\begin{pmatrix} u \\ v \end{pmatrix} = \begin{pmatrix} u_p e^{i\chi/2} \\ v_p e^{-i\chi/2} \end{pmatrix} \exp\left(\frac{i}{\hbar}\int \mathbf{p}\cdot d\mathbf{r}\right) \tag{1.52}$$

where χ is the order parameter phase. Substituting it into eqn (1.49) we find

$$p^2 = p_F^2 - p_s^2 \pm 2m\sqrt{(\epsilon_{\mathbf{p}} - \mathbf{p}_s \cdot \mathbf{v})^2 - |\Delta|^2} \tag{1.53}$$

with $\mathbf{v} = \mathbf{p}/m$. The energy spectrum becomes $\epsilon_K = \epsilon_{\mathbf{p}}$ where

$$\epsilon_{\mathbf{p}} = \tilde{\epsilon}_{\mathbf{p}} + \mathbf{p}_s \cdot \mathbf{v}; \quad \tilde{\epsilon}_{\mathbf{p}} = \sqrt{\tilde{\xi}_{\mathbf{p}}^2 + |\Delta|^2}. \tag{1.54}$$

In this equation,

$$\mathbf{p}_s = \frac{1}{2}\mathbf{P}_s = \frac{\hbar}{2}\nabla\chi - \frac{e}{c}\mathbf{A}$$

is the Cooper pair momentum per particle, and

$$\tilde{\xi}_{\mathbf{p}} = \xi_{\mathbf{p}} + \mathbf{p}_s^2/2m. \tag{1.55}$$

Usually, $\mathbf{p}_s^2/2m \ll |\Delta|$, so that the energy $\tilde{\epsilon}_{\mathbf{p}}$ is defined by eqn (1.1). Equation (1.54) determines the energy spectrum only for a constant magnitude of the order parameter. Moreover, one has to assume that the supercurrent (or $\mathbf{p}_s$) is also constant. The Bogoliubov–de Gennes wave functions are

$$u_{\mathbf{p}}^2 = \frac{1}{2}\left(1 + \frac{\tilde{\xi}_{\mathbf{p}}}{\tilde{\epsilon}_{\mathbf{p}}}\right), \; v_{\mathbf{p}}^2 = \frac{1}{2}\left(1 - \frac{\tilde{\xi}_{\mathbf{p}}}{\tilde{\epsilon}_{\mathbf{p}}}\right). \tag{1.56}$$

They are sometimes called the coherence factors.

In the momentum space, $\sum \to \int d^3p/(2\pi\hbar)^3$ and the self-consistency equation (1.51) becomes

$$\Delta_{\mathbf{p}} = -\int U_{\mathbf{p},\mathbf{p}'} \frac{\Delta_{\mathbf{p}'}}{2\tilde{\epsilon}_{\mathbf{p}'}} (1 - 2n_{\mathbf{p}'}) \frac{d^3p'}{(2\pi\hbar)^3}. \tag{1.57}$$

Equation (1.57) determines the order parameter as a function of temperature in a spatially homogeneous situation.

1.1.4 *Quasiclassical approximation*

If the order parameter and/or current vary in space, eqns (1.54) and (1.56) are no longer valid. Solution of the Bogoliubov–de Gennes equation (1.49) would become an impossible task if not for a very important observation.

In almost all known superconductors, the Fermi momentum p_F is much larger than the characteristic wave vectors associated with the order parameter variations $\hbar/\xi_0$ where ξ_0 is the zero-temperature coherence length defined by eqn (1.17). This implies that the particle wave length is much shorter than the characteristic scale of variations of the order parameter. It allows one to use a semi-classical approach for solving the Bogoliubov–de Gennes equations. One can write, for example,

$$\begin{pmatrix} u \\ v \end{pmatrix} = \begin{pmatrix} u_{\mathbf{p}} \\ v_{\mathbf{p}} \end{pmatrix} \exp\left(\frac{i}{\hbar}\int \mathbf{p}_F \cdot d\mathbf{r}\right) \tag{1.58}$$

where $u_{\mathbf{p}}$ and $v_{\mathbf{p}}$ vary in space on distances of the order of ξ. The Bogoliubov–de Gennes equations take the form

$$\begin{aligned} \mathbf{v}_F\left(-i\hbar\nabla - \frac{e}{c}\mathbf{A}\right) u_{\mathbf{p}} + \Delta v_{\mathbf{p}} &= \epsilon_p u_{\mathbf{p}}, \\ \mathbf{v}_F\left(-i\hbar\nabla + \frac{e}{c}\mathbf{A}\right) v_{\mathbf{p}} - \Delta^* u_{\mathbf{p}} &= -\epsilon_p v_{\mathbf{p}}. \end{aligned} \tag{1.59}$$

We denote $\mathbf{v}_F = \partial E_n/\partial \mathbf{p}$ and assign the index F to indicate that it is the velocity taken at the Fermi surface. Moreover, we neglect $\mathbf{p}_s^2/2m$ as compared to $|\Delta|$ because

$$\mathbf{p}_s^2/2m \sim \hbar^2/m\xi_0^2 \sim |\Delta|\,\hbar/p_F\xi_0$$

which is much smaller than $|\Delta|$ since $p_F\xi_0/\hbar \gg 1$. The self–consistency equation (1.51) becomes

$$\Delta_{\mathbf{p}}(\mathbf{r}) = -\int U_{\mathbf{p},\mathbf{p}'} u_{\mathbf{p}'}(\mathbf{r})\, v^*_{\mathbf{p}'}(\mathbf{r})\,(1 - 2n_{\mathbf{p}'})\,\frac{d^3p'}{(2\pi\hbar)^3}. \tag{1.60}$$

The possibility to separate and then exclude fast oscillating parts in the quasiparticle wave functions known as the quasiclassical approximation is provided by the relations between the magnitudes of the superconducting- and normal-state characteristic parameters of the superconducting material, namely $p_F\xi_0/\hbar \gg 1$. It is this fundamental property of most of superconductors which makes the BCS theory so successful in its practical applications. The major simplification is that the momentum dependence of the pairing potential is separated from the coordinate dependence of the wave functions $u_{\mathbf{p}}$ and $v_{\mathbf{p}}$ in the self-consistency equation (1.60). The momentum $\mathbf{p}$ enters the wave functions only as a parameter. Moreover, equations (1.59) are now each of the first order. Therefore, the mathematical complexity of the theory is considerably reduced.

The accuracy of the quasiclassical approximation depends on how well the inequality

$$\frac{\hbar}{p_F\xi_0} \sim \frac{\Delta}{E_F} \ll 1 \tag{1.61}$$

is satisfied. For conventional low-temperature superconductors, the parameter Δ/E_F is of the order of 10^{-3} thus the accuracy is very good. High temperature superconductors have $\Delta/E_F \sim 10^{-1}$ to 10^{-2} so that the quasiclassical approximation is less universal. However, it still has a reasonably solid base for validity, though description of some phenomena requires a more careful analysis. Using this approximation, one can formulate a very powerful *quasiclassical method* which is indispensable for solving spatially nonhomogeneous problems of the microscopic theory of superconductivity. It also makes the basis for the modern microscopic theory of nonstationary phenomena in superconductors. The great advantage of the quasiclassical nonstationary theory is that it can incorporate, in a coherent way, various relaxation mechanisms including interaction with impurities in superconducting alloys. It is this quasiclassical method which is the main subject of the present book.

1.2 Nonstationary phenomena

Nonstationary theory considers behavior of superconductors in a.c. external fields (electromagnetic fields, sound waves, etc.). It treats transport phenomena such as electric or thermal conductivity, and thermopower. Problems associated with a d.c. electric field should also be considered within the nonstationary theory. Indeed, a d.c. electric field accelerates electrons which should then relax and give away their energy to the superconductor. This produces a time-dependent or dissipative state where the absorbed power creates a nonequilibrium distribution of excitations. The nonstationary theory should provide a consistent approach to the whole class of such problems using the superconducting characteristics discussed above.

1.2.1 *Time-dependent Ginzburg–Landau theory*

The simplest description of nonstationary processes in superconductors is provided by the so-called time-dependent Ginzburg–Landau (TDGL) model which generalizes the usual Ginzburg–Landau (GL) theory to include relaxation processes. The TDGL model is widely used, it often gives a reasonable picture of superconducting dynamics. However, as distinct from its static counterpart, validity of the TDGL theory is much more limited. It is not enough just to be close to the critical temperature. The necessary condition requires also that deviations from equilibrium are small: the quasiparticle excitations should remain essentially in equilibrium with the heat bath. For real superconductors it is, in general, a very strong limitation. It can normally be fulfilled only for the so-called gapless superconductivity. The latter corresponds to a situation where mechanisms working to destroy Cooper pairs are almost successful: the energy gap in the spectrum disappears, but the order parameter retains the phase coherence, and the supercurrent is finite. These mechanisms are: interaction with magnetic impurities which act differently on the electrons with opposite spins in a Cooper pair (see Section 6.2), inelastic (not conserving energy) interactions with phonons (see Section 11.2), etc. We shall consider these examples later. For instance, inelastic scattering by phonons is characterized by mean free time $\tau_{\rm ph}$. The condition for applicability of the TDGL theory is $\tau_{\rm ph}\Delta \ll \hbar$. Due to a comparatively large magnitude of $\tau_{\rm ph}$ (for example, $\tau_{\rm ph} \sim 10^{-9}$ s as compared to $\Delta/\hbar \sim 10^{11}$ s^{-1} for Al), this condition is only satisfied in a very narrow vicinity of T_c. In this section we consider a phenomenological derivation of the TDGL model. Its microscopic justification will be given later in Chapter 11.

The TDGL model is constructed in the following way. In equilibrium, the GL energy eqn (1.3) has a minimum with respect to Δ and $\mathbf{A}$:

$$\frac{\delta \mathcal{F}_{sn}}{\delta \Delta} = 0; \ \frac{\delta \mathcal{F}_{sn}}{\delta \mathbf{A}} + \frac{1}{c}\mathbf{j}_s = 0. \tag{1.62}$$

If the superconductor is driven out of equilibrium, the order parameter should relax back to its equilibrium value. The rate of relaxation depends on the deviation from equilibrium:

$$-\Gamma \frac{\partial \Delta}{\partial t} = \frac{\delta \mathcal{F}_{sn}}{\delta \Delta^*} \tag{1.63}$$

where Γ is a positive constant. This equation, however, is not gauge invariant. For a time-dependent state, the gauge-invariance requires, in addition to eqn (1.7), that the equations describing the superconductor do not change under the transformation

$$\chi \to \chi + f \ , \ \mathbf{A} \to \mathbf{A} + \frac{\hbar c}{2e}\nabla f, \ \varphi \to \varphi - \frac{\hbar}{2e}\frac{\partial f}{\partial t} \tag{1.64}$$

where $f(\mathbf{r}, t)$ is an arbitrary function of coordinates and time. To preserve the gauge invariance of eqn (1.63) we add the scalar potential to the time-derivative in the form

$$-\Gamma\left(\frac{\partial \Delta}{\partial t} + \frac{2ie\varphi}{\hbar}\Delta\right) = \frac{\delta \mathcal{F}_{sn}}{\delta \Delta^*}. \tag{1.65}$$

The second equilibrium condition of eqn (1.62) is generalized in the following way. The current in eqn (1.62) being just a supercurrent in equilibrium is now replaced with $\mathbf{j}_s = \mathbf{j} - \mathbf{j}_n$ which accounts for the fact that a part of current can be produced by normal electrons in presence of an electric field:

$$\mathbf{j}_n = \sigma_n \mathbf{E} = -\sigma_n\left(\nabla\varphi + \frac{1}{c}\frac{\partial \mathbf{A}}{\partial t}\right). \tag{1.66}$$

Here σ_n is the conductivity in the normal state. As a result

$$\frac{1}{c}(\mathbf{j} - \mathbf{j}_n) = -\frac{\delta \mathcal{F}_{sn}}{\delta \mathbf{A}}. \tag{1.67}$$

Finally, the TDGL equations become

$$-\Gamma\left(\frac{\partial \Delta}{\partial t} + \frac{2ie\varphi}{\hbar}\Delta\right) = -|\alpha|\Delta + \beta|\Delta|^2\Delta + \gamma\left(-i\hbar\nabla - \frac{2e}{c}\mathbf{A}\right)^2\Delta, \tag{1.68}$$

The total current $\mathbf{j} = \mathbf{j}_s + \mathbf{j}_n$ has the form

$$\mathbf{j} = \sigma_n\left(-\frac{1}{c}\frac{\partial \mathbf{A}}{\partial t} - \nabla\varphi\right)$$
$$+2e\gamma\left[\Delta^*\left(-i\hbar\nabla - \frac{2e}{c}\mathbf{A}\right)\Delta + \Delta\left(i\hbar\nabla - \frac{2e}{c}\mathbf{A}\right)\Delta^*\right] \tag{1.69}$$

where $\mathbf{j}_s$ is defined by eqn (1.13).

Equations (1.68) and (1.69) define characteristic relaxation times of the order parameter

$$\tau_\Delta = \frac{\Gamma}{|\alpha|} \propto \left(1 - \frac{T}{T_c}\right)^{-1} \tag{1.70}$$

and of the vector potential (or current)

$$\tau_j = \frac{\beta\sigma_n}{8e^2\gamma|\alpha|} = \frac{\sigma_n}{8e^2\gamma\Delta_0^2} \propto \left(1 - \frac{T}{T_c}\right)^{-1} . \tag{1.71}$$

The ratio of these two relaxation times

$$u = \frac{\tau_\Delta}{\tau_j} = \frac{8\Gamma e^2\gamma}{\beta\sigma_n} \tag{1.72}$$

is independent of T.

Separating real and imaginary parts of eqn (1.68) we obtain

$$\tau_\Delta \frac{\partial|\Delta|}{\partial t} = |\Delta| - \frac{|\Delta|^3}{\Delta_\infty^2} + \xi^2\left(\nabla^2 - \frac{4e^2}{\hbar^2c^2}\mathbf{Q}^2\right)|\Delta| \tag{1.73}$$

and

$$c\tau_\Delta|\Delta|^2\Phi + \xi^2 \mathrm{div}\left(\mathbf{Q}|\Delta|^2\right) = 0 \tag{1.74}$$

where we introduce one more gauge-invariant (scalar) potential

$$\Phi = \varphi + \frac{\hbar}{2e}\frac{\partial\chi}{\partial t} \tag{1.75}$$

in addition to the vector potential $\mathbf{Q}$ defined in eqn (1.46).

Equation (1.74) can also be obtained by calculating the variation of the free energy with respect to the order parameter phase

$$\begin{aligned} 2\frac{\delta\mathcal{F}_{sn}}{\delta\chi} &\equiv i\left(\Delta\frac{\delta\mathcal{F}_{sn}}{\delta\Delta} - \Delta^*\frac{\delta\mathcal{F}_{sn}}{\delta\Delta^*}\right) \\ &= -i\hbar\gamma\left[\Delta\left(\nabla + \frac{2ie}{\hbar c}\mathbf{A}\right)^2\Delta^* - \Delta\left(\nabla - \frac{2ie}{\hbar c}\mathbf{A}\right)^2\Delta\right] \\ &= -\frac{\hbar}{2e}\mathrm{div}\,\mathbf{j}_s \end{aligned} \tag{1.76}$$

With help of eqn (1.65) we obtain from eqn (1.76)

$$\frac{\delta\mathcal{F}_{sn}}{\delta\chi} = -\Gamma|\Delta|^2\left(\frac{\partial\chi}{\partial t} + \frac{2e\phi}{\hbar}\right) = -\frac{\hbar}{4e}\mathrm{div}\,\mathbf{j}_s \tag{1.77}$$

which is nothing but eqn (1.74).

According to the charge neutrality of a superconductor as a metal

$$\text{div}\ (\mathbf{j}_n+\mathbf{j}_s) = 0 \tag{1.78}$$

eqn (1.77) gives

$$\frac{\hbar}{4e}\text{div}\,\mathbf{j}_s = -\frac{\hbar}{4e}\text{div}\,\mathbf{j}_n = \frac{2e\Gamma|\Delta|^2}{\hbar}\Phi. \tag{1.79}$$

Writing

$$\mathbf{j}_n = -\sigma_n\left(\nabla\Phi + \frac{1}{c}\frac{\partial\mathbf{Q}}{\partial t}\right)$$

we obtain

$$l_E^2\left(\nabla^2\Phi + \frac{1}{c}\text{div}\frac{\partial\mathbf{Q}}{\partial t}\right) = \frac{|\Delta|^2}{\Delta_\infty^2}\Phi. \tag{1.80}$$

This equation defines the new characteristic length

$$l_E = \sqrt{\frac{\hbar\sigma_n}{8e^2\Gamma\Delta_\infty^2}} = \xi/\sqrt{u} \tag{1.81}$$

which determines the scale of spatial variations of the gauge-invariant potential Φ. In equilibrium, when the electric field is absent, the potential Φ should vanish according to eqn (1.80) since a d.c. electric field exists as long as Φ varies in space. In other words, l_E is the distance over which a d.c. electric field decays into a superconductor; it is thus called the electric-field penetration length. At the same time, a nonzero Φ initiates a conversion of a normal current into a supercurrent according to eqn (1.79). We discuss this in more detail later in Chapter 11.

1.2.2 *Microscopic argumentation*

Let us discuss how one can justify the TDGL model microscopically. The TDGL equation was first derived by Schmid (1966) and by Abrahams and Tsuneto (1966) using a Green function formalism. We consider here a more simple method based on the BCS equation (1.57). Assume that $\mathbf{p}_s = 0$ and put

$$\frac{d^3p}{(2\pi\hbar)^3} = \nu\left(\xi_{\mathbf{p}}\right)d\xi_{\mathbf{p}}$$

where $\nu\left(\xi_{\mathbf{p}}\right)$ is the density of states in the normal metallic state. We assume that $\nu\left(\xi_{\mathbf{p}}\right)$ is approximately constant and equal to the density of states $\nu(0)$ at the Fermi surface, where

$$\nu\left(0\right) = mp_F/2\pi^2\hbar^3$$

for a parabolic normal-state spectrum. This assumption will be discussed in detail later in Chapter 5. Next, we transform to the energy integration in eqn (1.57) through $\xi_{\mathbf{p}}d\xi_{\mathbf{p}} = \epsilon_{\mathbf{p}}d\epsilon_{\mathbf{p}}$ using eqn (1.1). The function $\xi_{\mathbf{p}}$ changes its sign during integration over d^3p. To account for this, we note that $1-2n_{\mathbf{p}} = \tanh\left(\epsilon_{\mathbf{p}}/2T\right)$ is an odd function of $\epsilon_{\mathbf{p}}$. Let us define $\xi_{\mathbf{p}}$ as an analytical function in the complex plane of the variable $\epsilon_{\mathbf{p}}$ with a cut from $-|\Delta|$ to $+|\Delta|$ (see Fig. 5.1). We obtain

$$\xi_{\mathbf{p}}(\epsilon_{\mathbf{p}}) = \sqrt{\epsilon_{\mathbf{p}}^2 - |\Delta|^2}. \tag{1.82}$$

Now $\epsilon_{\mathbf{p}}$ can be both positive and negative provided $\epsilon_{\mathbf{p}}^2 > |\Delta|^2$. The function $\xi_{\mathbf{p}}$ changes its sign as $\epsilon_{\mathbf{p}}$ varies from $\epsilon_{\mathbf{p}} > |\Delta|$ to $\epsilon_{\mathbf{p}} < -|\Delta|$. Equation (1.82) correctly transforms into the normal state spectrum $\epsilon_{\mathbf{p}} = \xi_{\mathbf{p}}$ if $|\Delta| = 0$. The BCS equation (1.57) becomes

$$\frac{\Delta_{\mathbf{p}}}{\lambda} = \int V_{\mathbf{p},\mathbf{p}'}\Delta_{\mathbf{p}'}\frac{d\Omega_{p'}}{4\pi}\int \frac{1-2n_{\mathbf{p}'}}{2\xi_{\mathbf{p}'}}\Theta\left(\epsilon_{\mathbf{p}'}^2 - |\Delta_{\mathbf{p}'}|^2\right) d\epsilon_{\mathbf{p}'} \tag{1.83}$$

where we put $U_{\mathbf{p},\mathbf{p}'} = -|g|\, V_{\mathbf{p},\mathbf{p}'}$ and denote $\lambda = \nu(0)\,|g|$. The factor $V_{\mathbf{p},\mathbf{p}'}$ is the spherical harmonics of the pairing interaction. For s-wave pairing $V_{\mathbf{p},\mathbf{p}'} = 1$. The integration runs over positive and negative energies. It is limited from above by $|\epsilon| < \Omega_{\mathrm{BCS}}$. The upper BCS cut-off energy is determined by the particular pairing mechanism.

Consider a limit of small Δ and assume s-wave pairing for simplicity. With our definition of $\xi_{\mathbf{p}}$ as an analytical function, eqn (1.83) becomes

$$\frac{\Delta}{\lambda} = \int \left(\frac{\Delta}{\epsilon + i\delta} + \frac{\Delta}{\epsilon - i\delta}\right)[1 - 2n(\epsilon)]\,\frac{d\epsilon}{4}. \tag{1.84}$$

Here we drop the subscript $\mathbf{p}$ at ϵ for brevity. If the order parameter depends on time the energy ϵ becomes shifted due the external frequency ω. To obtain the correct shift, we notice that ϵ is a single-particle energy while ω refers to the frequency of a Cooper pair. Therefore, ϵ should be shifted by $\hbar\omega/2$. According to the casualty principle, we require that $\Delta(\omega)$ is an analytical function in the upper half-plane of complex ω. As a result, eqn (1.84) takes the form

$$\frac{\Delta(\omega)}{\lambda} = \int \Delta(\omega)\left(\frac{1}{\epsilon + \hbar\omega/2 + i\delta} + \frac{1}{\epsilon - \hbar\omega/2 - i\delta}\right)[1 - 2n(\epsilon)]\,\frac{d\epsilon}{4}.$$

Expanding it in a small ω we get

$$\begin{aligned}\frac{\Delta(\omega)}{\lambda} &= \Delta(\omega)\int_{-\Omega_{\mathrm{BCS}}}^{\Omega_{\mathrm{BCS}}} \frac{1-2n(\epsilon)}{2\epsilon}d\epsilon \\ &\quad + \frac{\hbar\omega\Delta(\omega)}{4}\int_{-\infty}^{\infty}\left(\frac{1}{\epsilon + i\delta} - \frac{1}{\epsilon - i\delta}\right)\frac{dn(\epsilon)}{d\epsilon}\,d\epsilon.\end{aligned} \tag{1.85}$$

Since

$$\frac{dn(\epsilon)}{d\epsilon} = -\frac{1}{4T}\cosh^{-2}\left(\frac{\epsilon}{2T}\right)$$

the last line of eqn (1.85) gives

$$\frac{\pi\hbar i}{8T}\omega\Delta(\omega) \quad \text{or} \quad -\frac{\pi\hbar}{8T}\frac{\partial\Delta}{\partial t}$$

in the time representation. As we shall see later, the first line gives the usual GL part of the equation. Comparing this with eqn (1.68) we find the microscopic value of the relaxation parameter

$$\Gamma = \frac{\nu(0)\,\pi\hbar}{8T_c}. \tag{1.86}$$

The derivation is based on simple intuitive arguments. However, it is not generally correct. Indeed, it treats nonstationary processes by including the external energy ω into the energy spectrum and ignoring the processes which are associated with relaxation. In fact, this derivation implicitly assumes that both the energy spectrum and the quasiparticle distribution remain the same as in equilibrium, i.e., that the relaxation is infinitely fast. We shall demonstrate later in Section 11.2, that this picture is appropriate only for gapless superconductors. The result eqn (1.86) itself is literally applicable only for gapless superconductors without magnetic impurities. We stress once again that general limitations of validity of the TDGL model come from nonequilibrium excitations: equation (1.85) becomes incorrect as soon as the deviation from equilibrium increases.

1.2.3 *Boltzmann kinetic equation*

Consider an alternative scheme of treating nonstationary processes which is capable of incorporating the quasiparticle relaxation. One would expect that it is the Boltzmann kinetic equation that is appropriate to describe nonequilibrium dynamics. The Boltzmann kinetic approach has been discussed in detail by Betbeder-Matibet and Nozières (1969), and by Aronov *et al.* (1981, 1986). We give only a brief outline here. A derivation of the Boltzmann equation is given later in Section 15.2.

With an account for a nonequilibrium scalar potential Φ the energy of the excitations in eqn (1.54) is defined with

$$\tilde{\xi}_{\mathbf{p}} = \xi_{\mathbf{p}} + e\Phi$$

where Φ is introduced by eqn (1.75). We neglect the term $p_s^2/2m$ in eqn (1.55). The canonical form of the Boltzmann equation for the distribution function for a situation where the order parameter, supercurrent, and external fields are slowly varying in space and in time, has the form

$$\frac{\partial n_{\mathbf{p}}}{\partial t} + \mathbf{v}_g \cdot \nabla n_{\mathbf{p}} + \mathbf{f}\frac{\partial n_{\mathbf{p}}}{\partial \mathbf{p}} = I\{n_{\mathbf{p}}\}. \tag{1.87}$$

The group velocity and the elementary force are defined as

$$\mathbf{v}_g = \frac{\partial \epsilon_{\mathbf{p}}}{\partial \mathbf{p}} = \frac{\tilde{\xi}_{\mathbf{p}}}{\tilde{\epsilon}_{\mathbf{p}}}\mathbf{v}_F,\ \mathbf{f} = -\frac{\partial \tilde{\epsilon}_{\mathbf{p}}}{\partial \mathbf{r}} + \mathbf{f}_L,\ \mathbf{f}_L = \frac{e}{c}\mathbf{v}_F \times \mathbf{H}. \tag{1.88}$$

The force $\mathbf{f}_L$ is the elementary Lorentz force acting on a particle in a magnetic field.

The r.h.s. of eqn (1.87) is the collision integral. The impurity part of the collision integral has the form (see Section 15.2)

$$I^{(\mathrm{imp})}\{n_{\mathbf{p}}\} = -\frac{1}{\nu(0)\,\tau_{\mathrm{imp}}}\int (u_{\mathbf{p}}u_{\mathbf{p}'} - v_{\mathbf{p}}v_{\mathbf{p}'})^2 \times (n_{\mathbf{p}} - n_{\mathbf{p}'})\,\delta(\epsilon_{\mathbf{p}} - \epsilon_{\mathbf{p}'})\frac{d^3p'}{(2\pi)^3}. \tag{1.89}$$

where the Bogoliubov–de Gennes coherence factors are determined by eqn (1.56). The expression for the phonon collision integral is rather complicated and we do not write it here.

The electric field enters the kinetic equation through the time-derivative of energy according to

$$\frac{\partial n_{\mathbf{p}}}{\partial t} = \frac{\partial n_0}{\partial \epsilon_{\mathbf{p}}}\frac{\partial \epsilon_{\mathbf{p}}}{\partial t} = \frac{\partial n_0}{\partial \epsilon_{\mathbf{p}}}\left(e\mathbf{v}_F\cdot\mathbf{E} + e\mathbf{v}_{\mathbf{F}}\cdot\nabla\Phi + \frac{\partial|\Delta|}{\partial t}\frac{|\Delta|}{\tilde{\epsilon}_{\mathbf{p}}}\right) \tag{1.90}$$

because

$$\frac{\partial \mathbf{p}_s}{\partial t} = e\mathbf{E} + e\nabla\Phi.$$

Solving the Boltzmann kinetic equation one can find the distribution function which should then be used in the BCS equation (1.57) for self-consistent determination of the order parameter.

Equation (1.87) is simple and transparent. However, its region of applicability is very limited because the order parameter and the supercurrent were assumed almost constant in space. In fact, attempts to use this equation together with the spectral characteristics of a superconductor for slow varying $|\Delta|$ and $\mathbf{p}_s$ may lead to incorrect results because the excitation spectrum eqn (1.54) is itself only defined for constant $|\Delta|$ and $\mathbf{p}_s$ and does not contain corrections which would appear if $|\Delta|$ and $\mathbf{p}_s$ become functions of coordinates. Moreover, the Boltzmann kinetic equation uses a concept of quasiparticle spectrum. The spectrum is well defined only for a clean superconductor such that $\Delta\tau \gg \hbar$. Dirty alloys can not be incorporated into the Boltzmann scheme. Again, we are in a situation where the use of a more general microscopic theory is vital.

The kinetic equation approach is not reduced to a TDGL scheme as T approaches T_c. In fact, these two descriptions do not overlap. Indeed, for the TDGL model it is important that the distribution function of excitations does not deviate from equilibrium. On the contrary, the kinetic equation approach implies that deviations from equilibrium do appear in a time-dependent state of a superconductor. The kinetic theory based on eqn (1.87) can give a hint on the limit of applicability of the TDGL model. If the order parameter gradients are slow on a length scale of order of the mean free path, the kinetic equation can be transformed into a diffusion-like equation which gives (compare with eqn (10.107) on page 211)

$$\left(D\nabla^2 - \frac{\epsilon_{\mathbf{p}}}{\tau_{\mathrm{ph}}\xi_{\mathbf{p}}}\right)\delta n_{\mathbf{p}} = \frac{\partial|\Delta|}{\partial t}\frac{|\Delta|}{\xi_{\mathbf{p}}}\frac{dn_0}{d\epsilon_{\mathbf{p}}} \tag{1.91}$$

where $\delta n_{\mathbf{p}} = n_{\mathbf{p}} - n_0$ and D is the diffusion constant. The inelastic relaxation is usually very slow. If the diffusion dominates, the deviation of distribution from equilibrium is

$$\delta n_{\mathbf{p}} \sim \left(Dk^2\right)^{-1} \frac{\partial |\Delta|}{\partial t} \frac{|\Delta|}{\xi_{\mathbf{p}}} \frac{dn_0}{d\epsilon_{\mathbf{p}}}.$$

The corresponding contribution to eqn (1.85) is much larger than the TDGL term

$$\frac{\hbar\pi}{8T_c} \frac{\partial |\Delta|}{\partial t}$$

due to a long-range diffusion relaxation. In the opposite limit, where the order parameter varies over distances shorter than the mean free path, i.e., when $\xi \ll \ell$, one generally has

$$\delta n_{\mathbf{p}} \sim \tau \frac{\partial |\Delta|}{\partial t} \frac{|\Delta|}{\epsilon_{\mathbf{p}}} \frac{dn_0}{d\epsilon_{\mathbf{p}}}.$$

The corresponding contribution is by a big factor $|\Delta| \tau \gg \hbar$ larger than the time-dependent Ginzburg–Landau term.

We observe from eqn (1.91) that a nonequilibrium correction to distribution is small compared to the TDGL term only for a hydrodynamic limit when $\tau_{\mathrm{ph}} |\Delta| \ll \hbar$. This is exactly the region of the so-called gapless superconductivity associated with inelastic scattering of electrons. The conclusion is that the TDGL model is expected to work only for gapless superconductors.

1.3 Outline of the contents

All the simple theories considered above have a limited applicability. The TDGL model can treat nonstationary problems in cases when the order parameter and magnetic field vary in space. Moreover, the effect of impurities can easily be taken into account. This model, however, works only for small deviations from equilibrium and only in a close vicinity of T_c. On the contrary, the kinetic scheme of eqn (1.87) can work at low temperatures and include large deviations from equilibrium. Unfortunately, it is not suitable for spatially inhomogeneous problems. In addition, it is not designed to deal with superconducting alloys where the mean free path is comparable to the coherence length, i.e., in cases when the excitation spectrum is not well defined.

A theory which is supposed to describe nonstationary properties of superconductors in the whole range of parameters and in a general, spatially inhomogeneous situation, which would include various types of relaxation processes in a self-consistent and universal way — this theory should provide answers at least to the following problems. (i) How can one define the order parameter and other physical observables in a nonstationary state? As an important part of this general issue, there emerges a problem of calculating the energy spectrum (or other equivalent characteristics of a superconductor if the spectrum is not well defined) in case when the superconducting state is not spatially uniform and varies in time. (ii) What are the processes which determine the distribution of excitations under nonstationary conditions? And, last but not least, (iii) what are the practical methods for solving nonstationary problems and for calculating the physical observables?

We shall discuss these problems throughout this book and describe a general approach which is able to deal with these problems in a coherent way. It is the Green function technique, and, in particular, the technique which uses the real-time Green functions. The important component of the general theory is the quasiclassical method which provides the easiest way of practical implementation of the Green function technique. The theory is based on general principles of the BCS scheme. In addition, it includes interactions of electrons with impurities, phonons, and electron–electron interaction. We start with the general definitions of the Green functions. We describe the BCS theory and derive the Gor'kov equations. Next, we introduce the quasiclassical approach. The real-time Green functions are introduced using two different but completely equivalent methods: the Keldysh formalism and the method of analytical continuation of the imaginary-time Matsubara Green functions onto the real-time domain. We obtain the general expressions for the order parameter and other quantities and derive transport-like and kinetic equations for the distribution functions of excitations. For practical implementations of the general formalism, we derive the TDGL equations and consider other examples of nonstationary phenomena. A considerable part of the book is devoted to the vortex dynamics as a problem which incorporates almost all nonstationary phenomena existing in a superconducting state. Discussion of the vortex dynamics for each particular system starts with an introductory section which elucidates the main processes characteristic for the system under consideration. Let us now turn to the general definition of the Green functions.

2

GREEN FUNCTIONS

We introduce the second quantization formalism and define the temperature and time-dependent Green functions which will be used throughout the book.

2.1 Second quantization

The Green function formalism uses the method of second quantization. We summarize its basic ideas (Landau and Lifshitz 1959 a, Abrikosov *et al.* 1965) to make the presentation self-consistent. We mostly follow the style of Abrikosov *et al.* (1965) in this chapter. The reader who is familiar with the general definition of the Green functions can proceed directly to the next chapter. From now on, we use the units where $\hbar = 1$.

Consider first a system of N Bose particles which can occupy quantum states with wave functions $\phi_1(x)$, $\phi_2(x)$, ... Here x is the set of general coordinates. We describe a state of such a system by a set of numbers N_1, N_2, ..., which show how many particles are in the states 1, 2, ... The wave function for the state with the occupation numbers N_1, N_2, ... is symmetric with respect to transposition of particles:

$$\Phi_{N_1 N_2 \dots} = \left(\frac{N_1! N_2! \dots}{N!}\right)^{1/2} \sum_P \phi_{i_1}(x_1)\phi_{i_2}(x_2)\dots\phi_{i_N}(x_N). \tag{2.1}$$

Here i_m are labels of the states, and the sum is taken over all possible transpositions of the labels i_m. The prefactor takes care of normalization. We introduce the operator a_i which decreases the occupation number in the state i by 1. According to eqn (2.1), its only nonzero matrix element is

$$(a_i)_{N_i}^{N_i - 1} = \sqrt{N_i}. \tag{2.2}$$

The conjugated operator increases the occupation number by 1:

$$(a^\dagger)_{N_i - 1}^{N_i} = \sqrt{N_i}. \tag{2.3}$$

Therefore

$$a_i^\dagger a_i = N_i, \; a_i a_i^\dagger = N_i + 1, \tag{2.4}$$

so that

$$a_i a_k^\dagger - a_k^\dagger a_i = \delta_{ik}, \tag{2.5}$$

and

$$a_i a_k - a_k a_i = a_i^\dagger a_k^\dagger - a_k^\dagger a_i^\dagger = 0. \tag{2.6}$$

For Fermi particles, the occupation numbers may be either 0 or 1 only, and the wave function is antisymmetric with respect to all variables:

$$\Phi_{N_1 N_2 \ldots} = \frac{1}{\sqrt{N!}} \sum_P \delta_P \phi_{i_1}(x_1)\phi_{i_2}(x_2)\ldots\phi_{i_N}(x_N). \tag{2.7}$$

Here δ_P takes +1 or -1 depending on whether the state $i_1, i_2, \ldots$ is obtained by an even or odd transposition of some initial configuration, respectively. Creation and annihilation operators now have the matrix elements

$$(a_i)_1^0 = (a_i^\dagger)_0^1 = (-1)^{\sum_{l=1}^{i-1} N_l}. \tag{2.8}$$

Therefore,

$$a_i^\dagger a_i = N_i, \; a_i a_i^\dagger = 1 - N_i. \tag{2.9}$$

The commutation rules are

$$a_i a_k^\dagger + a_k^\dagger a_i = \delta_{ik}, \tag{2.10}$$

$$a_i a_k + a_k a_i = a_i^\dagger a_k^\dagger + a_k^\dagger a_i^\dagger = 0. \tag{2.11}$$

Consider an operator of the form

$$\hat{L} = \sum_a L^{(1)}(x_a) + \sum_{a,b} L^{(2)}(x_a, x_b) + \ldots \tag{2.12}$$

where $L^{(1)}$ depends on a single coordinate (single-particle operator), $L^{(2)}$ depends on two coordinates (two-particle operator), etc. In terms of creation and annihilation operators, we can write

$$\hat{L} = \sum_{ik} L_{ik}^{(1)} a_i^\dagger a_k + \sum_{iklm} L_{lm}^{(2)\,ik} a_i^\dagger a_k^\dagger a_l a_m + \ldots \tag{2.13}$$

where the matrix elements are

$$L_{ik}^{(1)} = \int \phi_i^*(x) L^{(1)}(x) \phi_k(x)\, dx$$

and

$$L_{ik\,lm}^{(2)} = \int \phi_i^*(x_1)\phi_k^*(x_2) L^{(2)}(x_1, x_2)\phi_l(x_2)\phi_m(x_1)\, dx_1\, dx_2.$$

The operator in eqn (2.13) acts on the occupation numbers of particles. This definition holds for both Bose and Fermi particles.

2.1.1 *Schrödinger and Heisenberg operators*

One can introduce the operators

$$\psi(x) = \sum_i \phi_i(x) a_i, \; \psi^\dagger(x) = \sum_i \phi_i^*(x) a_i^\dagger. \tag{2.14}$$

Operator $\psi^\dagger(x)$ creates a particle at the point x while $\psi(x)$ annihilates it. These operators are called Schrödinger "particle-field" operators. We have

$$\psi(x)\psi^\dagger(x') \pm \psi^\dagger(x')\psi(x) = \delta(x - x'), \tag{2.15}$$

$$\psi(x)\psi(x') \pm \psi(x')\psi(x) = 0, \tag{2.16}$$

$$\psi^\dagger(x)\psi^\dagger(x') \pm \psi^\dagger(x')\psi^\dagger(x) = 0. \tag{2.17}$$

The upper sign is for Fermi particles while the lower is for Bose particles. The coordinate x may also include spin. The operator $\hat{L}$, which for definiteness is taken to be dependent on spatial coordinates, can be written as

$$\hat{L} = \sum_\alpha \int \psi_\alpha^\dagger(\mathbf{r}) L^{(1)}(\mathbf{r}) \psi_\alpha(\mathbf{r})\, d^3r$$
$$+ \sum_{\alpha\beta} \int \int \psi_\alpha^\dagger(\mathbf{r}) \psi_\beta^\dagger(\mathbf{r}') L^{(2)}(\mathbf{r}, \mathbf{r}') \psi_\beta(\mathbf{r}') \psi_\alpha(\mathbf{r})\, d^3r\, d^3r' \ldots$$

Here α and β are the spin indices.

Let the Hamiltonian be $\hat{\mathcal{H}}$. The wave function obeys the Schrödinger equation

$$i\frac{\partial \Phi}{\partial t} = \hat{\mathcal{H}}\Phi. \tag{2.18}$$

Its solution can be symbolically written as

$$\Phi(t) = e^{-i\hat{\mathcal{H}}t}\Phi_H \tag{2.19}$$

where Φ_H does not depend on t. A time-dependent matrix element of an operator F is

$$F_{nm}(t) = \int \Phi_n^*(t) F \Phi_m(t)\, d^3r = \int \Phi_{Hn}^*(t) e^{i\hat{\mathcal{H}}t} F e^{-i\hat{\mathcal{H}}t} \Phi_{Hm}(t)\, d^3r. \tag{2.20}$$

This is the matrix element of the operator

$$\tilde{F}(t) = e^{i\hat{\mathcal{H}}t} F e^{-i\hat{\mathcal{H}}t}. \tag{2.21}$$

This operator is called the Heisenberg operator. Its convenience is in that now one can use time-independent wave functions Φ_H to calculate time-dependent matrix elements. The time dependence is transferred to the operators. The time derivative of $\tilde{F}$ is

$$\frac{\partial \tilde{F}}{\partial t} = i(\hat{\mathcal{H}}\tilde{F} - \tilde{F}\hat{\mathcal{H}}). \tag{2.22}$$

Note that the commutation rules for Heisenberg operators taken at the same time are the same as for the corresponding operators in Schrödinger representation, since the exponential factors cancel in products of two operators.

2.2 Imaginary-time Green function

2.2.1 *Definitions*

The imaginary-time (Matsubara) Green function is defined for imaginary time $t = -i\tau$ within the interval

$$-\frac{1}{T} < \tau_1 - \tau_2 < \frac{1}{T}$$

as follows:

$$G_{\alpha\beta}(\mathbf{r}_1, \tau_1; \mathbf{r}_2, \tau_2) = \sum\nolimits^{(d)} \left[\exp\left(\frac{\Omega + \mu\hat{\mathcal{N}} - \hat{\mathcal{H}}}{T}\right) e^{(\hat{\mathcal{H}} - \mu\hat{\mathcal{N}})(\tau_1 - \tau_2)} \right.$$
$$\left. \times \psi_\alpha(\mathbf{r}_1)\, e^{-(\hat{\mathcal{H}} - \mu\hat{\mathcal{N}})(\tau_1 - \tau_2)} \psi_\beta^\dagger(\mathbf{r}_2) \right] \tag{2.23}$$

for $\tau_1 > \tau_2$ and

$$G_{\alpha\beta}(\mathbf{r}_1, \tau_1; \mathbf{r}_2, \tau_2) = \mp \sum\nolimits^{(d)} \left[\exp\left(\frac{\Omega + \mu\hat{\mathcal{N}} - \hat{\mathcal{H}}}{T}\right) e^{-(\hat{\mathcal{H}} - \mu\hat{\mathcal{N}})(\tau_1 - \tau_2)} \right.$$
$$\left. \times \psi_\beta^\dagger(\mathbf{r}_2)\, e^{(\hat{\mathcal{H}} - \mu\hat{\mathcal{N}})(\tau_1 - \tau_2)} \psi_\alpha(\mathbf{r}_1) \right] \tag{2.24}$$

if $\tau_1 < \tau_2$. Here $\psi_\alpha(\mathbf{r})$ and $\psi_\beta^\dagger(\mathbf{r})$ are Schrödinger operators; the signs $\mp$ refer to Fermi and Bose particles, respectively. Operators $\hat{\mathcal{H}}$ and $\hat{\mathcal{N}}$ are the Hamiltonian of the system and the particle-number operator. Operator $\sum^{(d)}$ means the sum over all diagonal matrix elements both for all quantum states for a given number of particles and for all numbers of particles. Thus G is a function of T and the chemical potential μ; Ω is the thermodynamic potential: $d\Omega = -\mathcal{S}\, dT - P\, dV - \mathcal{N}\, d\mu$. It is defined as

$$\Omega = \mathcal{F} - \mu\mathcal{N} \tag{2.25}$$

where $\mathcal{F}$ is the free energy,

$$\mathcal{F} = -T \ln \sum_n e^{-E_n/T}. \tag{2.26}$$

Therefore, the sum

$$\sum\nolimits^{(d)} \left[\exp\left(\frac{\Omega + \mu\hat{\mathcal{N}} - \hat{\mathcal{H}}}{T}\right) \ldots \right] = \langle \ldots \rangle_{st} \tag{2.27}$$

is the Gibbs statistical average. Note that

$$\langle 1 \rangle_{st} = 1$$

due to the definitions of eqns (2.25) and (2.26).

The phonon Green function is

$$D(\mathbf{r}_1, \tau_1; \mathbf{r}_2, \tau_2) = \sum{}^{(d)} \left[\exp\left(\frac{\Omega - \hat{\mathcal{H}}}{T}\right) e^{\hat{\mathcal{H}}(\tau_1 - \tau_2)} \phi(\mathbf{r}_1) e^{-\hat{\mathcal{H}}(\tau_1 - \tau_2)} \phi(\mathbf{r}_2) \right]$$

for $\tau_1 > \tau_2$ and

$$D(\mathbf{r}_1, \tau_1; \mathbf{r}_2, \tau_2) = \sum{}^{(d)} \left[\exp\left(\frac{\Omega - \hat{\mathcal{H}}}{T}\right) e^{-\hat{\mathcal{H}}(\tau_1 - \tau_2)} \phi(\mathbf{r}_2) e^{\hat{\mathcal{H}}(\tau_1 - \tau_2)} \phi(\mathbf{r}_1) \right]$$

for $\tau_1 < \tau_2$.

The function $G(\tau_1 - \tau_2)$ experiences a jump at $\tau = \tau_1 - \tau_2 = 0$. For Fermi particles,

$$\begin{aligned} [G(\tau) - G(-\tau)]_{\tau \to +0} &= \left\langle \left(\psi_\alpha(\mathbf{r}_1) \psi^\dagger_\beta(\mathbf{r}_2) + \psi^\dagger_\beta(\mathbf{r}_2) \psi_\alpha(\mathbf{r}_1) \right) \right\rangle_{st} \\ &= \delta_{\alpha\beta} \delta(\mathbf{r}_1 - \mathbf{r}_2). \end{aligned} \tag{2.28}$$

Let us introduce the "Heisenberg" particle operators which depend on time τ. In the representation with a given μ we have

$$\begin{aligned} \tilde{\psi}_\alpha(\mathbf{r}, \tau) &= e^{(\hat{\mathcal{H}} - \mu\hat{\mathcal{N}})\tau} \psi_\alpha(\mathbf{r}) e^{-(\hat{\mathcal{H}} - \mu\hat{\mathcal{N}})\tau}, \\ \tilde{\psi}^\dagger_\alpha(\mathbf{r}, \tau) &= e^{(\hat{\mathcal{H}} - \mu\hat{\mathcal{N}})\tau} \psi^\dagger_\alpha(\mathbf{r}) e^{-(\hat{\mathcal{H}} - \mu\hat{\mathcal{N}})\tau}, \end{aligned} \tag{2.29}$$

$$\tilde{\phi}(\mathbf{r}, \tau) = e^{\hat{\mathcal{H}}\tau} \phi(\mathbf{r}) e^{-\hat{\mathcal{H}}\tau}. \tag{2.30}$$

The Heisenberg operators satisfy the equations

$$\begin{aligned} \frac{\partial \tilde{\psi}_\alpha}{\partial \tau} &= \left(\hat{\mathcal{H}} - \mu\hat{\mathcal{N}}\right) \tilde{\psi}_\alpha - \tilde{\psi}_\alpha \left(\hat{\mathcal{H}} - \mu\hat{\mathcal{N}}\right), \\ \frac{\partial \tilde{\psi}^\dagger_\alpha}{\partial \tau} &= \left(\hat{\mathcal{H}} - \mu\hat{\mathcal{N}}\right) \tilde{\psi}^\dagger_\alpha - \tilde{\psi}^\dagger_\alpha \left(\hat{\mathcal{H}} - \mu\hat{\mathcal{N}}\right). \end{aligned} \tag{2.31}$$

Using this definition, we can write

$$\begin{aligned} G_{\alpha\beta}(\mathbf{r}_1, \tau_1; \mathbf{r}_2, \tau_2) &= \mathrm{Tr} \left[\exp\left(\frac{\Omega + \mu\hat{\mathcal{N}} - \hat{\mathcal{H}}}{T}\right) T_\tau \left(\tilde{\psi}_\alpha(\mathbf{r}_1, \tau_1) \tilde{\psi}^\dagger_\beta(\mathbf{r}_2, \tau_2) \right) \right] \\ &\equiv \left\langle T_\tau \left(\tilde{\psi}_\alpha(\mathbf{r}_1, \tau_1) \tilde{\psi}^\dagger_\beta(\mathbf{r}_2, \tau_2) \right) \right\rangle_{st}. \end{aligned} \tag{2.32}$$

Here T_τ means ordering in τ: the ψ-operators under T_τ are placed from left to right in order of decreasing "time" τ. For Fermi particles we have

$$T_\tau \left(\psi_1 \psi_2 \ldots \right) = \delta_P \psi_{i_1} \psi_{i_2} \cdots \tag{2.33}$$

In the r.h.s., the ψ-operators are put in the chronological order, and δ_P takes $+1$ or -1 depending on whether the transposition

$$1, 2, \ldots \to i_1, i_2, \ldots$$

is even or odd. For example,

$$T_\tau\left(\tilde{\psi}_1\tilde{\psi}_2^\dagger\right) = \begin{cases} \tilde{\psi}_1\tilde{\psi}_2^\dagger, & \tau_1 > \tau_2, \\ -\tilde{\psi}_2^\dagger\tilde{\psi}_1, & \tau_1 < \tau_2. \end{cases}$$

The definition eqn (2.32) can be generalized to the case when the Hamiltonian depends on time τ explicitly. The Heisenberg operators are defined as

$$\begin{aligned} \tilde{\psi}_\alpha(\mathbf{r},\tau) &= \hat{S}^{-1}(\tau,0)\,\psi_\alpha(\mathbf{r},0)\hat{S}(\tau,0)\,, \\ \tilde{\psi}_\alpha^\dagger(\mathbf{r},\tau) &= \hat{S}^{-1}(\tau,0)\,\psi_\alpha^\dagger(\mathbf{r},0)\hat{S}(\tau,0)\,, \end{aligned} \tag{2.34}$$

where we introduce the S-matrix

$$\hat{S}(\tau,0) = T_\tau \exp\left[-\int_0^\tau (\hat{\mathcal{H}} - \mu\hat{\mathcal{N}})d\tau'\right] \tag{2.35}$$

and

$$\hat{S}^{-1}(\tau,0) = \bar{T}_\tau \exp\left[-\int_\tau^0 (\hat{\mathcal{H}} - \mu\hat{\mathcal{N}})d\tau'\right]. \tag{2.36}$$

The operator $\bar{T}_\tau$ orders times as they appear in the integral, i.e., first comes the time τ then a time which is close to τ and lies between τ and 0, etc. Therefore,

$$\hat{S}^{-1}(\tau,0)\,\hat{S}(\tau,0) = 1.$$

We have for $\tau_1 > \tau_2$

$$\hat{S}(\tau_1,\tau_2) = T_\tau \exp\left[-\int_{\tau_2}^{\tau_1} (\hat{\mathcal{H}} - \mu\hat{\mathcal{N}})d\tau'\right] = \hat{S}(\tau_1,0)\,\hat{S}^{-1}(\tau_2,0)\,.$$

Moreover

$$\hat{S} \equiv \hat{S}\left(\frac{1}{T},0\right) = T_\tau \exp\left[-\int_0^{1/T} (\hat{\mathcal{H}} - \mu\hat{\mathcal{N}})d\tau\right].$$

We also have

$$\exp\left(-\frac{\Omega}{T}\right) = \sum\nolimits^{(d)}\left[\hat{S}\right]. \tag{2.37}$$

The Green function is now defined as

$$\begin{aligned} G_{\alpha\beta}(\mathbf{r}_1,\tau_1;\mathbf{r}_2,\tau_2) &= \frac{\sum^{(d)}\left[\hat{S}T_\tau\left(\tilde{\psi}_\alpha(\mathbf{r}_1,\tau_1)\tilde{\psi}_\beta^\dagger(\mathbf{r}_2,\tau_2)\right)\right]}{\sum^{(d)}\left[\hat{S}\right]} \\ &\equiv \left\langle T_\tau\left(\tilde{\psi}_\alpha(\mathbf{r}_1,\tau_1)\tilde{\psi}_\beta^\dagger(\mathbf{r}_2,\tau_2)\right)\right\rangle_{st} \end{aligned} \tag{2.38}$$

which coincides with eqn (2.32). It is very important to note that since the Hamiltonian of the system is also constructed of the ψ operators, it is the use of imaginary time which allows us to place all the ψ operators in proper positions consistent with the T_τ operation. This is possible if the Green function is defined within the time interval $-1/T < \tau < 1/T$ where $\tau = \tau_1 - \tau_2$. It is exactly the reason why we consider an imaginary time: for a real time it is impossible to define unambiguously the order of the ψ-operators.

There exists an important relation which couples $G(\tau < 0)$ with $G(\tau > 0)$. Indeed, if we make a cyclic transposition under the trace operator $\sum^{(d)}$, we obtain

$$\begin{aligned} G(\tau < 0) &= \mp \sum^{(d)} \left[\exp\left(\frac{\Omega + \mu\hat{\mathcal{N}} - \hat{\mathcal{H}}}{T} \right) \right. \\ &\quad \left. \times e^{-(\hat{\mathcal{H}} - \mu\hat{\mathcal{N}})\tau} \psi_\beta^\dagger(\mathbf{r}_2) e^{(\hat{\mathcal{H}} - \mu\hat{\mathcal{N}})\tau} \psi_\alpha(\mathbf{r}_1) \right] \\ &= \mp \sum^{(d)} \left[\exp\left(\frac{\Omega}{T} \right) e^{(\hat{\mathcal{H}} - \mu\hat{\mathcal{N}})\tau} \psi_\alpha(\mathbf{r}_1) e^{-(\hat{\mathcal{H}} - \mu\hat{\mathcal{N}})(\tau + 1/T)} \psi_\beta^\dagger(\mathbf{r}_2) \right] \\ &= \mp \sum^{(d)} \left[\exp\left(\frac{\Omega + \mu\hat{\mathcal{N}} - \hat{\mathcal{H}}}{T} \right) \right. \\ &\quad \left. \times e^{(\hat{\mathcal{H}} - \mu\hat{\mathcal{N}})(\tau + 1/T)} \psi_\alpha(\mathbf{r}_1) e^{-(\hat{\mathcal{H}} - \mu\hat{\mathcal{N}})(\tau + 1/T)} \psi_\beta^\dagger(\mathbf{r}_2) \right] . \end{aligned}$$

Comparing it with eqn (2.23) for $\tau > 0$, we see that

$$G(\tau < 0) = \mp G(\tau + 1/T). \tag{2.39}$$

For phonons we get

$$D(\tau < 0) = D(\tau + 1/T). \tag{2.40}$$

Equation (2.39) is of a great importance. Indeed, let us make a Fourier transformation in τ:

$$G(\tau) = T \sum_n e^{-i\omega_n \tau} G(\omega_n), \tag{2.41}$$

$$G(\omega_n) = \frac{1}{2} \int_{-1/T}^{1/T} e^{i\omega_n \tau} G(\tau)\, d\tau. \tag{2.42}$$

Because time τ is defined within a finite range, the frequency is discrete $\omega_n = n\pi T$. These frequencies are called the Matsubara frequencies. One has

$$\begin{aligned} G(\omega_n) &= \frac{1}{2} \int_0^{1/T} e^{i\omega_n \tau} G(\tau)\, d\tau + \frac{1}{2} \int_{-1/T}^{0} e^{i\omega_n \tau} G(\tau)\, d\tau \\ &= \frac{1}{2} \int_0^{1/T} e^{i\omega_n \tau} G(\tau)\, d\tau \mp \frac{1}{2} \int_{-1/T}^{0} e^{i\omega_n \tau} G(\tau + 1/T)\, d\tau \end{aligned}$$

$$= \frac{1}{2}\left(1 \mp e^{-i\omega_n/T}\right)\int_0^{1/T} e^{i\omega_n\tau}G(\tau)\,d\tau. \tag{2.43}$$

Therefore, the Fourier components

$$G(\omega_n) = \int_0^{1/T} e^{i\omega_n\tau}G(\tau)\,d\tau \tag{2.44}$$

are nonzero only for

$$\omega_n = \begin{cases} (2n+1)\pi T & \text{for Fermi particles,} \\ 2n\pi T & \text{for Bose particles.} \end{cases} \tag{2.45}$$

The Green function determines all properties of the system. For example, the particle-number operator is

$$\hat{\mathcal{N}} = \sum_\alpha \int \psi_\alpha^\dagger(\mathbf{r})\psi_\alpha(\mathbf{r})\,d^3r. \tag{2.46}$$

Thus, the particle density becomes

$$N = -\sum_\alpha G_{\alpha\alpha}(\mathbf{r},\tau;\mathbf{r},\tau+0). \tag{2.47}$$

Using this equation one can find μ as a function of temperature, etc. The momentum density operator is

$$\hat{\mathbf{P}} = -\frac{i}{2}\sum_\alpha \left(\psi_\alpha^\dagger\nabla\psi_\alpha - \nabla\psi_\alpha^\dagger\psi_\alpha\right).$$

Therefore, the momentum density is

$$\mathbf{P} = \frac{i\hbar}{2}\left[(\nabla_1 - \nabla_2)\sum_\alpha G_{\alpha\alpha}(\mathbf{r}_1,\tau;\mathbf{r}_2,\tau+0)\right]_{\mathbf{r}_1=\mathbf{r}_2}. \tag{2.48}$$

2.2.2 *Example: Free particles*

For free noninteracting particles the statistical average is taken over the states of every particle independently of other particles. The energy is

$$E_n^{(0)} = \sum_{\mathbf{p},\alpha} n_{\mathbf{p}\alpha}\epsilon_0(\mathbf{p}),\ \Omega_0 = \sum_{\mathbf{p},\alpha}\Omega_{\mathbf{p}\alpha}^{(0)} \tag{2.49}$$

and the Schrödinger operators are

$$\psi_\alpha(\mathbf{r}) = \frac{1}{\sqrt{V}}\sum_{\mathbf{p}} a_{\mathbf{p}\alpha}e^{i\mathbf{p}\mathbf{r}},\ \psi_\alpha^\dagger(\mathbf{r}) = \frac{1}{\sqrt{V}}\sum_{\mathbf{p}} a_{\mathbf{p}\alpha}^\dagger e^{-i\mathbf{p}\mathbf{r}}.$$

Since the Hamiltonian and the particle-number operators are

$$\hat{\mathcal{H}}_0 = \sum_{\mathbf{p},\alpha}\hat{n}_{\mathbf{p}\alpha}\epsilon_0(\mathbf{p}),\ \hat{\mathcal{N}} = \sum_{\mathbf{p},\alpha}\hat{n}_{\mathbf{p}\alpha} \tag{2.50}$$

the creation and annihilation operators satisfy

$$e^{(\hat{\mathcal{H}}_0-\mu\hat{\mathcal{N}})\tau}a_{\mathbf{p}\alpha}e^{-(\hat{\mathcal{H}}_0-\mu\hat{\mathcal{N}})\tau} = a_{\mathbf{p}\alpha}e^{-(\epsilon_0(\mathbf{p})-\mu)\tau},$$

$$e^{(\hat{\mathcal{H}}_0-\mu\hat{\mathcal{N}})\tau}a^{\dagger}_{\mathbf{p}\alpha}e^{-(\hat{\mathcal{H}}_0-\mu\hat{\mathcal{N}})\tau}=a^{\dagger}_{\mathbf{p}\alpha}e^{(\epsilon_0(\mathbf{p})-\mu)\tau}. \tag{2.51}$$

We have for $\tau>0$

$$G^{(0)}_{\alpha\beta}(\tau>0)=\frac{1}{V}\sum\nolimits^{(d)}\sum_{\mathbf{p}_1,\mathbf{p}_2}e^{i(\mathbf{p}_1\mathbf{r}_1-\mathbf{p}_2\mathbf{r}_2)}\exp\left(\frac{\Omega_0+\mu\hat{\mathcal{N}}-\hat{\mathcal{H}}_0}{T}\right)$$
$$\times\left[e^{(\hat{\mathcal{H}}_0-\mu\hat{\mathcal{N}})\tau}a_{\mathbf{p}_1\alpha}e^{-(\hat{\mathcal{H}}_0-\mu\hat{\mathcal{N}})\tau}a^{\dagger}_{\mathbf{p}_2\beta}\right].$$

As a result,

$$G^{(0)}_{\alpha\beta}(\tau>0)=\delta_{\alpha\beta}\frac{1}{V}\sum_{\mathbf{p}}e^{i\mathbf{p}(\mathbf{r}_1-\mathbf{r}_2)-\tau[\epsilon_0(\mathbf{p}_1)-\mu]}\left\langle a_{\mathbf{p}_1\alpha}a^{\dagger}_{\mathbf{p}_2\beta}\right\rangle_{st}$$

since

$$\left\langle a_{\mathbf{p}_1\alpha}a^{\dagger}_{\mathbf{p}_2\beta}\right\rangle_{st}=\delta_{\alpha\beta}\delta_{\mathbf{p}_1\mathbf{p}_2}\left\langle a_{\mathbf{p}\alpha}a^{\dagger}_{\mathbf{p}\alpha}\right\rangle_{st}.$$

The average for Fermi particles is

$$\left\langle a_{\mathbf{p}\alpha}a^{\dagger}_{\mathbf{p}\alpha}\right\rangle_{st}=1-n(\mathbf{p}),\ n(\mathbf{p})=\left[\exp\left(\frac{\epsilon_0(\mathbf{p})-\mu}{T}\right)+1\right]^{-1}.$$

For Bose particles it is

$$\left\langle a_{\mathbf{p}\alpha}a^{\dagger}_{\mathbf{p}\alpha}\right\rangle_{st}=1+n(\mathbf{p}),\ n(\mathbf{p})=\left[\exp\left(\frac{\epsilon_0(\mathbf{p})-\mu}{T}\right)-1\right]^{-1}.$$

In these equations, $n(\mathbf{p})$ is the average equilibrium occupation number or the distribution function which is different for Fermi and Bose particles.

We now take the limit $V\to\infty$ using

$$\frac{1}{V}\sum_{\mathbf{p}}\to\int\frac{d^3p}{(2\pi)^3}$$

and obtain

$$G^{(0)}_{\alpha\beta}(\mathbf{r},\tau>0)=\delta_{\alpha\beta}\int\frac{d^3p}{(2\pi)^3}e^{i\mathbf{p}\mathbf{r}-\tau[\epsilon_0(\mathbf{p}_1)-\mu]}[1\mp n(\mathbf{p})] \tag{2.52}$$

where $\mathbf{r}=\mathbf{r}_1-\mathbf{r}_2$. For $\tau<0$ we write

$$G^{(0)}_{\alpha\beta}(\mathbf{r},\tau<0)=\mp G^{(0)}_{\alpha\beta}(\mathbf{r},\tau+1/T)$$
$$=\mp\delta_{\alpha\beta}\int\frac{d^3p}{(2\pi)^3}e^{i\mathbf{p}\mathbf{r}-\tau[\epsilon_0(\mathbf{p}_1)-\mu]}n(\mathbf{p}). \tag{2.53}$$

The phonon Green function can be calculated in a similar way. The Schrödinger operator is

$$\phi(\mathbf{r}) = \frac{1}{\sqrt{V}} \sum_{\mathbf{k}} \sqrt{\frac{\omega_0(\mathbf{k})}{2}} \left(b_{\mathbf{k}} e^{i\mathbf{kr}} + b^{\dagger}_{\mathbf{k}} e^{-i\mathbf{kr}} \right)$$

where $\omega_0(\mathbf{k})$ is the phonon energy. The result is

$$D^{(0)}(\mathbf{r}, \tau) = \frac{1}{2} \int \frac{d^3k}{(2\pi)^3} \omega_0(\mathbf{k}) \left[(n(\mathbf{k}) + 1)\, e^{i\mathbf{kr} - \omega_0(\mathbf{k})|\tau|} + n(\mathbf{k}) e^{i\mathbf{kr} + \omega_0(\mathbf{k})|\tau|} \right] \tag{2.54}$$

where

$$n(\mathbf{k}) = \left[e^{\omega_0(\mathbf{k})/T} - 1 \right]^{-1}.$$

It is useful to look at the free-particle Green function in the Fourier representation. Consider a Fermi particle. To calculate $G(\mathbf{p}, \omega_n)$ we take $G(\tau > 0)$ and use eqns (2.44) and (2.52). We have

$$\begin{aligned}
G^{(0)}_{\alpha\beta}(\mathbf{p}, \omega_n) &= \int d^3r \int_0^{1/T} d\tau G^{(0)}_{\alpha\beta}(\mathbf{r}, \tau) e^{-i\mathbf{pr} + i\omega_n \tau} \\
&= \delta_{\alpha\beta}[1 - n(\mathbf{p})] \int_0^{1/T} d\tau e^{i\omega_n \tau - [\epsilon_0(\mathbf{p}) - \mu]\tau} \\
&= \frac{\delta_{\alpha\beta}}{i\omega_n - \epsilon_0(\mathbf{p}) + \mu} \left[e^{(2n+1)\pi i - [\epsilon_0(\mathbf{p}) - \mu]/T} - 1 \right] [1 - n(\mathbf{p})] \\
&= -\frac{\delta_{\alpha\beta}}{i\omega_n - \epsilon_0(\mathbf{p}) + \mu}, \quad \omega_n = (2n+1)\pi T. \qquad (2.55)
\end{aligned}$$

For Bose particles we get

$$G^{(0)}_{\alpha\beta}(\mathbf{p}, \omega_n) = -\frac{\delta_{\alpha\beta}}{i\omega_n - \epsilon_0(\mathbf{p}) + \mu}, \quad \omega_n = 2n\pi T. \tag{2.56}$$

For phonons we have

$$D^{(0)}(\mathbf{k}, \omega_n) = \frac{\omega_0^2(\mathbf{k})}{\omega_0^2(\mathbf{k}) + \omega_n^2}, \quad \omega_n = 2n\pi T. \tag{2.57}$$

2.2.3 *The Wick theorem*

Consider the statistical average

$$\left\langle T_\tau \left(\tilde{\psi}(1) \tilde{\psi}(2) \ldots \tilde{\psi}^\dagger(3) \tilde{\psi}^\dagger(4) \ldots \right) \right\rangle_{st}$$

taken over the states of noninteracting particles. It is the sum over all the quantum states. The wave functions of each state are normalized in such a way that the normalization constant is proportional to $1/\sqrt{V}$ where V is volume of the system. Therefore, the average has the form

$$\frac{1}{\sqrt{V}} \sum_n \frac{1}{\sqrt{V}} \sum_m \cdots \frac{1}{\sqrt{V}} \sum_{n'} \frac{1}{\sqrt{V}} \sum_{m'} \cdots$$

$$\times \left\langle T_\tau \left(a_n(\tau_1) a_m(\tau_2) \ldots a^\dagger_{n'}(\tau_3) a^\dagger_{m'}(\tau_4) \ldots \right) \right\rangle_{st} . \tag{2.58}$$

The matrix elements for the whole combination of a-operators are only nonzero if there are equal numbers of creation and annihilation operators belonging to the same state. For example, nonzero is the term

$$\frac{1}{V}\sum_n \frac{1}{V}\sum_m \ldots \langle T_\tau \left(a_n(\tau_1) a_m(\tau_2) \ldots a^\dagger_n(\tau_3) a^\dagger_m(\tau_4) \ldots \right) \rangle_{st} \tag{2.59}$$

where the states with $n \neq m$ are different from all other quantum states. Nonzero are also other terms which can be obtained from eqn (2.59) by transpositions of the indices n, m, etc. If there are four creation and annihilation operators belonging to the same state the nonzero terms have the form

$$\frac{1}{V}\frac{1}{V}\sum_n \frac{1}{V}\sum_m \ldots \langle T_\tau \left(a_n a_n a_m a_m \ldots a^\dagger_n a^\dagger_n a^\dagger_m a^\dagger_m \ldots \right) \rangle_{st} \tag{2.60}$$

where $n \neq m$. Expressions of the type of eqn (2.59) are different from all other nonzero terms like eqn (2.60) in that the former have as many summation as there are $1/V$ factors. All other terms have at least one extra factor $1/V$. Since the number of quantum states is proportional to the volume of the system, only the terms of the type of eqn (2.59) remain finite in the limit of a large volume of the system $V \to \infty$.

We come to the conclusion that, in the limit of $V \to \infty$, i.e., for a macroscopic system, the statistical average of a combination of ψ-operators is nonzero only if the pairs of creation and annihilation operators $a_n a^\dagger_n$ all belong to different quantum states. This implies that the statistical average can be calculated for each pair of operators $a_n a^\dagger_n$ independently. Indeed, the Hamiltonian and the particle-number operator of non-interacting particles, which both consist of pairs of the $a\, a^\dagger$-operators, commute with any pair of $a\, a^\dagger$ if they all belong to different states. Therefore,

$$\begin{aligned} &\sum\nolimits^{(d)} \left[T_\tau \exp\left(\frac{\Omega + \mu\hat{\mathcal{N}} - \hat{\mathcal{H}}}{T} \right) a_n a^\dagger_n a_m a^\dagger_m \right] \\ &\quad = \sum\nolimits^{(d)} \left[T_\tau \exp\left(\frac{\Omega_n + \mu N_n - H_n}{T} \right) a_n a^\dagger_n \right] \\ &\quad \times \sum\nolimits^{(d)} \left[T_\tau \exp\left(\frac{\Omega_m + \mu N_m - H_m}{T} \right) a_m a^\dagger_m \right] \end{aligned} \tag{2.61}$$

since Hamiltonian is an additive function for noninteracting particles. The fact that all τ are in the interval $-1/T < \tau < 1/T$ allows us to put the exponents again before all the ψ-operators.

In other words, the statistical average of a T_τ-product of a large number of ψ-and $\psi^\dagger$-operators is equal to the sum of T_τ-products of all possible $\psi\psi^\dagger$ pair averages. The sign of each term in the sum is + if the corresponding sequence of the ψ-operators is obtained from the initial sequence by an even transposition, and it is $-$ if the required transposition is odd. This is the contents of the Wick theorem. For example,

$$\left\langle T_\tau\left(\tilde{\psi}(1)\tilde{\psi}(2)\tilde{\psi}^\dagger(3)\tilde{\psi}^\dagger(4)\right)\right\rangle_{st} = \left\langle T_\tau\left(\tilde{\psi}(1)\tilde{\psi}^\dagger(4)\right)\right\rangle_{st}\left\langle T_\tau\left(\tilde{\psi}(2)\tilde{\psi}^\dagger(3)\right)\right\rangle_{st}$$
$$\mp\left\langle T_\tau\left(\tilde{\psi}(1)\tilde{\psi}^\dagger(3)\right)\right\rangle_{st}\left\langle T_\tau\left(\tilde{\psi}(2)\tilde{\psi}^\dagger(4)\right)\right\rangle_{st} .$$

The Wick theorem is of crucial importance for the Green function technique. Indeed, the Hamiltonian of the system is usually made up of various combinations of the particle operators. If the particles interact, the Hamiltonian contains the operators to powers higher than the second power. They come in the exponent, and a general expression for the Green function is thus hopelessly complicated. Here the Wick theorem saves the situation. We can expand the exponent into the power series in the interaction Hamiltonian and obtain the sum of terms which are various even powers of the particle operators. According to the Wick theorem, each term can be reduced to a product of several Green functions of noninteracting particles. As a result, we obtain an expression for the full Green function of interacting particles as a series of terms containing powers of the interaction strength and products of noninteracting Green functions (Feynman diagrams). This allows one to find the full Green function to any desirable accuracy in the interaction strength. In many cases, the summation of such a series can be performed analytically, which results in the so-called Dyson equation for the full Green function. The Dyson equation is the central result for the Green function technique because it provides the algorithm for calculation of the Green function and thus of any physical observable of a complicated system of interacting particles. A more detailed description of the diagram technique for various systems can be found in the book by Abrikosov *et al.* (1965).

2.3 The real-time Green functions

2.3.1 *Definitions*

The real-time Green function is defined as

$$G_{\alpha\beta}\left(\mathbf{r}_1, t_1; \mathbf{r}_2, t_2\right) = i\sum\nolimits^{(d)}\left[\exp\left(\frac{\Omega + \mu\hat{\mathcal{N}}\left(t_0\right) - \hat{\mathcal{H}}\left(t_0\right)}{T}\right)\right.$$
$$\left.\times\, T_t\left(\tilde{\psi}_\alpha\left(\mathbf{r}_1, t_1\right)\tilde{\psi}^\dagger_\beta\left(\mathbf{r}_2, t_2\right)\right)\right] . \tag{2.62}$$

The Heisenberg operators are determined through the real-time S-matrix which is defined similarly to eqn (2.35)

$$\hat{S}\left(t, t_0\right) = T_t \exp\left[-i\int_{t_0}^{t}(\hat{\mathcal{H}} - \mu\hat{\mathcal{N}})\, dt'\right] . \tag{2.63}$$

The real-time S matrix can be obtained from eqn (2.35) by a formal substitution $\tau \to it$.

The statistical average in eqn (2.62) is taken at a time instant t_0. This definition implies that, at the time t_0, the system was in thermodynamic equilibrium; the field operators evolve since then according to the Schrödinger equation or as determined by the S matrix in eqn (2.63). Note that the Hamiltonian can now be a function of time. The real-time Green function defines the density matrix and thus it is this function which determines the physical quantities for a time-dependent system. For example, the particle density is

$$N(\mathbf{r},t) = \pm i \lim_{t_1 \to t_2 - 0} \sum_{\alpha} G_{\alpha\alpha}(\mathbf{r},t_1;\mathbf{r},t_2) . \tag{2.64}$$

We define also the so-called retarded G^R and advanced G^A Green functions:

$$\begin{aligned} G^R_{\alpha\beta}(1;2) = i\sum^{(d)} & \left[\exp\left(\frac{\Omega + \mu\hat{\mathcal{N}}(t_0) - \hat{\mathcal{H}}(t_0)}{T}\right)\right. \\ & \left.\times \left(\tilde{\psi}_\alpha(1)\tilde{\psi}^\dagger_\beta(2) \pm \tilde{\psi}^\dagger_\beta(2)\tilde{\psi}_\alpha(1)\right)\right] , \end{aligned} \tag{2.65}$$

for $t_1 > t_2$ and

$$G^R_{\alpha\beta}(\mathbf{r}_1,t_1;\mathbf{r}_2,t_2) = 0$$

for $t_1 < t_2$. Similarly,

$$\begin{aligned} G^A_{\alpha\beta}(1;2) = -i\sum^{(d)} & \left[\exp\left(\frac{\Omega + \mu\hat{\mathcal{N}}(t_0) - \hat{\mathcal{H}}(t_0)}{T}\right)\right. \\ & \left.\times \left(\tilde{\psi}_\alpha(1)\tilde{\psi}^\dagger_\beta(2) \pm \tilde{\psi}^\dagger_\beta(2)\tilde{\psi}_\alpha(1)\right)\right] , \end{aligned} \tag{2.66}$$

for $t_1 < t_2$ and

$$G^A_{\alpha\beta}(1;2) = 0$$

for $t_1 > t_2$.

2.3.2 *Analytical properties*

Consider the Fourier transformed retarded Green function

$$G^R(t_1,t_2) = \int G^R(\epsilon)e^{-i\epsilon(t_1-t_2)}\,\frac{d\epsilon}{2\pi} . \tag{2.67}$$

We assume for simplicity that the Hamiltonian does not depend on time. The Fourier transformed Green function $G^R(\epsilon)$ is an analytical function of ϵ in the upper half-plane of complex ϵ. Indeed, in this case we can calculate the integral in eqn (2.67) for $t_1 - t_2 < 0$ by shifting the contour of integration into the upper half-plane of ϵ. Since $G^R(\epsilon)$ is analytical, the integral vanishes, and we have $G^R(t_1,t_2) = 0$ for $t_1 < t_2$ as it should be according to the definition. Similarly, $G^A(\epsilon)$ is an analytical function of ϵ in the lower half-plane.

We now want to prove that the Matsubara functions are coupled to the retarded and advanced functions through

$$G(\omega_n) = G^R(i\omega_n),\ \omega_n > 0, \tag{2.68}$$

$$G(\omega_n) = G^A(i\omega_n),\ \omega_n < 0. \tag{2.69}$$

To do this we note that the matrix elements of the Heisenberg operators are

$$\tilde{\psi}_{nm}(\mathbf{r},t) = \psi_{nm}(\mathbf{r})e^{i\omega_{nm}t},$$
$$\tilde{\psi}^\dagger_{nm}(\mathbf{r},t) = \psi^\dagger_{nm}(\mathbf{r})e^{-i\omega_{nm}t}, \tag{2.70}$$

where

$$\omega_{nm} = E_n - E_m - \mu(N_n - N_m) \tag{2.71}$$

with $N_n = N_m \pm 1$. Therefore, for $t > 0$ according to eqn (2.65)

$$\begin{aligned} G^R(t) &= i\sum_{n,m} e^{(\Omega+\mu N_n - E_n)/T}\left(\psi_{nm}(\mathbf{r}_1)\psi^\dagger_{mn}(\mathbf{r}_2)\right)e^{i\omega_{nm}t} \\ &\quad \pm i\sum_{n,m} e^{(\Omega+\mu N_m - E_m)/T}\left(\psi^\dagger_{mn}(\mathbf{r}_2)\psi_{nm}(\mathbf{r}_1)\right)e^{-i\omega_{mn}t} \\ &= i\sum_{n,m} e^{(\Omega+\mu N_n - E_n)/T}\left(\psi_{nm}(\mathbf{r}_1)\psi^\dagger_{mn}(\mathbf{r}_2)\right)e^{-i\omega_{mn}t} \\ &\quad \times\left[1 \pm e^{(E_n - E_m - \mu N_n + \mu N_m)/T}\right], \end{aligned} \tag{2.72}$$

and $G^R(t) = 0$ for $t < 0$. For the Fourier transforms we have

$$\begin{aligned} G^R(\omega) &= \int_{-\infty}^{\infty} G^R(t)e^{i\omega t}\,dt \\ &= \sum_{n,m} e^{(\Omega+\mu N_n - E_n)/T}\left(\psi_{nm}(\mathbf{r}_1)\psi^\dagger_{mn}(\mathbf{r}_2)\right) \\ &\quad \times\left(1 \pm e^{-\omega_{mn}/T}\right)\left[i\pi\delta(\omega - \omega_{mn}) - \frac{1}{\omega - \omega_{mn}}\right]. \end{aligned}$$

Therefore,

$$G^R(\omega) = \int_{-\infty}^{\infty} \frac{\rho(x)\,dx}{x - \omega - i\delta} \tag{2.73}$$

where $\delta \to 0$ and

$$\begin{aligned} \rho(\omega) &= \sum_{n,m} e^{(\Omega+\mu N_n - E_n)/T}\left(\psi_{nm}(\mathbf{r}_1)\psi^\dagger_{mn}(\mathbf{r}_2)\right) \\ &\quad \times\left(1 \pm e^{-\omega_{mn}/T}\right)\delta(\omega - \omega_{mn}). \end{aligned}$$

During the derivation we use the identity

$$\int_0^\infty e^{i\alpha x}\,dx = \left[\int_0^\infty e^{(i\alpha-\delta)x}\,dx\right]_{\delta\to 0} = \left[\frac{i}{\alpha+i\delta}\right]_{\delta\to 0} = \pi\delta(\alpha) + \frac{i}{\alpha}.$$

In the same way we get

$$G^A(\omega) = \int_{-\infty}^{\infty} \frac{\rho(x)\,dx}{x-\omega+i\delta}. \tag{2.74}$$

For the Matsubara Green function we have for $\tau > 0$

$$G(\tau) = \sum_{n,m} e^{(\Omega+\mu N_n - E_n)/T}\,(\psi_{nm}(\mathbf{r}_1)\psi^\dagger_{mn}(\mathbf{r}_2))\,e^{\omega_{nm}\tau}. \tag{2.75}$$

This gives

$$\begin{aligned} G(\omega_k) &= \int_0^{1/T} G(\tau)e^{i\omega_n\tau}\,d\tau \\ &= \sum_{n,m} e^{(\Omega+\mu N_n - E_n)/T}\,(\psi_{nm}(\mathbf{r}_1)\psi^\dagger_{mn}(\mathbf{r}_2))\,\frac{1\pm e^{-\omega_{mn}/T}}{\omega_{mn}-i\omega_k} \\ &= \int_{-\infty}^{\infty}\frac{\rho(x)\,dx}{x-i\omega_k} \end{aligned} \tag{2.76}$$

where $\omega_k = (2k+1)\pi T$ for Fermi particles and $\omega_k = 2k\pi T$ for Bose particles. Comparing eqns (2.73) and (2.74) with (2.76) we arrive at eqn (2.69).

3

THE BCS MODEL

We apply the Green function formalism to the Bardeen–Cooper–Schrieffer theory of superconductivity and derive the Gor'kov equations which make the basis for our analysis. We couple the Green functions to such physical quantities as the order parameter of the superconductor, the electric current, the electron density, and the thermodynamic potential. We derive the Bogoliubov–de Gennes equations from the Gor'kov equations. Several applications of the Gor'kov theory are considered.

3.1 BCS theory and Gor'kov equations

The BCS Hamiltonian (Bardeen *et al.* 1957) contains the attractive interaction between electrons. This attraction is needed for electrons to form Cooper pairs which then condense into a superconducting state. We consider a spin-independent interaction. This leads to a zero-spin (spin-singlet) superconducting state which is what one usually has in real superconductors. The Fermi statistics of electrons requires a spin-singlet state to have an even parity with respect to the transposition of the particle coordinates or (which is the same) with respect to inversion of the relative momentum of particles. This means that the superconducting state should be of either s-wave or d-wave (or of other higher-order even) symmetry in the relative momentum of particles. We start our consideration with the s-wave superconducting state and will consider some specifics of a d-wave state later.

There can be various physical mechanisms of attraction between electrons. In the phonon model, for example, the attraction is mediated by an exchange of phonons. The pairing interaction usually works in a restricted energy range and vanishes for energy transfer larger than some cut-off value $\Omega_{\mathbf{BCS}}$. In case of phonons, the interaction is effective only when the exchange of energy is less than the characteristic Debye frequency Ω_D. If the interaction is relatively weak the characteristic energies of particles participating in the superconducting phenomena, are much smaller than both the Fermi energy and the cut-off frequency $\Omega_{\mathbf{BCS}}$. In this case, a point-like interaction between the electrons in the form

$$U(\mathbf{r}_1 - \mathbf{r}_2) = \frac{g}{2}\delta(\mathbf{r}_1 - \mathbf{r}_2)$$

is a reasonable approximation. An attraction corresponds to $g < 0$. The limit of small interaction corresponds to $|g|\nu \ll 1$ where ν is the density of states at the Fermi level in the normal state. It is called a weak-coupling approximation.

Let us derive the equations for the Green functions for the BCS model following Gor'kov (1958) and Abrikosov *et al.* (1965). The BCS Hamiltonian is

$$\hat{\mathcal{H}}_{\rm BCS} = \int \left[-\psi^\dagger_\alpha \frac{\nabla^2}{2m}\psi_\alpha + \frac{g}{2}\psi^\dagger_\beta\psi^\dagger_\alpha\psi_\alpha\psi_\beta \right] d^3r. \tag{3.1}$$

The Hamiltonian and the particle number operator

$$\hat{\mathcal{N}} = \int \psi^\dagger_\alpha \psi_\alpha \, d^3 r$$

can be written through the Heisenberg operators which are defined in eqn (2.29). Since $\hat{\mathcal{N}}$ commutes with the Hamiltonian, the operators $\hat{\mathcal{H}}$ and $\hat{\mathcal{N}}$ in this representation will have the same form as in eqns (3.1), (2.46).

Using eqn (2.31) we can calculate the time-derivative of the Heisenberg operators $\tilde{\psi}$. Commuting $\tilde{\psi}$ with the Hamiltonian and the particle-number operator taken at the same time we get

$$\begin{aligned}
\frac{\partial \tilde{\psi}_\alpha(x)}{\partial \tau} &= \left(\frac{\nabla^2}{2m} + \mu\right)\tilde{\psi}_\alpha(x) - g\tilde{\psi}^\dagger_\gamma(x)\tilde{\psi}_\gamma(x)\tilde{\psi}_\alpha(x),\\
\frac{\partial \tilde{\psi}^\dagger_\alpha(x)}{\partial \tau} &= -\left(\frac{\nabla^2}{2m} + \mu\right)\tilde{\psi}^\dagger_\alpha(x) + g\tilde{\psi}^\dagger_\alpha(x)\tilde{\psi}^\dagger_\gamma(x)\tilde{\psi}_\gamma(x).
\end{aligned} \tag{3.2}$$

Here $x = (\tau, \mathbf{r})$. The time-derivative of the Green function is

$$\begin{aligned}
\frac{\partial G_{\alpha\beta}(x_1,x_2)}{\partial \tau_1} &= \frac{\partial}{\partial \tau_1}(\Delta G) + \left\langle T_\tau \frac{\partial \tilde{\psi}_\alpha(x_1)}{\partial \tau_1}\tilde{\psi}^\dagger_\beta(x_2)\right\rangle_{st}\\
&= \frac{\partial}{\partial \tau_1}[\delta_{\alpha\beta}\delta(\mathbf{r}_1 - \mathbf{r}_2)\Theta(\tau_1-\tau_2)]\\
&\quad + \left\langle T_\tau \left[\left(\frac{\nabla^2}{2m}+\mu\right)\tilde{\psi}_\alpha(x_1)\tilde{\psi}^\dagger_\beta(x_2) \right.\right.\\
&\quad \left.\left. - \, g\tilde{\psi}^\dagger_\gamma(x_1)\tilde{\psi}_\gamma(x_1)\tilde{\psi}_\alpha(x_1)\tilde{\psi}^\dagger_\beta(x_2)\right]\right\rangle_{st}\\
&= \delta_{\alpha\beta}\delta(\mathbf{r}_1-\mathbf{r}_2)\delta(\tau_1-\tau_2) + \left(\frac{\nabla^2}{2m}+\mu\right)G_{\alpha\beta}(x_1,x_2)\\
&\quad - g\left\langle T_\tau \tilde{\psi}^\dagger_\gamma(x_1)\tilde{\psi}_\gamma(x_1)\tilde{\psi}_\alpha(x_1)\tilde{\psi}^\dagger_\beta(x_2)\right\rangle_{st}.
\end{aligned}$$

(ΔG) denotes the Green function jump at $\tau_1 = \tau_2$ according to eqn (2.28). The last line can be simplified using the Wick theorem:

$$\begin{aligned}
&\left\langle T_\tau \tilde{\psi}^\dagger_\gamma(x_1)\tilde{\psi}_\gamma(x_1)\tilde{\psi}_\alpha(x_1)\tilde{\psi}^\dagger_\beta(x_2)\right\rangle_{st}\\
&\quad = -\left\langle T_\tau \tilde{\psi}_\gamma(x_1)\tilde{\psi}^\dagger_\gamma(x_1)\right\rangle \left\langle T_\tau \tilde{\psi}_\alpha(x_1)\tilde{\psi}^\dagger_\beta(x_2)\right\rangle_{st}\\
&\qquad + \left\langle T_\tau \tilde{\psi}_\alpha(x_1)\tilde{\psi}^\dagger_\gamma(x_1)\right\rangle \left\langle T_\tau \tilde{\psi}_\gamma(x_1)\tilde{\psi}^\dagger_\beta(x_2)\right\rangle_{st}
\end{aligned}$$

$$-\left\langle T_\tau\tilde{\psi}_\alpha(x_1)\tilde{\psi}_\gamma(x_1)\right\rangle\left\langle T_\tau\tilde{\psi}^\dagger_\gamma(x_1)\tilde{\psi}^\dagger_\beta(x_2)\right\rangle_{st}. \tag{3.3}$$

Here we need some discussion: First, we note that the decomposition of the four-operator average into a product of two averages of pairs of the operators is strictly true for noninteracting particles according to the Wick theorem. Here we perform the decomposition for real, and thus interacting, particles. There are two major types of interaction. Interactions of one type are such that do not result in the superconducting behavior. We neglect them. This is a model approximation; the quality of the model should be considered separately for each particular case. For example, it is not correct, of course, for a strongly correlated electron system. However, for usual superconductors, where electrons form a Fermi liquid with weak correlations, the BCS model proved to be very realistic. The interaction of the second kind is the one which results in the superconducting pairing. What we should have done with this interaction is to expand the average of the four particle operators into the power series in the interaction strength g and then to average each term using the Wick theorem. We would obtain the series of diagrams containing the non-superconducting Green functions. Performing summation of this series we would then obtain the full Green function as it appears in eqn (3.3).

The second major feature of the model is as follows. The averages in the third line in the r.h.s of eqn (3.3) do not disappear in the limit of a macroscopic volume $V \to \infty$ as it was assumed earlier during the derivation of the Wick theorem. The reason is that now the average

$$\left\langle T_\tau\tilde{\psi}_\alpha(x_1)\tilde{\psi}_\gamma(x_1)\right\rangle_{st} \tag{3.4}$$

contains a macroscopic number of particles because of the Cooper pairing effect: due to a small attraction between electrons they form pairs which then condense into a ground state. This is the basic assumption of the BCS theory. The average of eqn (3.4) has the form of a wave function of the condensed Cooper pairs, i.e., of the superconducting electrons.

The first and the second terms in the r.h.s. of eqn (3.3) have the form

$$-\Sigma_{\gamma\gamma}(x_1)G_{\alpha\beta}(x_1,x_2)+\Sigma_{\alpha\gamma}(x_1)G_{\gamma\beta}(x_1,x_2)$$

where

$$\Sigma_{\alpha\beta}(x)=\left\langle T_\tau\tilde{\psi}_\alpha(x_1)\tilde{\psi}^\dagger_\beta(x_1)\right\rangle_{st}\equiv G_{\alpha\beta}(x,x)$$

is called the self-energy. The self-energy leads to a renormalization of the chemical potential μ. In addition, it introduces a small imaginary part proportional to the small g. In the BCS model, this relaxation term is ignored, and the only effect of the interaction is assumed to be the pairing leading to a nonzero average of the type of $\langle\psi\psi\rangle$. In the phonon model to be discussed in Section 8.2, the self-energy is of the order of T^3/Ω_D^2; it is much smaller than the characteristic superconducting energy in the so-called weak-coupling limit to which we restrict our consideration. However, the relaxation part can be important for nonstationary processes; we shall incorporate it where appropriate while considering nonstationary phenomena throughout the book.

Dealing with the BCS model, we omit the self-energy from the expression for the time-derivative of the Green function. Finally, we obtain

$$\left(\frac{\partial}{\partial\tau_1} - \frac{\nabla_1^2}{2m} - \mu\right) G_{\alpha\beta}(x_1, x_2) - g\left\langle T_\tau \tilde{\psi}_\alpha(x_1)\tilde{\psi}_\gamma(x_1)\right\rangle_{st}\left\langle T_\tau \tilde{\psi}_\gamma^\dagger(x_1)\tilde{\psi}_\beta^\dagger(x_2)\right\rangle_{st}$$
$$= \delta_{\alpha\beta}\delta(x_1 - x_2)$$

where $\delta(x_1 - x_2) = \delta(\tau_1 - \tau_2)\delta(\mathbf{r}_1 - \mathbf{r}_2)$. We define the new Green functions

$$F_{\alpha\beta}^\dagger(x_1, x_2) = \left\langle T_\tau \tilde{\psi}_\alpha^\dagger(x_1)\tilde{\psi}_\beta^\dagger(x_2)\right\rangle_{st}, \quad (3.5)$$

$$F_{\alpha\beta}(x_1, x_2) = \left\langle T_\tau \tilde{\psi}_\alpha(x_1)\tilde{\psi}_\beta(x_2)\right\rangle_{st}. \quad (3.6)$$

They are called the "anomalous" or Gor'kov functions. We also define

$$\Delta_{\alpha\beta}(x) = |g| F_{\alpha\beta}(x, x). \quad (3.7)$$

Here we take into account that $g < 0$. The equation for the Green function becomes

$$\left(\frac{\partial}{\partial\tau_1} - \frac{\nabla_1^2}{2m} - \mu\right) G_{\alpha\beta}(x_1, x_2) + \Delta_{\alpha\gamma}(x_1)F_{\gamma\beta}^\dagger(x_1, x_2) = \delta_{\alpha\beta}\delta(x_1 - x_2). \quad (3.8)$$

The function $F_{\alpha\beta}(x_1, x_2)$ is odd in transposition of the particle coordinates and spin indices because of the Fermi statistics of electrons. For a pairing interaction which has an even parity in the orbital space such as an s-wave (or d-wave) interaction, the Cooper pairing can only occur between the electrons with opposite spin projections into a singlet state and the pair wave function is antisymmetric in spin indices:

$$\Delta_{\alpha\beta}(x) = -\Delta_{\beta\alpha}(x).$$

We can write

$$\Delta_{\alpha\beta} = i\sigma_{\alpha\beta}^{(2)}\Delta(x)\ ;\ \Delta_{\alpha\beta}^\dagger = -i\sigma_{\alpha\beta}^{(2)}\Delta^*(x)$$

where

$$\sigma_{\alpha\beta}^{(2)} = \begin{pmatrix} 0 & -i \\ i & 0 \end{pmatrix}$$

is the Pauli spin matrix. We also denote

$$F_{\alpha\beta}^\dagger(x_1, x_2) = -i\sigma_{\alpha\beta}^{(2)}F^\dagger(x_1, x_2)\ ;\ F_{\alpha\beta}(x_1, x_2) = i\sigma_{\alpha\beta}^{(2)}F(x_1, x_2).$$

Since our interaction does not depend on spin, the Green function G is proportional to the unit matrix in the spin indices:

$$G_{\alpha\beta}(x_1, x_2) = \delta_{\alpha\beta} G(x_1, x_2).$$

As a result, we obtain from eqn (3.8)

$$\left(\frac{\partial}{\partial \tau_1} - \frac{\nabla_1^2}{2m} - \mu \right) G(x_1, x_2) + \Delta(x_1) F^\dagger(x_1, x_2) = \delta(x_1 - x_2). \tag{3.9}$$

This equation contains an unknown function $F^\dagger$ and the quantity $\Delta(x)$ which is expressed through another unknown function F. To find these functions we need more equations. The equation for $F^\dagger$ can be obtained from the second eqn (3.2) in a way similar to that used to derive eqn (3.9). We have

$$\left(\frac{\partial}{\partial \tau_1} + \frac{\nabla_1^2}{2m} + \mu \right) F^\dagger(x_1, x_2) + \Delta^*(x_1) G(x_1, x_2) = 0. \tag{3.10}$$

The equation for F contains the function

$$\bar{G}_{\alpha\beta}(x_1, x_2) = -\left\langle T_\tau \bar{\psi}_\alpha^\dagger(x_1) \psi_\beta(x_2) \right\rangle = \delta_{\alpha\beta} \bar{G}(x_1, x_2). \tag{3.11}$$

It is easy to see that

$$\bar{G}(x_1, x_2) = G(x_2, x_1). \tag{3.12}$$

We have for F and $\bar{G}$:

$$-\left(\frac{\partial}{\partial \tau_1} + \frac{\nabla_1^2}{2m} + \mu \right) \bar{G}(x_1, x_2) + \Delta^*(x_1) F(x_1, x_2) = \delta(x_1 - x_2) \tag{3.13}$$

and

$$\left(-\frac{\partial}{\partial \tau_1} + \frac{\nabla_1^2}{2m} + \mu \right) F(x_1, x_2) + \Delta(x_1) \bar{G}(x_1, x_2) = 0. \tag{3.14}$$

Equations (3.9), (3.10), (3.13) and (3.14) form the system of generalized Dyson equations for the BCS model. These equations are known as the Gor'kov equations (Gor'kov 1958). The quantity Δ is called the order parameter. We have already encountered it in Chapter 1. It has exactly the same meaning as in the Landau theory of second-order phase transitions: it is zero in the non-superconducting (normal) state and nonzero in a superconducting state. It is proportional to the wave function of superconducting electrons and thus is a complex function. According to the definition of eqn (3.7), we have

$$\Delta(x) = |g| F(x, x) \; ; \; \Delta^*(x) = |g| F^\dagger(x, x). \tag{3.15}$$

This is a self-consistency equation: it couples the order parameter with the function F which has to be found from equations containing Δ.

The function $G(x_1, x_2)$ describes a particle moving from point x_1 to point x_2 while $\bar{G}(x_1, x_2)$ describes a particle moving from point x_2 to point x_1. The latter particle is equivalent to a hole. According to eqn (3.12), the fact that the Gor'kov equations contain $G(x_1, x_2)$ and $\bar{G}(x_1, x_2)$ on equal footing is a manifestation of the particle–hole symmetry of the BCS theory.

We can introduce the matrix Green function

$$\check{G}(x_1,x_2)=\begin{pmatrix} G(x_1,x_2) & F(x_1,x_2)\\ -F^{\dagger}(x_1,x_2) & \bar{G}(x_1,x_2)\end{pmatrix} \tag{3.16}$$

and the matrix operator

$$\check{G}^{-1}=\check{\tau}_3\frac{\partial}{\partial\tau}+\check{\mathcal{H}}$$

where

$$\check{\mathcal{H}}=\begin{pmatrix} -\frac{\nabla^2}{2m}-\mu & -\Delta\\ \Delta^* & -\frac{\nabla^2}{2m}-\mu\end{pmatrix},\quad \check{\tau}_3=\begin{pmatrix}1 & 0\\ 0 & -1\end{pmatrix}. \tag{3.17}$$

These are called the matrices in the Nambu space. The Hamiltonian $\check{\mathcal{H}}$ is now measured from the chemical potential μ. It is equivalent to $\mathcal{H}-\mu\mathcal{N}$ of the previous chapter. The four equations for the four Green functions can be written in the compact matrix form

$$\check{G}^{-1}(x_1)\check{G}(x_1,x_2)=\check{1}\delta(x_1-x_2). \tag{3.18}$$

We can apply the operator to act on the second coordinate x_2. To find the corresponding equations, we multiply the derivatives of eqn (3.2) taken from the functions of the coordinate x_2 by $\psi(x_1)$ or $\psi^{\dagger}(x_1)$ from the left. After decoupling of the averages using the Wick theorem we find for the BCS model

$$\check{G}(x_1,x_2)\overline{\check{G}}^{-1}(x_2)=\check{1}\delta(x_1-x_2) \tag{3.19}$$

where the conjugated operator is

$$\overline{\check{G}}^{-1}=-\check{\tau}_3\frac{\partial}{\partial\tau}+\overline{\check{\mathcal{H}}}. \tag{3.20}$$

Here $\overline{\check{\mathcal{H}}}$ coincides with $\check{\mathcal{H}}$.

3.1.1 *Magnetic field*

In the presence of a magnetic field, the momentum operator should be replaced with

$$-i\nabla-\frac{e}{c}\mathbf{A},\ \text{or}\ i\nabla-\frac{e}{c}\mathbf{A} \tag{3.21}$$

when it acts on ψ or $\psi^{\dagger}$, respectively. For a stationary electromagnetic field, the scalar potential φ can be incorporated into the chemical potential μ. Therefore, it does not appear in stationary problems. The operator $\check{G}^{-1}$ becomes

$$\check{G}^{-1}=\check{\tau}_3\frac{\partial}{\partial\tau}+\check{\mathcal{H}} \tag{3.22}$$

where now

$$\check{\mathcal{H}}=\begin{pmatrix} -\frac{1}{2m}\left(\nabla-\frac{ie}{c}\mathbf{A}\right)^2-\mu & -\Delta\\ \Delta^* & -\frac{1}{2m}\left(\nabla+\frac{ie}{c}\mathbf{A}\right)^2-\mu\end{pmatrix}. \tag{3.23}$$

The conjugated operator is

$$\bar{\check{G}}^{-1} = -\check{\tau}_3 \frac{\partial}{\partial \tau} + \bar{\check{\mathcal{H}}}$$

where

$$\bar{\check{\mathcal{H}}} = \begin{pmatrix} -\frac{1}{2m}\left(\nabla + \frac{ie}{c}\mathbf{A}\right)^2 - \mu & -\Delta \\ \Delta^* & -\frac{1}{2m}\left(\nabla - \frac{ie}{c}\mathbf{A}\right)^2 - \mu \end{pmatrix}. \tag{3.24}$$

The electron-number density is

$$N = -2G(\mathbf{r}, \tau; \mathbf{r}, \tau + 0) = -2\bar{G}(\mathbf{r}, \tau + 0; \mathbf{r}, \tau). \tag{3.25}$$

The electric current can be expressed through the Green functions using the momentum density of eqn (2.48):

$$\begin{aligned} \mathbf{j} &= -\frac{e}{m}\left[\left[\left(-i\nabla_1 - \frac{e}{c}\mathbf{A}\right) + \left(i\nabla_2 - \frac{e}{c}\mathbf{A}\right)\right] G(\mathbf{r}_1, \tau; \mathbf{r}_2, \tau + 0)\right]_{\mathbf{r}_1 = \mathbf{r}_2} \\ &= \frac{ie}{m}\left[(\nabla_1 - \nabla_2) G(\mathbf{r}_1, \tau; \mathbf{r}_2, \tau + 0)\right]_{\mathbf{r}_1 = \mathbf{r}_2} - \frac{Ne^2}{mc}\mathbf{A} \end{aligned} \tag{3.26}$$

or

$$\begin{aligned} \mathbf{j} &= -\frac{e}{m}\left[\left[\left(i\nabla_1 - \frac{e}{c}\mathbf{A}\right) + \left(-i\nabla_2 - \frac{e}{c}\mathbf{A}\right)\right] \bar{G}(\mathbf{r}_1, \tau + 0; \mathbf{r}_2, \tau)\right]_{\mathbf{r}_1 = \mathbf{r}_2} \\ &= -\frac{ie}{m}\left[(\nabla_1 - \nabla_2) \bar{G}(\mathbf{r}_1, \tau; \mathbf{r}_2, \tau + 0)\right]_{\mathbf{r}_1 = \mathbf{r}_2} - \frac{Ne^2}{mc}\mathbf{A}. \end{aligned} \tag{3.27}$$

The factor $1/2$ is canceled here due to summation over spin indices implied in the definition, eqn (2.48).

The fact that the spatial derivatives enter the inverse operators for the Green functions and the expressions for the electric current in combinations $-i\nabla - (e/c)\mathbf{A}$ and $i\nabla - (e/c)\mathbf{A}$ is a manifestation of the gauge invariance of the theory of superconductivity. Indeed, a change in the phase $\chi/2$ of the single-electron wave function ψ by an arbitrary function $f/2$ results in the f-change in the order parameter phase; it can be compensated by the variation of the vector potential by $(c/2e)\nabla f$. Since the magnetic field $H = \operatorname{curl} \mathbf{A}$ does not change, the superconducting state is invariant under the transformation of eqn (1.7) on page 4. Therefore, the vector potential should always appear in the combination with the order parameter phase $\mathbf{A} - (c/2e)\nabla\chi$ to preserve the gauge invariance. This gives a convenient test for checking whether the calculations which we are doing are correct.

3.1.2 *Frequency and momentum representation*

Equations for the Green functions are more convenient in the frequency representation. We denote $\tau = \tau_1 - \tau_2$ and write

$$\check{G}(\mathbf{r}_1, \mathbf{r}_2; \tau) = T \sum_n e^{-i\omega_n \tau} \check{G}_{\omega_n}(\mathbf{r}_1, \mathbf{r}_2) \ ,$$

$$\check{G}_{\omega_n}(\mathbf{r}_1, \mathbf{r}_2) = \int_0^{1/T} e^{i\omega_n \tau} \check{G}(\mathbf{r}_1, \mathbf{r}_2; \tau)\, d\tau,$$

where $\omega_n = (2n+1)\pi T$. The operators are

$$\check{G}^{-1}_{\omega_n}(\mathbf{r}) = -i\omega_n \check{\tau}_3 + \check{\mathcal{H}}, \tag{3.28}$$

$$\overline{\check{G}}^{-1}_{\omega_n}(\mathbf{r}) = i\omega_n \check{\tau}_3 + \overline{\check{\mathcal{H}}}. \tag{3.29}$$

The matrix Gor'kov equation in the frequency representation is

$$\check{G}^{-1}_{\omega_n}(\mathbf{r}_1)\check{G}_{\omega_n}(\mathbf{r}_1, \mathbf{r}_2) = \check{1}\delta(\mathbf{r}_1 - \mathbf{r}_2), \tag{3.30}$$

$$\check{G}_{\omega_n}(\mathbf{r}_1, \mathbf{r}_2)\overline{\check{G}}^{-1}_{\omega_n}(\mathbf{r}_2) = \check{1}\delta(\mathbf{r}_1 - \mathbf{r}_2) \tag{3.31}$$

The self-consistency equation for the order parameter becomes

$$\frac{\Delta(\mathbf{r})}{|g|} = T\sum_n F_{\omega_n}(\mathbf{r}, \mathbf{r})\ , \quad \frac{\Delta^*(\mathbf{r})}{|g|} = T\sum_n F^\dagger_{\omega_n}(\mathbf{r}, \mathbf{r})\ , \tag{3.32}$$

since we have to take $\tau = \tau_1 - \tau_2 = 0$. The electron density is obtained in a similar way

$$\begin{aligned} N &= -2T \lim_{\tau \to -0} \sum_n G_{\omega_n}(\mathbf{r}, \mathbf{r}) e^{-i\omega_n \tau} \\ &= -2T \lim_{\tau \to +0} \sum_n \bar{G}_{\omega_n}(\mathbf{r}, \mathbf{r}) e^{-i\omega_n \tau}. \end{aligned} \tag{3.33}$$

The limits in eqn (3.33) need to be taken because $G(\mathbf{r}_1, t_1; \mathbf{r}_2, t_2)$ is discontinuous at $(\mathbf{r}_1, t_1) = (\mathbf{r}_2, t_2)$ and the sums for $\tau = 0$ do not converge.

The electric current becomes

$$\mathbf{j} = \frac{ie}{m}\left[(\nabla_1 - \nabla_2)\, T\sum_n G_{\omega_n}(\mathbf{r}_1, \mathbf{r}_2)\right]_{\mathbf{r}_1 = \mathbf{r}_2} - \frac{Ne^2}{mc}\mathbf{A} \tag{3.34}$$

or

$$\mathbf{j} = -\frac{ie}{m}\left[(\nabla_1 - \nabla_2)\, T\sum_n \bar{G}_{\omega_n}(\mathbf{r}_1, \mathbf{r}_2)\right]_{\mathbf{r}_1 = \mathbf{r}_2} - \frac{Ne^2}{mc}\mathbf{A}. \tag{3.35}$$

The limiting procedure is not necessary for eqn (3.34) since the discontinuous part vanishes as a result of action of the differential operators.

The order parameter equation and the expressions for the current and electron density can be written in terms of the real-time retarded and advanced Green functions, as well. We start with the order parameter equation. The sum over frequencies $\omega_n = (2n+1)\pi T$ can be presented as

$$T\sum_n F_{\omega_n} = T\sum_{n>0} F_{\omega_n} + T\sum_{n<0} F_{\omega_n}$$

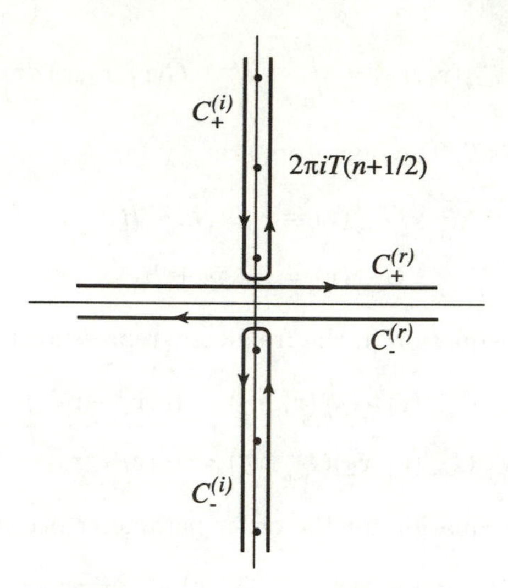

FIG. 3.1. Contours of integration over complex frequency. Contours $C_+^{(i)}$ and $C_-^{(i)}$ encircle the poles of $\tanh(\epsilon/2T)$ in the upper (lower) half plane, respectively. Contours $C_\pm^{(r)}$ are obtained by shifting the respective contours $C_\pm^{(i)}$ onto the real frequency axis.

$$= \int_{C_+^{(i)}} \tanh\left(\frac{\epsilon}{2T}\right) F_\epsilon^R \frac{d\epsilon}{4\pi i} + \int_{C_-^{(i)}} \tanh\left(\frac{\epsilon}{2T}\right) F_\epsilon^A \frac{d\epsilon}{4\pi i}.$$

The function $\tanh(\epsilon/2T)$ has the poles at $\epsilon = (2n+1)\pi iT$, i.e., at $\epsilon = i\omega_n$. The contour $C_+^{(i)}$ encircles all the poles of $\tanh(\epsilon/2T)$ in the upper complex half-plane of ϵ while the contour $C_-^{(i)}$ encircles all the poles of $\tanh(\epsilon/2T)$ in the lower complex half-plane of ϵ (see Fig. 3.1). In the upper half plane, the function F_ϵ^R is an analytical function and coincides with F_{ω_n} for $\epsilon = i\omega_n$ when $\omega_n > 0$. In the same way, the function F_ϵ^A is an analytical function in the lower half-plane and coincides with F_{ω_n} for $\epsilon = i\omega_n$ when $\omega_n < 0$. We can now shift the contours in such a way that $C_+^{(i)}$ transforms into $C_+^{(r)}$ and goes along the real axis of ϵ from $-\infty$ to $+\infty$ and $C_-^{(i)}$ transforms into $C_-^{(r)}$ which goes along the real axis of ϵ from $+\infty$ to $-\infty$. Finally, we have

$$T\sum_n F_{\omega_n} = \int_{-\infty}^{+\infty} \tanh\left(\frac{\epsilon}{2T}\right)\left(F_\epsilon^R - F_\epsilon^A\right)\frac{d\epsilon}{4\pi i}. \tag{3.36}$$

The order parameter equation becomes

$$\frac{\Delta}{|g|} = \int_{-\infty}^{+\infty} \tanh\left(\frac{\epsilon}{2T}\right)\left(F_\epsilon^R - F_\epsilon^A\right)\frac{d\epsilon}{4\pi i}. \tag{3.37}$$

The electric current is

$$\mathbf{j} = \frac{ie}{m}\left[(\nabla_1 - \nabla_2)\int_{-\infty}^{+\infty} \tanh\left(\frac{\epsilon}{2T}\right)\right.$$

$$\times \left(G^R_\epsilon(\mathbf{r}_1,\mathbf{r}_2) - G^A_\epsilon(\mathbf{r}_1,\mathbf{r}_2)\right)\Big]_{\mathbf{r}_1=\mathbf{r}_2} \frac{d\epsilon}{4\pi i} - \frac{Ne^2}{mc}\mathbf{A}, \tag{3.38}$$

or

$$\mathbf{j} = -\frac{ie}{m}\Big[(\nabla_1 - \nabla_2)\int_{-\infty}^{+\infty} \tanh\left(\frac{\epsilon}{2T}\right)$$
$$\times \left(\bar{G}^R_\epsilon(\mathbf{r}_1,\mathbf{r}_2) - \bar{G}^A_\epsilon(\mathbf{r}_1,\mathbf{r}_2)\right)\Big]_{\mathbf{r}_1=\mathbf{r}_2} \frac{d\epsilon}{4\pi i} - \frac{Ne^2}{mc}\mathbf{A}.$$

The transformation to the real frequencies for the electron density cannot be performed straightforwardly because of the discontinuity of the corresponding Green functions. The discontinuity results in a divergence of the integrals over the remote contours. Let us first separate the contribution to the density from the normal state. The remaining part of the Green functions δG then decreases quickly enough for large ϵ, and the integrals over remote contours vanish. As a result we obtain

$$N = N_0 - 2\int_{-\infty}^{+\infty} \tanh\left(\frac{\epsilon}{2T}\right)\delta\left[G^R_\epsilon(\mathbf{r},\mathbf{r}) - G^A_\epsilon(\mathbf{r},\mathbf{r})\right]\frac{d\epsilon}{4\pi i}, \tag{3.39}$$

or

$$N = N_0 - 2\int_{-\infty}^{+\infty} \tanh\left(\frac{\epsilon}{2T}\right)\delta\left[\bar{G}^R_\epsilon(\mathbf{r},\mathbf{r}) - \bar{G}^A_\epsilon(\mathbf{r},\mathbf{r})\right]\frac{d\epsilon}{4\pi i},$$

where N_0 is the electron density in the normal state.

The retarded and advanced Green functions satisfy the same equations as the Matsubara functions with the replacement $\omega_n = -i\epsilon$:

$$\check{G}^{-1}_\epsilon(\mathbf{r}_1)\check{G}^{R(A)}_\epsilon(\mathbf{r}_1,\mathbf{r}_2) = \hat{1}\delta(\mathbf{r}_1 - \mathbf{r}_2)\ ,$$
$$\check{G}^{R(A)}_\epsilon(\mathbf{r}_1,\mathbf{r}_2)\overline{\check{G}}^{-1}_\epsilon(\mathbf{r}_2) = \hat{1}\delta(\mathbf{r}_1 - \mathbf{r}_2)\ , \tag{3.40}$$

where the inverse operators are

$$\check{G}^{-1}_\epsilon(\mathbf{r}) = -\epsilon\check{\tau}_3 + \check{\mathcal{H}};\ \overline{\check{G}}^{-1}_\epsilon(\mathbf{r}) = -\epsilon\check{\tau}_3 + \overline{\check{\mathcal{H}}}. \tag{3.41}$$

The factor $\tanh(\epsilon/2T)$ appears as a result of the replacement of the summation over imaginary frequencies with integration over the real frequencies. It has a simple physical meaning. Indeed,

$$\tanh(\epsilon/2T) = 1 - 2n(\epsilon) \tag{3.42}$$

where

$$n(\epsilon) = [e^{\epsilon/T} + 1]^{-1}$$

is the Fermi distribution function expressed in terms of the energy ϵ measured from the chemical potential.

One can introduce the momentum representation of the Green function,

$$\check{G}(\mathbf{r}_1,\mathbf{r}_2) = \int \frac{d^3p}{(2\pi)^3}\frac{d^3k}{(2\pi)^3}\check{G}(\mathbf{p},\mathbf{p}-\mathbf{k})e^{i\mathbf{p}\mathbf{r}_1 - i(\mathbf{p}-\mathbf{k})\mathbf{r}_2}. \tag{3.43}$$

The particle–hole symmetry equation (3.12), in particular, implies

$$G_{\omega_n}(\mathbf{p}_1,\mathbf{p}_2) = \bar{G}_{-\omega_n}(-\mathbf{p}_2,-\mathbf{p}_1) \tag{3.44}$$

or

$$G^{R(A)}_{\epsilon}(\mathbf{p}_1,\mathbf{p}_2) = \bar{G}^{A(R)}_{-\epsilon}(-\mathbf{p}_2,-\mathbf{p}_1) \tag{3.45}$$

In the momentum–frequency representation, the order parameter equation becomes

$$\frac{\Delta(\mathbf{k})}{|g|} = T\sum_n \int \frac{d^3p}{(2\pi)^3}F_{\omega_n}(\mathbf{p},\mathbf{p}-\mathbf{k}) \tag{3.46}$$

or

$$\frac{\Delta(\mathbf{k})}{|g|} = \int \frac{d^3p}{(2\pi)^3}\int_{-\infty}^{+\infty} \tanh\left(\frac{\epsilon}{2T}\right)\left[F^R_\epsilon(\mathbf{p},\mathbf{p}-\mathbf{k}) - F^A_\epsilon(\mathbf{p},\mathbf{p}-\mathbf{k})\right]\frac{d\epsilon}{4\pi i}. \tag{3.47}$$

The current is

$$\mathbf{j}(\mathbf{k}) = -\frac{e}{m}T\sum_n\int\frac{d^3p}{(2\pi)^3}(2\mathbf{p}-\mathbf{k})\,G_{\omega_n}(\mathbf{p},\mathbf{p}-\mathbf{k}) - \frac{Ne^2}{mc}\mathbf{A} \tag{3.48}$$

or

$$\mathbf{j}(\mathbf{k}) = -\frac{e}{m}\int\frac{d^3p}{(2\pi)^3}\int\frac{d\epsilon}{4\pi i}\tanh\left(\frac{\epsilon}{2T}\right) \times (2\mathbf{p}-\mathbf{k})\left[G^R_\epsilon(\mathbf{p},\mathbf{p}-\mathbf{k}) - G^A_\epsilon(\mathbf{p},\mathbf{p}-\mathbf{k})\right] - \frac{Ne^2}{mc}\mathbf{A}. \tag{3.49}$$

The electron density

$$N = -2T\lim_{\tau\to -0}\sum_n\int\frac{d^3p}{(2\pi)^3}G_{\omega_n}(\mathbf{p},\mathbf{p}-\mathbf{k})e^{-i\omega_n\tau} \tag{3.50}$$

or

$$N = N_0 - 2\int\frac{d^3p}{(2\pi)^3}\int\frac{d\epsilon}{4\pi i}\tanh\left(\frac{\epsilon}{2T}\right)\delta\left[G^R_\epsilon(\mathbf{p},\mathbf{p}-\mathbf{k}) - G^A_\epsilon(\mathbf{p},\mathbf{p}-\mathbf{k})\right]. \tag{3.51}$$

3.1.3 *Order parameter of a d-wave superconductor*

Until now we assumed a point-like interaction between electrons independent of particle momenta. In general, the pairing interaction may have a dependence on directions of the momenta of participating particles. In such a case, the order parameter will also be momentum-dependent. In general, one can write

$$\Delta_{\mathbf{p}}(\mathbf{k}) = -T\sum_n U(\hat{\mathbf{p}},\hat{\mathbf{p}}_1)F_{\omega_n}(\mathbf{p}_1,\mathbf{p}_1-\mathbf{k})\,\frac{d^3p_1}{(2\pi)^3} \tag{3.52}$$

where $U(\hat{\mathbf{p}},\hat{\mathbf{p}}_1)$ is a Fourier transform of the (attractive) interaction matrix element. For simplicity we consider systems which have the inversion symmetry. For such systems, the order parameter has a definite symmetry in the momentum $\mathbf{p}$ because of the general parity of Fermi particles. The full order parameter including the spin degrees of freedom is an odd function with respect to transposition of particles. If the pairing occurs into a spin-singlet state which is odd in spin indices, the remaining orbital part should be even in the transposition of the particle coordinates, i.e., in the substitution $\mathbf{p}\to-\mathbf{p}$. This requires that the potential $U(\mathbf{p},\mathbf{p}_1)$ is even in $\mathbf{p}\to-\mathbf{p}$ and, of course, in $\mathbf{p}_1\to-\mathbf{p}_1$, as well, since it is symmetric in $\mathbf{p}\to\mathbf{p}_1$.

One can expand the pairing potential in terms of orthogonal eigenfunctions of angular motion

$$U(\hat{\mathbf{p}},\hat{\mathbf{p}}_1) = U_s + U_d(\hat{\mathbf{p}},\hat{\mathbf{p}}_1) + \dots$$

The s-wave component which we denoted earlier as $U_s = -|g|$ is independent of the momentum directions. The next expansion term represents the so-called d-wave pairing potential. In some cases it can be favorable, for example, because of the crystalline symmetry. If the crystal has a symmetry axis around the axis c the d-wave potential has the from $U_d(\hat{\mathbf{p}},\hat{\mathbf{p}}_1) = -|g_d|\,V_d(\hat{\mathbf{p}},\hat{\mathbf{p}}_1)$ where

$$V_d(\hat{\mathbf{p}},\hat{\mathbf{p}}_1) = 2(\hat{p}_x^2-\hat{p}_y^2)(\hat{p}_{1\,x}^2-\hat{p}_{1\,y}^2) = 2\cos(2\alpha)\cos(2\alpha_1) \tag{3.53}$$

where α is the angle between the momentum in the crystalline (ab) plane and, say, the a axis. The factor 2 is introduced for normalization. It is easy to see that the order parameter has a form

$$\Delta_{\mathbf{p}} = \Delta_0\cos(2\alpha)$$

with the amplitude satisfying the self-consistency order parameter equation

$$\frac{\Delta_0(\mathbf{k})}{|g_d|} = T\sum_n 2\cos(2\alpha)\,F_{\omega_n}(\mathbf{p},\mathbf{p}-\mathbf{k})\,\frac{d^3p}{(2\pi)^3}. \tag{3.54}$$

3.2 Derivation of the Bogoliubov–de Gennes equations

The Green functions can be expanded in terms of the eigenfunctions of the matrix Schrödinger equation. Let us introduce the functions $u(\mathbf{r})$ and $v(\mathbf{r})$ and construct a vector in the Nambu space

$$\mathcal{U} = \begin{pmatrix} u \\ -v \end{pmatrix}.$$

Let the functions u and v satisfy the matrix Schrödinger equation

$$\check{\mathcal{H}}\mathcal{U} = E_K \check{\tau}_3 \mathcal{U} \tag{3.55}$$

where E_K are the eigenvalues, i.e., energy spectrum of the Hamiltonian $\mathcal{H}$. The two equations

$$\left[-\frac{1}{2m}\left(\nabla - \frac{ie}{c}\mathbf{A}\right)^2 - \mu\right] u + \Delta v = E_K u,$$

$$\left[-\frac{1}{2m}\left(\nabla + \frac{ie}{c}\mathbf{A}\right)^2 - \mu\right] v - \Delta^* u = -E_K v, \tag{3.56}$$

arising from matrix equation (3.55) are called the Bogoliubov–de Gennes equations. We have encountered them already in Section 1.1.3. The functions u and v coincide with the Bogoliubov–de Gennes wave functions (Bogoliubov 1958) discussed earlier in Section 1.1.3. The conjugated vector

$$\mathcal{U}^\dagger = (\, u^*, \, v^* \,)$$

satisfies the conjugated equation

$$\mathcal{U}^\dagger \overline{\check{\mathcal{H}}} = \mathcal{U}^\dagger \check{\tau}_3 E_K \tag{3.57}$$

where $\overline{\check{\mathcal{H}}}$ acts on the left. In components,

$$\left[-\frac{1}{2m}\left(\nabla + \frac{ie}{c}\mathbf{A}\right)^2 - \mu\right] u^* + \Delta^* v^* = E_K u^*,$$

$$\left[-\frac{1}{2m}\left(\nabla - \frac{ie}{c}\mathbf{A}\right)^2 - \mu\right] v^* - \Delta u^* = -E_K v^*, \tag{3.58}$$

which are the complex conjugated eqns (3.56).

Since u and v are eigenfunctions of a linear operator, the functions belonging to different states are orthogonal. We normalize them in such a way that

$$\int d^3r \mathcal{U}_K^\dagger(\mathbf{r}) \check{\tau}_3 \mathcal{U}_{K'}(\mathbf{r}) = \delta_{K,K'}. \tag{3.59}$$

Moreover, the functions u and v constitute the full basis thus

$$\sum_K \check{\tau}_3 \mathcal{U}_K(\mathbf{r}_1) \mathcal{U}_K^\dagger(\mathbf{r}_2) = \check{1}\delta(\mathbf{r}_1 - \mathbf{r}_2). \tag{3.60}$$

Using the wave functions u and v we can write

$$\check{G}_\epsilon^{R(A)}(\mathbf{r}_1, \mathbf{r}_2) = \sum_K \frac{\mathcal{U}_K(\mathbf{r}_1)\mathcal{U}_K^\dagger(\mathbf{r}_2)}{E_K - \epsilon \mp i\delta}. \tag{3.61}$$

It easy to check that the function defined by eqn (3.61) indeed satisfies eqn (3.40) with the operators defined by eqn (3.41). The infinitely small imaginary term in the denominator is chosen in accordance with the analytical properties of the retarded and advanced functions.

We can see that the zeroes in the denominator, i.e., the poles of the Green functions G and F are at $\epsilon = E_K$; they thus correspond to the energy spectrum of excitations. Moreover, the quantity

$$\nu_s(\epsilon) = \frac{1}{4\pi i}\mathrm{Tr}\int \check{\tau}_3 \left[\check{G}^R_\epsilon(\mathbf{r},\mathbf{r}) - \check{G}^A_\epsilon(\mathbf{r},\mathbf{r})\right] dV = \sum_K \delta\left(\epsilon - E_K\right) \tag{3.62}$$

is the density of states in a superconductor per one spin projection. The operator Tr is the trace over the Nambu indices.

3.3 Thermodynamic potential

Here we derive a useful relation between the thermodynamic potential and the Green functions for a BCS system (Kopnin 1987). We first recall the definition of the thermodynamic potential. According to eqn (2.37)

$$\Omega_{sn} = -T \ln \sum\nolimits^{(d)} \hat{S}, \tag{3.63}$$

where

$$\hat{S} = T_\tau \exp\left[-\int_0^{1/T}\left[\hat{\mathcal{H}} - \mu\hat{\mathcal{N}}\right] d\tau\right], \tag{3.64}$$

and T_τ is the time-ordering operator in the imaginary "time" τ. The sum in eqn (3.63) denotes the sum of the diagonal matrix elements over the quantum states. In the BCS theory formulated through the Green functions which satisfy the Gor'kov equations, the order parameter plays the part of an external field. In this context, the BCS Hamiltonian measured from the chemical potential should be written as

$$\begin{aligned}\hat{\mathcal{H}}_{\mathrm{BCS}} = \sum_{\alpha,\gamma}\Bigg\{&\int d^3r \left[\tilde{\psi}^\dagger_\alpha(\mathbf{r})\left(-\frac{1}{2m}\left(\nabla - \frac{ie}{c}\mathbf{A}\right)^2 - \mu\right)\tilde{\psi}_\alpha(\mathbf{r})\right] \\ &-\frac{1}{2}\int d^3r \left[\tilde{\psi}^\dagger_\alpha(\mathbf{r})\tilde{\psi}^\dagger_\gamma(\mathbf{r})\Delta_{\gamma\alpha}(\mathbf{r})\right. \\ &\left.+\Delta^*_{\alpha\gamma}(\mathbf{r})\tilde{\psi}_\gamma(\mathbf{r})\tilde{\psi}_\alpha(\mathbf{r})\right] + \frac{\Delta^*_{\alpha\gamma}(\mathbf{r})\Delta_{\gamma\alpha}(\mathbf{r})}{2\,|g|}\Bigg\}. \end{aligned} \tag{3.65}$$

It plays the part of $\hat{\mathcal{H}} - \mu\hat{\mathcal{N}}$ in eqn (3.64). We consider an s-wave pairing for definiteness. Here $\tilde{\psi}_\alpha(\mathbf{r})$ is a single-particle operator. The sum is over the spin indices. The Hamiltonian can have also terms describing interactions with other external fields (including impurities) and interparticle interactions. With the Hamiltonian (3.65), one can directly reproduce equations (3.9), (3.10), (3.13) and (3.14) for the Green functions. The last term is added to ensure that the thermodynamic potential has a minimum for the value of the order parameter which satisfies the self-consistency equation.

To see this, let us calculate the variation of the thermodynamic potential Ω with respect to the fields $\delta\hat{\Delta}$, $\delta\hat{\Delta}^*$, and $\mathbf{A}$:

$$\delta\Omega_{sn} = -\frac{T\sum^{(d)}\delta\hat{S}}{\sum^{(d)}\hat{S}}. \tag{3.66}$$

The variation of the S-matrix is expressed through variations of the Hamiltonian (3.65) and contains the Green functions defined according to eqns (2.38) and (3.6):

$$F_{\alpha\beta}(x,x') = \frac{\sum^{(d)}\left(T_\tau\tilde{\psi}_\alpha(x)\tilde{\psi}_\beta(x')\hat{S}\right)}{\sum^{(d)}\hat{S}} \equiv \left\langle T_\tau\tilde{\psi}_\alpha(x)\tilde{\psi}_\beta(x')\right\rangle_{st}, \tag{3.67}$$

$$G_{\alpha\beta}(x,x') = \frac{\sum^{(d)}\left(T_\tau\tilde{\psi}_\alpha(x)\tilde{\psi}^\dagger_\beta(x')\hat{S}\right)}{\sum^{(d)}\hat{S}} \equiv \left\langle T_\tau\tilde{\psi}_\alpha(x)\tilde{\psi}^\dagger_\beta(x')\right\rangle_{st}. \tag{3.68}$$

One has from eqn (3.66)

$$\delta\Omega_{sn} = -T\sum_n \mathrm{Tr}\int\left(\left[\delta\check{H}\check{G}_{\omega_n}(\mathbf{r},\mathbf{r}')\right]_{\mathbf{r}=\mathbf{r}'} + \frac{\delta\left(\Delta\Delta^*\right)}{|g|}\right)dV \tag{3.69}$$

where $\check{H}$ is the effective Hamiltonian

$$\check{H} = \begin{pmatrix} \frac{ie}{2mc}\left(\mathbf{A}\nabla + \mathbf{A}\nabla\right) - \frac{e^2}{2mc^2}A^2 & -\Delta \\ \Delta^* & -\frac{ie}{2mc}\left(\mathbf{A}\nabla + \mathbf{A}\nabla\right) - \frac{e^2}{2mc^2}A^2 \end{pmatrix}. \tag{3.70}$$

It is easy to see that a minimization of Ω_{sn} with respect to $\delta\Delta^*$ leads to the self-consistency equation (3.32).

Let us perform transformation into momentum representation. The variation of the thermodynamic potential can now be written as

$$\begin{aligned}\delta\Omega_{sn} &= -T\sum_n\int \mathrm{Tr}\left[\delta\check{H}(-\mathbf{k})\check{G}_{\omega_n}(\mathbf{p}_+,\mathbf{p}_-)\right]\frac{d^3p}{(2\pi)^3}\frac{d^3k}{(2\pi)^3} \\ &\quad +\frac{1}{|g|}\int\delta\left(\Delta(\mathbf{k})\Delta^*(-\mathbf{k})\right)\frac{d^3k}{(2\pi)^3}.\end{aligned}$$

Here we omit the spin indices and denote $\mathbf{p}_\pm = \mathbf{p}\pm\mathbf{k}/2$.

Including the magnetic energy, we find for the total potential

$$\begin{aligned}&\delta\left(\Omega_{sn} + \int\frac{H^2}{8\pi}d^3r\right) \\ &\quad = -T\sum_n\int \mathrm{Tr}\left[\delta\check{H}(-\mathbf{k})\check{G}_{\omega_n}(\mathbf{p}_+,\mathbf{p}_-)\right]\frac{d^3p}{(2\pi)^3}\frac{d^3k}{(2\pi)^3} \\ &\qquad + \int\left[\frac{1}{|g|}\delta\left(\Delta(\mathbf{k})\Delta^*(-\mathbf{k})\right) + \frac{1}{c}\mathbf{j}(\mathbf{k})\cdot\delta\mathbf{A}(-\mathbf{k})\right]\frac{d^3k}{(2\pi)^3}.\end{aligned} \tag{3.71}$$

The total thermodynamic potential has a minimum also with respect to variation of the vector potential $\mathbf{A}$ because of equation (3.34) for the current.

One can check that eqn (3.71) holds also for a nontrivial pairing interaction determined, for example, by eqn (3.53). One only needs to replace the order parameter Δ with

$$\int V(\mathbf{p}, \mathbf{p}')\Delta_{\mathbf{p}'} \frac{d\Omega_{\mathbf{p}'}}{4\pi}$$

in the first term in the r.h.s. of eqn (3.71) and introduce the momentum–direction average $\int d\Omega_{\mathbf{p}}/4\pi$ together with $\Delta_{\mathbf{p}}(\mathbf{k})\Delta^*_{\mathbf{p}}(-\mathbf{k})$ in the last term in the r.h.s. of eqn (3.71).

3.4 Example: Homogeneous state

3.4.1 *Green functions*

Consider a homogeneous state where the order parameter does not depend on coordinates, and the magnetic field is absent. The simplest way to solve the Gor'kov equations, eqns (3.28), (3.30), is to use the momentum representation of eqn (3.43). We have for a homogeneous case

$$\check{G}_{\omega_n}(\mathbf{r}_1, \mathbf{r}_2) = \int e^{i\mathbf{p}(\mathbf{r}_1 - \mathbf{r}_2)} \check{G}_{\omega_n}(\mathbf{p}) \frac{d^3p}{(2\pi)^3}.$$

For the Fourier transformed Green functions we obtain

$$\begin{aligned} (-i\omega_n + \xi_{\mathbf{p}})G + \Delta F^\dagger &= 1, \\ (-i\omega_n - \xi_{\mathbf{p}})F^\dagger + \Delta^* G &= 0. \end{aligned} \tag{3.72}$$

where $\xi_{\mathbf{p}} = p^2/2m - \mu$ according to eqn (1.2). In a metal, the chemical potential $\mu \approx E_F$. The solution of the Gor'kov equations is

$$G = \frac{\xi_{\mathbf{p}} + i\omega_n}{\xi_{\mathbf{p}}^2 + \omega_n^2 + |\Delta|^2}\ , \ F^\dagger = \frac{\Delta^*}{\xi_{\mathbf{p}}^2 + \omega_n^2 + |\Delta|^2}. \tag{3.73}$$

Similarly,

$$\bar{G} = \frac{\xi_{\mathbf{p}} - i\omega_n}{\xi_{\mathbf{p}}^2 + \omega_n^2 + |\Delta|^2}\ , \ F = \frac{\Delta}{\xi_{\mathbf{p}}^2 + \omega_n^2 + |\Delta|^2}. \tag{3.74}$$

The retarded and advanced real-time functions are

$$\begin{aligned} G^{R(A)} &= \frac{\xi_{\mathbf{p}} + \epsilon}{\xi_{\mathbf{p}}^2 - (\epsilon \pm i\delta)^2 + |\Delta|^2}\ , \ F^{\dagger\, R(A)} = \frac{\Delta^*}{\xi_{\mathbf{p}}^2 - (\epsilon \pm i\delta)^2 + |\Delta|^2}, \\ \bar{G}^{R(A)} &= \frac{\xi_{\mathbf{p}} - \epsilon}{\xi_{\mathbf{p}}^2 - (\epsilon \pm i\delta)^2 + |\Delta|^2}\ , \ F^{R(A)} = \frac{\Delta}{\xi_{\mathbf{p}}^2 - (\epsilon \pm i\delta)^2 + |\Delta|^2}, \end{aligned}$$

which comply with the particle–hole symmetry of eqns (3.44) and (3.45).

One can also write

$$G^{R(A)} = -\frac{\xi_{\mathbf{p}} + \epsilon}{(\epsilon - \epsilon_{\mathbf{p}} \pm i\delta)(\epsilon + \epsilon_{\mathbf{p}} \pm i\delta)} \tag{3.75}$$

where

$$\epsilon_{\mathbf{p}} = \sqrt{\xi_{\mathbf{p}}^2 + |\Delta|^2}. \tag{3.76}$$

This equation coincides with eqn (1.1) in Chapter 1. The Green functions have poles at $\epsilon = \pm\epsilon_{\mathbf{p}}$. This means that $\pm\epsilon_{\mathbf{p}}$ is the energy spectrum of excitations in a superconductor. There are two branches: one with positive and another with negative energies. We also notice that energy of an excitation cannot be less than $|\Delta|$: there are no excitations with energies below the energy gap $|\Delta|$.

3.4.2 *Gap equation for an s-wave superconductor*

Inserting our solution for F into the self-consistency equation we can find the order parameter as a function of temperature. We have

$$\begin{aligned} \frac{\Delta}{|g|} &= T\sum_n \int \frac{d^3p}{(2\pi)^3} \frac{\Delta}{\xi_{\mathbf{p}}^2 + \omega_n^2 + |\Delta|^2} \\ &= \nu(0)2T\sum_{n\geq 0}\int d\xi_{\mathbf{p}} \frac{\Delta}{\xi_{\mathbf{p}}^2 + \omega_n^2 + |\Delta|^2}. \end{aligned} \tag{3.77}$$

We integrate over the momentum with the help of

$$\frac{d^3p}{(2\pi)^3} = \frac{p^2\,dp}{2\pi^2}\frac{d\Omega_{\mathbf{p}}}{4\pi} = \nu(0)d\xi_{\mathbf{p}}\frac{d\Omega_{\mathbf{p}}}{4\pi} \tag{3.78}$$

where $d\xi_{\mathbf{p}} = v_F\,dp$ and $d\Omega_{\mathbf{p}}$ is the increment of the solid angle in the momentum space.

$$\nu(0) = \frac{mp_F}{2\pi^2} \tag{3.79}$$

is the density of states for one spin projection in the normal state. Indeed, the characteristic values of $\xi_{\mathbf{p}}$ are of the order of Δ or T which is much smaller than the Fermi energy E_F. Therefore, the momentum is always close to the Fermi momentum, and we can replace p with p_F everywhere. At this stage, we explicitly employ the quasiclassical approximation introduced in Section 1.1.4. We see that the quasiclassical approximation results in a substantial simplification of calculations even at this initial level.

The function under the integral in eqn (3.77) has poles at $\xi_{\mathbf{p}} = \pm i\sqrt{\omega_n^2 + |\Delta|^2}$. Shifting the contour of integration into the upper half-plane of the complex variable $\xi_{\mathbf{p}}$ (see Fig. 3.2) we find

$$\frac{1}{\lambda} = 2\pi T\sum_{n\geq 0}\frac{1}{\sqrt{\omega_n^2 + |\Delta|^2}} \tag{3.80}$$

where $\lambda = \nu(0)|g|$ is the interaction constant. Inspection of this equation shows that the summation over n diverges logarithmically at large frequencies. This

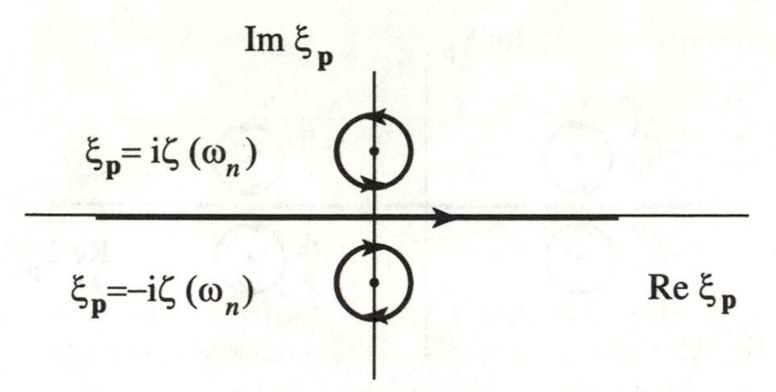

FIG. 3.2. Contour of integration over $d\xi_{\mathbf{p}}$. We denote $\zeta(\omega_n) = \sqrt{\omega_n^2 + |\Delta|^2}$.

is because the pairing interaction was assumed energy independent. As we have already discussed, however, the pairing interaction is restricted from above by a characteristic energy $\Omega_{\rm BCS}$. For a weak coupling approximation, the cut-off frequency is much larger than T_c but considerably smaller than E_F. The energy cut-off is equivalent to truncating the sum at a limiting number $N_0 = \Omega_{\rm BCS}/2\pi T \gg 1$:

$$\frac{1}{\lambda} = 2\pi T \sum_{n=0}^{N_0} \frac{1}{\sqrt{\omega_n^2 + |\Delta|^2}}. \tag{3.81}$$

This equation determines the temperature dependence of the order parameter.

The order parameter vanishes at the critical temperature $T = T_c$. To find T_c we put $\Delta = 0$ in eqn (3.81) which becomes

$$\frac{1}{\lambda} = 2\pi T_c \sum_{n=0}^{N_0} \frac{1}{(2n+1)\pi T_c} = \sum_{n=0}^{N_0} \frac{1}{n+1/2} \approx \ln(\Omega_{\rm BCS}/2\pi T_c) + 2\ln 2 + C \tag{3.82}$$

where $C = 0.5772\ldots$ is the Euler constant. The critical temperature is found to be

$$T_c = \frac{2\Omega_{\rm BCS}\gamma}{\pi} e^{-1/\lambda} \tag{3.83}$$

where $\gamma = e^C \approx 1.78$. The critical temperature is thus much lower than Ω_0 in the weak-coupling approximation $\lambda \ll 1$. It is due to the divergence of the sum in eqn (3.81) that the order parameter does not vanish: a small λ is compensated by a large logarithm $\ln(\Omega_{\rm BCS}/T_c)$ which introduces a new energy scale $\Omega_{\rm BCS}$ for the temperature. Equation (3.83) also applies to a d-wave superconductor because we omitted the angle-dependent $\Delta_{\mathbf{p}}$ in the denominator of the self-consistency equation.

Equation (3.81) can be presented in terms of real frequencies. Using expressions for retarded and advanced functions, we obtain

$$\frac{1}{\lambda} = \int d\xi_{\mathbf{p}} \int_0^{\Omega_{\rm BCS}} \frac{d\epsilon}{2\pi i} \tanh\left(\frac{\epsilon}{2T}\right)$$

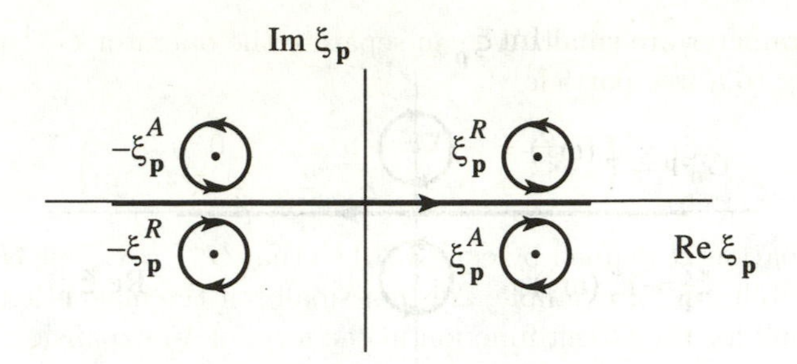

FIG. 3.3. Contour of integration over $d\xi_{\mathbf{p}}$. After shifting the contour into the upper half-plane, the integrals are determined by poles.

$$\times\left(\frac{1}{\xi_{\mathbf{p}}^2-(\epsilon+i\delta)^2+|\Delta|^2}-\frac{1}{\xi_{\mathbf{p}}^2-(\epsilon-i\delta)^2+|\Delta|^2}\right). \quad (3.84)$$

Integration over $d\xi_{\mathbf{p}}$ can be performed by shifting the contours of integration into the upper half-plane of the complex variable $\xi_{\mathbf{p}}$. For $|\epsilon>|\Delta|$, the contribution to the integrals come from the pole $\xi_{\mathbf{p}}=\xi_{\mathbf{p}}^R=\sqrt{(\epsilon+i\delta)^2-|\Delta|^2}$ for the retarded function and from the pole $\xi_{\mathbf{p}}=-\xi_{\mathbf{p}}^A=-\sqrt{(\epsilon-i\delta)^2-|\Delta|^2}$ for the advanced function (see Fig. 3.3). If $|\epsilon|<|\Delta|$, the poles are at $\xi_{\mathbf{p}}=\pm i\sqrt{|\Delta|^2-\epsilon^2}$ for both the retarded and advanced functions. Therefore, for $|\epsilon|<|\Delta|$, the retarded and advanced functions are equal to each other, and the integral vanishes. We have

$$\begin{aligned}\frac{1}{\lambda}&=\int_0^{\Omega_{\rm BCS}}\tanh\left(\frac{\epsilon}{2T}\right)\frac{d\epsilon}{\sqrt{\epsilon^2-|\Delta|^2}}\Theta\left[\epsilon^2-|\Delta|^2\right]\\&=\int_{|\Delta|}^{\Omega_{\rm BCS}}\tanh\left(\frac{\epsilon}{2T}\right)\frac{d\epsilon}{\sqrt{\epsilon^2-|\Delta|^2}}.\end{aligned} \quad (3.85)$$

Equation (3.85) is nothing but eqn (1.57). To see this we use eqn (3.42) and express ξ_p in terms of ϵ through eqn (3.76). Equation (3.85) gives the same value of T_c as eqn (3.81), of course. It can also be used to get the order parameter value at zero temperature:

$$\frac{1}{\lambda}=\int_{|\Delta|}^{\Omega_{\rm BCS}}\frac{d\epsilon}{\sqrt{\epsilon^2-|\Delta|^2}}=\ln\left(\frac{2\Omega_{\rm BCS}}{|\Delta|}\right). \quad (3.86)$$

Therefore, at $T=0$,

$$|\Delta|=(\pi/\gamma)T_c\approx 1.76T_c. \quad (3.87)$$

3.5 Perturbation theory

3.5.1 *Diagram technique*

Let us assume that the operator $\check{G}^{-1}$ consists of a main part plus a small correction. For example, we can consider the case when the magnetic field and the order parameter are small. We can separate the operator $\check{G}^{-1}$ into the part corresponding to a free particle

$$\check{G}_0^{-1} = \begin{pmatrix} -\frac{\partial}{\partial \tau} - \frac{1}{2m}\nabla^2 - \mu & 0 \\ 0 & \frac{\partial}{\partial \tau} - \frac{1}{2m}\nabla^2 - \mu \end{pmatrix} \tag{3.88}$$

and a perturbation $\check{H}$ defined by eqn (3.70) so that $\check{G}^{-1} = \check{G}_0^{-1} + \check{H}$. There may be other possibilities; for example, Δ is not small but the magnetic field is small, etc. Let us look for the Green function in the form of an expansion in powers of perturbation. We write

$$\check{G} = \check{G}^{(0)} + \check{G}^{(1)} + \check{G}^{(2)} + \dots$$

where $\check{G}^{(0)}$ is the Green function of a free particle:

$$\check{G}_0^{-1}\check{G}^{(0)} = \check{1}\delta(x_1 - x_2)$$

and the correction $\check{G}^{(k)}$ satisfies the equation

$$\check{G}_0^{-1}\check{G}^{(k)} + \check{H}\check{G}^{(k-1)} = 0.$$

The full Green function can be written as a series

$$\begin{aligned} \check{G}(x,x') = \check{G}^{(0)}(x,x') &- \int d^4x_1\, \check{G}^{(0)}(x,x_1)\check{H}(x_1)\check{G}^{(0)}(x_1,x') \\ &+ \int d^4x_1 d^4x_2\, \check{G}^{(0)}(x,x_1)\check{H}(x_1)\check{G}^{(0)}(x_1,x_2)\check{H}(x_2)\check{G}^{(0)}(x_2,x') \\ &- \dots \end{aligned}$$

or in the form of the Dyson equation

$$\check{G}(x,x') = \check{G}^{(0)}(x,x') - \int d^4x_1\, \check{G}^{(0)}(x,x_1)\check{H}(x_1)\check{G}(x_1,x'). \tag{3.89}$$

The Dyson equation (3.89) is equivalent to the Gor'kov equation (3.18). The same expansion holds also for retarded and advanced functions, as can be seen from eqns (3.40).

The series expansion and the Dyson equation can also be represented as graphs. Let us ascribe a single line to the Green function $\check{G}^{(0)}$, a double line to the full Green function $\check{G}$. Two ends represent two coordinates x_1 and x_2. We denote the interaction $\check{H}$ by a wavy line with one coordinate. As a result, the full Green function can be presented either as an infinite series of diagrams, Fig. 3.4 (a), or as an graphical Dyson equation, Fig. 3.4 (b).

3.5.2 *Electric current*

The electric current can be calculated using eqn (3.38) and the Green function which can be found, for example, by means of the perturbation theory. We have

$$\mathbf{j} = \frac{ie\hbar}{m}\int_{-\infty}^{+\infty} \tanh\left(\frac{\epsilon}{2T}\right)\frac{d\epsilon}{4\pi i}$$

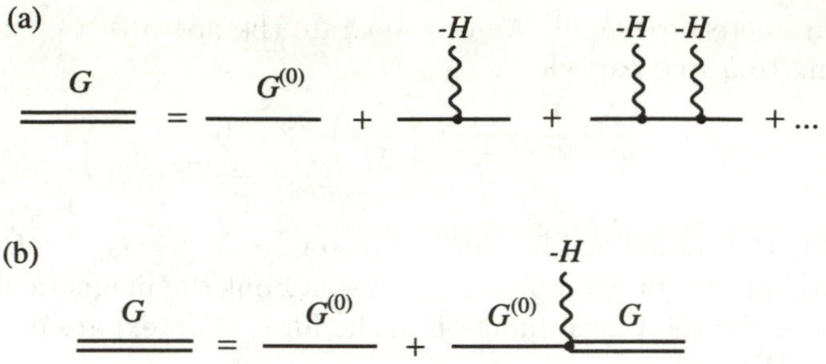

(a)

$$G = G^{(0)} + \; \overset{-H}{\sim} \; + \; \overset{-H\;-H}{\sim\;\sim} \; + \ldots$$

(b)

$$G = G^{(0)} + G^{(0)} \overset{-H}{\sim} G$$

FIG. 3.4. Diagrammatic representations for the Green function. (a) The perturbation series. (b) The Dyson equation.

$$\times \left[(\nabla_1 - \nabla_2) \left(G_\epsilon^R(\mathbf{r}_1, \mathbf{r}_2) - G_\epsilon^A(\mathbf{r}_1, \mathbf{r}_2) \right) \right]_{\mathbf{r}_1 = \mathbf{r}_2} - \frac{Ne^2}{mc}\mathbf{A}.$$

In the momentum representation we have

$$\mathbf{j}(\mathbf{k}) = -\frac{2e}{m} \int_{-\infty}^{+\infty} \tanh\left(\frac{\epsilon}{2T}\right) \frac{d\epsilon}{4\pi i} \int \frac{d^3p}{(2\pi)^3}$$
$$\times \mathbf{p} \left[G_\epsilon^R(\mathbf{p}_+, \mathbf{p}_-) - G_\epsilon^A(\mathbf{p}_+, \mathbf{p}_-) \right] - \frac{Ne^2}{mc}\mathbf{A}. \tag{3.90}$$

The last term is called the diamagnetic term. It should disappear in the normal state if one neglects the Landau quantization in a magnetic field. We prove this using the Green function of the normal state. In the presence of magnetic field

$$\check{H} = \begin{pmatrix} -(e/c)\mathbf{v}\cdot\mathbf{A} & 0 \\ 0 & (e/c)\mathbf{v}\cdot\mathbf{A} \end{pmatrix}$$

in the leading approximation in $\mathbf{A}$. Since $\xi_{\mathbf{p}-\mathbf{k}} \approx \xi_p - \mathbf{v}_F\mathbf{k}$, we get

$$G_\epsilon^{R(A)}(\mathbf{p}_+, \mathbf{p}_-) = (2\pi)^3 \delta(\mathbf{k}) \frac{1}{\xi_\mathbf{p} - \epsilon \mp i\delta}$$
$$- \frac{e}{c} \frac{\mathbf{v}\cdot\mathbf{A}}{(\xi_\mathbf{p} - \epsilon \mp i\delta)(\xi_\mathbf{p} - \mathbf{v}\mathbf{k} - \epsilon \mp i\delta)}.$$

The first term does not give a contribution to the current. Since $p \gg k$, we have

$$G_\epsilon^{R(A)}(\mathbf{p}_+, \mathbf{p}_-) = -\frac{e}{c} \frac{\mathbf{v}\mathbf{A}}{\mathbf{v}\mathbf{k}} \left(\frac{1}{\epsilon - \xi_\mathbf{p} \pm i\delta} - \frac{1}{\epsilon - \xi_\mathbf{p} + \mathbf{v}\mathbf{k} \pm i\delta} \right).$$

We use the identity

$$\frac{1}{\epsilon - \xi_\mathbf{p} \pm i\delta} = \mp \pi i \delta(\epsilon - \xi_\mathbf{p}) + \frac{1}{\epsilon - \xi_\mathbf{p}}.$$

Therefore,

$$\int_{-\infty}^{+\infty} \tanh\left(\frac{\epsilon}{2T}\right) \left[G(\mathbf{p}_+, \mathbf{p}_-) - G_\epsilon^A(\mathbf{p}_+, \mathbf{p}_-) \right] \frac{d\epsilon}{4\pi i}$$

$$= -\frac{e\mathbf{v}\mathbf{A}}{2c}\frac{1}{\mathbf{v}\mathbf{k}}\left[\tanh\left(\frac{\xi_{\mathbf{p}}}{2T}\right) - \tanh\left(\frac{\xi_{\mathbf{p}} - \mathbf{v}\mathbf{k}}{2T}\right)\right]$$
$$= -\frac{e\mathbf{v}\mathbf{A}}{2c}\frac{1}{2T}\cosh^{-2}\left(\frac{\xi_{\mathbf{p}}}{2T}\right).$$

The corresponding contribution to the current is

$$\frac{e^2}{c}A_k\int\frac{d^3p}{(2\pi)^3}\frac{\mathbf{v}v_k}{2T}\cosh^{-2}\left(\frac{\xi_{\mathbf{p}}}{2T}\right)$$
$$= \frac{e^2}{c}\mathbf{A}\frac{\nu(0)v_F^2}{3}\int d\xi_{\mathbf{p}}\frac{1}{2T}\cosh^{-2}\left(\frac{\xi_{\mathbf{p}}}{2T}\right) = \frac{Ne^2}{mc}\mathbf{A}.$$

This cancels the diamagnetic term.

4

SUPERCONDUCTING ALLOYS

We learn how to incorporate scattering by random impurity atoms into the general Green function formalism of the theory of superconductivity. The cross-diagram technique is derived using the Born approximation for the scattering amplitude.

4.1 Averaging over impurity positions

We describe a simple and powerful method how one can incorporate the interaction between electrons and random impurity atoms into the general BCS scheme. The method has been developed by Abrikosov and Gor'kov (1958). It is assumed that physical properties of a superconductor containing a large amount of random impurities can be obtained by averaging over realizations of the disordered impurity potentials. Another assumption is the Born approximation which implies that the scattering potential is small compared to the characteristic atomic potential the latter being of the order of the Fermi energy of the host material.

Each impurity interacts with electrons via a potential $u(\mathbf{r} - \mathbf{r}_a)$ where $\mathbf{r}_a$ is the position of an impurity atom. Let us consider this interaction as perturbation and calculate the Green function. We assume for simplicity a homogeneous order parameter. Generalization for a spatially dependent order parameter is straightforward. The Green function has the form

$$\check{G}_{\omega_n}(\mathbf{r}, \mathbf{r}') = \check{G}^{(0)}_{\omega_n}(\mathbf{r} - \mathbf{r}') - \sum_a \int d^3r_1\, \check{G}^{(0)}_{\omega_n}(\mathbf{r} - \mathbf{r}_1)u(\mathbf{r}_1 - \mathbf{r}_a)\check{G}^{(0)}_{\omega_n}(\mathbf{r}_1 - \mathbf{r}')$$
$$+\sum_{a,b} \int d^3r_1\, d^3r_2\, \check{G}^{(0)}_{\omega_n}(\mathbf{r} - \mathbf{r}_1)u(\mathbf{r}_1 - \mathbf{r}_a)$$
$$\times\check{G}^{(0)}_{\omega_n}(\mathbf{r}_1 - \mathbf{r}_2)u(\mathbf{r}_2 - \mathbf{r}_b)\check{G}^{(0)}_{\omega_n}(\mathbf{r}_2 - \mathbf{r}') - \ldots$$

Here $\check{G}^{(0)}$ refers to a superconductor without impurities. Note that the Green functions on the both sides of each impurity potential have the same frequency ω_n because the impurity scattering is elastic. We introduce the Fourier transformed Green functions and find

$$\check{G}(\mathbf{r}, \mathbf{r}') = \int \frac{d^3p}{(2\pi)^3} \check{G}^{(0)}(\mathbf{p})e^{i\mathbf{p}(\mathbf{r}-\mathbf{r}')}$$
$$-\sum_a \int \frac{d^3p\, d^3p'}{(2\pi)^6} e^{i(\mathbf{p}\mathbf{r}-\mathbf{p}'\mathbf{r}')} e^{-i(\mathbf{p}-\mathbf{p}')\mathbf{r}_a} \check{G}^{(0)}(\mathbf{p})u(\mathbf{p} - \mathbf{p}')\check{G}^{(0)}(\mathbf{p}')$$

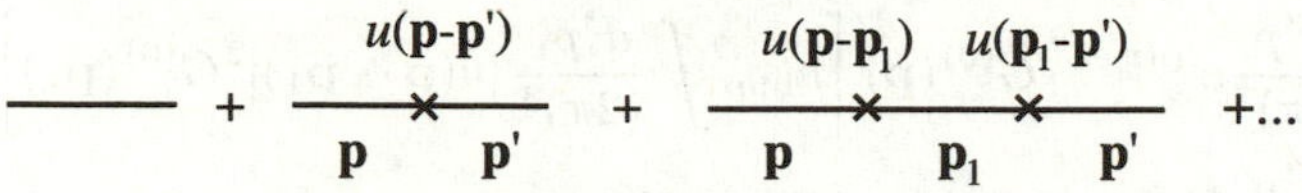

$u(\mathbf{p}\text{-}\mathbf{p}')$ $u(\mathbf{p}\text{-}\mathbf{p}_1)$ $u(\mathbf{p}_1\text{-}\mathbf{p}')$

+ $\mathbf{p}$ $\mathbf{p}'$ + $\mathbf{p}$ $\mathbf{p}_1$ $\mathbf{p}'$ +...

FIG. 4.1. Graphic representation of interaction with impurities.

$$+\sum_{a,b}\int \frac{d^3p\,d^3p'\,d^3p_1}{(2\pi)^9} e^{i(\mathbf{pr}-\mathbf{p}'\mathbf{r}')} e^{-i(\mathbf{p}-\mathbf{p}_1)\mathbf{r}_a} e^{-i(\mathbf{p}_1-\mathbf{p}')\mathbf{r}_b}$$

$$\times \check{G}^{(0)}(\mathbf{p})u(\mathbf{p}-\mathbf{p}_1)\check{G}^{(0)}(\mathbf{p}_1)u(\mathbf{p}_1-\mathbf{p}')\check{G}^{(0)}(\mathbf{p}') - \ldots \tag{4.1}$$

We omit the index ω_n for brevity. This series can be presented graphically as lines separated by crosses, where each cross designates the scattering potential u, Fig. 4.1

Since the impurity atoms are distributed randomly, summation over positions of impurity atoms can be substituted with integration over their coordinates, i.e., with the averaging over positions of impurities and multiplication by the number of impurity atoms

$$\sum_a \Rightarrow n_{\rm imp}\int d^3r_a$$

where $n_{\rm imp}$ is the concentration of impurity atoms. The integration results in δ-functions of the corresponding differences of momenta in eqn (4.1). The first-order correction in eqn (4.1) gives

$$n_{\rm imp}\int \check{G}^{(0)}(\mathbf{p})u(0)\check{G}^{(0)}(\mathbf{p})e^{i\mathbf{p}(\mathbf{r}-\mathbf{r}')}\frac{d^3p}{(2\pi)^3}. \tag{4.2}$$

In the second-order correction we separate first the terms with $\mathbf{r}_a \neq \mathbf{r}_b$, i.e., the terms where the impurity scattering occurs on different atoms. The average gives the δ-functions $\delta(\mathbf{p}-\mathbf{p}_1)$ and $\delta(\mathbf{p}'-\mathbf{p}_1)$. The second-order term becomes

$$n_{\rm imp}^2\int \check{G}^{(0)}(\mathbf{p})u(0)\check{G}^{(0)}(\mathbf{p})u(0)\check{G}^{(0)}(\mathbf{p})e^{i\mathbf{p}(\mathbf{r}-\mathbf{r}')}\frac{d^3p}{(2\pi)^3}. \tag{4.3}$$

If we now collect the terms of all orders in u where scattering occurs on different atoms we obtain the effective correction to the Hamiltonian in the form

$$\check{H}_1 = \begin{pmatrix} n_{\rm imp}u(0) & 0 \\ 0 & n_{\rm imp}u(0)\end{pmatrix}.$$

This term is not important: it can be incorporated in the chemical potential. We shall omit this contribution in what follows.

Consider now the second-order term where scattering occurs at the same atom $\mathbf{r}_a = \mathbf{r}_b$. The sum gives

$$n_{\rm imp}\int \frac{d^3p\,d^3p_1}{(2\pi)^6} e^{i\mathbf{p}(\mathbf{r}-\mathbf{r}')}\check{G}^{(0)}(\mathbf{p})u(\mathbf{p}-\mathbf{p}_1)\check{G}^{(0)}(\mathbf{p}_1)u(\mathbf{p}_1-\mathbf{p})\check{G}^{(0)}(\mathbf{p})$$

$$= \int \frac{d^3p}{(2\pi)^3} e^{i\mathbf{p}(\mathbf{r}-\mathbf{r}')} \check{G}^{(0)}(\mathbf{p}) \left[n_{\rm imp} \int \frac{d^3p_1}{(2\pi)^3} |u(\mathbf{p}-\mathbf{p}_1)|^2 \check{G}^{(0)}(\mathbf{p}_1) \right] \check{G}^{(0)}(\mathbf{p})$$
$$\equiv \int \frac{d^3p}{(2\pi)^3} e^{i\mathbf{p}(\mathbf{r}-\mathbf{r}')} \check{G}^{(0)}(\mathbf{p}) \check{\Sigma}^{(0)}(\mathbf{p}) \check{G}^{(0)}(\mathbf{p}). \tag{4.4}$$

Here we introduce the "self-energy"

$$\check{\Sigma}^{(0)}_{\omega_n}(\mathbf{p}) = n_{\rm imp} \int \frac{d^3p_1}{(2\pi)^3} |u(\mathbf{p}-\mathbf{p}_1)|^2 \check{G}^{(0)}_{\omega_n}(\mathbf{p}_1). \tag{4.5}$$

Since both $\mathbf{p}$ and $\mathbf{p}_1$ are close to the Fermi surface, the scattering amplitude depends only on the angle θ between $\mathbf{p}$ and $\mathbf{p}_1$: $|u(\mathbf{p}-\mathbf{p}_1)|^2 = |u(\theta)|^2$.

The scattering amplitude determines the scattering mean free time. In the normal state it is

$$\frac{1}{\tau_{\rm imp}} = 2\pi\nu(0) n_{\rm imp} \int |u(\theta)|^2 \frac{d\Omega_{\mathbf{p}}}{4\pi}. \tag{4.6}$$

One can introduce the scattering cross section $\sigma_{\mathbf{p},\mathbf{p}_1}$. In the Born approximation

$$|u(\mathbf{p}-\mathbf{p}_1)|^2 = \frac{2v_F}{\nu(0)} \sigma_{\mathbf{p},\mathbf{p}_1}.$$

The scattering mean free time becomes

$$\frac{1}{\tau_{\rm imp}} = 4\pi v_F n_{\rm imp} \int \sigma_{\mathbf{p},\mathbf{p}_1} \frac{d\Omega_{\mathbf{p}}}{4\pi}.$$

For an isotropic scattering by impurities $|u(\theta)|^2 = const$, the self-energy is independent of the momentum direction:

$$\check{\Sigma}^{(0)}_{\omega_n} = n_{\rm imp}|u|^2 \int \frac{d^3p_1}{(2\pi)^3} \check{G}^{(0)}_{\omega_n}(\mathbf{p}_1) = \frac{1}{2\pi\nu(0)\tau_{\rm imp}} \int \frac{d^3p_1}{(2\pi)^3} \check{G}^{(0)}_{\omega_n}(\mathbf{p}_1). \tag{4.7}$$

Here the scattering mean free time is

$$\frac{1}{\tau_{\rm imp}} = 2\pi\nu(0)|u|^2 n_{\rm imp}. \tag{4.8}$$

The corresponding contribution to the Green function can be shown on a diagram as a dashed line connecting two crosses belonging to the same atom (see Fig. 4.2).

Consider now higher-order terms in u. The terms where all the atoms are different have already been taken into account. The rest can be separated into two groups. The first contains such scattering processes that each atom only participates twice. On a diagram this corresponds to various combinations of dashed lines connecting two crosses. The second group involves processes where each atom participates more than twice. One can show that the contribution from more than two crosses per atom is smaller than that from two crosses per

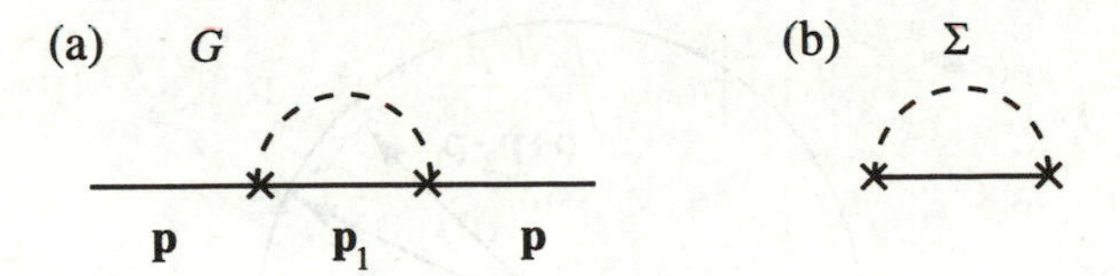

FIG. 4.2. (a) Averaging over positions of one atom. (b) First-order self-energy.

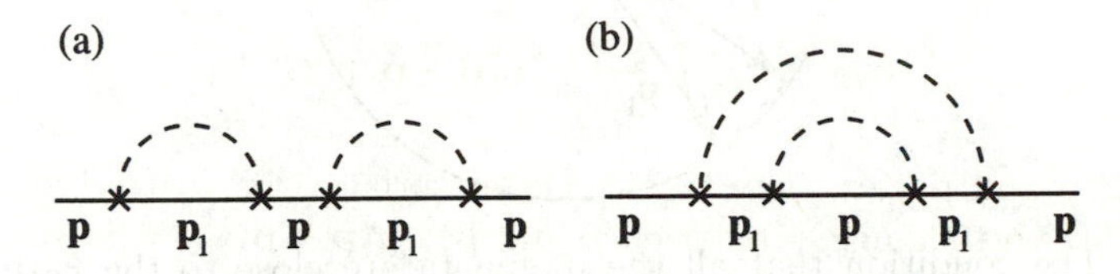

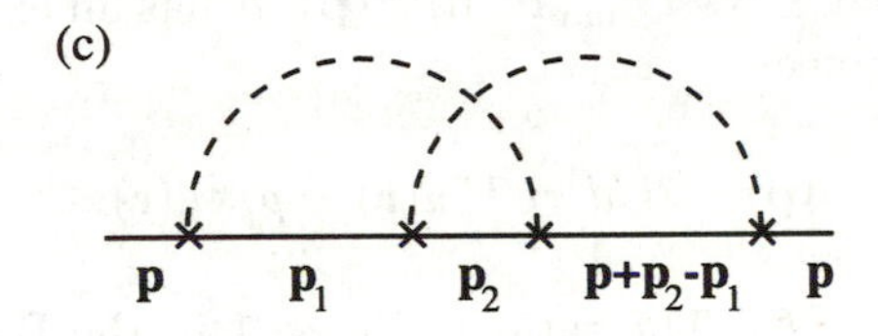

FIG. 4.3. Averaging over positions of two atoms. (a) and (b) Dashed lines do not cross. (c) Dashed lines cross, which restricts the interval of the momentum directions.

atom by a factor u/E_F for each extra cross. Indeed, the diagram with three crosses gives

$$\sum_{a,b,c} \int \frac{d^3p\, d^3p'\, d^3p_1\, d^3p_2}{(2\pi)^{12}} e^{i(\mathbf{pr}-\mathbf{p'r'})} e^{-i(\mathbf{p}-\mathbf{p}_1)\mathbf{r}_a - i(\mathbf{p}_1-\mathbf{p}_2)\mathbf{r}_b - i(\mathbf{p}_2-\mathbf{p}')\mathbf{r}_c}$$
$$\times \check{G}^{(0)}(\mathbf{p}) u(\mathbf{p}-\mathbf{p}_1) \check{G}^{(0)}(\mathbf{p}_1) u(\mathbf{p}_1-\mathbf{p}_2) \check{G}^{(0)}(\mathbf{p}_2) u(\mathbf{p}_2-\mathbf{p}') \check{G}^{(0)}(\mathbf{p}').$$

We put $\mathbf{r}_a = \mathbf{r}_b = \mathbf{r}_c = \mathbf{r}$ and integrate over d^3r. We get

$$n_{\text{imp}} \int \frac{d^3p\, d^3p_1\, d^3p_2}{(2\pi)^9} e^{i\mathbf{p}(\mathbf{r}-\mathbf{r}')} \check{G}^{(0)}_{\omega_n}(\mathbf{p})$$
$$\times u(\mathbf{p}-\mathbf{p}_1) \check{G}^{(0)}(\mathbf{p}_1) u(\mathbf{p}_1-\mathbf{p}_2) \check{G}^{(0)}(\mathbf{p}_2) u(\mathbf{p}_2-\mathbf{p}) \check{G}^{(0)}(\mathbf{p})$$
$$= \int \frac{d^3p}{(2\pi)^3} e^{i\mathbf{p}(\mathbf{r}-\mathbf{r}')} \check{G}^{(0)}(\mathbf{p}) \check{\Sigma}^{(3)}(\mathbf{p}) \check{G}^{(0)}(\mathbf{p})$$

where

$$\check{\Sigma}^{(3)}_{\omega_n}(\mathbf{p}) = n_{\text{imp}} \int \frac{d^3p_1\, d^3p_2}{(2\pi)^6} u(\mathbf{p}-\mathbf{p}_1) u(\mathbf{p}_1-\mathbf{p}_2) u(\mathbf{p}_2-\mathbf{p})$$
$$\times \check{G}^{(0)}_{\omega_n}(\mathbf{p}_1) \check{G}^{(0)}_{\omega_n}(\mathbf{p}_2).$$

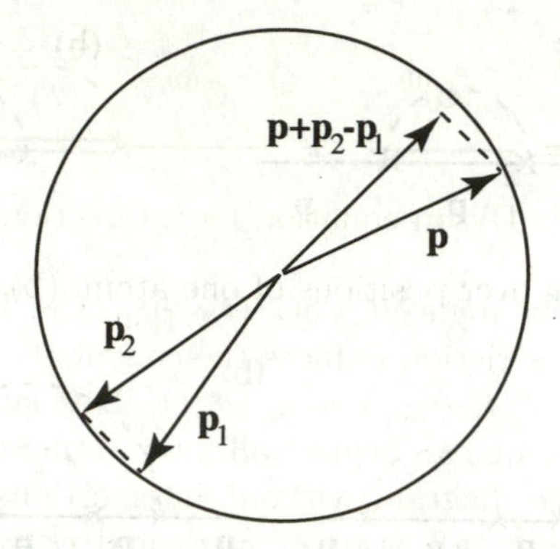

FIG. 4.4. The condition that all the momenta are close to the Fermi surface imposes a restriction on the momentum directions.

The estimate shows that $\Sigma^{(3)} \sim n_{\rm imp}\nu^2(0)u^3(\mathbf{p})$. It has an extra factor $\nu(0)u(\mathbf{p})$ as compared to $\Sigma^{(0)}$. Since

$$u(\mathbf{p}) = \int d^3r e^{i\mathbf{pr}}u(\mathbf{r}) \sim p_F^{-3}u(\mathbf{r})$$

the factor $\nu(0)u(\mathbf{p}) \sim u/E_F$ The ratio $u/E_F \ll 1$ in the Born approximation. It is this approximation which we adopt for our consideration of the impurity scattering. Therefore, the processes with two crosses per atom are sufficient for us.

The processes where each atom participates only twice can be of two different types shown in Fig. 4.3. The first type has dashed lines which do not intersect, Fig. 4.3 (a,b). The second type has intersecting lines, Fig. 4.3 (c). For example, the diagram with four crosses gives

$$\sum_{a,b,c,d} \int \frac{d^3p\, d^3p'\, d^3p_1\, d^3p_2\, d^3p_3}{(2\pi)^{15}} e^{i(\mathbf{pr}-\mathbf{p}'\mathbf{r}')}$$
$$\times e^{-i(\mathbf{p}-\mathbf{p}_1)\mathbf{r}_a - i(\mathbf{p}_1-\mathbf{p}_2)\mathbf{r}_b - i(\mathbf{p}_2-\mathbf{p}_3)\mathbf{r}_c - i(\mathbf{p}_3-\mathbf{p}')\mathbf{r}_d}$$
$$\times \check{G}^{(0)}(\mathbf{p})u(\mathbf{p}-\mathbf{p}_1)\check{G}^{(0)}(\mathbf{p}_1)u(\mathbf{p}_1-\mathbf{p}_2)\check{G}^{(0)}(\mathbf{p}_2)u(\mathbf{p}_2-\mathbf{p}_3)$$
$$\times \check{G}^{(0)}(\mathbf{p}_3)u(\mathbf{p}_3-\mathbf{p}')\check{G}^{(0)}(\mathbf{p}').$$

The intersecting lines correspond to $\mathbf{r}_a = \mathbf{r}_c$ and $\mathbf{r}_b = \mathbf{r}_d$. After averaging we obtain

$$n_{\rm imp}^2 \int \frac{d^3p\, d^3p_1\, d^3p_2}{(2\pi)^9} e^{i\mathbf{p}(\mathbf{r}-\mathbf{r}')}\check{G}^{(0)}(\mathbf{p})|u(\mathbf{p}-\mathbf{p}_1)|^2$$
$$\times \check{G}^{(0)}(\mathbf{p}_1)\check{G}^{(0)}(\mathbf{p}_2)|u(\mathbf{p}_1-\mathbf{p}_2)|^2\check{G}^{(0)}(\mathbf{p}+\mathbf{p}_2-\mathbf{p}_1)\check{G}^{(0)}(\mathbf{p}).$$

All the momenta which are arguments of the Green functions should be close to the Fermi surface, $|\mathbf{p}_i| \approx p_F$. This condition imposes a restriction on the integration range over the angles of the momenta (see Fig. 4.4). For example,

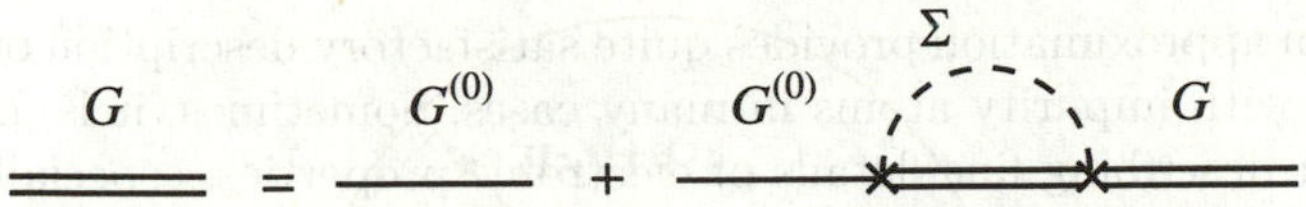

FIG. 4.5. Graphic Dyson equation for interaction with impurities.

the range of the azimuthal angle $\delta\theta \sim \delta G/G \sim \delta\xi_{\mathbf{p}}/E_F$. The characteristic $\delta\xi_{\mathbf{p}} \sim 1/\tau_{\text{imp}}$, therefore, this restriction reduces the contribution from the intersecting diagram by the factor $1/(E_F\tau_{\text{imp}}) \sim 1/(p_F\ell)$. This factor is small for metals $1/(p_F\ell) \ll 1$. Therefore, we can neglect all intersecting diagrams.

On the contrary, the diagram without intersections corresponds to either $\mathbf{r}_a = \mathbf{r}_b$ and $\mathbf{r}_c = \mathbf{r}_d$ (Fig. 4.3 a) or $\mathbf{r}_a = \mathbf{r}_d$ and $\mathbf{r}_b = \mathbf{r}_c$ (Fig. 4.3 b). These diagrams do not have restrictions on the integration over the angles, therefore, they have to be taken into account. After summation of all the diagram, we obtain an expression of the type of eqn (4.4) where the function $\check{G}^{(0)}(\mathbf{p}_1)$ in the self-energy and the function $\check{G}^{(0)}(\mathbf{p})$ from the right are replaced with the total Green function containing all the diagrams:

$$\check{G}_{\omega_n}(\mathbf{p}) = \check{G}^{(0)}_{\omega_n}(\mathbf{p}) + \check{G}^{(0)}_{\omega_n}(\mathbf{p})\check{\Sigma}_{\omega_n}(\mathbf{p})\check{G}_{\omega_n}(\mathbf{p}) \tag{4.9}$$

where the total self-energy is

$$\check{\Sigma}_{\omega_n}(\mathbf{p}) = n_{\text{imp}} \int \frac{d^3p_1}{(2\pi)^3} |u(\theta)|^2 \check{G}_{\omega_n}(\mathbf{p}_1) \tag{4.10}$$

For isotropic scattering, we have

$$\check{\Sigma}_{\omega_n} = \frac{1}{2\pi\nu(0)\tau_{\text{imp}}} \int \frac{d^3p_1}{(2\pi)^3} \check{G}_{\omega_n}(\mathbf{p}_1). \tag{4.11}$$

The Dyson equation (4.9) is presented graphically in Fig. 4.5. It is equivalent to the equation for the total Green function in the form

$$\check{G}^{-1}_{\omega_n}(\mathbf{r}_1)\check{G}_{\omega_n}(\mathbf{r}_1, \mathbf{r}_2) = \hat{1}\delta(\mathbf{r}_1 - \mathbf{r}_2) \tag{4.12}$$

with the operator

$$\check{G}^{-1}_{\omega_n}(\mathbf{r}_1) = \check{G}^{-1}_{\text{clean}}(\mathbf{r}_1) - \check{\Sigma}_{\omega_n}$$

where $\check{G}^{-1}_{\text{clean}}(\mathbf{r}_1)$ is defined by eqn (3.28). The matrix self-energy can be written in components:

$$\check{\Sigma} = \begin{pmatrix} \Sigma_1 & \Sigma_2 \\ -\Sigma_2^\dagger & \bar{\Sigma}_1 \end{pmatrix}. \tag{4.13}$$

The method of averaging over the impurity atoms which leads to the Dyson equation in the form of Fig. 4.5 is called the cross-diagram technique. Its characteristic feature is the absence of intersecting lines (impurity averaging) which connect different crosses (impurity atoms).

The Born approximation provides quite satisfactory description of interaction of electrons with impurity atoms in many cases. Sometimes, it is, however, not sufficient for describing fine details of electronic properties, especially in d-wave superconductors. Beyond the Born approximation, one has to take into account various processes with more than two crosses per atom. This results in a substitution of the Born amplitude $u(\mathbf{p})$ with the full scattering amplitude. For a strong scattering potential, the so-called unitary limit, the exact amplitude may be substantially different from that in the Born approximation. We do not consider such problems here. Some examples of full scattering amplitude analysis have been discussed by Buchholtz and Zwicknagl (1981), Graf *et al.* (1995), and Graf *et al.* (1996).

4.1.1 *Magnetic impurities*

Considering interaction of electrons with impurities one has to distinguish between the scattering by nonmagnetic impurities and that by impurity atoms having a finite magnetic moment (Abrikosov and Gor'kov 1960). We already noticed that magnetic impurities scatter differently the two electrons which make a Cooper pair because of different spin states of the constituent electrons in the potential of the magnetic impurity. Recall that the functions G, on one hand, and F, $F^\dagger$, on the other hand, have different spin structures; G is proportional to the unit matrix in the spin state, while F is proportional to the antisymmetric matrix $i\sigma_y$. Therefore, the scattering amplitudes contain different projections of the impurity spin operators. One can write, therefore,

$$\Sigma_1 = n_{\text{imp}} \int \frac{d^3p_1}{(2\pi)^3} |u_1(\theta)|^2 G_{\omega_n}(\mathbf{p}_1) \tag{4.14}$$

and similarly for $\bar{\Sigma}_1$ in terms of $\bar{G}$. On the other hand,

$$\Sigma_2 = n_{\text{imp}} \int \frac{d^3p_1}{(2\pi)^3} |u_2(\theta)|^2 F_{\omega_n}(\mathbf{p}_1) \tag{4.15}$$

with a similar expression for $\Sigma_2^\dagger$ in terms of $F^\dagger$.

The corresponding scattering times are

$$\frac{1}{\tau_1} = 2\pi\nu(0)n_{\text{imp}} \int \frac{d\Omega_{\mathbf{p}_1}}{4\pi} |u_1(\theta)|^2, \tag{4.16}$$

$$\frac{1}{\tau_2} = 2\pi\nu(0)n_{\text{imp}} \int \frac{d\Omega_{\mathbf{p}_1}}{4\pi} |u_2(\theta)|^2. \tag{4.17}$$

One can define the difference of these scattering rates as

$$\frac{2}{\tau_s} = \frac{1}{\tau_1} - \frac{1}{\tau_2} \tag{4.18}$$

where τ_s is the spin-flip scattering time. It can be shown (Abrikosov and Gor'kov 1960) that

$$\frac{1}{\tau_s} = 4\pi n_{\text{imp}}\nu(0)\frac{S(S+1)}{(2S+1)^2}\int \frac{d\Omega_{\mathbf{p}_1}}{4\pi}|u_+(\theta) - u_-(\theta)|^2 > 0. \tag{4.19}$$

Here u_+ is the scattering amplitude for an electron in the state with the total spin $S+1/2$, and u_- is the scattering amplitude for an electron in the state with the total spin $S-1/2$; S is the spin of an impurity atom. This is the so-called spin-flip time: the average time between collisions resulting in a change of the spin state of an electron.

The scattering amplitudes can be parametrized by two different scattering cross sections σ_1 and σ_2 associated with the diagonal and off-diagonal components (i.e., with either the functions G and $\bar{G}$ or the functions F and $F^\dagger$) of the matrix Green function $\check{G}$, respectively,

$$|u_{1,2}(\theta)|^2 = \frac{2v_F}{\nu(0)}\sigma_{1,2}(\theta).$$

4.2 Homogeneous state of an s-wave superconductor

Consider a homogeneous s-wave superconductor without magnetic impurities. We assume also that the magnetic field is absent. In the Fourier representation, the Gor'kov equations are

$$(-i\omega_n + \xi_{\mathbf{p}} - \Sigma_1)G_{\omega_n}(\mathbf{p}) + (\Delta + \Sigma_2)F^\dagger_{\omega_n}(\mathbf{p}) = 1, \tag{4.20}$$

$$(i\omega_n + \xi_{\mathbf{p}} - \bar{\Sigma}_1)F^\dagger_{\omega_n}(\mathbf{p}) - (\Delta^* + \Sigma_2^\dagger)G_{\omega_n}(\mathbf{p}) = 0. \tag{4.21}$$

We shall see later that $\Sigma_1 = -\bar{\Sigma}_1$. Moreover, for a homogeneous state, one can take Δ to be real and thus $F = F^\dagger$ and $\Sigma_2 = \Sigma_2^\dagger$. The solution becomes

$$G = \frac{i\omega_n + \xi_{\mathbf{p}} + \Sigma_1}{\xi_{\mathbf{p}}^2 + (\omega_n - i\Sigma_1)^2 + (\Delta + \Sigma_2)^2}, \tag{4.22}$$

$$F = \frac{\Delta + \Sigma_2}{\xi_{\mathbf{p}}^2 + (\omega_n - i\Sigma_1)^2 + (\Delta + \Sigma_2)^2}. \tag{4.23}$$

The self-energies are

$$\Sigma_1 = n_{\text{imp}}\int \frac{d^3p_1}{(2\pi)^3}\,|u(\theta)|^2 G_{\omega_n}(\mathbf{p}_1),$$

$$\Sigma_2 = n_{\text{imp}}\int \frac{d^3p_1}{(2\pi)^3}\,|u(\theta)|^2 F_{\omega_n}(\mathbf{p}_1). \tag{4.24}$$

The Green functions for a homogeneous state depend only on the magnitude of the momentum. We shall see that the momentum integrals here are determined by the region near the Fermi surface. Therefore,

$$\Sigma_1 = \nu(0)n_{\text{imp}}\int \frac{d\Omega_{\mathbf{p}}}{4\pi}\,|u(\theta)|^2 \int d\xi_{\mathbf{p}}G_{\omega_n}(\mathbf{p}) = \frac{1}{2\pi\tau_{\text{imp}}}\int d\xi_{\mathbf{p}}G_{\omega_n}(\mathbf{p}),$$

$$\Sigma_2 = \nu(0) n_{\rm imp} \int \frac{d\Omega_{\mathbf{p}}}{4\pi} |u(\theta)|^2 \int d\xi_{\mathbf{p}} F_{\omega_n}(\mathbf{p}) = \frac{1}{2\pi\tau_{\rm imp}} \int d\xi_{\mathbf{p}} F_{\omega_n}(\mathbf{p}),$$

since the Green functions do not depend on the momentum direction. We have

$$\int d\xi_{\mathbf{p}} G_{\omega_n}(\mathbf{p}) = \int d\xi_{\mathbf{p}} \frac{i\omega_n + \xi_{\mathbf{p}} + \Sigma_1}{\xi_{\mathbf{p}}^2 + (\omega_n - i\Sigma_1)^2 + (\Delta + \Sigma_2)^2}$$
$$= \int \frac{d\xi_{\mathbf{p}}}{2} \frac{1}{\xi_{\mathbf{p}} - i\sqrt{(\omega_n - i\Sigma_1)^2 + (\Delta+\Sigma_2)^2}}$$
$$+ \int \frac{d\xi_{\mathbf{p}}}{2} \frac{1}{\xi_{\mathbf{p}} + i\sqrt{(\omega_n - i\Sigma_1)^2 + (\Delta+\Sigma_2)^2}}$$
$$+ \int d\xi_{\mathbf{p}} \frac{i\omega_n + \Sigma_1}{\left(\xi_{\mathbf{p}} - i\sqrt{(\omega_n - i\Sigma_1)^2 + (\Delta+\Sigma_2)^2}\right)}$$
$$\times \frac{1}{\left(\xi_{\mathbf{p}} + i\sqrt{(\omega_n - i\Sigma_1)^2 + (\Delta+\Sigma_2)^2}\right)}.$$

We shall see that Σ_1 is imaginary while Σ_2 is real. Therefore, the first two integrals result in real quantities which are determined by large ξ_p and thus depend only on the normal properties of the system. We can incorporate these contributions to the chemical potential and disregard them in further calculations. The third integral can be calculated by residues. It gives

$$\Sigma_1 = \frac{1}{2\pi\tau_{\rm imp}} \int d\xi_{\mathbf{p}} G_{\omega_n}(\mathbf{p}) = \frac{i}{2\tau_{\rm imp}} \frac{\omega_n - i\Sigma_1}{\sqrt{(\omega_n - i\Sigma_1)^2 + (\Delta+\Sigma_2)^2}} \tag{4.25}$$

For Σ_2, the normal-state part is absent since F decreases quickly enough as a function of ξ_p. Calculating it by residues, we find

$$\Sigma_2 = \frac{i}{2\tau_{\rm imp}} \frac{\Delta + \Sigma_2}{i\sqrt{(\omega_n - i\Sigma_1)^2 + (\Delta+\Sigma_2)^2}}. \tag{4.26}$$

Equations (4.25) and (4.26) imply that

$$1 + \frac{\Sigma_1}{i\omega_n} = 1 + \frac{\Sigma_2}{\Delta} = \eta_n \tag{4.27}$$

where the function η_n is

$$\eta_n = 1 + \frac{1}{2\tau_{\rm imp}} \frac{1}{\sqrt{\omega_n^2 + \Delta^2}}. \tag{4.28}$$

Finally, the Green functions become

$$G = \frac{i\omega_n\eta_n + \xi_{\mathbf{p}}}{\xi_{\mathbf{p}}^2 + \eta_n^2(\omega_n^2 + \Delta^2)}, \quad F = \frac{\Delta\eta_n}{\xi_{\mathbf{p}}^2 + \eta_n^2(\omega_n^2 + \Delta^2)}. \tag{4.29}$$

The gap equation takes the form

$$\frac{\Delta}{\lambda} = 2T \sum_{n \geq 0}^{N_0} \int d\xi_{\mathbf{p}} F_{\omega_n}(\mathbf{p}) = 2\pi T \sum_{n \geq 0}^{N_0} \frac{\Delta}{\sqrt{\omega_n^2 + \Delta^2}}. \tag{4.30}$$

The function η drops out after the integration over $d\xi_{\mathbf{p}}$! The gap equation for a superconducting alloy with an s-wave pairing in a spatially homogeneous state is exactly the same as for a clean superconductor. This means that neither the critical temperature nor the order parameter is affected by impurities. This statement holds for an s-wave superconductor and is known as the Anderson theorem.

Part II

Quasiclassical method

5

GENERAL PRINCIPLES OF THE QUASICLASSICAL APPROXIMATION

The quasiclassical Green functions integrated over the energy near the Fermi surface are introduced. Eilenberger equations and the normalization condition are derived for the quasiclassical Green functions. It is shown how to reduce the Eilenberger equations to diffusion-like Usadel equations in case of superconducting alloys. The boundary conditions for the quasiclassical Green functions at a rough interface between a superconductor and an insulator are derived.

5.1 Quasiclassical Green functions

As we already discussed in the Introduction, all presently known superconductors have the Fermi energy considerably larger than Δ. This condition is especially well satisfied for conventional, low-temperature superconductors where the ratio E_F/Δ can be as high as 10^3. For high-temperature superconductors, the ratio E_F/Δ is not that high; nevertheless, it still ranges by the order of magnitude from 10 to 10^2. The BCS theory appears to be so successful mostly because it benefits essentially from the situation when the superconducting order parameter magnitude is much smaller than the characteristic energy of the normal state, i.e., the Fermi energy. In such a case, the Fermi momentum is much larger than the inverse coherence length $\xi^{-1} \sim \Delta/v_F$ such that the condition eqn (1.61) is satisfied. This fact offers a possibility to develop a relatively simple and powerful method of dealing with the Green functions in the BCS theory based on the quasiclassical approximation. The quasiclassical method is now commonly accepted; it is widely used for practical calculations especially for problems where the order parameter and electromagnetic fields are functions of coordinates and time. In the present chapter we introduce the quasiclassical method for stationary properties of superconductors. Nonstationary problems are considered in the following parts of the book.

The definitions of physical quantities involve the Green functions whose spatial coordinates are taken in the limit $\mathbf{r}_1 \to \mathbf{r}_2$. In the momentum representation, the limit $\mathbf{r}_1 \to \mathbf{r}_2$ corresponds to the integration over the momentum space. The definitions of physical quantities would now contain Green functions integrated with various powers of the quasiparticle momenta. For example, the order parameter of an s-wave superconductor contains the function $F(\mathbf{p}_+, \mathbf{p}_-)$ integrated with the momentum to the zeroth power. The order parameter of a d-wave superconductor involves second powers of components of $\mathbf{p}$. The current is expressed through $G(\mathbf{p}_+, \mathbf{p}_-)$ integrated with the first power of $\mathbf{p}$, etc.

The fastest function under these integrals is the Green function itself: as a function of $\xi_{\mathbf{p}}$, it varies strongly near the Fermi surface when $\xi_{\mathbf{p}}$ changes by an amount $\delta\xi_{\mathbf{p}}$ of the order of Δ. Since this variation $\delta\xi_{\mathbf{p}}$ is much smaller than E_F the magnitude of the quasiparticle momentum $\mathbf{p}$ remains close to the Fermi momentum. Indeed, the variation of the momentum magnitude is $\delta p \sim \delta\xi_{\mathbf{p}}/v_F \sim \xi^{-1}$ such that $\delta p/p_F \sim (p_F\xi)^{-1} \ll 1$. Here we use again the quasiclassical condition eqn (1.61), namely that $p_F\xi$ is a large number. It means that, in all the prefactors in front of the Green function, we can put the momentum magnitude to be at the Fermi surface. In particular, one can parameterize the momentum-space integral using the Fermi surface defined for the normal state:

$$\frac{d^3p}{(2\pi)^3} = d\xi_{\mathbf{p}} \frac{dS_F}{(2\pi)^3 v_F} \tag{5.1}$$

where $d\xi_{\mathbf{p}}/v_F$ is the momentum increment in the direction perpendicular to the Fermi surface while dS_F is the Fermi-surface area element. For simplicity, we shall consider mostly a spherical Fermi surface. If the Fermi surface is a sphere, $dS_F = p_F^2 d\Omega_{\mathbf{p}}$, and we recover eqn (3.78). Generalizations for more complicated Fermi surfaces are straightforward. Of course, in the superconducting state, the Fermi surface is not defined in a strict sense because a gap in the energy spectrum opens near the original position of Fermi surface in the normal state. However, if the energy gap is much smaller than the original Fermi energy, the latter can still be used as a reference energy for the variable

$$\xi_{\mathbf{p}} = E_n(\mathbf{p}) - E_F.$$

Recall that E_n is the electronic spectrum in the normal state.

All the expressions for physical quantities now contain the Green functions integrated over $d\xi_{\mathbf{p}}$ near the Fermi surface and then multiplied with various functions of the particle momentum directions. Since these expressions do not use energies of the order of the Fermi energy, the question arises: Is it possible to develop such a formalism that operates only with the Green functions taken near the Fermi surface? Or, what would be even better, can one develop a method which uses only Green functions already integrated over $d\xi_{\mathbf{p}}$ near the Fermi surface? The answer is yes; we show how this should be done in the present chapter. The new formulation of the theory of superconductivity was elaborated first by Eilenberger (1968), Eliashberg (1971), and Larkin and Ovchinnikov (1977). It is much simpler to handle than the original approach which uses the full Green functions. Fundamentals of the quasiclassical method can be found in the review by Serene and Rainer (1983).

We turn now to definitions of the quasiclassical Green functions. Integrating over the energy near the Fermi surface we note that it is easy to define the integrals

$$\int \frac{d\xi_{\mathbf{p}}}{\pi i} F_{\omega_n}(\mathbf{p}_+, \mathbf{p}_-)\ , \quad \int \frac{d\xi_{\mathbf{p}}}{\pi i} F^{\dagger}_{\omega_n}(\mathbf{p}_+, \mathbf{p}_-). \tag{5.2}$$

They converge rather quickly for $\xi_{\mathbf{p}} \gg T, \Delta$ since F and $F^{\dagger}$ decrease as $\xi_{\mathbf{p}}^{-2}$. This is due to the fact that the functions F and $F^{\dagger}$ exist only in the superconducting state. The integrals are, of course, determined by contributions from the quasiparticles having energies near the Fermi surface. These quasiparticle states correspond to the poles of the Green functions. We denote this fact by

$$\int \frac{d\xi_{\mathbf{p}}}{\pi i} F_{\omega_n}(\mathbf{p}_+, \mathbf{p}_-) = \oint \frac{d\xi_{\mathbf{p}}}{\pi i} F_{\omega_n}(\mathbf{p}_+, \mathbf{p}_-) \equiv f_{\omega_n}(\hat{\mathbf{p}}, \mathbf{k}), \tag{5.3}$$

$$\int \frac{d\xi_{\mathbf{p}}}{\pi i} F^{\dagger}_{\omega_n}(\mathbf{p}_+, \mathbf{p}_-) = \oint \frac{d\xi_{\mathbf{p}}}{\pi i} F^{\dagger}_{\omega_n}(\mathbf{p}_+, \mathbf{p}_-) \equiv f^{\dagger}_{\omega_n}(\hat{\mathbf{p}}, \mathbf{k}), \tag{5.4}$$

where $\oint$ shows that we take the contributions from poles close to the Fermi surface. The particle momentum lies at the Fermi surface, the only variable is the momentum direction; we use this fact and denote $\hat{\mathbf{p}}$ the unit vector in the direction of $\mathbf{p}$.

However, the full integral over $d\xi_{\mathbf{p}}$ diverges for the function G. This is because G contains the normal part which decays slowly for large $\xi_{\mathbf{p}}$. We can, however, introduce the integral

$$\oint \frac{d\xi_{\mathbf{p}}}{\pi i} G_{\omega_n}(\mathbf{p}_+, \mathbf{p}_-) \equiv g_{\omega_n}(\hat{\mathbf{p}}, \mathbf{k}) \tag{5.5}$$

which only takes the contribution from poles near the Fermi surface. With help of eqn (3.78) the full integral over the magnitude of the momentum can be written as

$$\int \frac{p^2\, dp}{2\pi^2} G_{\omega_n}(\mathbf{p}_+, \mathbf{p}_-) = \int \frac{p^2\, dp}{2\pi^2} G^{(n)}_{\omega_n} + \nu(0) \oint d\xi_{\mathbf{p}} \left[G_{\omega_n} - G^{(n)}_{\omega_n} \right] \tag{5.6}$$

where $G^{(n)}_{\omega_n}$ is the zero-field Green function in the normal state. In the last line, the integral over $d\xi_{\mathbf{p}}$ converges; it is thus determined by poles near the Fermi surface. The accuracy of this approximation is again of the order of Δ/E_F.

Since

$$G^{(n)}_{\omega_n}(\mathbf{p}_+, \mathbf{p}_-) = G^{(n)}_{\omega_n}(\mathbf{p})(2\pi)^3 \delta(\mathbf{k})$$

and

$$G^{(n)}_{\omega_n}(\mathbf{p}) = \frac{1}{\xi_{\mathbf{p}} - i\omega_n} = \frac{1}{\xi_{\mathbf{p}}} + \pi i \operatorname{sign}(\omega_n) \delta(\xi_{\mathbf{p}})$$

we have

$$\oint \frac{d\xi_{\mathbf{p}}}{\pi i} G^{(n)}_{\omega}(\mathbf{p}) = \operatorname{sign}(\omega_n). \tag{5.7}$$

Therefore,

$$\int \frac{p^2\, dp}{2\pi^2} G_{\omega_n}(\mathbf{p}_+, \mathbf{p}_-) = \int \frac{p^2\, dp}{2\pi^2} G^{(n)}_{\omega_n}(\mathbf{p}_+, \mathbf{p}_-)$$

$$-\nu(0)\pi i \mathrm{sign}\,(\omega_n)(2\pi)^3\delta(\mathbf{k})$$
$$+\nu(0)\pi i g_\omega(\hat{\mathbf{p}},\mathbf{k}) \tag{5.8}$$

so that

$$\int \frac{p^2\,dp}{2\pi^2} G_{\omega_n}(\mathbf{p}_+,\mathbf{p}_-) = (2\pi)^3\delta(\mathbf{k})\mathcal{P}\int \frac{p^2\,dp}{2\pi^2}\frac{1}{\xi_{\mathbf{p}}} + \nu(0)\pi i g_\omega(\hat{\mathbf{p}},\mathbf{k}) \tag{5.9}$$

where $\mathcal{P}\int$ means the principal value integral.

In the same way we introduce

$$\oint \frac{d\xi_p}{\pi i}\bar{G}_\omega(\mathbf{p}_+,\mathbf{p}_-) \equiv \bar{g}_\omega(\hat{\mathbf{p}},\mathbf{k}) \tag{5.10}$$

so that

$$\int \frac{p^2\,dp}{2\pi^2}\bar{G}_{\omega_n}(\mathbf{p}_+,\mathbf{p}_-) = \int \frac{p^2 dp}{2\pi^2}\bar{G}^{(n)}_{\omega_n}(\mathbf{p}_+,\mathbf{p}_-)$$
$$+\nu(0)\pi i\,\mathrm{sign}\,(\omega_n)(2\pi)^3\delta(\mathbf{k})$$
$$+\nu(0)\pi i\bar{g}_\omega(\hat{\mathbf{p}},\mathbf{k}) \tag{5.11}$$

and

$$\int \frac{p^2\,dp}{2\pi^2}\bar{G}_\omega(\mathbf{p}_+,\mathbf{p}_-) = (2\pi)^3\delta(\mathbf{k})\mathcal{P}\int \frac{p^2\,dp}{2\pi^2}\frac{1}{\xi_{\mathbf{p}}} + \nu(0)\pi i\bar{g}_\omega(\hat{\mathbf{p}},\mathbf{k}). \tag{5.12}$$

We can see that the large-$\xi_{\mathbf{p}}$ contribution to the full momentum integral of G only depends on the properties of the material in the normal state. All the superconducting properties are included into the functions g, f, etc., determined by a vicinity of the Fermi surface. The functions

$$g,\ f,\ f^\dagger,\ \text{and}\ \bar{g}$$

are called quasiclassical Green functions. The large-$\xi_{\mathbf{p}}$ contributions are important, for example, when we calculate the density of electrons in the superconducting state. We consider this issue in the next section.

We have defined the quasiclassical function in a symmetric way with respect to incoming $\mathbf{p}+\mathbf{k}/2$ and outgoing $\mathbf{p}-\mathbf{k}/2$ momenta in eqns (5.3), (5.4), and (5.5). Making shift in the momentum $\mathbf{p}$, one observes that also, with the accuracy of the quasiclassical approximation,

$$g_{\omega_n}(\hat{\mathbf{p}},\mathbf{k}) = \oint \frac{d\xi_{\mathbf{p}}}{\pi i}G_{\omega_n}(\mathbf{p},\mathbf{p}-\mathbf{k}) \tag{5.13}$$

and similarly for f and $f^\dagger$.

Now, after the $\xi_{\mathbf{p}}$-integration, the Green function only depends on the momentum direction and on the center-of-mass coordinate Fourier-transformed into the $\mathbf{k}$ space. The coordinate dependence of the Green function contains

$$e^{i(\mathbf{p}+\mathbf{k}/2)\mathbf{r}_1 - i(\mathbf{p}-\mathbf{k}/2)\mathbf{r}_2} = e^{i\mathbf{p}(\mathbf{r}_1-\mathbf{r}_2)+i\mathbf{k}(\mathbf{r}_1+\mathbf{r}_2)/2}. \tag{5.14}$$

The integration over $d\xi_p$ is the integration over the magnitude of the momentum; it is equivalent to the limit $\mathbf{r}_1 \to \mathbf{r}_2$ taken along the direction of the quasiparticle momentum $\mathbf{p}$. The result thus depends on the unit vector $\hat{\mathbf{p}}$.

The term "quasiclassical Green function" means that the fast oscillations of the Green function associated with variations of the relative coordinate $\mathbf{r}_1 - \mathbf{r}_2$ on a scale of p_F^{-1} are excluded. They are not relevant to the superconducting properties. Instead, there remains only a slow dependence on the center-of-mass coordinate

$$(\mathbf{r}_1 + \mathbf{r}_2)/2 = \mathbf{r}$$

on a scale of ξ^{-1} characteristic to the superconducting state. Sometimes we use a mixed Fourier-coordinate representation

$$\check{g}_{\omega_n}(\hat{\mathbf{p}}, \mathbf{r}) = \int \frac{d^3k}{(2\pi)^3} e^{i\mathbf{k}\mathbf{r}} \check{g}_{\omega_n}(\hat{\mathbf{p}}, \mathbf{k}) \tag{5.15}$$

which deals with the functions of the center-of-mass coordinate and of the particle momentum direction.

To make the formulas more compact, we introduce the matrix notation

$$\check{g}_{\omega_n}(\hat{\mathbf{p}}, \mathbf{k}) = \begin{pmatrix} g_{\omega_n} & f_{\omega_n} \\ -f^{\dagger}_{\omega_n} & \bar{g}_{\omega_n} \end{pmatrix} \tag{5.16}$$

where the matrices in the Nambu space $\check{g}$ are constructed out of g, f, etc., in the same way as the matrix $\check{G}$ is made of G, F, etc.

5.2 Density, current, and order parameter

Using the momentum–space volume element in the form of eqn (3.78) the electron density can be written as

$$\begin{aligned} N(\mathbf{k}) &= -2T \lim_{\tau \to -0} \sum_n \int \frac{d^3p}{(2\pi)^3} G_{\omega_n}(\mathbf{p}_+, \mathbf{p}_-) e^{-i\omega_n \tau} \\ &= -2T \lim_{\tau \to -0} \sum_n \left[\int \frac{d^3p}{(2\pi)^3} G^{(n)}_{\omega_n}(\mathbf{p}_+, \mathbf{p}_-) e^{-i\omega_n \tau} \right. \\ &\quad \left. -\nu(0)\pi i \,\mathrm{sign}\,(\omega_n)(2\pi)^3 \delta(\mathbf{k}) + \nu(0)\pi i \int g_{\omega_n}(\hat{\mathbf{p}}, \mathbf{k}) \frac{d\Omega_{\mathbf{p}}}{4\pi} \right] \end{aligned}$$

where

$$\mathbf{p}_{\pm} = \mathbf{p} \pm \mathbf{k}/2.$$

Introducing the normal-state electron density

$$N_0 = -2T \lim_{\tau\to-0} \sum_n \int \frac{d^3p}{(2\pi)^3} G^{(n)}_{\omega_n}(\mathbf{p}_+, \mathbf{p}_-) e^{-i\omega_n \tau}$$

we find

$$\begin{aligned} N &= N_0 (2\pi)^3 \delta(\mathbf{k}) \\ &\quad -2\nu(0)\pi i T \sum_n \left[\int g_{\omega_n}(\hat{\mathbf{p}}, \mathbf{k}) \frac{d\Omega_{\mathbf{p}}}{4\pi} - \operatorname{sign}(\omega_n)(2\pi)^3 \delta(\mathbf{k}) \right]. \end{aligned} \tag{5.17}$$

Note that the procedure of taking the limit is only important for the normal-state part of the Green function: there is no singularity in the last two terms for large ω_n. Indeed, the sum in eqn (5.17) is

$$\sum_n \operatorname{sign}(\omega_n) = 0 \tag{5.18}$$

because the function under the sum is odd. Therefore,

$$N(\mathbf{k}) = N_0(2\pi)^3\delta(\mathbf{k}) - 2\nu(0)\pi i T \sum_n \int g_{\omega_n}(\hat{\mathbf{p}}, \mathbf{k}) \frac{d\Omega_{\mathbf{p}}}{4\pi}. \tag{5.19}$$

Similarly,

$$\begin{aligned} N(\mathbf{k}) &= -2T \lim_{\tau\to+0} \sum_n \int \frac{d^3p}{(2\pi)^3} \bar{G}_{\omega_n}(\mathbf{p}_+, \mathbf{p}_-) e^{-i\omega_n\tau} \\ &= -2T \lim_{\tau\to+0} \sum_n \left[\int \frac{d^3p}{(2\pi)^3} \bar{G}^{(n)}_{\omega_n}(\mathbf{p}_+, \mathbf{p}_-) e^{-i\omega_n\tau} \right. \\ &\quad \left. +\nu(0)\pi i \operatorname{sign}(\omega_n)(2\pi)^3\delta(\mathbf{k}) + \nu(0)\pi i \int \bar{g}_{\omega_n}(\hat{\mathbf{p}}, \mathbf{k}) \frac{d\Omega_{\mathbf{p}}}{4\pi} \right] \\ &= N_0(2\pi)^3\delta(\mathbf{k}) - 2\nu(0)\pi i T \sum_n \int \bar{g}_{\omega}(\hat{\mathbf{p}}, \mathbf{k}) \frac{d\Omega_{\mathbf{p}}}{4\pi}. \end{aligned} \tag{5.20}$$

Combining eqns (5.19) and (5.20) we get

$$N(\mathbf{k}) = N_0(2\pi)^3\delta(\mathbf{k}) - \nu(0)\pi i T \sum_n \int [g_{\omega_n}(\hat{\mathbf{p}}, \mathbf{k}) + \bar{g}_{\omega_n}(\hat{\mathbf{p}}, \mathbf{k})] \frac{d\Omega_{\mathbf{p}}}{4\pi}. \tag{5.21}$$

Note that the sums of the type of eqn (5.18) in eqn (5.21) cancel each other.

We shall see later that always

$$g_{\omega_n}(\hat{\mathbf{p}}, \mathbf{k}) + \bar{g}_{\omega_n}(\hat{\mathbf{p}}, \mathbf{k}) = 0 \tag{5.22}$$

in a static situation. This is a consequence of the particle–hole symmetry of a superconductor as a Fermi-liquid, i.e., the symmetry between "particles" defined as excitations in the part of the spectrum with $\xi_{\mathbf{p}} > 0$ and "holes" with $\xi_{\mathbf{p}} < 0$.

This property holds only for quasiparticle states near the Fermi surface and is in addition to the general relation (3.44) or (3.45). The particle–hole symmetry is conserved in expressions for the order parameter, the current and the particle density within the quasiclassical approximation which assumes a linear dependence of the phase–space volume on $\xi_{\mathbf{p}}$ and uses a fixed value of the particle momentum magnitude $|\mathbf{p}| = p_F$. If the density of states and/or the momentum magnitude are energy dependent, the particle–hole symmetry is broken in expressions for the physical observables. In particular, the electron density in the quasiclassical approximation does not change after a transition to an equilibrium superconducting state:

$$N = N_0.$$

The expression for the electric current can be transformed as follows:

$$\begin{aligned}
\mathbf{j}(\mathbf{k}) &= -2eT\sum_n \int \frac{d^3p}{(2\pi)^3}\mathbf{v}_F G_{\omega_n}(\mathbf{p}_+,\mathbf{p}_-) - \frac{Ne^2}{mc}\mathbf{A} \\
&= -2eT\sum_n \left[\int \frac{d^3p}{(2\pi)^3}\mathbf{v}_F G^{(n)}_{\omega_n}(\mathbf{p}_+,\mathbf{p}_-)\right. \\
&\quad \left. -(2\pi)^3\delta(\mathbf{k})\nu(0)\pi i\,\mathrm{sign}\,(\omega_n)\int \frac{d\Omega_{\mathbf{p}}}{4\pi}\mathbf{v}_F\right] \\
&\quad -\frac{Ne^2}{mc}\mathbf{A} - 2e\nu(0)\pi i T\sum_n \int \frac{d\Omega_{\mathbf{p}}}{4\pi}\mathbf{v}_F g_{\omega_n}(\hat{\mathbf{p}},\mathbf{k}).
\end{aligned}$$

The diamagnetic term in this equation is canceled by the large-$\xi_{\mathbf{p}}$ contribution from the normal-state Green function $G^{(n)}_{\omega}(\mathbf{p}_+,\mathbf{p}_-)$ in the same way as it happened in Section 3.4. Finally,

$$\mathbf{j}(\mathbf{k}) = -2e\nu(0)\pi i T\sum_n \int \frac{d\Omega_p}{4\pi}\mathbf{v}_F g_{\omega_n}(\hat{\mathbf{p}},\mathbf{k}). \tag{5.23}$$

Similarly,

$$\begin{aligned}
\mathbf{j}(\mathbf{k}) &= 2eT\sum_n \int \frac{d^3p}{(2\pi)^3}\mathbf{v}_F \bar{G}_{\omega_n}(\mathbf{p}_+,\mathbf{p}_-) - \frac{Ne^2}{mc}\mathbf{A} \\
&= 2eT\sum_n \left[\int \frac{d^3p}{(2\pi)^3}\mathbf{v}_F \bar{G}^{(n)}_{\omega_n}(\mathbf{p}_+,\mathbf{p}_-)\right. \\
&\quad \left. +(2\pi)^3\delta(\mathbf{k})\nu(0)\pi i\,\mathrm{sign}\,(\omega_n)\int \frac{d\Omega_p}{4\pi}\mathbf{v}_F\right] \\
&\quad -\frac{Ne^2}{mc}\mathbf{A} + 2e\nu(0)\pi i T\sum_n \int \frac{d\Omega_{\mathbf{p}}}{4\pi}\mathbf{v}_F \bar{g}_{\omega_n}(\hat{\mathbf{p}},\mathbf{k}) \\
&= 2e\nu(0)\pi i T\sum_n \int \frac{d\Omega_{\mathbf{p}}}{4\pi}\mathbf{v}_F \bar{g}_{\omega_n}(\hat{\mathbf{p}},\mathbf{k}).
\end{aligned} \tag{5.24}$$

Therefore,

$$\mathbf{j}(\mathbf{k}) = -e\nu(0)\pi i T \sum_n \int \frac{d\Omega_{\mathbf{p}}}{4\pi} \mathbf{v}_F \left[g_{\omega_n}(\hat{\mathbf{p}}, \mathbf{k}) - \bar{g}_{\omega_n}(\hat{\mathbf{p}}, \mathbf{k}) \right]. \tag{5.25}$$

The order parameter can also be written in terms of a $\xi_{\mathbf{p}}$-integrated Green function. For an s-wave superconductor,

$$\frac{\Delta(\mathbf{k})}{|g|} = T \sum_n \int \frac{d^3p}{(2\pi)^3} F_{\omega_n}(\mathbf{p}_+, \mathbf{p}_-) = \nu(0)\pi i T \sum_n \int f_{\omega_n}(\hat{\mathbf{p}}, \mathbf{k}) \frac{d\Omega_{\mathbf{p}}}{4\pi} \tag{5.26}$$

or

$$\frac{\Delta(\mathbf{k})}{\lambda} = \pi i T \sum_n \int f_{\omega_n}(\hat{\mathbf{p}}, \mathbf{k}) \frac{d\Omega_{\mathbf{p}}}{4\pi}. \tag{5.27}$$

Similarly,

$$\frac{\Delta^*(\mathbf{k})}{\lambda} = \pi i T \sum_n \int f^{\dagger}_{\omega_n}(\hat{\mathbf{p}}, \mathbf{k}) \frac{d\Omega_{\mathbf{p}}}{4\pi}. \tag{5.28}$$

For nontrivial pairing, the order parameter equations contains the corresponding normalized moment of the pairing potential. For example, in case of d-wave pairing, the order parameter is

$$\frac{\Delta_{\mathbf{p}}(\mathbf{k})}{\lambda} = \pi i T \sum_n \int V(\hat{\mathbf{p}}, \hat{\mathbf{p}}') f_{\omega_n}(\hat{\mathbf{p}}', \mathbf{k}) \frac{d\Omega_{\mathbf{p}'}}{4\pi} \tag{5.29}$$

where $V(\hat{\mathbf{p}}, \hat{\mathbf{p}}') = 2\left(p_x^2 - p_y^2\right)\left(p_x'^2 - p_y'^2\right)$ or $V(\hat{\mathbf{p}}, \hat{\mathbf{p}}') = 4\left(p_x p_y\right)\left(p_x' p_y'\right)$, etc. For simplicity, we assume an s-wave pairing if not specified otherwise.

Using the quasiclassical representation, one can write the expression for the variation of the thermodynamic potential eqn (3.71) in the form

$$\begin{aligned} \delta\left(\Omega_{sn} + \int \frac{H^2}{8\pi} dV\right) &= -\nu(0)\pi i T \sum_n \int \mathrm{Tr}\left[\delta \check{H} \check{g}_{\omega_n}(\mathbf{p}, \mathbf{r})\right] \frac{d\Omega_{\mathbf{p}}}{4\pi} dV \\ &\quad + \frac{1}{|g|} \int \delta |\Delta|^2 \, dV + \frac{1}{c} \int \mathbf{j} \delta \mathbf{A} \, dV \end{aligned} \tag{5.30}$$

where now

$$\check{H} = \begin{pmatrix} -(e/c)\mathbf{v}_F \cdot \mathbf{A} & -\Delta \\ \Delta^* & (e/c)\mathbf{v}_F \cdot \mathbf{A} \end{pmatrix} \tag{5.31}$$

instead of eqn (3.70).

5.3 Homogeneous state

For further use we calculate the functions g, f, etc., for a homogeneous case. We have

$$f_{\omega_n} = \int \frac{d\xi_{\mathbf{p}}}{\pi i} \frac{\Delta}{(\xi_{\mathbf{p}} + i\sqrt{\omega_n^2 + |\Delta|^2})(\xi_{\mathbf{p}} - i\sqrt{\omega_n^2 + |\Delta|^2})}$$

$$= \frac{\Delta}{i\sqrt{\omega_n^2 + |\Delta|^2}} \tag{5.32}$$

and, similarly,

$$f_{\omega_n}^\dagger = \int \frac{d\xi_{\mathbf{p}}}{\pi i} \frac{\Delta^*}{(\xi_{\mathbf{p}} + i\sqrt{\omega_n^2 + |\Delta|^2})(\xi_{\mathbf{p}} - i\sqrt{\omega_n^2 + |\Delta|^2})}$$
$$= \frac{\Delta^*}{i\sqrt{\omega_n^2 + |\Delta|^2}}. \tag{5.33}$$

The integration is performed by shifting the contour of integration to either upper or lower half-plane of $\xi_{\mathbf{p}}$, see Fig. 3.2.

To calculate g we write

$$g_{\omega_n} = \oint \frac{d\xi_{\mathbf{p}}}{\pi i} \frac{\xi_{\mathbf{p}} + i\omega_n}{(\xi_{\mathbf{p}} + i\sqrt{\omega_n^2 + |\Delta|^2})(\xi_p - i\sqrt{\omega_n^2 + |\Delta|^2})}$$
$$= \frac{1}{2}\oint \frac{d\xi_{\mathbf{p}}}{\pi i}\left(\frac{1}{\xi_{\mathbf{p}} - i\sqrt{\omega_n^2 + |\Delta|^2}} + \frac{1}{\xi_{\mathbf{p}} + i\sqrt{\omega_n^2 + |\Delta|^2}}\right)$$
$$+ \oint \frac{d\xi_{\mathbf{p}}}{\pi i} \frac{i\omega_n}{(\xi_{\mathbf{p}} + i\sqrt{\omega_n^2 + |\Delta|^2})(\xi_{\mathbf{p}} - i\sqrt{\omega_n^2 + |\Delta|^2})}. \tag{5.34}$$

The two terms in the second line cancel each other. Using the contour of Fig. 3.2 we obtain from the third line

$$g_{\omega_n} = \frac{\omega_n}{\sqrt{\omega_n^2 + |\Delta|^2}}. \tag{5.35}$$

We find in exactly the same way

$$\bar{g}_{\omega_n} = \oint \frac{d\xi_{\mathbf{p}}}{\pi i} \frac{\xi_{\mathbf{p}} - i\omega_n}{(\xi_{\mathbf{p}} + i\sqrt{\omega_n^2 + |\Delta|^2})(\xi_{\mathbf{p}} - i\sqrt{\omega_n^2 + |\Delta|^2})}$$
$$= -\frac{\omega_n}{\sqrt{\omega_n^2 + |\Delta|^2}}. \tag{5.36}$$

We see from these equations that

$$g_{\omega_n} + \bar{g}_{\omega_n} = 0. \tag{5.37}$$

This is in agreement with the general relation eqn (5.22). Another useful relation is the so-called normalization condition which is satisfied by the functions g, f, and $f^\dagger$:

$$g_{\omega_n}^2 - f_{\omega_n} f_{\omega_n}^\dagger = 1. \tag{5.38}$$

Equations (5.37) and (5.38) can be combined into one matrix equation

$$\check{g}_{\omega_n}\check{g}_{\omega_n} = \check{1} \tag{5.39}$$

Indeed, since

$$\check{g}_{\omega_n}\check{g}_{\omega_n} = \begin{pmatrix} gg - ff^\dagger & (g+\bar{g})f \\ -f^\dagger(g+\bar{g}) & \bar{g}^2 - ff^\dagger \end{pmatrix},$$

eqn (5.39) is satisfied because of eqns (5.37, 5.38). We shall prove later that eqn (5.39) holds also in a general spatially nonhomogeneous case for any stationary situation.

5.4 Real-frequency representation

Equations for the order parameter, the current, and the particle density can be presented in terms of real-frequency Green functions, as well, using the recipe

$$\pi i T \sum_{\omega_n} \check{g}_{\omega_n} \Rightarrow \int \frac{d\epsilon}{4} \tanh\left(\frac{\epsilon}{2T}\right) \left[\check{g}^R_\epsilon - \check{g}^A_\epsilon\right]. \tag{5.40}$$

This transformation can also be used for the particle density because the normal-state contribution vanishes due to an odd parity of $\tanh(\epsilon/2T)$. The quasiclassical retarded and advanced functions $\check{g}^{R(A)}_\epsilon$ can be obtained from the Matsubara functions $\check{g}_{\omega_n}$ by an analytical continuation using the relation

$$\check{g}^R_{i\omega_n} = \check{g}_{\omega_n} \text{ for } \omega_n > 0, \tag{5.41}$$

$$\check{g}^A_{i\omega_n} = \check{g}_{\omega_n} \text{ for } \omega_n < 0. \tag{5.42}$$

The retarded function is the Matsubara function continued analytically from the positive imaginary axis while the advanced function is its continued from the negative imaginary axis.

The functions g^R and g^A, in particular, determine the density of states. Indeed, we know from eqn (3.62) that

$$\begin{aligned}\nu_s(\epsilon) &= \frac{1}{2\pi i}\left[G^R_\epsilon(\mathbf{r},\mathbf{r}) - G^A_\epsilon(\mathbf{r},\mathbf{r})\right] \\ &= \frac{1}{2\pi i}\int \frac{d^3p}{(2\pi)^3}\left[G^R_\epsilon(\mathbf{p}_+,\mathbf{p}_-) - G^A_\epsilon(\mathbf{p}_+,\mathbf{p}_-)\right]\end{aligned}$$

is the density of states per one spin projection. We observe now that

$$\nu_s(\epsilon) = \frac{1}{2}\nu(0)\left[g^R_\epsilon(\mathbf{p}) - g^A_\epsilon(\mathbf{p})\right] \tag{5.43}$$

is the density of states near the Fermi surface (per one spin) for a given momentum direction.

5.4.1 *Example: Homogeneous state*

First of all, we note that, in the normal state

$$g_{\omega_n} = \text{sign}(\omega_n).$$

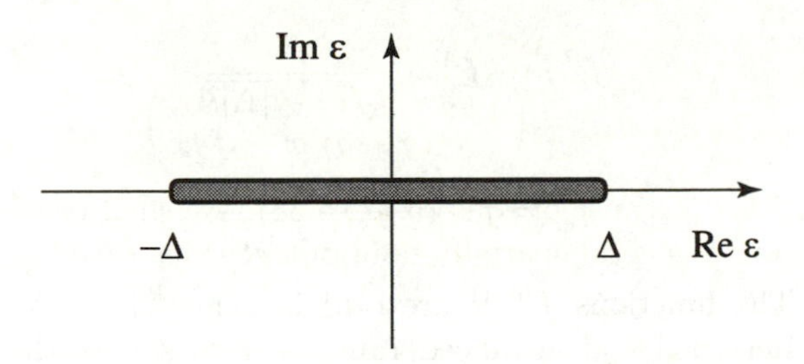

FIG. 5.1. The plane of complex ϵ with the cut. It is this plane where the retarded and advanced functions for a homogeneous state are defined.

Therefore,

$$g_\epsilon^R = -g_\epsilon^A = 1 \tag{5.44}$$

in the normal state. In a homogeneous superconductor, we obtain from eqn (5.35)

$$g_\epsilon^R = \frac{\epsilon}{\sqrt{(\epsilon + i\delta)^2 - |\Delta|^2}}\ ,\ g_\epsilon^A = -\frac{\epsilon}{\sqrt{(\epsilon - i\delta)^2 - |\Delta|^2}}. \tag{5.45}$$

The infinitely small complex part $i\delta$ is introduced to demonstrate the analytical properties of the functions. We can also define $\sqrt{\epsilon^2 - |\Delta|^2}$ as an analytical function on the plane of complex ϵ with the cut connecting the points $-|\Delta|$ and $+|\Delta|$ (see Fig. 5.1). This is the definition which results in eqn (5.44) in the limit $|\Delta| \to 0$. With this definition, we have for $|\epsilon| > |\Delta|$

$$g_\epsilon^R = -g_\epsilon^A = \frac{\epsilon}{\sqrt{\epsilon^2 - |\Delta|^2}}. \tag{5.46}$$

To get the function g^R in the region $|\epsilon| < |\Delta|$, we continue the square root $\sqrt{\epsilon^2 - |\Delta|^2}$ to the upper edge of the cut; it becomes $i\sqrt{|\Delta|^2 - \epsilon^2}$. To get g^A, we continue the square root to the lower edge of the cut; it becomes $-i\sqrt{|\Delta|^2 - \epsilon^2}$. Therefore, we have for $|\epsilon| < |\Delta|$

$$g_\epsilon^R = g_\epsilon^A = \frac{-i\epsilon}{\sqrt{|\Delta|^2 - \epsilon^2}}. \tag{5.47}$$

The functions $g^{R(A)}$ are thus "even" in ϵ for $|\epsilon| > |\Delta|$ and odd in ϵ for $|\epsilon| < |\Delta|$. This contrasts to g_{ω_n} which is always odd in ω_n. Of course, the functions $g_\epsilon^{R(A)}$ and $\bar{g}_\epsilon^{R(A)}$ are coupled through eqn (5.37), i.e.,

$$\bar{g}_\epsilon^{R(A)} = -g_\epsilon^{R(A)} \tag{5.48}$$

Similarly,

$$f_\epsilon^R = -f_\epsilon^A = \frac{\Delta}{\sqrt{(\epsilon \pm i\delta)^2 - |\Delta|^2}}. \tag{5.49}$$

One has for $|\epsilon| > |\Delta|$

$$f_\epsilon^R = -f_\epsilon^A = \frac{\Delta}{\sqrt{\epsilon^2 - |\Delta|^2}} \tag{5.50}$$

and

$$f_\epsilon^R = f_\epsilon^A = \frac{-i\Delta}{\sqrt{|\Delta|^2 - \epsilon^2}} \tag{5.51}$$

for $|\epsilon| < |\Delta|$. The functions $f^{R(A)}$ are odd in ϵ for $|\epsilon| > |\Delta|$ and even in ϵ for $|\epsilon| < |\Delta|$. Retarded and advanced function also satisfy the normalization condition of eqn (5.39).

One can find the density of states in a homogeneous superconductor. One has from eqn (5.43)

$$\nu_s(\epsilon) = \nu(0)\frac{\epsilon}{\sqrt{\epsilon^2 - |\Delta|^2}}\Theta(\epsilon^2 - |\Delta|^2).$$

It diverges near the gap edge and is zero for $|\epsilon| < |\Delta|$: there are no states within the energy gap.

5.5 Eilenberger equations

To derive equations for the quasiclassical Green functions we consider the Gor'kov equations in the momentum representation

$$\left[\check{G}^{-1}\check{G}\right]_{\mathbf{p}+\mathbf{k}/2,\mathbf{p}-\mathbf{k}/2} = \check{1}. \tag{5.52}$$

Here we use the shortcut notation

$$\left[\check{A}\check{B}\right]_{\mathbf{p}_1,\mathbf{p}_2} = \int \check{A}(\mathbf{p}_1,\mathbf{p}')\check{B}(\mathbf{p}',\mathbf{p}_2)\frac{d^3p'}{(2\pi)^3}. \tag{5.53}$$

We write the inverse operator as

$$\check{G}^{-1} = \check{G}_0^{-1} + \check{H} - \check{\Sigma}. \tag{5.54}$$

Here $\check{\Sigma}$ is the self-energy matrix with the components

$$\check{\Sigma} = \begin{pmatrix} \Sigma_1 & \Sigma_2 \\ -\Sigma_2^\dagger & \bar{\Sigma}_1 \end{pmatrix}. \tag{5.55}$$

The zero-order inverse operator is

$$\check{G}_0^{-1}(\mathbf{p}+\mathbf{k}/2) = \begin{pmatrix} -i\omega + \xi_\mathbf{p} + \frac{\mathbf{vk}}{2} + \frac{k^2}{8m} & 0 \\ 0 & i\omega + \xi_\mathbf{p} + \frac{\mathbf{vk}}{2} + \frac{k^2}{8m} \end{pmatrix}, \tag{5.56}$$

and the effective Hamiltonian is [compare with eqn (3.70)]

$$\check{H} = \begin{pmatrix} -\frac{e}{c}\mathbf{v}_F \cdot \mathbf{A}(\mathbf{k}) + e\varphi & -\Delta(\mathbf{k}) \\ \Delta^*(\mathbf{k}) & \frac{e}{c}\mathbf{v}_F \cdot \mathbf{A}(\mathbf{k}) + e\varphi \end{pmatrix}. \tag{5.57}$$

We include the scalar electromagnetic potential φ into the Hamiltonian eqn (5.57) because $\mu = E_F - e\varphi$ while we measure energy from the Fermi level:

$\xi_{\mathbf{p}} = E_n - E_F$. We neglect $\mathbf{k}$ as compared to $\mathbf{p}_F$ in the Hamiltonian since $k/p_F \sim 1/(p_F\xi)$ and omit the term

$$\frac{e^2}{mc^2}A^2 \sim v_F\frac{e}{c}A\frac{eH_{c2}\xi}{cp_F} \sim v_F\frac{e}{c}A\frac{1}{p_F\xi}$$

which is also small in the same parameter, eqn (1.61), that determines the accuracy of the quasiclassical approximation.

The left-sided Gor'kov equation becomes

$$\left[\left(\check{G}_0^{-1} + \check{H} - \check{\Sigma}\right)\check{G}\right]_{\mathbf{p}_+,\mathbf{p}_-} = \check{1}. \tag{5.58}$$

Similarly, the right-sided equation is

$$\left[\check{G}\left(\check{G}_0^{-1} + \check{H} - \check{\Sigma}\right)\right]_{\mathbf{p}_+,\mathbf{p}_-} = \check{1} \tag{5.59}$$

where

$$\check{G}_0^{-1}(\mathbf{p} - \mathbf{k}/2) = \begin{pmatrix} i\omega + \xi_{\mathbf{p}} - \frac{\mathbf{vk}}{2} + \frac{k^2}{8m} & 0 \\ 0 & -i\omega + \xi_{\mathbf{p}} - \frac{\mathbf{vk}}{2} + \frac{k^2}{8m} \end{pmatrix}. \tag{5.60}$$

Let us now subtract two eqns (5.58) and (5.59). We have

$$\mathbf{v}_F\mathbf{k}\check{G} - i\omega_n\left(\check{\tau}_3\check{G} - \check{G}\check{\tau}_3\right) + \left[\left(\check{H} - \check{\Sigma}\right)\check{G}\right] - \left[\check{G}\left(\check{H} - \check{\Sigma}\right)\right] = 0. \tag{5.61}$$

The term with ξ_p drops out, being proportional to the unit matrix which commutes with any matrix. We can now integrate this equation over $d\xi_p$ near the Fermi surface to obtain

$$\mathbf{v}_F\mathbf{k}\check{g} - i\omega_n\left(\check{\tau}_3\check{g} - \check{g}\check{\tau}_3\right) + \left[\check{H}\check{g}\right] - \left[\check{g}\check{H}\right] = \check{I}. \tag{5.62}$$

With the quasiclassical accuracy, we can put everywhere $|\mathbf{p}| = p_F$ so that $\check{g} \equiv \check{g}_{\omega_n}(\hat{\mathbf{p}}, \mathbf{k})$ and $\check{H} \equiv \check{H}(\hat{\mathbf{p}}, \mathbf{k})$. The product $\left[\check{H}\check{g}\right]$ is a usual product in the Fourier representation:

$$\left[\check{H}\check{g}\right] = \int \check{H}(\hat{\mathbf{p}}, \mathbf{k}_1)\check{g}(\hat{\mathbf{p}}, \mathbf{k} - \mathbf{k}_1)\frac{d^3k_1}{(2\pi)^3}.$$

The "collision integral" is

$$\check{I} = \left[\check{\Sigma}\check{g}\right] - \left[\check{g}\check{\Sigma}\right]. \tag{5.63}$$

It is the matrix in the Nambu space with the components

$$\check{I} = \begin{pmatrix} I_1 & I_2 \\ -I_2^\dagger & \bar{I}_1 \end{pmatrix}.$$

The equation for the real-frequency Green functions can be obtained by continuation of all Green functions in eqn (5.62) onto the real frequency axis from the positive or from the negative imaginary axis for retarded or advanced Green functions, respectively. We have

$$\mathbf{v}_F \mathbf{k} \check{g}^{R(A)} - \epsilon \left(\check{\tau}_3 \check{g}^{R(A)} - \check{g}^{R(A)} \check{\tau}_3 \right) + \left[\check{H} \check{g}^{R(A)} \right] - \left[\check{g}^{R(A)} \check{H} \right] = \check{I}^{R(A)} \quad (5.64)$$

where

$$\check{I}^{R(A)} = \left[\check{\Sigma}^{R(A)} \check{g}^{R(A)} \right] - \left[\check{g}^{R(A)} \check{\Sigma}^{R(A)} \right]. \quad (5.65)$$

Equations (5.62) or (5.64) are the famous Eilenberger equation (Eilenberger 1968) for quasiclassical Green functions. Note that the scalar potential $e\varphi$ disappears from the Eilenberger equations in the stationary case because it enters $\check{H}$ with the unit matrix and commutes with any matrix $\check{g}$.

5.5.1 *Self-energy*

The self-energies can also be presented in terms of the $\xi_{\mathbf{p}}$-integrated functions. For example, the impurity self-energy eqn (4.10) is

$$\check{\Sigma}(\mathbf{p}, \mathbf{k}) = n_{\mathrm{imp}} \int \frac{d^3 p'}{(2\pi)^3} |u(\theta)|^2 \check{G}(\mathbf{p}'_+, \mathbf{p}'_-). \quad (5.66)$$

For isotropic scattering,

$$\check{\Sigma} = \frac{1}{2\pi \nu(0) \tau_{\mathrm{imp}}} \int \check{G}(\mathbf{p}_+, \mathbf{p}_-) \frac{d^3 p}{(2\pi)^3}. \quad (5.67)$$

The momentum integral can be written as

$$\int \check{G}_{\omega_n}(\mathbf{p}_+, \mathbf{p}_-) \frac{d^3 p}{(2\pi)^3} = \nu(0) \pi i \int \frac{d\Omega_{\mathbf{p}}}{4\pi} \check{g}_{\omega_n}(\hat{\mathbf{p}}, \mathbf{k}) + \check{1} \mathcal{P} \int \frac{p^2\, dp}{2\pi^2} \frac{1}{\xi_{\mathbf{p}}}.$$

The last term gives equal positive contributions to both Σ_1 and $\bar{\Sigma}_1$; being proportional to the unit matrix, it can thus be incorporated into the chemical potential. It then drops out of the Eilenberger equation. One can write finally

$$\check{\Sigma}_{\omega_n}(\hat{\mathbf{p}}, \mathbf{k}) = \pi i \nu(0) n_{\mathrm{imp}} \int \frac{d\Omega_{\mathbf{p}_1}}{4\pi} |u(\hat{\mathbf{p}} - \hat{\mathbf{p}}_1)|^2 \check{g}_{\omega_n}(\hat{\mathbf{p}}_1, \mathbf{k}), \quad (5.68)$$

or

$$\check{\Sigma}^{R(A)}_{\epsilon}(\hat{\mathbf{p}}, \mathbf{k}) = \pi i \nu(0) n_{\mathrm{imp}} \int \frac{d\Omega_{\mathbf{p}_1}}{4\pi} |u(\hat{\mathbf{p}} - \hat{\mathbf{p}}_1)|^2 \check{g}^{R(A)}_{\epsilon}(\hat{\mathbf{p}}_1, \mathbf{k}). \quad (5.69)$$

For isotropic scattering, we obtain

$$\check{\Sigma}_{\omega_n} = \frac{i}{2\tau_{\mathrm{imp}}} \int \frac{d\Omega_{\mathbf{p}}}{4\pi} \check{g}_{\omega_n}(\hat{\mathbf{p}}, \mathbf{k}) = \frac{i}{2\tau_{\mathrm{imp}}} \left\langle \check{g}_{\omega_n} \right\rangle, \quad (5.70)$$

or

$$\check{\Sigma}^{R(A)}_{\epsilon} = \frac{i}{2\tau_{\mathrm{imp}}} \int \frac{d\Omega_{\mathbf{p}}}{4\pi} \check{g}^{R(A)}_{\epsilon}(\hat{\mathbf{p}}, \mathbf{k}) = \frac{i}{2\tau_{\mathrm{imp}}} \left\langle \check{g}^{R(A)}_{\epsilon} \right\rangle. \quad (5.71)$$

We introduce the average over the Fermi surface

$$\langle A(\mathbf{p})\rangle = \int A(\mathbf{p})\frac{d\Omega_{\mathbf{p}}}{4\pi} \tag{5.72}$$

or, more generally,

$$\langle A(\mathbf{p})\rangle = \frac{1}{\nu(0)}\int A(\mathbf{p})\frac{dS_F}{(2\pi)^3 v_F} \tag{5.73}$$

for a nonspherical Fermi surface. The quantity $\nu(0)$ here is the integrated density of states at the Fermi surface

$$\nu(0) = \int \frac{dS_F}{(2\pi)^3 v_F}.$$

For magnetic impurities one has from eqns (4.14) and (4.15)

$$\Sigma_1(\hat{\mathbf{p}},\mathbf{k}) = \pi i\nu(0)n_{\text{imp}}\int \frac{d\Omega_{\mathbf{p}_1}}{4\pi}|u_1(\hat{\mathbf{p}}-\hat{\mathbf{p}}_1)|^2 g_{\omega_n}(\hat{\mathbf{p}}_1,\mathbf{k}), \tag{5.74}$$

$$\Sigma_2(\hat{\mathbf{p}},\mathbf{k}) = \pi i\nu(0)n_{\text{imp}}\int \frac{d\Omega_{\mathbf{p}_1}}{4\pi}|u_2(\hat{\mathbf{p}}-\hat{\mathbf{p}}_1)|^2 f_{\omega_n}(\hat{\mathbf{p}}_1,\mathbf{k}). \tag{5.75}$$

The scattering amplitudes can be parametrized by two different scattering cross sections σ_1 and σ_2 associated with the diagonal and off-diagonal components (i.e., with either the functions g and $\bar{g}$ or the functions f and $f^\dagger$) of the matrix Green function $\check{g}$, respectively. As in eqns (5.74) and (5.75),

$$\Sigma_1(\mathbf{p}_F,\mathbf{k}) = \frac{i}{2}v_F n_{\text{imp}}\int d\Omega_{\mathbf{p}'}\sigma_1(\theta)\, g(\mathbf{p}'_F,\mathbf{k}),$$
$$\Sigma_2(\mathbf{p}_F,\mathbf{k}) = \frac{i}{2}v_F n_{\text{imp}}\int d\Omega_{\mathbf{p}'}\sigma_2(\theta)\, f(\mathbf{p}'_F,\mathbf{k}).$$

One can define the scattering times according to eqns (4.16), (4.17) together with the spin-flip time eqn (4.18). Let us introduce the cross sections

$$\sigma_{\mathbf{pp}'} = \frac{1}{2}\left[\sigma_1(\theta)+\sigma_2(\theta)\right],\ \sigma^{(s)}_{\mathbf{pp}'} = \frac{1}{2}\left[\sigma_1(\theta)-\sigma_2(\theta)\right].$$

The spin-flip scattering cross section is positive according to eqn (4.19) (Abrikosov and Gor'kov 1960), however, it is usually much smaller than the nonmagnetic cross section $\sigma^{(s)}_{\mathbf{pp}'} \ll \sigma_{\mathbf{pp}'}$. We can write

$$\check{\Sigma}(\mathbf{p}_F,\mathbf{k}) = \frac{i}{2}v_F n_{\text{imp}}\int d\Omega_{\mathbf{p}'}\sigma_{\mathbf{pp}'}\check{g}(\mathbf{p}'_F,\mathbf{k}) + \frac{i}{2}v_F n_{\text{imp}}\int d\Omega_{\mathbf{p}'}\sigma^{(s)}_{\mathbf{pp}'}\check{\tau}_3\check{g}(\mathbf{p}'_F,\mathbf{k})\check{\tau}_3. \tag{5.76}$$

The first term describes the scattering without changes in the spin state of electrons and coincides with the scattering by nonmagnetic impurities while the second term refers to the scattering accompanied by a spin-flip process. We shall attribute the notation $\check{\Sigma}_{\rm imp}$ only to the nonmagnetic part of the self-energy, i.e., to the term in the first line of eqn (5.76), and denote the spin-flip self-energy in the second line of eqn (5.76) by $\check{\Sigma}_s$.

The impurity collision integral for nonmagnetic scattering eqn (5.63) or (5.65) is expressed through nonmagnetic self-energies defined by eqn (5.76). The collision integral vanishes after averaging over the momentum directions:

$$\langle \check{I}_{\rm imp}(\mathbf{p})\rangle = \left\langle \check{I}^{R(A)}_{\rm imp}(\mathbf{p})\right\rangle = 0. \tag{5.77}$$

This is because

$$\begin{aligned}\left\langle \langle \check{g}(\hat{\mathbf{p}}')\,\sigma_{\mathbf{pp}'}\rangle_{\mathbf{p}'}\,\check{g}(\hat{\mathbf{p}})\right\rangle_{\mathbf{p}} &= \left\langle \check{g}(\hat{\mathbf{p}})\,\langle \sigma_{\mathbf{pp}'}\check{g}(\hat{\mathbf{p}}')\rangle_{\mathbf{p}'}\right\rangle_{\mathbf{p}} \\ &= \int\!\!\int \frac{d\Omega_{\mathbf{p}}d\Omega_{\mathbf{p}'}}{(4\pi)^2}\check{g}(\hat{\mathbf{p}})\,\sigma_{\mathbf{pp}'}\check{g}(\hat{\mathbf{p}}').\end{aligned} \tag{5.78}$$

Equation (5.77) does not hold, however, for the spin-flip scattering integral:

$$\langle \check{I}_s(\mathbf{p})\rangle \neq 0 \tag{5.79}$$

because the matrices $\check{g}$ and $\check{\tau}_3\check{g}\check{\tau}_3$ do not commute.

Another useful relation is

$$\langle \sigma_{\mathbf{pp}'}\rangle_{\mathbf{p}'}\,\hat{\mathbf{p}} - \langle \sigma_{\mathbf{pp}'}\hat{\mathbf{p}}'\rangle_{\mathbf{p}'} = \langle [1-(\hat{\mathbf{p}}\hat{\mathbf{p}}')]\,\sigma_{\mathbf{pp}'}\rangle_{\mathbf{p}'}\,\hat{\mathbf{p}} \equiv \sigma_{\rm tr}\hat{\mathbf{p}} \tag{5.80}$$

where

$$\sigma_{\rm tr} = \frac{1}{2}\int_0^\pi (1-\cos\theta)\,\sigma(\theta)\sin\theta\,d\theta$$

is the transport cross section. To prove it we note that $\langle \hat{\mathbf{p}}'\sigma_{\mathbf{pp}'}\rangle_{\mathbf{p}'}$ should be directed along $\hat{\mathbf{p}}$: $\langle \hat{\mathbf{p}}'\sigma_{\mathbf{pp}'}\rangle_{\mathbf{p}'} = a\hat{\mathbf{p}}$. Multiplying this by $\hat{\mathbf{p}}$, we find $a = \langle (\hat{\mathbf{p}}\hat{\mathbf{p}}')\,\sigma_{\mathbf{pp}'}\rangle_{\mathbf{p}'}$ which proves eqn (5.80). We shall also use the transport mean free time defined as

$$\frac{1}{\tau_{\rm tr}} = 4\pi\sigma_{\rm tr}v_F n_{\rm imp}. \tag{5.81}$$

We have considered the impurity scattering in detail here. The phonon self-energy can also be expressed through the quasiclassical Green functions. We shall discuss this later in Section 8.2.

5.5.2 *Normalization*

In the mixed Fourier-coordinate representation of eqn (5.15), the Eilenberger equation for $\check{g}(\hat{\mathbf{p}},\mathbf{r})$ becomes

$$-i\mathbf{v}_F\nabla\check{g} - i\omega_n(\check{\tau}_3\check{g} - \check{g}\check{\tau}_3) + \check{H}\check{g} - \check{g}\check{H} = \check{I}. \tag{5.82}$$

Here the product $\check{H}\check{g}$ is just the usual product of two matrices. All the functions here are functions of $\hat{\mathbf{p}}$ and $\mathbf{r}$. The matrix Eilenberger equation can be written in components. We have

$$-i\mathbf{v}_F\nabla g + \Delta f^\dagger - f\Delta^* = I_1,$$

$$-i\mathbf{v}_F\nabla\bar{g} + \Delta^* f - f^\dagger\Delta = \bar{I}_1,$$

$$-i\mathbf{v}_F\left(\nabla - \frac{2ie}{c}\mathbf{A}\right)f - 2i\omega_n f - \Delta\bar{g} + g\Delta = I_2,$$

$$i\mathbf{v}_F\left(\nabla + \frac{2ie}{c}\mathbf{A}\right)f^\dagger - 2i\omega_n f^\dagger + \Delta^* g - \bar{g}\Delta^* = I_2^\dagger.$$

We observe in particular that the Eilenberger equations together with the quasi-classical functions for a stationary superconductor are invariant under the gauge transformation of eqn (1.7).

From the first two equations we see immediately that g and $\bar{g}$ are not independent:

$$g + \bar{g} = const\ .$$

To find the constant we assume that our superconducting state transforms gradually into a homogeneous state without impurities at large enough distances along the direction of $\mathbf{v}_F$ from the region which is under consideration for a given problem. Moreover, we assume that the magnetic field vanished at these distances. In a homogeneous case, the constant is zero according to eqn (5.37); this implies that $g + \bar{g} = 0$ everywhere along $\mathbf{v}_F$, i.e., in any inhomogeneous situation, as well. Using this condition, we see that there are only three independent functions, i.e., g, f, and $f^\dagger$ which satisfy three equations

$$-i\mathbf{v}_F\nabla g + \Delta f^\dagger - f\Delta^* = I_1, \tag{5.83}$$

$$-i\mathbf{v}_F\left(\nabla - \frac{2ie}{c}\mathbf{A}\right)f - 2i\omega_n f + 2g\Delta = I_2, \tag{5.84}$$

$$i\mathbf{v}_F\left(\nabla + \frac{2ie}{c}\mathbf{A}\right)f^\dagger - 2i\omega_n f^\dagger + 2\Delta^* g = I_2^\dagger. \tag{5.85}$$

However, they are still not completely independent due to the normalization condition of eqn (5.38). It was obtained for a homogeneous superconductor. To prove it for a general case, we multiply the matrix Eilenberger equation (5.82) by $\check{g}$ first from the left and then from the right, and add the two thus obtained equations. We have

$$-i\mathbf{v}_F\nabla(\check{g}\check{g}) - i\omega_n\left[\check{\tau}_3(\check{g}\check{g}) - (\check{g}\check{g})\check{\tau}_3\right] + \left(\check{H} - \check{\Sigma}\right)(\check{g}\check{g}) - (\check{g}\check{g})\left(\check{H} - \check{\Sigma}\right) = 0.$$

The solution of this equation has the form

$$(\check{g}\check{g}) = A\check{1} + B\check{g} \tag{5.86}$$

where A and B are constants. At large distances, i.e., in a homogeneous state without impurities, we have $A = 1$ and $B = 0$. Therefore, they should have the same values also for a nonhomogeneous case. This proves eqn (5.39) together with eqns (5.37) and (5.38).

Equations (5.83–5.85) can be written in a matrix form. Let us introduce the matrices

$$\hat{\nabla}\check{g} = \begin{pmatrix} \nabla g & \left(\nabla - \frac{2ie}{c}\mathbf{A}\right) f \\ -\left(\nabla + \frac{2ie}{c}\mathbf{A}\right) f^{\dagger} & -\nabla g \end{pmatrix}, \quad \check{H}_0 = \begin{pmatrix} -i\omega_n & -\Delta \\ \Delta^* & i\omega_n \end{pmatrix}. \tag{5.87}$$

The Eilenberger equations take the form

$$-i\mathbf{v}_F \hat{\nabla}\check{g} + \check{H}_0\check{g} - \check{g}\check{H}_0 = \check{I}. \tag{5.88}$$

The Eilenberger equations for the real-time retarded and advanced functions are obtained from eqns (5.83)–(5.85) by substitution $i\omega_n = \epsilon$:

$$-i\mathbf{v}_F \nabla g^{R(A)} + \Delta f^{\dagger R(A)} - f^{R(A)}\Delta^* = I_1^{R(A)}, \tag{5.89}$$

$$-i\mathbf{v}_F \left(\nabla - \frac{2ie}{c}\mathbf{A}\right) f^{R(A)} - 2\epsilon f^{R(A)} + 2g^{R(A)}\Delta = I_2^{R(A)}, \tag{5.90}$$

$$i\mathbf{v}_F \left(\nabla + \frac{2ie}{c}\mathbf{A}\right) f^{\dagger R(A)} - 2\epsilon f^{\dagger R(A)} + 2\Delta^* g^{R(A)} = I_2^{\dagger R(A)}. \tag{5.91}$$

Retarded and advanced Green functions also satisfy the normalization condition (5.39). Therefore, only two of the equations (5.89), (5.90) and (5.91) are independent.

5.6 Dirty limit. Usadel equations

The Eilenberger equations can be simplified considerably in the dirty limit (Usadel 1970), when the impurity scattering rate or the mean free path ℓ satisfies the condition

$$1/\tau \gg T_c, \; i.e., \; \ell \ll \xi_0.$$

This limit applies only for s-wave superconductors where Δ is independent of the momentum direction.

A strong scattering by impurities produces averaging over momentum directions. Therefore, in the first approximation, the Green function does not depend on the momentum directions. The first-order correction is linear in $\mathbf{v}_F$. We can thus write

$$\check{g} = g_0 + \hat{\mathbf{v}}_F \check{\mathbf{g}} \tag{5.92}$$

where $\check{g}_0$ does not depend on directions of $\mathbf{v}_F$, and $|\mathbf{g}| \ll g_0$; $\hat{\mathbf{v}}_F$ is the unit vector in the direction of momentum. Therefore

$$\langle \check{g} \rangle = \check{g}_0.$$

The normalization condition gives

$$\check{g}_0\check{g}_0 = \check{1}\,, \quad \check{g}_0\check{\mathbf{g}} + \check{\mathbf{g}}\check{g}_0 = 0. \tag{5.93}$$

Consider a superconductor without magnetic impurities. The nonmagnetic collision integral is determined by eqn (5.63). As we know, it vanishes after averaging over the momentum directions, eqn (5.77). Let us average eqn (5.88) over directions of $\mathbf{v}_F$. We obtain

$$-i\frac{v_F}{3}\hat{\nabla}\check{\mathbf{g}} + \check{H}_0\check{g}_0 - \check{g}_0\check{H}_0 = 0. \tag{5.94}$$

We now multiply eqn (5.88) by $\hat{\mathbf{v}}_F$ and average it over momenta. We have

$$-iv_F\hat{\nabla}\check{g}_0 + (\check{H}_0\check{\mathbf{g}} - \check{\mathbf{g}}\check{H}_0) - \frac{i}{2\tau_{\mathrm{tr}}}(\check{g}_0\check{\mathbf{g}} - \check{\mathbf{g}}\check{g}_0) = 0. \tag{5.95}$$

We use here eqns (5.80) and (5.81). In the dirty limit, we can neglect the terms with $\check{H}_0$. Multiplying this equation by $\check{g}_0$ and using the normalization condition eqn (5.93), we get

$$\check{\mathbf{g}} = -\ell_{\mathrm{tr}}\check{g}_0\hat{\nabla}\check{g}_0 \tag{5.96}$$

where $\ell_{\mathrm{tr}} = v_F\tau_{\mathrm{tr}}$.

Equation for the averaged functions becomes

$$iD\hat{\nabla}(\check{g}_0\hat{\nabla}\check{g}_0) + \check{H}_0\check{g}_0 - \check{g}_0\check{H}_0 = 0 \tag{5.97}$$

where $D = v_F\ell_{\mathrm{tr}}/3$ is the diffusion constant. In components, eqn (5.97) has the form

$$-iD\left(\nabla - \frac{2ie}{c}\mathbf{A}\right)\left[g_0\left(\nabla - \frac{2ie}{c}\mathbf{A}\right)f_0 - f_0\nabla g_0\right] = 2\Delta g_0 - 2i\omega_n f_0, \tag{5.98}$$

and

$$-iD\left(\nabla + \frac{2ie}{c}\mathbf{A}\right)\left[g_0\left(\nabla + \frac{2ie}{c}\mathbf{A}\right)f_0^\dagger - f_0^\dagger\nabla g_0\right] = 2\Delta^* g_0 - 2i\omega_n f_0^\dagger. \tag{5.99}$$

These equations are known as the Usadel equations.

We can easily get the expression for the current from eqn (5.96):

$$\begin{aligned}\mathbf{j} &= -\frac{e\nu(0)v_F^2}{3}\pi iT\sum_n \mathrm{Tr}\,(\tau_3\check{\mathbf{g}}) = \frac{\sigma_n}{e}\pi iT\sum_{n\geq 0}\mathrm{Tr}\,(\tau_3\check{g}_0\hat{\nabla}\check{g}_0)\\ &= \frac{\sigma_n}{e}\pi iT\sum_{n\geq 0}\left[f_0^\dagger\left(\nabla - \frac{2ie}{c}\mathbf{A}\right)f_0 - f_0\left(\nabla + \frac{2ie}{c}\mathbf{A}\right)f_0^\dagger\right].\end{aligned} \tag{5.100}$$

The normal-state Drude conductivity is

$$\sigma_n = 2\nu\,(0)\,De^2. \tag{5.101}$$

In a homogeneous case, $|\Delta| = const$, we have

$$\mathbf{j} = -\frac{i\sigma_n}{e}\left[\Delta^*\left(\nabla - \frac{2ie}{c}\mathbf{A}\right)\Delta - \Delta\left(\nabla + \frac{2ie}{c}\mathbf{A}\right)\Delta^*\right]\sum_{n\geq 0}\frac{\pi T}{\omega_n^2 + |\Delta|^2}. \tag{5.102}$$

5.7 Boundary conditions

To solve the Eilenberger equations one needs some boundary conditions. They are imposed either at the surfaces of the superconductor or in the bulk, i.e., at large distances from those regions where the spatial dependence of the Green functions is nontrivial. The boundary conditions in the bulk are more or less obvious: one has to match the obtained solution with the Green functions for a homogeneous situation. The boundary conditions at the surfaces depend on the characteristics of the particular surface and vary from one surface to another. In general, boundary conditions can be formulated in terms of the scattering T-matrix (Serene and Rainer 1983). However, in their general form, these conditions are rather complicated. We discuss briefly a few simple examples of quasiclassical conditions at various boundaries.

The boundary conditions for dirty superconductors reduce to the boundary conditions for diffusion-like Usadel equations (5.97). They are usually formulated in terms of continuity of quasiparticle currents. For example, at the interface between the superconductor and vacuum (non-conducting medium), the current through the surface should vanish. According to eqn (5.100) this requires

$$\mathbf{n}\cdot\left(\nabla-\frac{2ie}{c}\mathbf{A}\right)f_0=\mathbf{n}\cdot\left(\nabla+\frac{2ie}{c}\mathbf{A}\right)f_0^{\dagger}=0$$

at the boundary. Here $\mathbf{n}$ is the unit vector normal to the surface.

For clean superconductors, the boundary conditions are more complicated. The whole set of conditions at interfaces between superconductors and metals or other conducting media was derived by Zaitsev (1984).

In this section, we discuss the boundary conditions which can be imposed for a clean superconductor at the boundary with vacuum. We consider two kinds of surfaces. First is a completely smooth surface which is characterized by specular reflection of incident quasiparticles. Specular reflections require that the Green function of the reflected particle taken at the surface $\mathbf{r}=\mathbf{r}_w$ is equal to that before the reflection:

$$\check{g}\left(\mathbf{p}_{\parallel},\mathbf{p}_{\perp};\mathbf{r}_w\right)=\check{g}\left(\mathbf{p}_{\parallel},-\mathbf{p}_{\perp};\mathbf{r}_w\right) \tag{5.103}$$

where $\mathbf{p}_{\parallel}$ and $\mathbf{p}_{\perp}$ are the parallel and perpendicular to the surface components of the particle momentum. Equation (5.103) automatically results in vanishing of the current through the surface. Another example is a diffusive surface.

5.7.1 *Diffusive surface*

Diffusive surface reflects the incident particles randomly. The conditions at a diffusive surface are much more complicated (Serene and Rainer 1983). First of all, they strongly depend on the particular characteristic of the surface, for example, on the degree of diffusivity. In certain cases, a diffusively scattering boundary with an non-conducting medium (vacuum) is modeled by an impurity layer which covers the surface. The thickness of the layer and the quasiparticle mean free path inside the layer are assumed to be much shorter than the coherence length ξ_0. The specular boundary conditions are applied at the outer side of the layer with respect to the superconductor (superfluid). To describe various degrees of diffusivity, the ratio of the layer thickness to the mean free path, $\rho = d/l$, varies from zero (specular wall) to infinity (diffusive wall). The equations for the Green functions ought to be solved first inside the impurity layer, and then the solutions should have to be matched to those in the bulk.

In this section we describe another approach where the limit $\rho \to \infty$ is accomplished analytically for a completely diffusive surface using the method developed by Ovchinnikov (1969). It gives the boundary conditions for the Green functions which are to be applied directly at the boundary of the superconductor.

For a derivation of the boundary conditions it is useful to expand the quasiclassical Green functions

$$\check{g} = \begin{pmatrix} g & f \\ -f^\dagger & -g \end{pmatrix} \tag{5.104}$$

into the three Pauli matrices in the Nambu space

$$\check{\tau}_1 = \begin{pmatrix} 0 & 1 \\ 1 & 0 \end{pmatrix}, \ \check{\tau}_2 = \begin{pmatrix} 0 & -i \\ i & 0 \end{pmatrix}, \check{\tau}_3 = \begin{pmatrix} 1 & 0 \\ 0 & -1 \end{pmatrix} \tag{5.105}$$

which constitute the vector $\check{\tau}$. We can write

$$\check{g} = \mathbf{n} \cdot \check{\tau} , \tag{5.106}$$

where $\mathbf{n}$ are the three expansion coefficients constituting a vector. Generally, the components n_α are functions of the coordinates and of the directions of the quasiparticle momenta, $\hat{\mathbf{p}}$.

The vector n_α can be expanded into three constant basis vectors $\mathbf{n}(N)$, where $N = 1, 2, 3$, which satisfy the orthogonality conditions

$$[\mathbf{n}(N_1) \cdot \check{\tau},\ \mathbf{n}(N_2) \cdot \check{\tau}]_+ = 2\delta_{N_1 N_2} \check{1}, \tag{5.107}$$

where $[\ ,\ ]_\pm$ is an anticommutator (commutator).

One can easily check that, for any two non-equal vectors $\mathbf{n}(N_1)$ and $\mathbf{n}(N_2)$, the commutator $[\mathbf{n}(N_1) \cdot \check{\tau},\ \mathbf{n}(N_2) \cdot \check{\tau}]_-$ is proportional to the product of $\check{\tau}$ and the third $\mathbf{n}$-vector:

$$[\mathbf{n}(N_1) \cdot \check{\tau},\ \mathbf{n}(N_2) \cdot \check{\tau}]_- = 2i\mathbf{n}(N_3) \cdot \check{\tau}. \tag{5.108}$$

Note that the numbers N_1, N_2, and N_3 constitute an even permutation of 1, 2, and 3.

We can now write the Green function as follows:

$$\check{g} = \sum_N C_N \cdot \mathbf{n}(N) \cdot \check{\tau}. \tag{5.109}$$

Above, the coefficients C_N are functions of $\mathbf{r}$ and $\mathbf{p}$. Due to the normalization

$$\check{g}^2 = \check{1}$$

the coefficients C_N satisfy the condition

$$\sum_N C_N^2 = 1. \tag{5.110}$$

The function $\check{g}$ in eqns (5.109) can be either retarded (advanced) or Matsubara Green function.

The quasiparticle self-energy due to impurity scattering inside the layer is

$$\check{\Sigma} = \frac{icv_F}{2} \int \sigma_{\mathbf{pp}'}\ \check{g}(\mathbf{p}', \mathbf{r})\, d\Omega_{\mathbf{p}'}, \tag{5.111}$$

where c is the impurity concentration in the layer, and $\sigma_{\mathbf{pp}'}$ is the scattering cross section. The mean free time is

$$\frac{1}{\tau_{\text{imp}}} = cv_F \int \sigma_{\mathbf{pp}'}\, d\Omega_{\mathbf{p}'}. \tag{5.112}$$

The Eilenberger equation for the Green function inside the layer has the form

$$-i\mathbf{v}_F \cdot \nabla \check{g}(\mathbf{p}, \mathbf{r}) = \check{\Sigma}\check{g}(\mathbf{p}, \mathbf{r}) - \check{g}(\mathbf{p}, \mathbf{r})\ \check{\Sigma}. \tag{5.113}$$

From the definition of the impurity self-energy one has, identically,

$$\int \left[\check{\Sigma}\check{g}(\mathbf{p}, \mathbf{r}) - \check{g}(\mathbf{p}, \mathbf{r})\ \check{\Sigma}\right]\, d\Omega_{\mathbf{p}'} \equiv 0. \tag{5.114}$$

Deep in the boundary layer, the Green function depends neither on the distance form the surface of the layer nor on the momentum direction due to isotropization in the course of the scattering. Therefore, it can be presented in the form $\check{g} = \mathbf{n}(1) \cdot \check{\tau}$ with the vector $\mathbf{n}(1)$ independent of $\mathbf{p}$ and of the distance from the surface of the layer. However, the vector $\mathbf{n}(1)$ may have a smooth dependence on the position at the layer surface, $\mathbf{r}_w$, if the boundary is not planar. We choose this vector $\mathbf{n}(1)$ as one of the basis vectors for the $\check{g}$-expansion.

One can present the Green function near the layer surface in the form

$$\check{g} = \mathbf{n}(1) \cdot \check{\tau} + C_2\mathbf{n}(2) \cdot \check{\tau} + iC_3\mathbf{n}(3) \cdot \check{\tau}, \tag{5.115}$$

where the vectors $\mathbf{n}(2)$ and $\mathbf{n}(3)$ are the other two constant basis vectors. Due to the normalization condition of eqn (4) we have

$$C_2^2 - C_3^2 = 0. \tag{5.116}$$

The coefficients C_2 and C_3 now contain all the information on the coordinate and momentum dependencies of the Green function.

The self-energy becomes in the boundary layer, according to eqn (5.115),

$$\check{\Sigma} = \frac{i}{2\tau_{\text{imp}}}\mathbf{n}(1)\cdot\check{\tau} + \frac{icv_F}{2}[\mathbf{n}(2)\cdot\check{\tau}\int\sigma_{\mathbf{pp}'}\, C_2(\mathbf{p}')\, d\Omega_{\mathbf{p}'}$$
$$+i\mathbf{n}(3)\cdot\check{\tau}\int\sigma_{\mathbf{pp}'}\, C_3(\mathbf{p}')\, d\Omega_{\mathbf{p}'}\,]\,. \tag{5.117}$$

Using eqn (5.117) and the commutation rules of eqns (5.107,5.108), one gets

$$[\check{\Sigma}\,,\,\check{g}]_- = \frac{i}{\tau_{\text{imp}}}\,[iC_2\,\mathbf{n}(3)\cdot\check{\tau} + C_3\,\mathbf{n}(2)\cdot\check{\tau}]$$
$$+\left[cv_F\int\sigma_{\mathbf{pp}'}\, C_2(\mathbf{p}')\, d\Omega_{\mathbf{p}'}\right]\,\mathbf{n}(3)\cdot\check{\tau}$$
$$-\left[icv_F\int\sigma_{\mathbf{pp}'}\, C_3(\mathbf{p}')\, d\Omega_{\mathbf{p}'}\right]\,\mathbf{n}(2)\cdot\check{\tau}\,. \tag{5.118}$$

This results in the equations

$$(v_F)_z\frac{\partial}{\partial z}\;[C_2(\mathbf{p}')\pm C_3(\mathbf{p}')] = \mp\frac{1}{\tau_{\text{imp}}}\;[C_2(\mathbf{p}')\pm C_3(\mathbf{p}')]$$
$$\pm\, cv_F\int\sigma_{\mathbf{pp}'}\;[C_2(\mathbf{p}')\pm C_3(\mathbf{p}')]\;d\Omega_{\mathbf{p}'}$$

for the coefficients C_2 and C_3 in the layer. Here z is measured along the layer-surface normal directed inwards the superconductor so that $z < 0$ corresponds to a position inside the layer.

Like Ovchinnikov (1969), we assume that the scattering cross section $\sigma_{\mathbf{pp}'}$ is nonzero only for $p_z p'_z > 0$. Using this assumption one can eliminate the solutions which increase exponentially as $z \to -\infty$, by requiring

$$(C_2 + C_3)_{p_z>0} = 0\;;\;(C_2 - C_3)_{p_z<0} = 0 \tag{5.119}$$

at the surface of the layer.

There are two more conditions for the coefficients $C_{2,3}$ which follow from eqns (5.114), (5.119) upon integration over the momentum directions:

$$\int_{p_z<0} |\,p_z\,|\,(C_2 + C_3)\,d\Omega_{\mathbf{p}'} = 0\;,\;\int_{p_z>0} p_z\;(C_2 - C_3)\,d\Omega_{\mathbf{p}'} = 0. \tag{5.120}$$

Equations (5.119) and (5.120) ensure that there is no current through the interface.

Using eqns (5.115), (5.119), and (5.120) one obtains for the basis vector

$$\mathbf{n}(1)\cdot\check{\tau} = \frac{1}{\pi}\int_{p_z>0}\hat{p}_z\;\check{g}(\mathbf{p},\mathbf{r}_w)\,d\Omega_{\mathbf{p}'} = \frac{1}{\pi}\int_{p_z<0}|\,\hat{p}_z\,|\;\check{g}(\mathbf{p},\mathbf{r}_w)\,d\Omega_{\mathbf{p}'}$$

$$\equiv \langle \hat{p}_z \check{g} \rangle_+ \tag{5.121}$$

where $\hat{\mathbf{p}}$ is the unit vector in the momentum direction. Since

$$C_2 \mathbf{n}(2) \cdot \check{\tau} + iC_3 \mathbf{n}(3) \cdot \check{\tau} = \check{g} - \mathbf{n}(1) \cdot \check{\tau} \,, \tag{5.122}$$

and

$$C_3 \mathbf{n}(2) \cdot \check{\tau} + iC_2 \mathbf{n}(3) \cdot \check{\tau} = \frac{1}{2} [\mathbf{n}(1) \cdot \check{\tau}, \, \check{g}]_- \,, \tag{5.123}$$

one can exclude the unknown vectors $\mathbf{n}(2)$ and $\mathbf{n}(3)$ using eqn (5.116):

$$C_2 [\check{g} - \mathbf{n}(1) \cdot \check{\tau}] = \frac{C_3}{2} [\mathbf{n}(1) \cdot \check{\tau}, \, \check{g}]_- .$$

Since the coefficients C_2 and C_3 are coupled by eqn (5.119): $C_2 = -C_3$ for $\hat{p}_z > 0$, and $C_2 = C_3$ for $\hat{p}_z < 0$, one obtains

$$2[\check{g} - \langle \hat{p}_z \check{g} \rangle_+] = -\mathrm{sign}\,(\hat{p}_z) \, [\langle \hat{p}_z \check{g} \rangle_+, \, \check{g}]_- \; . \tag{5.124}$$

The conditions of eqn (5.124) can also be written directly for the functions g, f, and $f^\dagger$:

$$\begin{aligned} 2\left(g - \langle \hat{p}_z g \rangle_+\right) &= -\mathrm{sign}\,(\hat{p}_z) \left(f \langle \hat{p}_z f^\dagger \rangle_+ - f^\dagger \langle \hat{p}_z f \rangle_+\right) \,, \\ f - \langle \hat{p}_z f \rangle_+ &= -\mathrm{sign}\,(\hat{p}_z) \left(\langle \hat{p}_z g \rangle_+ f - g \langle \hat{p}_z f \rangle_+\right) \,, \\ f^\dagger - \langle \hat{p}_z f^\dagger \rangle_+ &= -\mathrm{sign}\,(\hat{p}_z) \left(g \langle \hat{p}_z f^\dagger \rangle_+ - \langle \hat{p}_z g \rangle_+ f^\dagger\right) . \end{aligned} \tag{5.125}$$

These three equations determine the three functions g, f, and $f^\dagger$.

These conditions hold for both the real-frequency (retarded and advanced) and Matsubara Green functions. They are linear with respect to the functions for a given momentum direction. This fact may be very important for practical use. The coefficients, however, are the integrals of the functions over momentum directions, so the boundary conditions of eqns (5.125) are nonlinear in a strict sense.

6

QUASICLASSICAL METHODS IN STATIONARY PROBLEMS

We demonstrate the potentialities of the quasiclassical approach for selected problems in the theory of stationary superconductivity: We derive the Ginzburg–Landau equations, calculate the upper critical field of dirty superconductors at arbitrary temperatures, and discuss the gapless regime in superconductors with magnetic impurities. We consider effects of impurities on the density of states in d-wave superconductors. Finally, we find the energy spectrum of excitations in vortex cores of s-wave and d-wave superconductors.

6.1 s-wave superconductors with impurities

6.1.1 *Small currents in a uniform state*

In this chapter we consider several useful applications of the quasiclassical methods to stationary properties of superconductors. We start with a homogeneous state of an s-wave superconductor. Without a magnetic field, the Green functions do not depend on coordinates and on the momentum direction. This fact tells us that

$$\check{\Sigma} = \frac{i}{2\tau}\check{g}.$$

We denote the impurity scattering time simply by τ for brevity. Therefore, the self-energies commute with the Green functions and drop out of the Eilenberger equations. As a result, the homogeneous state does not depend on impurity concentration. We know this already as the Anderson theorem. This consideration clearly demonstrates how much simpler the quasiclassical method is as compared to the full Green function technique: without any calculations we have arrived to the conclusion which took two pages of algebra in Section 4.2. Eilenberger equations (5.84, 5.85) give

$$f = \frac{\Delta}{i\omega_n} g \ , \ f^\dagger = \frac{\Delta^*}{i\omega_n} g.$$

Equation (5.83) vanishes. To find g and f we need to use the normalization condition eqn (5.38). We obtain

$$g = \frac{\omega_n}{\sqrt{\omega_n^2 + |\Delta|^2}} \ , \quad \text{and} \ \ f = \frac{\Delta}{i\sqrt{\omega_n^2 + |\Delta|^2}}.$$

The sign was chosen by comparison with our previous result. For retarded and advanced functions we recover eqns (5.46), (5.47), (5.50), and (5.51).

Consider now a superconducting state where the magnetic field and currents are small, and the gradients are slow. The condition is

$$(2e/c)\,A \ll \xi_0^{-1} \text{ i.e., } H \ll \frac{c}{2e\xi^2} = H_{c2}$$

and $v_F k \ll T_c$. The latter means that variations are slow on the scale of the coherence length ξ_0. We also assume an isotropic scattering by impurities.

We solve the Eilenberger equations by perturbations. Let the Green function be

$$\check{g} = \check{g}_0 + \check{g}_1$$

where $\check{g}_0$ is the solution for a homogeneous state and $\check{g}_1$ is a small correction proportional to gradients and the vector potential. The first-order correction is proportional either to the dot product of ∇ and $\mathbf{v}_F$ or to the dot product of $\mathbf{A}$ and $\mathbf{v}_F$. It is thus linear in $\mathbf{v}_F$. We have therefore

$$\langle \check{g} \rangle = \check{g}_0.$$

The normalization condition (5.38) gives

$$2g_0 g_1 = f_0 f_1^\dagger + f_0^\dagger f_1. \tag{6.1}$$

The corrections f_1 and $f_1^\dagger$ are found from eqns (5.84) and (5.85)

$$-i\mathbf{v}_F\left(\nabla - \frac{2ie}{c}\mathbf{A}\right) f_0 - 2i\omega_n f_1 + 2g_1\Delta - \frac{i}{\tau}\left(g_0 f_1 - g_1 f_0\right) = 0, \tag{6.2}$$

$$i\mathbf{v}_F\left(\nabla + \frac{2ie}{c}\mathbf{A}\right) f_0^\dagger - 2i\omega_n f_1^\dagger + 2\Delta^* g_1 - \frac{i}{\tau}\left(g_0 f_1^\dagger - g_1 f_0^\dagger\right) = 0. \tag{6.3}$$

We multiply the first equation by $f_0^\dagger$, the second equation by f_0 and add the two equations. Using the normalization condition and eqn (6.1), we obtain

$$\frac{i}{2}\mathbf{v}_F\left[f_0^\dagger\left(\nabla - \frac{2ie}{c}\mathbf{A}\right) f_0 - f_0\left(\nabla + \frac{2ie}{c}\mathbf{A}\right) f_0^\dagger\right]$$
$$= -\left[2i\omega_n g_0 - \left(\Delta f_0^\dagger + \Delta^* f_0\right) + \frac{i}{\tau}\right] g_1.$$

We find from this equation

$$g_1 = \frac{\mathbf{v}_F\left[\Delta^*\left(\nabla - (2ie/c)\mathbf{A}\right)\Delta - \Delta\left(\nabla + (2ie/c)\mathbf{A}\right)\Delta^*\right]}{4(\omega_n^2 + |\Delta|^2)[\sqrt{\omega_n^2 + |\Delta|^2} + (1/2\tau)]}. \tag{6.4}$$

The current becomes

$$\mathbf{j} = -\frac{c}{4\pi\lambda_L^2}\left(\mathbf{A} - \frac{c}{2e}\nabla\chi\right). \tag{6.5}$$

where the London penetration length is found from

$$\lambda_L^{-2} = \frac{16\pi^2 \nu(0)\, v_F^2 e^2 |\Delta|^2 T}{3c^2} \sum_{n\geq 0} \frac{1}{(\omega_n^2 + |\Delta|^2)\left[\sqrt{\omega_n^2 + |\Delta|^2} + (1/2\tau)\right]}. \tag{6.6}$$

Note that the order parameter does depend on the impurity concentration in presence of magnetic field or current and/or in presence of order parameter gradients because the Green function now contains the mean free time τ.

The London penetration depth in the dirty case, $1/\tau \gg T_c$, is determined by

$$\lambda_L^{-2} = \frac{16\pi^2 \sigma_n |\Delta|^2 T}{c^2} \sum_{n\geq 0} \frac{1}{\omega_n^2 + |\Delta|^2}. \tag{6.7}$$

This sum can be evaluated using

$$\sum_{n=0}^{\infty} \frac{1}{(2n+1)^2 + x^2} = \frac{\pi}{4x} \tanh \frac{\pi x}{2}. \tag{6.8}$$

We find

$$\lambda_L^{-2} = \frac{4\pi^2 \sigma_n}{c^2} |\Delta| \tanh \frac{|\Delta|}{2T}.$$

λ_L is proportional to $1/\sqrt{\ell}$ and increases with a decrease in ℓ.

One can express the density of superconducting electrons through λ_L using eqn (1.25). In the clean limit, $1/\tau \ll T_c$, the superconducting density is

$$N_s = N|\Delta|^2\, 2\pi T \sum_{n\geq 0} \frac{1}{(\omega_n^2 + |\Delta|^2)^{3/2}}. \tag{6.9}$$

In the dirty case we have

$$N_s = \frac{\pi m \sigma_n |\Delta|}{e^2} \tanh \frac{|\Delta|}{2T}. \tag{6.10}$$

6.1.2 *Ginzburg–Landau theory*

The Ginzburg–Landau equations can easily be derived using the quasiclassical method. Let us assume that Δ and the magnetic field are small and spatial variations of all the quantities are slow. Under these conditions, we can expand the Eilenberger equations in small gradients and in Δ. Within the zero approximation in gradients, we have for an s-wave superconductor

$$\tilde{g}_0 = \frac{\omega_n}{\sqrt{\omega_n^2 + |\Delta|^2}}, \tag{6.11}$$

$$\tilde{f}_0 = \frac{\Delta}{i\sqrt{\omega_n^2 + |\Delta|^2}} \approx \frac{\Delta}{i\omega_n} - \frac{|\Delta|^2 \Delta}{2i\omega_n^3}, \tag{6.12}$$

$$\tilde{f}_0^{\dagger} = \frac{\Delta^*}{i\sqrt{\omega_n^2 + |\Delta|^2}} \approx \frac{\Delta^*}{i\omega_n} - \frac{|\Delta|^2 \Delta^*}{2i\omega_n^3}. \tag{6.13}$$

We need the function f up to the terms which are simultaneously of the third-order in Δ and zero-order in gradients, or of the second-order in gradients and of the first-order in Δ. Within the first-order in Δ, the function $g = 1$. Note that the function g contains only even powers of Δ. Therefore, a first-order correction in gradients for the function g is simultaneously of the second order in Δ.

The first-order corrections both in the gradient and in Δ are found from eqn (6.2)

$$f_1 = -\frac{\mathbf{v}_F \left(\nabla - \frac{2ie}{c}\mathbf{A}\right) f_0}{2\omega_n + 1/\tau}; \quad g_1 = 0. \tag{6.14}$$

The correction of the second-order in gradients and first-order in Δ is found from

$$-i\mathbf{v}_F \left(\nabla - \frac{2ie}{c}\mathbf{A}\right) f_1 - 2i\omega_n f_2 + 2g_2\Delta - \frac{i}{\tau}\left(f_2 - \langle f_2 \rangle\right) = 0.$$

Since g_2 is zero again, we get

$$f_2 = -\frac{\mathbf{v}_F \left(\nabla - 2ie\mathbf{A}/c\right) f_1}{2\omega_n + 1/\tau} + \frac{\langle f_2 \rangle}{\left(2\omega_n + 1/\tau\right)\tau}. \tag{6.15}$$

Performing the angular averaging we solve the resulting equation for $\langle f_2 \rangle$ whence

$$\langle f_2 \rangle = \frac{v_F^2}{3} \frac{\left(\nabla - 2ie\mathbf{A}/c\right)^2 f_0}{\left(2\omega_n + 1/\tau\right) 2\omega_n} = -\frac{iD\left(\nabla - 2ie\mathbf{A}/c\right)^2 \Delta}{2\omega_n^2 \left(2\omega_n\tau + 1\right)}$$

where we use

$$\int \frac{d\Omega_{\mathbf{p}}}{4\pi} v_i v_k = \frac{v^2}{3}\delta_{ik}$$

for a spherical Fermi surface. For the third-order terms in Δ we do not need gradients. Therefore, the final expression for the averaged Green function is

$$\langle f \rangle = \frac{\Delta}{i\omega_n} - \frac{|\Delta|^2 \Delta}{2i\omega_n^3} - \frac{iD\left(\nabla - 2ie\mathbf{A}/c\right)^2 \Delta}{2\omega_n^2 \left(2\omega_n\tau + 1\right)}. \tag{6.16}$$

The gradient-independent terms do not contain τ in accordance with the Anderson theorem.

6.1.2.1 *Order parameter* The self-consistency equation for the order parameter (5.27) gives

$$\begin{aligned} \frac{\Delta(\mathbf{r})}{\lambda} &= 2\pi T \sum_{n=0}^{N_0(T)} \frac{\Delta(\mathbf{r})}{\omega_n} - \pi T|\Delta|^2\Delta \sum_{n=0}^{\infty} \frac{1}{\omega_n^3} \\ &+ \pi T D \left(\nabla - \frac{2ie}{c}\mathbf{A}\right)^2 \Delta(\mathbf{r}) \sum_{n=0}^{\infty} \frac{1}{\omega_n^2 \left(2\omega_n\tau + 1\right)}. \end{aligned} \tag{6.17}$$

In the first sum, we introduce the cut-off value as in eqn (3.81)

$$N_0(T) = \frac{\Omega_{\mathrm{BCS}}}{2\pi T} > \frac{\Omega_{\mathrm{BCS}}}{2\pi T_c} = N_0(T_c). \tag{6.18}$$

Therefore

$$2\pi T \sum_{n=0}^{N_0(T)} \frac{1}{\omega_n} = 2\pi T \sum_{n=0}^{N_0(T_c)} \frac{1}{\omega_n} + 2\pi T \sum_{N_0(T_c)}^{N_0(T)} \frac{1}{\omega_n}$$

$$= \frac{1}{\lambda} + \sum_{N_0(T_c)}^{N_0(T)} \frac{1}{n+1/2} = \frac{1}{\lambda} + \ln\left(\frac{T_c}{T}\right) \tag{6.19}$$

where we use the expression for the critical temperature eqn (3.83) through the interaction constant. Next,

$$2\pi T \sum_{n=0}^{\infty} \frac{1}{\omega_n^3} = \frac{1}{4\pi^2 T^2} \sum_{n=0}^{\infty} \frac{1}{(n+1/2)^3} = \frac{7\zeta(3)}{4\pi^2 T^2}. \tag{6.20}$$

Here we use

$$\sum_{n=0}^{\infty} \frac{1}{(n+1/2)^z} = (2^z - 1)\zeta(z)$$

where

$$\zeta(z) = \sum_{n=1}^{\infty} \frac{1}{n^z}$$

is the Riemann ζ function. The equation for the order parameter becomes

$$\Delta \ln\left(\frac{T_c}{T}\right) - \frac{7\zeta(3)}{8\pi^2 T^2}|\Delta|^2\Delta + \frac{\pi\tilde{D}}{8T_c}\left(\nabla - \frac{2ie}{c}\mathbf{A}\right)^2 \Delta = 0 \tag{6.21}$$

where

$$\tilde{D} = D\,\Lambda\,(T_c\tau)$$

and

$$\Lambda\,(x) = \frac{8}{\pi^2}\sum_{n=1}^{\infty} \frac{1}{(2n+1)^2\left[(2n+1)\,2\pi x + 1\right]}. \tag{6.22}$$

The function

$$\Lambda\,(\tau T_c) = \begin{cases} 1 & \text{for } \tau T_c \ll 1, \\ 7\zeta\,(3)\,/2\pi^3\tau T_c & \text{for } \tau T_c \gg 1. \end{cases}$$

For clean superconductors $\tau T_c \gg 1$ the coefficient

$$\tilde{D} = \frac{7\zeta\,(3)\,v_F^2}{6\pi^3 T_c}.$$

The condition of a small Δ is $\Delta \ll T$. This is satisfied for $T \to T_c$. Therefore, $\ln(T_c/T) \approx 1 - T/T_c$. We finally obtain

$$\left(1 - \frac{T}{T_c}\right)\Delta - \frac{7\zeta(3)}{8\pi^2 T_c^2}|\Delta|^2\Delta + \frac{\pi\tilde{D}}{8T_c}\left(\nabla - \frac{2ie}{c}\mathbf{A}\right)^2 \Delta = 0. \tag{6.23}$$

Equation (6.23) is identical to the famous Ginzburg–Landau equation (1.12) for the order parameter. It was derived microscopically by Gor'kov (1959 a, 1959 c).

Equation (6.23) determines the constants α, β, and γ introduced on page 4. The constants α and β were already defined by eqn (1.8). The constant γ is

$$\gamma = \frac{\pi\nu(0)\,\tilde{D}}{8T_c}. \tag{6.24}$$

The equilibrium order parameter is

$$\Delta_\infty(T) = T_c\sqrt{\frac{8\pi^2}{7\zeta(3)}}\,\sqrt{1-\frac{T}{T_c}}. \tag{6.25}$$

The GL equation determines the temperature-dependent coherence length, i.e., characteristic scale of variations of Δ:

$$\xi(T) = \sqrt{\frac{\pi^3 T \Lambda D}{7\zeta(3)\,\Delta_\infty^2}} = \sqrt{\frac{\pi\tilde{D}}{8(T_c-T)}} \propto \left(1-\frac{T}{T_c}\right)^{-1/2}. \tag{6.26}$$

We have already encountered it on page 6. For clean superconductors

$$\xi(T) = \frac{v_F}{\sqrt{6}\Delta(T)} \sim \xi_0\left(1-\frac{T}{T_c}\right)^{-1/2}$$

where $\xi_0 = v_F/2\pi T_c$ is the zero-temperature coherence length. For dirty superconductors

$$\xi(T) \sim \sqrt{\xi_0\ell_0}\left(1-\frac{T}{T_c}\right)^{-1/2}.$$

The condition of slow variations of the order parameter implies that

$$\xi(T) \gg \xi_0.$$

This is always satisfied for $T \to T_c$.

6.1.2.2 *Current* To find the current, we can use the results obtained earlier. We take eqn (6.5) and put $\Delta \to 0$ in the denominator. We reproduce eqn (1.13) which can be written as

$$\mathbf{j} = -\frac{\pi\tilde{D}\nu(0)\,e^2|\Delta|^2}{cT_c}\left(\mathbf{A}-\frac{c}{2e}\nabla\chi\right) \tag{6.27}$$

using the microscopic value for γ. With this expression for the supercurrent we find, in particular, that the density of superconducting electrons is

$$N_s = 2N\,(1-T/T_c)$$

in clean superconductors. For dirty superconductors

$$N_s = \frac{\pi N\tau|\Delta|^2}{2T_c}. \tag{6.28}$$

This agrees, of course, with eqn (6.6).

Note that in addition to $\xi(T) \gg \xi_0$, the condition of slow variations implies also

$$\lambda_L(T) \gg \xi_0.$$

It is fulfilled for both type I and type II superconductors when temperatures are close enough to T_c because λ_L increases as T approaches T_c.

6.1.2.3 *Free energy* One can use equation (5.30) for the thermodynamic potential to find the Ginzburg–Landau free energy of a superconductor. For a constant chemical potential we have

$$\delta\Omega_{sn} = \nu(0)\int\left[-2\pi iT\sum_n\left(\delta\Delta\left\langle f^{\dagger}_{\omega_n}\right\rangle + \delta\Delta^*\left\langle f_{\omega_n}\right\rangle\right) + \frac{\delta\left(\hat{\Delta}\hat{\Delta}^*\right)}{\lambda}\right]d^3r. \quad (6.29)$$

The same expression also holds for variation of the free energy for a constant particle density. Calculating the sums we find from eqn (6.16)

$$\mathcal{F}_{sn} = \nu(0)\int\left[-\left(1-\frac{T}{T_c}\right)|\Delta|^2 + \frac{7\zeta(3)}{16\pi^2T_c^2}|\Delta|^4 + \frac{\pi\tilde{D}}{8T_c}\left|\left(\nabla - \frac{2ie}{c}\mathbf{A}\right)\Delta\right|^2\right]d^3r. \quad (6.30)$$

It coincides, of course, with eqn (1.3) for given α, β, and γ.

6.1.3 *The upper critical field in a dirty alloy*

Consider the second-order phase transition of a dirty alloy from normal into the superconducting state in a decreasing magnetic field. As in Section 1.1.2, we are going to calculate the critical magnetic field below which the superconducting state first appears, i.e., the upper critical field H_{c2}. At this time, however, we do not restrict ourselves to temperatures close to T_c but shall use the quasiclassical method and demonstrate that it is able to treat this problem for arbitrary temperatures.

Near the transition, the order parameter is small. We can again expand the Green function in a small Δ. In the zero-order approximation, $g_0 = 1$, and

$$f = \frac{\Delta}{i\alpha_n}$$

where α_n has yet to be found. The Usadel eqn (5.98) gives

$$D\left(\nabla - \frac{2ie}{c}\mathbf{A}\right)^2\Delta + 2(\alpha_n - \omega_n)\Delta = 0. \quad (6.31)$$

This equation has exactly the same form as the Ginzburg–Landau equation for the order parameter near the upper critical magnetic field, eqn (1.31). We again take the vector potential in the form

$$\mathbf{A} = (0, Hx, 0)$$

where the magnetic field is nearly homogeneous and is directed along the z-axis and put

$$\Delta = e^{iky} Y(x). \tag{6.32}$$

The equation for $Y(x)$ becomes the oscillator equation

$$\frac{\partial^2 Y}{\partial x^2} - \left(k - \frac{2eH_{c2}x}{c}\right)^2 Y + 2\left(\frac{\alpha_n - \omega_n}{D}\right) Y = 0. \tag{6.33}$$

The solution which appears first with decreasing the magnetic field corresponds to the lowest level of eqn (6.33) and has the form of eqn (1.35) with

$$\alpha_n = \omega_n + \frac{DeH_{c2}}{c} \tag{6.34}$$

The full expression for Δ is a linear combination of functions (6.32) with various k. A periodic solution has the form of a vortex lattice, eqn (1.36), obtained by Abrikosov (1957).

The Green function becomes

$$f_0 = \frac{\Delta}{i(\omega_n + \epsilon_0)} \tag{6.35}$$

where

$$\epsilon_0 = DeH_{c2}/c. \tag{6.36}$$

Using the expression for T_c we can write the order parameter equation as

$$\frac{\Delta}{\lambda} = \sum_{n=0}^{N_0(T_c)} \frac{\Delta}{n+\frac{1}{2}} + \sum_{N_0(T_c)}^{N_0(T)} \frac{\Delta}{n+\frac{1}{2}} + \sum_{n=0}^{\infty}\left[\frac{\Delta}{n+\frac{1}{2}+\frac{\epsilon_0}{2\pi T}} - \frac{\Delta}{n+\frac{1}{2}}\right]$$
$$= \frac{\Delta}{\lambda} + \Delta \ln\left(\frac{T_c}{T}\right) + \Delta\left[\psi\left(\frac{1}{2}\right) - \psi\left(\frac{1}{2} + \frac{\epsilon_0}{2\pi T}\right)\right].$$

Here

$$\psi(z) = \frac{d}{dz}\left[\ln \Gamma(z)\right] \tag{6.37}$$

is the so-called digamma function coupled to the Euler gamma function

$$\Gamma(z) = \int_0^\infty e^{-t} t^{z-1}\, dt.$$

Finally, we obtain the equation which determines the upper critical magnetic field (Maki 1969)

$$\ln\left(\frac{T_c}{T}\right) = \psi\left(\frac{1}{2} + \frac{\epsilon_0}{2\pi T}\right) - \psi\left(\frac{1}{2}\right). \tag{6.38}$$

For low temperatures, $T \to 0$, we can use the asymptotic expression for $\psi(z)$ for large z:

$$\psi(z) \approx \ln z - \frac{1}{2z}.$$

Since $\psi(1/2) = -C - 2\ln 2$ we obtain

$$\epsilon_0 = \pi T_c / 2\gamma$$

where $\gamma = e^C \approx 1.78$. Equation (6.36) gives the upper critical field

$$H_{c2} = \frac{\pi c T_c}{2\gamma D e}. \tag{6.39}$$

For high temperatures, $T \to T_c$, we have

$$\psi\left(\frac{1}{2} + \frac{\epsilon_0}{2\pi T}\right) - \psi\left(\frac{1}{2}\right) \approx \frac{d\psi(1/2)}{dz}\frac{\epsilon_0}{2\pi T_c}.$$

Since $\psi'(1/2) = \pi^2/2$, we obtain

$$H_{c2} = \frac{4cT}{\pi D e}\left(1 - \frac{T}{T_c}\right) = \frac{c}{2e\xi^2(T)}. \tag{6.40}$$

This result coincides, of course, with the solution of the Ginzburg–Landau equation (1.34) since the coherence length in the dirty limit is

$$\xi^2(T) = \frac{\pi D}{8T_c}\left(1 - \frac{T}{T_c}\right)^{-1}. \tag{6.41}$$

6.2 Gapless s-wave superconductivity

Consider an alloy with magnetic impurities (Abrikosov and Gor'kov 1960). Scattering on an impurity atom depends on the spin state of an electron. The scattering process breaks the spin coherence of paired electrons and leads to a suppression of superconductivity. The key point is that the self-energies for g and f contain different scattering times since the functions g, on one hand, and f, $f^\dagger$, on the other hand, have different spin structures, For a homogeneous case we have from eqns (5.74) and (5.75)

$$\Sigma_1 = \frac{i}{2\tau_1} g_{\omega_n}, \text{ while } \Sigma_2 = \frac{i}{2\tau_2} f_{\omega_n}, \; \Sigma_2^\dagger = \frac{i}{2\tau_2} f_{\omega_n}^\dagger \tag{6.42}$$

where the mean free times are defined by eqns (4.16) and (4.17). The Eilenberger equation (5.84) gives

$$-2i\omega_n f + 2\Delta g - \frac{2i}{\tau_s} g f = 0.$$

where the spin-flip time is determined by eqn (4.18). Therefore,

$$f = \frac{\Delta g}{i(\omega_n + g/\tau_s)}. \tag{6.43}$$

From the normalization condition $g^2 - f^2 = 1$ we find

$$g^2 = \frac{(\omega_n + g/\tau_s)^2}{(\omega_n + g/\tau_s)^2 + \Delta^2} \tag{6.44}$$

and

$$f = \frac{\Delta}{i\sqrt{(\omega_n + g/\tau_s)^2 + \Delta^2}}. \tag{6.45}$$

6.2.1 *Critical temperature*

The order parameter equation (5.27) takes the form

$$\frac{1}{\lambda} = 2\pi T \sum_{n\geq 0} \frac{1}{\sqrt{(\omega_n + g/\tau_s)^2 + \Delta^2}}. \tag{6.46}$$

Let us find the critical temperature. Putting $\Delta = 0$, we get $g = 1$ and

$$\begin{aligned}\frac{1}{\lambda} &= 2\pi T \sum_{n=0}^{N_0(T_c)} \frac{1}{\omega_n} + 2\pi T \sum_{n\geq 0} \left(\frac{1}{\omega_n + 1/\tau_s} - \frac{1}{\omega_n} \right) \\ &= 2\pi T \sum_{n=0}^{N_0(T_{c0})} \frac{1}{\omega_n} + 2\pi T \sum_{N_0(T_{c0})}^{N_0(T_c)} \frac{1}{\omega_n} \\ &\quad + 2\pi T \sum_{n=0}^{\infty} \left(\frac{1}{\omega_n + 1/\tau_s} - \frac{1}{\omega_n} \right).\end{aligned} \tag{6.47}$$

Here T_{c0} is the critical temperature without magnetic impurities, eqn (3.83):

$$T_{c0} = (2\Omega_{\text{BCS}}\gamma/\pi)e^{-1/\lambda}.$$

We obtain

$$\ln\left(\frac{T_{c0}}{T_c}\right) + \sum_{n=0}^{\infty} \left(\frac{1}{n + 1/2 + 1/2\pi T_c \tau_s} - \frac{1}{n + 1/2} \right) = 0$$

or

$$\ln\left(\frac{T_{c0}}{T_c}\right) + \psi\left(\frac{1}{2}\right) - \psi\left(\frac{1}{2} + \frac{1}{2\pi T_c \tau_s}\right) = 0. \tag{6.48}$$

It is the familiar equation: we have seen it on page 108 when we calculated the upper critical field for a dirty superconductor.

Let us consider a limiting case when the spin-flip scattering rate is large: $\tau_s T_c \ll 1$. In this limit, the critical temperature decreases down to zero, $T_c \ll T_{c0}$ and $1/2\pi T_c \tau_s$ increases. For large arguments, we can use the asymptotics

$$\psi(z) = \ln z - \frac{1}{2z} - \frac{1}{12z^2}.$$

The value $\psi(1/2) = -\ln(4\gamma)$ where $\gamma \approx 1.78$. We have

$$\psi\left(\frac{1}{2} + \frac{1}{2\pi T_c \tau_s}\right) - \psi\left(\frac{1}{2}\right) = \ln\left(\frac{2\gamma}{\pi T_c \tau_s}\right) + \frac{(\pi T_c \tau_s)^2}{6}.$$

As the result, the critical temperature is

$$T_c^2 = \frac{6}{\pi^2 \tau_s^2} \ln\left(\frac{\pi T_{c0} \tau_s}{2\gamma}\right). \tag{6.49}$$

The critical temperature vanishes when the scattering rate is

$$\frac{1}{\tau_s} = \pi T_{c0}/2\gamma.$$

For low concentration of magnetic impurities, $1/\tau_s \to 0$, we have $\psi(1/2 + x) - \psi(1/2) = \psi'(1/2)x$ and

$$1 - \frac{T_c}{T_{c0}} = \frac{\pi}{4 T_{c0} \tau_s}.$$

The critical temperature is slightly reduced from its value without magnetic impurities.

6.2.2 *Gap in the energy spectrum*

The real-frequency (retarded and advanced) Green functions are found from

$$g^2 = \frac{(\epsilon + ig/\tau_s)^2}{(\epsilon + ig/\tau_s)^2 - \Delta^2} \tag{6.50}$$

which is obtained from eqn (6.44) by replacing ω_n with $-i\epsilon$. Retarded function g^R is defined as the solution of eqn (6.50) which has no singularities in the upper half-plane of ϵ, while g^A should not have singularities in the lower half-plane.

Let us look for a solution in the form $g^R = g^A = -ia$ where a is a real function. If such a solution exists for a given energy ϵ, the density of states $g^R - g^A = 0$. This means that the corresponding energy is below the energy gap. We have for a

$$a^2 = \frac{(\epsilon + a/\tau_s)^2}{\Delta^2 - (\epsilon + a/\tau_s)^2}. \tag{6.51}$$

This gives

$$\frac{\epsilon}{\Delta} = -\frac{a}{\tau_s \Delta} \pm \frac{a}{\sqrt{a^2 + 1}}. \tag{6.52}$$

We choose the solution with the + sign since it gives the correct behavior in the case without magnetic impurities $1/\tau_s \to 0$. Indeed, without magnetic impurities we would have

$$g = -i\frac{\epsilon}{\sqrt{\Delta^2 - \epsilon^2}}$$

for $|\epsilon| < \Delta$. It results in the asymptotics $a \to +\infty$ for $\epsilon/\Delta \to 1 - 0$ which should be reproduced by eqn (6.52). Equation (6.52) for a takes the form

$$\epsilon = \Delta F(a)$$

where we define the function

$$F(a) = -\frac{a}{\tau_s \Delta} + \frac{a}{\sqrt{a^2 + 1}}. \tag{6.53}$$

For $\tau_s \Delta > 1$, the function $F(a)$ has a positive maximum. Calculating the derivative we find that the maximum is reached at

$$a = \sqrt{(\tau_s \Delta)^{2/3} - 1}.$$

The function at maximum is equal to

$$F_{max} = \left[1 - \frac{1}{(\tau_s \Delta)^{2/3}}\right]^{3/2}.$$

Therefore, a real solution for a exists if $\epsilon < \Delta F_{max}$. Thus

$$\epsilon_0 = \Delta \left[1 - \frac{1}{(\tau_s \Delta)^{2/3}}\right]^{3/2}$$

is the gap in the energy spectrum. For $\epsilon > \epsilon_0$, there is no real solution for a, thus g becomes complex: $g^R = -ia + b$ and $g^A = -ia - b$ with a finite density of states $g^R - g^A = 2b$.

The energy gap disappears when the order parameter satisfies the gapless condition $\tau_s \Delta(T) \leq 1$. Note that there is a temperature range where superconductivity exists without an energy gap in the excitation spectrum. Indeed, superconductivity exists for temperatures below T_c if $\pi T_{c0} \tau_s / 2\gamma > 1$. For low enough τ_s^{-1}, the condition $\tau_s \Delta(T = 0) > 1$ is obviously fulfilled, and there is an energy gap for low temperatures. However, the gap vanishes at higher temperatures when $\tau_s \Delta(T)$ decreases down to 1. In the opposite case, $\tau_s \Delta(T = 0) < 1$, the energy gap is absent for all temperatures.

We note that the vicinity of the upper critical field considered in the previous section also belongs to the gapless regime. Indeed, for the real frequency $\epsilon = 0$ one has from eqn (6.35)

$$f^R = f^A = \frac{\Delta}{i\epsilon_0};\; g^R = -g^A = 1 + \frac{\Delta^2}{2\epsilon_0^2}.$$

The density of states is thus finite. The gapless situation can be created also by an inelastic electron–phonon scattering within certain range of temperatures (see Section 11.2).

6.3 Aspects of d-wave superconductivity

Many problems of d-wave superconductivity can be successfully solved using the quasiclassical methods. Their applications for stationary and nonequilibrium properties of d-wave superconductors can be found in many publications (see, for example, Graf *et al.* 1995, Graf *et al.* 1996); they belong to a rapidly growing field of research. We consider in this section only several selected examples.

6.3.1 *Impurities and d-wave superconductivity*

The order parameter of a d-wave superconductor satisfies the self-consistency equation (5.29). It has the form

$$\frac{\Delta_{\mathbf{p}}(\mathbf{k})}{\lambda} = 2\pi i T \sum_{n\geq 0} \left\langle V_d(\hat{\mathbf{p}}, \hat{\mathbf{p}}_1) f_{\omega_n}(\mathbf{p}_1 + \mathbf{k}/2, \mathbf{p}_1 - \mathbf{k}/2) \right\rangle_{\mathbf{p}_1} \tag{6.54}$$

in the quasiclassical approximation. Here we average over the directions of $\mathbf{p}_1$ at the Fermi surface according to eqn (5.73) having in mind that the Fermi surface can be nonspherical. For example, in d-wave superconductors which have uniaxial crystal symmetry, its shape is more close to a cylinder. The d-wave pairing potential has the structure of eqn (3.53). It is normalized in such a way that

$$\left\langle V_d(\hat{\mathbf{p}}, \hat{\mathbf{p}}_1) \Delta_{\mathbf{p}_1} \right\rangle_{\mathbf{p}_1} = \Delta_{\mathbf{p}}. \tag{6.55}$$

If $V_d = 2\cos(2\alpha)\cos(2\alpha_1)$ the order parameter in a d-wave superconductor is

$$\Delta_{\mathbf{p}} = \Delta_0 \cos(2\alpha) \tag{6.56}$$

where the amplitude satisfies the equation

$$\frac{\Delta_0(\mathbf{r})}{\lambda} = 2\pi i T \sum_{n\geq 0} \left\langle 2\cos(2\alpha) f_{\omega_n}(\hat{\mathbf{p}}, \mathbf{r}) \right\rangle . \tag{6.57}$$

The order parameter has nodes, i.e., it vanishes for momentum directions $\alpha = \pi/4 + (\pi/2)n$. Note that

$$\langle \Delta_{\mathbf{p}} \rangle = 0. \tag{6.58}$$

We restrict ourselves to a homogeneous case. The quasiclassical Green functions are found from the Eilenberger equations. We assume an isotropic scattering by impurities. We have from eqn (5.84)

$$-2i\omega_n f + 2\Delta_{\mathbf{p}} g - \frac{i}{\tau} \left(\langle g \rangle f - \langle f \rangle g \right) = 0.$$

As we shall see later, in a homogeneous state,

$$\langle f(\hat{\mathbf{p}}) \rangle = 0. \tag{6.59}$$

Under this condition, the self-energies do not drop out of the Eilenberger equation eqns (5.83) – (5.85) any more. Instead, we obtain with help of the normalization condition $g^2 - f^2 = 1$

$$f = \frac{\Delta_{\mathbf{p}}}{i\sqrt{(\omega_n + \langle g \rangle/2\tau)^2 + |\Delta_{\mathbf{p}}|^2}}, \; g = \frac{\omega_n + \langle g \rangle/2\tau}{\sqrt{(\omega_n + \langle g \rangle/2\tau)^2 + |\Delta_{\mathbf{p}}|^2}}. \tag{6.60}$$

With this solution, it easy to see that eqn (6.59) holds.

Equation (6.60) resembles the expressions for an alloy with magnetic impurities. Indeed, properties of a d-wave superconductor with usual nonmagnetic impurities are very much similar to those for magnetic s-wave alloys. First of all, the presence of impurities suppresses the d-wave superconductivity. The reason is that the scattering by impurities destroys the momentum coherence of the paired state in a way similar to that by which the spin-dependent scattering destroys the spin coherence of the paired state in an s-wave superconductor. We thus can anticipate that there exists a critical scattering rate $1/\tau$ such that the critical temperature of the d-wave superconductor vanishes.

To find the critical temperature as a function of the scattering rate, we put $\Delta = 0$ in eqn (6.60) and insert thus obtained f into eqn (6.57). We have

$$\frac{1}{\lambda} = 2\pi T \sum_{n\geq 0} \frac{1}{\omega_n + 1/2\tau}. \tag{6.61}$$

This is exactly the equation (6.47) for the critical temperature of a magnetic alloy with the substitution $1/\tau_s \to 1/2\tau$. Therefore, we obtain

$$\ln\left(\frac{T_{c0}}{T_c}\right) + \psi\left(\frac{1}{2}\right) - \psi\left(\frac{1}{2} + \frac{1}{4\pi T_c \tau}\right) = 0. \tag{6.62}$$

Here T_{c0} is the critical temperature without impurities

$$T_{c0} = (2\Omega_{\mathrm{BCS}}\gamma/\pi)e^{-1/\lambda}.$$

The critical temperature vanishes for

$$\frac{1}{\tau} = \pi T_{c0}/\gamma$$

This shows that d-wave superconductivity can exists only in rather clean compounds.

6.3.2 *Impurity-induced gapless excitations*

Presence of impurities broadens the order parameter nodes into gapless low-energy states (Gor'kov and Kalugin 1985). In a strict sense, there is no energy gap in a d-wave alloy. To see that, let us find the real-frequency Green function. The retarded function is from eqn (6.60)

$$g^R = \frac{\epsilon + \frac{i}{2\tau}\langle g^R \rangle}{\sqrt{\left(\epsilon + \frac{i}{2\tau}\langle g^R \rangle\right)^2 - |\Delta_{\mathbf{p}}|^2}}. \tag{6.63}$$

Let us find the average

$$\langle g^R \rangle = \int \frac{d\alpha}{2\pi} \frac{\epsilon + \frac{i}{2\tau}\langle g^R \rangle}{\sqrt{\left(\epsilon + \frac{i}{2\tau}\langle g^R \rangle\right)^2 - |\Delta_{\mathbf{p}}|^2}}.$$

We have for $\epsilon = 0$

$$1 = \frac{1}{2\tau} \int \frac{d\alpha}{2\pi} \frac{1}{\sqrt{\Delta_0^2 |\cos(2\alpha)|^2 + \left(\frac{1}{2\tau}\langle g^R \rangle\right)^2}}. \tag{6.64}$$

We shall see that $(1/2\tau)\langle g^R \rangle \ll \Delta_0$ therefore, the angles with a small $\cos(2\alpha)$ are important. We put $x = \pi/4 - \alpha$. Substituting $\cos(2\alpha) \approx 2x$ for small x, we get

$$\frac{1}{\pi\tau} \int_0^\infty \frac{dx}{\sqrt{\Delta_0^2 x^2 + (\langle g^R \rangle / 4\tau)^2}} = 1 \tag{6.65}$$

since there are 4 nodes. Calculating the integral we find

$$\ln\left(\frac{4\tau\Delta_0}{\langle g^R \rangle}\right) = \pi\Delta_0\tau. \tag{6.66}$$

The averaged Green function is

$$\langle g^R \rangle = 4\Delta_0\tau \exp(-\pi\Delta_0\tau). \tag{6.67}$$

It is a real quantity. Thus there exists a finite density of states at zero energy

$$\nu = 4\nu(0)\Delta_0\tau \exp(-\pi\Delta_0\tau).$$

It is exponentially small in the clean limit $\Delta_0\tau \gg 1$ in the Born approximation. However, the density of states becomes a power-law function if the impurity scattering is strong such that it should be treated within the full scattering amplitude (Graf *et al.* 1996).

We see that there is no energy gap: a small but finite density of states exists down to zero energy. The angle-resolved density of states for $\epsilon = 0$ is

$$\frac{1}{2}\left(g^R - g^A\right) = \frac{\epsilon_0}{\sqrt{\epsilon_0^2 + [\Delta_0 \cos(2\alpha)]^2}}$$

where ϵ_0 is

$$\epsilon_0 = \langle g^R \rangle / 2\tau = 2\Delta_0 \exp(-\pi\Delta_0\tau).$$

One can say that, instead of gap nodes, there opens "an impurity band" with the bandwidth $2\epsilon_0$.

6.3.3 *The Ginzburg–Landau equations*

Here we consider an ideal case where the pairing potential has only a d-wave component. A more complete version of the Ginzburg–Landau equation including both d-wave and s-wave components of the pairing potential and the order parameter is given by Heeb *et al.* (1996). Since a d-wave superconductor can only be clean, we have from eqns (6.12, 6.14, 6.15) for $\tau \to \infty$

$$f = \frac{\Delta_{\mathbf{p}}}{i\omega_n} - \frac{|\Delta_{\mathbf{p}}|^2 \Delta_{\mathbf{p}}}{2i\omega_n} + \frac{1}{4i\omega_n} v_i v_k \hat{\nabla}_i \hat{\nabla}_k \Delta_{\mathbf{p}}$$

where

$$\hat{\nabla} = \nabla \mp \frac{2ie}{c}\mathbf{A}$$

for Δ and Δ^*, respectively. Result of integration over the Fermi surface depends on its shape. d-wave superconductors are usually anisotropic. We assume for simplicity that the Fermi surface has an axial symmetry around the crystal c axis. Performing integration in the self-consistency equation over the angle α within the (ab) plane we obtain

$$\frac{\Delta_0}{\lambda} = 2\pi T \sum_{n>0} \left\langle \frac{\Delta_0}{\omega_n} + \frac{1}{8\omega_n^3} \left[\left(v_{ab}^2 \hat{\nabla}_{ab}^2 + 2v_c^2 \hat{\nabla}_c^2 \right) \Delta_0 - 3 |\Delta_0|^2 \Delta_0 \right] \right\rangle$$

where v_{ab} and v_c are the Fermi velocity projections on the ab plane and the c axis, respectively. The average now is over the momentum projection on the z axis. Using the sums calculated earlier, we find the Ginzburg–Landau equation for uniaxial anisotropic d-wave superconductor

$$\alpha \Delta_0 + \beta |\Delta_0|^2 \Delta_0 + \gamma_{ab} \left(\nabla_{ab} - \frac{2ie}{c} \mathbf{A}_{ab} \right)^2 \Delta_0 + \gamma_c \left(\nabla_c - \frac{2ie}{c} \mathbf{A}_c \right)^2 \Delta_0 = 0, \tag{6.68}$$

where α is defined by eqn (1.8). The constant β now is

$$\beta = \frac{21 \nu(0) \zeta(3)}{32\pi^2 T_c^2},$$

while

$$\gamma_{ab} = \frac{7\zeta(3) \nu(0) \langle v_{ab}^2 \rangle}{32\pi^2 T_c^2}, \ \gamma_c = \frac{7\zeta(3) \nu(0) \langle v_c^2 \rangle}{16\pi^2 T_c^2}.$$

Equation (6.68) is isotropic in the (ab) plane.

We can reconstruct the GL thermodynamic potential (free energy) using eqn (6.29):

$$\mathcal{F}_{sn} = \int \left[\alpha |\Delta_0|^2 + \frac{\beta}{2} |\Delta_0|^4 + \gamma_{ab} \left| \left(\nabla - \frac{2ie}{c} \mathbf{A} \right)_{ab} \Delta_0 \right|^2 + \gamma_c \left| \left(\nabla - \frac{2ie}{c} \mathbf{A} \right)_c \Delta_0 \right|^2 \right] d^3 r. \tag{6.69}$$

The supercurrent can be found from

$$\mathbf{j}_s = -c\frac{\delta \mathcal{F}}{\delta \mathbf{A}}.$$

We obtain

$$\mathbf{j}_{ab} = -2ie\gamma_{ab}\left[\Delta_0^*\left(\nabla - \frac{2ie}{c}\mathbf{A}\right)_{ab}\Delta_0 - \Delta_0\left(\nabla + \frac{2ie}{c}\mathbf{A}\right)_{ab}\Delta_0^*\right],$$
$$j_c = -2ie\gamma_c\left[\Delta_0^*\left(\nabla - \frac{2ie}{c}\mathbf{A}\right)_c\Delta_0 - \Delta_0\left(\nabla + \frac{2ie}{c}\mathbf{A}\right)_c\Delta_0^*\right] \quad (6.70)$$

in a full analogy to eqn (1.13).

6.4 Bound states in vortex cores

We turn now to another important problem associated with the structure of a vortex in the mixed state of clean type II superconductors. We already know from Section 1.1.2 that each vortex is a singular line at which the order parameter goes to zero; the order parameter phase winds around this line by 2π. Each vortex carries one magnetic flux quantum $\Phi_0 = \pi\hbar c/|e|$. The magnetic field decays away from vortex at a distance of the order of the London penetration depth λ_L due to the screening supercurrent which circulates around the vortex within the distance on order λ_L. The order parameter recovers its bulk magnitude at a distance of the order of the coherence length ξ from the vortex axis. The region where the order parameter is essentially suppressed is called the vortex core. The structure of a single vortex is shown schematically in Fig. 1.1 for the case $\kappa \gg 1$.

The order parameter magnitude profile $|\Delta(\rho)|$ forms a potential well near the vortex axis. Quasiparticles with energies below the bulk gap Δ_∞ become localized and occupy discrete levels in the vortex core. The energy spectrum of excitations in the vortex core for a single vortex was first found by Caroli *et al.* (1964). These authors have calculated the low energy levels $\epsilon \ll \Delta_\infty$ by solving the Bogoliubov–de Gennes equations (3.56) (see also de Gennes 1966).

Here we consider the same problem using the quasiclassical scheme according to the method developed by Kramer and Pesch (1974). We assume that the Ginzburg–Landau parameter of the superconductor $\kappa \gg 1$; it is the condition under which vortices can be treated separately from each other. Indeed, the intervortex distance which is of the order of the radius of the Bravais unit cell r_0 is such that the flux quantum is $\Phi_0 = \pi r_0^2 B$ therefore, $r_0 \sim \xi\sqrt{H_{c2}/H}$. It is much longer than the core size ξ if $H \ll H_{c2}$. At the same time, magnetic field should be larger than the lower critical field H_{c1} which is only possible if $H_{c1} \ll H_{c2}$. The latter is fulfilled for $\kappa \gg 1$ since $H_{c1} \sim H_{c2}/\kappa^2$.

The problem of quasiparticle states in the vortex core is important for many applications. One needs their energy spectrum, for example, to calculate the density of states in the mixed state of superconductors which determines the low-temperature behavior of the specific heat and of the London penetration depth, etc. We shall use the energy spectrum of the core states later for the vortex dynamics.

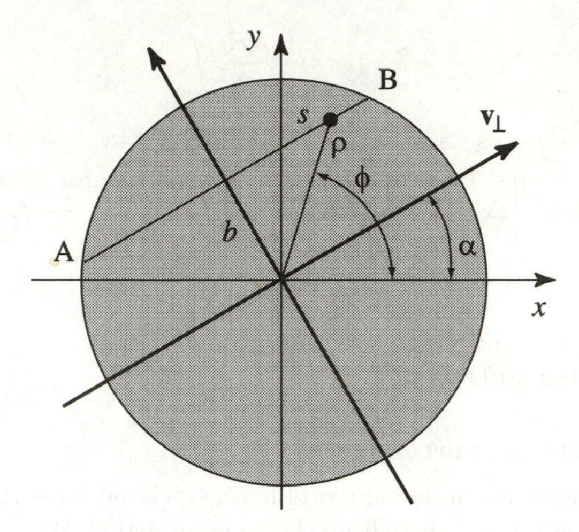

FIG. 6.1. The coordinate frame associated with the particle localized in the vortex core. The line AB is the quasiparticle trajectory passing by the vortex axis at an impact parameter b through the position point (ρ, ϕ) shown by the black dot.

As usually, the problem starts with calculating the quasiclassical Green functions. We need them for $\epsilon \ll \Delta_\infty$. Let us first choose the proper reference frame. We take the direction of the z-axis of the cylindrical frame in such a way that the vortex has a positive circulation. The z-axis is thus parallel the magnetic field for positive charge carriers, while it is antiparallel to it for negative carriers: $\hat{\mathbf{z}} = (\mathbf{H}/H)\,\mathrm{sign}\,(e)$. Let $\mathbf{v}_\perp$ be the projection of the quasiparticle (Fermi) velocity $\mathbf{v}_F$ on the (x, y) plane. The vector $\mathbf{v}_\perp$ makes an angle α with the x-axis. The quasiparticle trajectory which passes through the position point is characterized by an impact parameter b with respect to the vortex axis. We take the impact parameter and the distance s along the trajectory as the new coordinate frame. The cylindrical coordinates of the position point (ρ, ϕ) are connected to the impact parameter and the coordinate along the trajectory through $\rho^2 = b^2 + s^2$ where (see Fig. 6.1)

$$b = \rho \sin(\phi - \alpha) \; ; \; s = \rho \cos(\phi - \alpha). \tag{6.71}$$

As a function of s, the angle ϕ is $\pi + \alpha$ for $s \to -\infty$, while $\phi = \pi/2 + \alpha$ for $s = 0$, and $\phi = \alpha$ for $s \to \infty$.

6.4.1 *Superconductors with s-wave pairing*

Let us consider first an s-wave superconductor. We put $\Delta = \Delta_0 \exp(i\phi)$ and

$$f = F_0 \exp(i\phi + i\eta); \; f^\dagger = F_0^\dagger \exp(-i\phi - i\eta), \tag{6.72}$$

where

$$\eta = -\int_0^s \left(\frac{\partial \phi}{\partial s} - \frac{2e}{c} A_s \right) ds$$

$$= \frac{\pi}{2} + \alpha - \phi(s) - \frac{2eb}{c} \int_0^s A_\phi \, \frac{ds'}{\rho'}. \tag{6.73}$$

Note that $\eta(s) = -\eta(-s)$. Moreover, $\eta \to \pm\pi/2$ for $s \to \pm s_0$ where $s_0 \gg \xi$.

Now we write F_0 and $F_0^\dagger$ in terms of the symmetric and antisymmetric functions $\theta(s) = -\theta(-s)$; $\zeta(s) = \zeta(-s)$:

$$F_0 = -\theta(s) + i\zeta(s); \; F_0^\dagger = \theta(s) + i\zeta(s). \tag{6.74}$$

The normalization condition eqn (5.39) requires $g^2 + \theta^2 + \zeta^2 = 1$. Eilenberger equations (5.90) and (5.91) give

$$v_\perp \frac{\partial \zeta}{\partial s} + 2\epsilon\theta - 2i\Delta_0 g \sin\eta = 0, \tag{6.75}$$

$$v_\perp \frac{\partial \theta}{\partial s} - 2\epsilon\zeta - 2i\Delta_0 g \cos\eta = 0. \tag{6.76}$$

The boundary conditions for $s \to \pm\infty$ are:

$$f^{R(A)} = -i \frac{|\Delta_\infty| e^{i\phi}}{\sqrt{|\Delta_\infty|^2 - \epsilon^2}}, \; g^{R(A)} = -i \frac{\epsilon}{\sqrt{|\Delta_\infty|^2 - \epsilon^2}}$$

where $\epsilon \ll |\Delta_\infty|$. Therefore, $g^{R(A)}(s) \to 0$, $f^{R(A)} \to -ie^{i\phi}$ at large distances. This requires $\theta \to \pm 1$, $\zeta \to 0$. For energies close to the bound state energy ϵ_0, i.e., close to the pole $\epsilon - \epsilon_0(b) \ll \epsilon_0(b)$, the functions g and f, $f^\dagger$ are large near the vortex. We assume that $\zeta^2 \gg \theta^2 - 1$, so that $g = i\zeta$. The plus sign here is chosen to satisfy the condition of vanishing of g at large distances according to eqn (6.75).

The solution of eqns (6.75, 6.76) is

$$\zeta^{R(A)} = \frac{v_\perp \exp[-K(s)]}{2C(v_\perp)[\epsilon - \epsilon_0(b) \pm i\delta]}, \tag{6.77}$$

$$\theta^{R(A)} = \frac{\int_0^s (\epsilon - \Delta_0 \cos\eta) \exp[-K(s')] \, ds'}{C(v_\perp)[\epsilon - \epsilon_0(b) \pm i\delta]}. \tag{6.78}$$

Here

$$K(s) = \left| \frac{2}{v_\perp} \int_0^s \Delta_0 \sin\eta \, ds' \right|. \tag{6.79}$$

The functions f and g have poles at $\epsilon = \epsilon_0(b)$, i.e., $\epsilon_0(b)$ is the energy of a bound state,

$$\epsilon_0(b) = C^{-1}(v_\perp) \int_0^\infty \Delta_0(s) \cos\eta \exp[-K(s)] \, ds, \tag{6.80}$$

where

$$C(v_\perp) = \int_0^\infty \exp[-K(s)] \, ds, \tag{6.81}$$

and

$$\cos\eta = \frac{b}{\rho}\left(1 + \frac{2es}{c}\int_0^s A_\phi \frac{ds'}{\rho'}\right). \tag{6.82}$$

The Green functions are

$$g^{R(A)} = \frac{iv_\perp e^{-K}}{2C\left[\epsilon - \epsilon_0 \pm i\delta\right]} \tag{6.83}$$

and $F_0 = F_0^\dagger = g$.

For small impact parameters, $b \ll \xi$, one has $\mid s \mid = \rho$. Therefore,

$$\begin{aligned}\epsilon_0(b) &= C^{-1}(v_\perp)b \\ &\quad\times \int_0^\infty \left[\left(\frac{1}{\rho} - \frac{2e}{c}A_\phi\right)\Delta_0 + \frac{ev_\perp H}{c}\right] e^{-K(\rho)}\, d\rho \\ &= C^{-1}(v_\perp)b\int_0^\infty \frac{\Delta_0}{\rho} e^{-K(\rho)}\, d\rho + \frac{\omega_c b p_\perp}{2}.\end{aligned} \tag{6.84}$$

where

$$\omega_c = \frac{|e|\, H}{mc}$$

is the cyclotron frequency. Modulus of charge appears due to the choice of the z axis.

In the quasiclassical approximation, the energy $\epsilon_0\,(b)$ is a continuous function of the impact parameter. However, the impact parameter is coupled to the angular momentum μ so that $b = -\mu/p_\perp$; the minus sign appears because a positive impact parameter corresponds to a negative angular momentum as can be seen from Fig. 6.1. The angular momentum is quantized such that $\mu = m + 1/2$ where m is an integer (de Gennes 1966). Therefore, the energy can be written in the form

$$\epsilon_0 = -\omega_0\left(m + \frac{1}{2}\right) \tag{6.85}$$

where

$$\omega_0 = p_\perp^{-1}\frac{\partial\epsilon_0}{\partial b} = p_\perp^{-1}C^{-1}(v_\perp)\int_0^\infty \frac{\Delta_0}{\rho}e^{-K(\rho)}\, d\rho + \frac{\omega_c}{2}. \tag{6.86}$$

For a superconductor with a large Ginzburg–Landau parameter λ_L/ξ, the contributions from H is small for fields $H \ll H_{c2}$. Indeed, the first term in eqn (6.86) is of the order of

$$\omega_0 \sim \frac{\Delta_\infty^2}{E_F}$$

which is much larger than the second term because $\omega_c/\omega_0 \sim H/H_{c2}$. Equation (6.85) coincides with the result by Caroli *et al.* (1964). Equation (6.86) with the account of magnetic field was obtained by Hansen (1968).

The energy spectrum (6.85) depends on two quantum numbers: the angular momentum and the momentum along the vortex axis. In principle, the spectrum

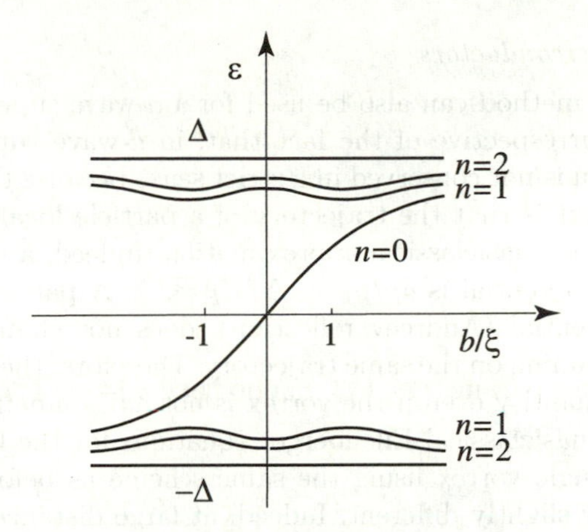

FIG. 6.2. Energy levels $\epsilon_n(b)$ as function of the impact parameter for various radial quantum numbers n. The anomalous branch with $n = 0$ crosses zero of energy and is odd in b. Other branches with $n \neq 0$ lie considerably higher than the anomalous branch.

should depend on three quantum numbers in a three-dimensional problem. The third is the radial quantum number. It does not appear in eqn (6.85) because the energies $\epsilon \ll \Delta_\infty$ are only accessible for particles having the radial quantum number zero. The levels with nonzero radial quantum numbers are located at energies of the order of Δ_∞ and cannot be calculated using this approach. Numerical calculations (Gygi and Schlüter 1991) show that states with $n \neq 0$ are practically indistinguishable from continuum $|\epsilon| > \Delta_\infty$. The general feature of levels with $n \neq 0$ is that they make the spectral branches which do not cross zero of energy as functions of μ as distinct from the branch with $n = 0$. The latter is chiral: it is odd in μ and crosses zero of energy. We shall see that it is of great importance for vortex dynamics. The energy spectrum $\epsilon_n(b)$ is shown schematically in Fig. 6.2.

6.4.1.1 *Density of states* Equation (6.83) can be used to calculate the density of states in the vortex core. Integrating the local density of states, eqn (5.43), over the vortex core area we find

$$\nu_s(\epsilon) = \frac{1}{2}\nu(0)\int_{-\infty}^{\infty}\left(g^R - g^A\right) ds\, db = \pi\nu(0)\, v_\perp \int_{-\infty}^{\infty}\delta(\epsilon - \epsilon_0)\, db$$
$$= \pi\nu(0)\, v_\perp \left(\frac{\partial \epsilon_0}{\partial b}\right)^{-1} = \frac{\pi\nu(0)}{m\omega_0} \tag{6.87}$$

for a given direction of the momentum with respect to the z-axis. Therefore, the density of states for low energies is finite

$$\nu_s(\epsilon) \sim \nu(0)\,\pi\xi_0^2.$$

It is independent of energy and has the same order of magnitude as the density of states in the normal region of the radius of the vortex core.

6.4.2 *d-wave superconductors*

The quasiclassical method can also be used for a d-wave superconductor within the same scheme irrespective of the fact that, in d-wave superconductors, the angular momentum is not conserved in a strict sense because the axial symmetry is broken. The point is that the trajectory of a particle localized in the core is a straight line in the quasiclassical approximation. Indeed, a momentum change due to the vortex potential is $\delta p/p_F \sim \Delta/E_F \ll 1$. A particle, being reflected by the vortex potential (Andreev reflection), does not change its momentum direction thus remaining on the same trajectory. Therefore, the impact parameter is a well defined quantity even if the vortex is not axisymmetric.

We solve the quasiclassical Eilenberger equations for the Green functions g, f, and $f^\dagger$ for a single vortex using the same scheme as before. The boundary conditions are now slightly different. Indeed, at large distances from the vortex axis, the Green function is

$$f^R = -i\frac{\Delta_{\mathbf{p}}}{\sqrt{|\Delta_{\mathbf{p}}|^2 - \tilde{\epsilon}^2}}. \tag{6.88}$$

For simplicity, we measure the angle α from one of the gap nodes such that $\Delta_{\mathbf{p}} = \Delta_0 \sin(2\alpha)\, e^{i\phi}$. In combination with eqn (6.74), it gives the boundary condition

$$\zeta \to 0;\ \theta \to \pm\mathrm{sign}\,[\sin(2\alpha)] \text{ for } s \to \pm\infty. \tag{6.89}$$

The solution of eqns (6.75) and (6.76) is

$$\zeta^{R(A)} = \frac{\mathrm{sign}\,[\sin(2\alpha)]\, v_\perp e^{-\tilde{K}}}{\int_{-\infty}^{\infty} [\epsilon - \Delta_0|\sin(2\alpha)|\cos\eta \pm i\delta]\, e^{-\tilde{K}}\, ds}$$

$$\theta^{R(A)} = 2\mathrm{sign}\,[\sin(2\alpha)] \times \left(\frac{\int_{-\infty}^{s} [\epsilon - \Delta_0|\sin(2\alpha)|\cos\eta]\, e^{-\tilde{K}}\, ds'}{\int_{-\infty}^{\infty} [\epsilon - \Delta_0|\sin(2\alpha)|\cos\eta \pm i\delta]\, e^{-\tilde{K}}\, ds} - \frac{1}{2}\right)$$

where

$$\tilde{K} = \frac{2}{v_\perp}\int_0^s \Delta_0|\sin(2\alpha)|\sin\eta\, ds'. \tag{6.90}$$

As a result,

$$g^{R(A)} = \frac{iv_\perp e^{-\tilde{K}}}{\tilde{C}\,[\epsilon - \epsilon_0 \pm i\delta]} \tag{6.91}$$

where now

$$\epsilon_0(\alpha, b) = \tilde{C}^{-1}|\sin(2\alpha)| \int_{-\infty}^{\infty} \Delta_0 \cos\eta e^{-\tilde{K}}\, ds \tag{6.92}$$

with

$$\tilde{C} = \int_{-\infty}^{\infty} e^{-\tilde{K}}\, ds. \tag{6.93}$$

For impact parameters, $b \ll \xi$, one has $|s| = \rho$. Therefore,

$$\epsilon_0(b) = b\left(\tilde{C}^{-1}|\sin(2\alpha)|\int_0^\infty \Delta_0 e^{-\tilde{K}(\rho)}\frac{d\rho}{\rho} + p_\perp\omega_c/2\right). \tag{6.94}$$

Equation (6.94) can be simplified for the momentum directions near the gap nodes, where $\sin(2\alpha) \ll 1$. Significant values of $\tilde{K}$ are determined by $\rho \gg \xi$. Therefore, one can put $\Delta_0 = \Delta_\infty$:

$$\begin{aligned}\tilde{K}(s) &= \frac{2\Delta_\infty|\sin(2\alpha)|}{v_\perp}\int_0^s \frac{s'\,ds'}{\rho'}\\ &= \frac{2\Delta_\infty|\sin(2\alpha)|}{v_\perp}(\rho - b).\end{aligned}$$

The normalization constant becomes

$$\tilde{C} = 2e^\gamma \int_b^\infty e^{-\gamma\rho/b}\frac{\rho\,d\rho}{\sqrt{\rho^2 - b^2}} = be^\gamma K_1(\gamma) \tag{6.95}$$

where K_n is the Bessel function of an imaginary argument with $\gamma = 2\Delta_\infty b \times|\sin(2\alpha)|/v_\perp$. The energy takes the form

$$\epsilon_0(b) = \Delta_0|\sin 2\alpha|\frac{K_0(\gamma)}{K_1(\gamma)} + bp_\perp\omega_c/2. \tag{6.96}$$

As a result, for $\xi \ll b \ll \xi/|\sin 2\alpha|$, we get

$$\epsilon_0(\alpha, b) = \frac{2\Delta_\infty^2\sin^2(2\alpha)Lb}{v_\perp} + \frac{p_\perp b\omega_c}{2} \tag{6.97}$$

where $L = \ln[1/|\sin(2\alpha)|]$ for $b \ll \xi$ and $L = \ln[\xi v_\perp/bv_F|\sin(2\alpha)|]$ for $\xi \ll b \ll \xi/|\sin(2\alpha)|$. The states with energies much below the gap at infinity, $\epsilon \ll \Delta_0|\sin 2\alpha|$ correspond to $b \ll \xi/|\sin 2\alpha|$.

Equation (6.97) defines the energy of a particle that moves along the trajectory passing by a vortex at an impact parameter b. However, it is not the true quantum mechanical state because its energy depends on the momentum direction α. For a fixed energy, the particle trajectory starts to precess. We can find the true quantum states for low energies using the semiclassical Bohr–Sommerfeld quantization rule (Kopnin and Volovik 1997). Let us define the angular momentum as $\mu = -b/p_\perp$. The angular momentum $\mu(\alpha)$ is expressed through the quasiparticle energy via $E = \epsilon_0(\alpha, b)$ which determines $\mu(\alpha)$ as a function of α. The quantization rule requires

$$\oint \mu(\alpha)\,d\alpha = 2\pi\left(m + \frac{1}{2}\right) \tag{6.98}$$

where m is an integer, $\frac{1}{2}$ appears because the single-particle wave function changes its sign after encircling a single-quantum vortex.

Consider an energy $\epsilon \ll \Delta_\infty \sqrt{H/H_{c2}}$. We have from eqn (6.97)

$$-\oint \frac{E\, d\alpha}{\omega_0(\alpha)} = 2\pi\left(m + \frac{1}{2}\right) \tag{6.99}$$

where

$$\omega_0(\alpha) = 8\tilde{\omega}_0\alpha^2 + \omega_c/2 \tag{6.100}$$

with $\tilde{\omega}_0 = (\Delta_\infty^2/v_\perp p_\perp)L$. The integral (6.99) converges and is determined by angles $\alpha \sim \sqrt{\omega_c E_F/\Delta_\infty^2} \sim \sqrt{H/H_{c2}}$. The characteristic impact parameters are of the order of $b \sim \epsilon/p_\perp\omega_c$, i.e., $b \ll r_0$ where $r_0 \sim \xi\sqrt{H_{c2}/H}$ is of the order of the distance between vortices. We obtain

$$E_0 = -\Omega_0(m + 1/2) \tag{6.101}$$

where

$$\Omega_0 = \sqrt{\tilde{\omega}_0\omega_c}. \tag{6.102}$$

The logarithmic factor in $\tilde{\omega}$ becomes

$$L = \ln[(\Delta_\infty/\epsilon_0)\sqrt{H/H_{c2}}].$$

The states defined by eqn (6.101) have much smaller interlevel spacing compared to those in an s-wave superconductor, eqn (6.85). Moreover, we see that only the states with very low energies $\epsilon \ll \Delta_\infty\sqrt{H/H_{c2}}$ are truly localized while the states with energies $\epsilon \sim \Delta_\infty\sqrt{H/H_{c2}}$ and above are not localized in a strict quantum mechanical sense. Indeed, these particles spend most of their time having such momentum directions $\alpha \sim \sqrt{H/H_{c2}}$ that their energies are above the gap along this direction $|\Delta_{\mathbf{p}}| \sim \Delta_\infty\sqrt{H/H_{c2}}$. Thus a particle with an energy $\epsilon \sim \Delta_\infty\sqrt{H/H_{c2}}$ escapes from the vortex core along the gap nodes.

The singular behavior of the energy spectrum in the directions of gap nodes results in a nontrivial magnetic field dependence of the density of states in the vortex core. Indeed, eqn (6.87) determines now a density of states which depends not only on p_z but also on the momentum direction within the ab plane. Averaging it over the angles α we obtain

$$\langle\nu(\epsilon)\rangle = \frac{\nu(0)\, v_\perp}{2p_\perp}\int_0^{2\pi}\frac{d\alpha}{\omega_0(\alpha)} = \frac{\pi\nu(0)}{m\Omega_0}. \tag{6.103}$$

The density of states per vortex is considerably larger than in an s-wave case. It can be understood as an increase in the effective size of the vortex core due to a larger extension of the wave function in the direction of the gap nodes. Moreover, the density of states depends on the magnetic field $\langle\nu(\epsilon)\rangle \propto \sqrt{H_{c2}/H}$. Multiplied with the number of vortices, it gives an energy-independent density of states proportional to $\sqrt{H/H_{c2}}$. This behavior of the low-energy density of states in d-wave superconductors in presence of vortices was first predicted by Volovik (1988, 1993).

7

QUASICLASSICAL METHOD FOR LAYERED SUPERCONDUCTORS

We generalize the quasiclassical scheme for layered superconductors. The Ginzburg–Landau theory for layered superconductors known as the Lawrence–Doniach model is derived. Using this model we discuss the behavior of the upper critical field and investigate interaction of vortices with the underlying crystalline structure (intrinsic pinning).

7.1 Quasiclassical Green functions

The quasiclassical method described in Chapter 5 can be directly applied to anisotropic superconductors, as well as to two-dimensional systems. However, its generalization for quasi-two-dimensional (layered) superconductors is more complicated. The difficulty arises because a quasi-two-dimensional system, in fact, is not fully quasiclassical: the quasiclassical approximation is not fulfilled when the coherence length in the direction perpendicular to the layers is comparable with the distance between the layers. Nevertheless, if the coherence length in other directions is still larger than the interatomic distance within the layers, one can construct a generalization of the above approach which can sometimes be used successfully for reducing the full Green function technique to a more manageable form.

Consider a system which consists of layers with a good conductivity in the crystallographic *ab* plane; these layers are stacked along the *c*-axis with a weak conductivity in the *c* direction. A strong anisotropy of conductivity in a metallic state is associated with a very anisotropic normal-state electronic spectrum which can be written in the form

$$\xi_p = E_n^{(2)}(\mathbf{p}) + \psi(p_z) - E_F \tag{7.1}$$

where $\psi \ll E_n^{(2)}$. The bold letters in this chapter denote two-dimensional vectors in the *ab* plane, and p_z is the momentum along the *c* axis chosen as the z coordinate axis. The function ψ can also depend on $\mathbf{p}$. If the distance between the layers is s, the energy $\psi(p_z)$ is periodic in p_z with the period $2\pi/s$. A small energy $\psi(p_z)$ is usually associated with a narrow energy band produced by a weak tunneling between the conducting layers. It is thus natural to assume the spectrum in the form

$$\psi(p_z) = J\cos(p_z s). \tag{7.2}$$

Characteristic superconducting wave vectors along the c axis are $k_z \sim \xi_c^{-1}$ where ξ_c is the coherence length in the c direction. If it were $\xi_c \gg s$, we would have $k_z \ll p_z$ and the energy $\psi(p_z + k_z)$ could be expanded as

$$\psi(p_z + k_z) = \psi(p_z) + \frac{\partial \psi}{\partial p_z} k_z = \psi(p_z) + v_{Fz}(p_z) k_z.$$

This would allow us to determine the Fermi momentum $(\mathbf{p}_F, p_{Fz})$ from the condition $E_n^{(2)}(\mathbf{p}) + \psi(p_z) - E_F = 0$ and define the quasiclassical Green functions in a usual way as

$$\check{g}(\mathbf{p}_F, p_{Fz}; \mathbf{k}, k_z) = \oint \check{G}(\mathbf{p}, p_z; \mathbf{p} - \mathbf{k}, p_z - k_z) \frac{d\xi_p}{\pi i}.$$

For such definition it is very important that all the components of both $(\mathbf{p}, p_z)$ and $(\mathbf{p} - \mathbf{k}, p_z - k_z)$ are close to the corresponding components of the Fermi momentum $(\mathbf{p}_F, p_{Fz})$. One then returns to the usual quasiclassical scheme with an anisotropic Fermi surface.

The situation changes when $\xi_c \sim s$. Now one can not determine a three-dimensional Fermi momentum in such a way that both $(\mathbf{p}, p_z)$ and $(\mathbf{p} - \mathbf{k}, p_z - k_z)$ are close to the corresponding components $(\mathbf{p}_F, p_{Fz})$. Nevertheless, one can introduce a *two-dimensional* Fermi momentum $\mathbf{p}_F$ through the condition $\xi_{\mathbf{p}}^{(2)} = 0$ where

$$\xi_{\mathbf{p}}^{(2)} = E_n(\mathbf{p}_F) - E_F$$

such that the momenta $\mathbf{p}$ and $\mathbf{p} - \mathbf{k}$ are close to $\mathbf{p}_F$. The Fermi momentum $\mathbf{p}_F$ belongs to a two-dimensional (cylindrical) Fermi surface and is specified by an angle α in the ab plane. Let us define the "quasiclassical" Green function as

$$\check{g}(\mathbf{p}_F, \mathbf{k}; p_z, p_z') = \oint \check{G}(\mathbf{p}, p_z; \mathbf{p} - \mathbf{k}, p_z') \frac{d\xi_{\mathbf{p}}^{(2)}}{\pi i}.$$

The energy integration is carried out along the normal

$$\mathbf{v}_F = \frac{\partial \xi_{\mathbf{p}}^{(2)}}{\partial \mathbf{p}}$$

to the cylindrical Fermi surface. The function $\check{g}$ depends on the Fermi-momentum direction α, on a two-dimensional center-of-mass coordinate $\mathbf{r}$ through the wave vector $\mathbf{k}$, and on two momenta p_z and p_z' in the z direction. The difference from the traditional quasiclassical Green function is that the dependence on the center-of-mass coordinate $(z + z')/2$ through $p_z - p_z'$ no longer separates from the relative momentum dependence $(p_z + p_z')/2$. This is the price which one has to pay for the fact that the interlayer distance is comparable with the coherence length ξ_c.

The momentum-space integral becomes

$$\int \frac{d^3p}{(2\pi)^3} = s\nu^{(2)} \int d\xi_{\mathbf{p}}^{(2)} \int_{-\pi/s}^{\pi/s} \frac{dp_z}{2\pi} \int \frac{d\alpha}{2\pi}.$$

Here $\nu^{(2)}$ is the two-dimensional density of states. It is

$$\nu^{(2)} = \frac{m}{2\pi s}$$

for the spectrum $E_n^{(2)} = \mathbf{p}^2/2m$.

The order parameter equation (5.27) for the Matsubara representation becomes

$$\frac{\Delta(\mathbf{k}, k_z)}{\lambda^{(2)}} = \pi i T \sum_j \int_{-\pi/s}^{\pi/s} \frac{s\,dp_z}{2\pi} \int \frac{d\alpha}{2\pi} f_{\omega_j}(\mathbf{p}_F, \mathbf{k}; p_+, p_-) \tag{7.3}$$

where the interaction constant $\lambda^{(2)} = \nu^{(2)}\,|g|$ and

$$p_\pm = p_z \pm \frac{k_z}{2}.$$

In the real-frequency representation, one has

$$\frac{\Delta(\mathbf{k}, k_z)}{\lambda^{(2)}} = \int_{-\pi/s}^{\pi/s} \frac{s\,dp_z}{2\pi} \int \frac{d\alpha}{2\pi} \int \frac{d\epsilon}{4} \tanh\left(\frac{\epsilon}{2T}\right) \times \left[f_\epsilon^R(\mathbf{p}_F, \mathbf{k}; p_+, p_-) - f_\epsilon^A(\mathbf{p}_F, \mathbf{k}; p_+, p_-)\right]. \tag{7.4}$$

The current in the ab plane has its usual form

$$\mathbf{j}(\mathbf{k}, k_z) = -2e\nu^{(2)}\pi i T \sum_j \int_{-\pi/s}^{\pi/s} \frac{s\,dp_z}{2\pi} \int \frac{d\alpha}{2\pi} \mathbf{v}_F\, g_{\omega_j}(\mathbf{p}_F, \mathbf{k}; p_+, p_-)$$

for the Matsubara representation or

$$\mathbf{j}(\mathbf{k}, k_z) = -2e\nu^{(2)} \int \frac{d\epsilon}{4} \tanh\left(\frac{\epsilon}{2T}\right) \int_{-\pi/s}^{\pi/s} \frac{s\,dp_z}{2\pi} \int \frac{d\alpha}{2\pi} \mathbf{v}_F \times \left[g_\epsilon^R(\mathbf{p}_F, \mathbf{k}; p_+, p_-) - g_\epsilon^A(\mathbf{p}_F, \mathbf{k}; p_+, p_-)\right] \tag{7.5}$$

for the real-frequency representation.

The current in the c direction is defined as follows

$$j_z(\mathbf{k}, k_z) = -2e\nu^{(2)} \int \frac{d\epsilon}{4} \tanh\left(\frac{\epsilon}{2T}\right) \int_{-\pi/s}^{\pi/s} \frac{s\,dp_z}{2\pi} \int \frac{d\alpha}{2\pi} V_z \times \left[g_\epsilon^R(\mathbf{p}_F, \mathbf{k}; p_+, p_-) - g_\epsilon^A(\mathbf{p}_F, \mathbf{k}; p_+, p_-)\right] \tag{7.6}$$

where

$$V_z = \frac{1}{2}\left[\frac{\partial\psi(p_+ - eA_z/c)}{\partial p_z} + \frac{\partial\psi(p_- - eA_z/c)}{\partial p_z}\right].$$

This definition complies with the condition of minimum of the thermodynamic potential $\delta\Omega/\delta A_z = 0$ where the variation of Ω is expressed through the Green function according to eqn (5.30). For a layered system, the Hamiltonian should be taken in the form

$$\check{H}(\mathbf{r},z) = \check{H}^{(2)}(\mathbf{r},z) + \psi\left(p_z - \frac{e}{c}A_z\check{\tau}_3\right) \tag{7.7}$$

where

$$\check{H}^{(2)}(\mathbf{r},z) = \begin{pmatrix} -(e/c)\,\mathbf{v}_F\mathbf{A}(\mathbf{r},z) & -\Delta(\mathbf{r},z) \\ \Delta^*(\mathbf{r},z) & (e/c)\,\mathbf{v}_F\mathbf{A}(\mathbf{r},z) \end{pmatrix}. \tag{7.8}$$

7.2 Eilenberger equations for layered systems

Equations for the retarded and advanced (or Matsubara) quasiclassical Green functions can be obtained in a way similar to that used earlier in Section 5.5. It is more convenient to formulate them for the functions in the z-coordinate representation

$$\check{g}(\mathbf{p}_F,\mathbf{r};z,z') = \int \check{g}(\mathbf{p}_F,\mathbf{k};p_z,p_z')\,e^{i\mathbf{kr}+ip_z z-ip_z' z'}\frac{d^3k}{(2\pi)^3}\frac{dp_z dp_z'}{(2\pi)^2}.$$

We obtain for the real-frequency functions

$$\begin{aligned}
&\left[-i\mathbf{v}_F\frac{\partial}{\partial\mathbf{r}} + \psi\left(\hat{p}_z - \frac{e}{c}A_z\check{\tau}_3\right) - \psi\left(\hat{p}_{z'}^{\dagger} - \frac{e}{c}A_{z'}\check{\tau}_3\right)\right]\check{g}_\epsilon(\mathbf{p}_F,\mathbf{r};z,z') \\
&-\epsilon\left[\check{\tau}_3\check{g}_\epsilon(\mathbf{p}_F,\mathbf{r};z,z') - \check{g}_\epsilon(\mathbf{p}_F,\mathbf{r};z,z')\check{\tau}_3\right] \\
&+\check{H}^{(2)}(\mathbf{r},z)\,\check{g}_\epsilon(\mathbf{p}_F,\mathbf{r};z,z') - \check{g}_\epsilon(\mathbf{p}_F,\mathbf{r};z,z')\,\check{H}^{(2)}(\mathbf{r},z') \\
&-\check{\Sigma}_\epsilon(\mathbf{r},z)\,\check{g}_\epsilon(\mathbf{p}_F,\mathbf{r};z,z') + \check{g}_\epsilon(\mathbf{p}_F,\mathbf{r};z,z')\,\check{\Sigma}_\epsilon(\mathbf{r},z') = 0.
\end{aligned} \tag{7.9}$$

Here $\check{g}$ stands for $\check{g}^{R(A)}$,

$$\hat{p}_z = -i\frac{\partial}{\partial z},\ \hat{p}_{z'}^{\dagger} = i\frac{\partial}{\partial z'},$$

and the z component of the vector potential $A_{z'}$ is a function of $(\mathbf{r},z')$. The effective Hamiltonian is determined by eqn (7.8). It is a two-dimensional part of the "full" Hamiltonian of the type of eqn (5.57) used to derive the Eilenberger equations for the three-dimensional case in Section 5.5. $\check{\Sigma}$ is the self-energy. For example, for isotropic impurity scattering it is

$$\check{\Sigma} = n_{\mathrm{imp}}u^2\int \check{G}\frac{d^3p}{(2\pi)^3} = \pi i\nu^{(2)}u^2 n_{\mathrm{imp}}\int \check{g}\frac{s\,dp_z}{2\pi}\frac{d\alpha}{2\pi} = \frac{i}{2\tau}\int \check{g}\frac{s\,dp_z}{2\pi}\frac{d\alpha}{2\pi}.$$

Let us derive now the normalization condition for quasiclassical Green functions in layered superconductors using the same approach as we did for a three-dimensional case on page 93. Multiplying eqn (7.9) with $\check{g}(\mathbf{p}_F,\mathbf{r};z,z')$ from the left and adding to the equation multiplied from the right we obtain that the combination

$$\int \check{g}_\epsilon^{R(A)}(\mathbf{p}_F,\mathbf{r};z,z'')\,\check{g}_\epsilon^{R(A)}(\mathbf{p}_F,\mathbf{r};z'',z')\,dz''$$

again satisfies eqn (7.9). Now we employ the same argumentation as for the three-dimensional case. Let us assume that, at large distances (in the direction of the *ab* planes) from the region under consideration, the order parameter becomes independent of coordinates, and the magnetic field vanishes. We have in this case

$$g(p_z, p_z') = \frac{\epsilon}{\sqrt{\epsilon^2 - \Delta^2}} 2\pi\delta(p_z - p_z'),\ f(p_z, p_z') = \frac{\Delta}{\sqrt{\epsilon^2 - \Delta^2}} 2\pi\delta(p_z - p_z').$$

It follows that, in the momentum representation, one has at large distances

$$\int \check{g}^{R(A)}(\mathbf{p}_F, \mathbf{r}; p_z, p_z'')\, \check{g}^{R(A)}(\mathbf{p}_F, \mathbf{r}; p_z'', p_z') \frac{dp_z''}{2\pi} = \check{1} \cdot 2\pi\delta(p_z - p_z'). \quad (7.10)$$

Since the expression in eqn (7.10) is an integral of motion, we find that eqn (7.10) holds also everywhere including the region where Δ has its actual spatial dependence, and the magnetic field is finite. Equation (7.10) in the coordinate representation reads

$$\int \check{g}_\epsilon^{R(A)}(\mathbf{p}_F, \mathbf{r}; z, z'')\, \check{g}_\epsilon^{R(A)}(\mathbf{p}_F, \mathbf{r}; z'', z')\, dz'' = \check{1}\delta(z - z'). \quad (7.11)$$

Eilenberger equations (7.9) and the normalization eqn (7.10) are not local in the center-of-mass coordinate $(z + z')/2$. This reduces considerably the potentialities of the quasiclassical methods for quasi-two-dimensional systems.

A layered superconductor can also be described by introducing the coupling between two-dimensional layers via mechanisms other than the electron tunneling between layers based on eqn (7.1). For example, an interaction between layers which, by themselves, form a "good" two-dimensional quasiclassical environment can be mediated via scattering by impurities (Graf *et al.* 1993).

7.3 Lawrence–Doniach model

One important example when the Eilenberger equations (7.9) can be solved is the Lawrence–Doniach model (Lawrence and Doniach 1971) which is an analogue of the Ginzburg–Landau theory for layered superconductors. The basic assumptions are as follows. First, it is required that the temperature is close to the critical temperature T_c. This allows expansion of the Green function in powers of Δ/T_c and in slow gradients in the *ab* plane in the same way as was done to derive the usual Ginzburg–Landau theory in Section 6.1.2. The second requirement is that the corrugation of the Fermi surface is small compared to the critical temperature, $\psi \ll T_c$. We shall also assume that the vector potential varies slowly at distances of the order of ξ_0 and s which is the case for $T \to T_c$. The approach which we describe below applies to either *s*-wave or *d*-wave pairing: In these cases the pairing interaction is independent of the momentum p_z, and the order parameter only depends on the center-of-mass coordinate $(z + z')/2$ through $p_z - p_z' = k_z$, i.e. $\Delta = \Delta(k_z)$. We shall see that this form of the p_z dependence is essential: the situation would be different if Δ depended on p_z and p_z' separately. For simplicity, we consider here only the *s*-wave case.

7.3.1 *Order parameter*

Assume an isotropic scattering by impurities and solve the Eilenberger equation in the momentum representation with respect to the coordinate z using the perturbation expansion in Δ/T_c and ψ/T_c:

$$g = g_0 + g_1; \; f = f_0 + f_1 + f_2.$$

Denote

$$\hat{L} = -i\mathbf{v}_F\left(\nabla - \frac{2ie}{c}\mathbf{A}\right) + \psi\left(p_+ - \frac{e}{c}A_z\right) - \psi\left(p_- + \frac{e}{c}A_z\right).$$

We find using the normalization eqn (7.10)

$$g_0^{R(A)}(p_+, p_-) = \pm 2\pi\delta(k_z), \tag{7.12}$$

$$g_1^{R(A)}(p_+, p_-) = \frac{|\Delta|^2}{2\epsilon^2} + \frac{\Delta(k_z')}{2\epsilon} f_1^{\dagger R(A)}(p_+ - k_z', p_-) + f_1^{R(A)}(p_+, p_- + k_z')\frac{\Delta(k_z')}{2\epsilon}, \tag{7.13}$$

and

$$f_0^{R(A)}(p_+, p_-) = \pm\frac{\Delta(k_z)}{\epsilon}, \tag{7.14}$$

$$f_1^{R(A)}(p_+, p_-) = \frac{\hat{L}f_0^{R(A)} \pm (i/2\tau\epsilon)\left\langle \hat{L}f_0^{R(A)}\right\rangle}{2\epsilon \pm i/\tau}, \tag{7.15}$$

$$f_2^{R(A)}(p_+, p_-) = \frac{\hat{L}f_1^{R(A)} \pm (i/2\tau\epsilon)\left\langle \hat{L}f_1^{R(A)}\right\rangle}{2\epsilon \pm i/\tau} \pm \frac{|\Delta|^2\Delta}{2\epsilon^3}. \tag{7.16}$$

The average over the Fermi surface for a layered system is defined as

$$\langle \cdots \rangle = \int \frac{s dp_z}{2\pi}\frac{d\alpha}{2\pi}(\cdots).$$

We obtain

$$\left\langle f^{R(A)}\right\rangle = \pm\left[\frac{\Delta}{\epsilon} + \frac{\left\langle \hat{L}^2\right\rangle \Delta}{2\epsilon^2(2\epsilon \pm i/\tau)} + \frac{|\Delta|^2\Delta}{2\epsilon^3}\right]. \tag{7.17}$$

We assume here that the spectrum has the inversion symmetry $\psi(p_z) = \psi(-p_z)$ such that $\left\langle \hat{L}\right\rangle = 0$.

Consider for simplicity an uniaxial superconductor isotropic in the *ab* plane. The order parameter equation (7.4) yields

$$\left[\frac{T_c - T}{T_c}\Delta - \frac{7\zeta(3)}{8\pi^2 T_c^2}|\Delta|^2\Delta + \frac{\pi v_F^2 \tau\Lambda}{16 T_c}\left(\nabla - \frac{2ie}{c}\mathbf{A}\right)^2\Delta\right]_{k_z}$$

$$-\frac{\pi\tau\Lambda}{8T_c}\int\frac{s\,dp_z}{2\pi}\left[\psi\left(p_+-\frac{e}{c}A_z\right)-\psi\left(p_-+\frac{e}{c}A_z\right)\right]^2\Delta\left(k_z\right)=0, \quad (7.18)$$

where $\Lambda\left(T_c\tau\right)$ is defined by eqn (6.22). The first line is a Fourier transform at the momentum k_z.

In the z-coordinate representation, the order parameter is determined only on the conducting layers. We specify the layers with the number n such that $z_n = ns$. We have

$$\Delta_n\left(\mathbf{r}\right)=s\int_{-\pi/s}^{\pi/s}\Delta\left(k_z,\mathbf{r}\right)e^{ik_zsn}\frac{dk_z}{2\pi};\ \Delta\left(k_z,\mathbf{r}\right)=\sum_n\Delta_n\left(\mathbf{r}\right)e^{-ik_zsn}. \quad (7.19)$$

Equations (7.18) and (7.19) completely determine the order parameter in a layered superconductor. However, eqn (7.18) can be simplified further using eqn (7.2). We have

$$\begin{aligned}
&\int_{-\pi/s}^{\pi/s}\frac{s\,dp_z}{2\pi}\left[\psi\left(p_z+\frac{k_z}{2}-\frac{e}{c}A_z\right)-\psi\left(p_z-\frac{k_z}{2}+\frac{e}{c}A_z\right)\right]^2\\
&=4J^2\sin^2\left(\frac{sk_z-\chi_A}{2}\right)\int_{-\pi/s}^{\pi/s}\frac{s\,dp_z}{2\pi}\sin^2\left(p_zs\right)\\
&=2J^2\sin^2\left(\frac{sk_z-\chi_A}{2}\right)
\end{aligned}$$

where

$$\chi_A=\frac{2e}{c}\int_{ns}^{(n+1)s}A_zdz=\frac{2e}{c}\int_{(n-1)s}^{ns}A_zdz.$$

We use the fact that the vector potential varies slowly at distances of the order of the interlayer distance. One has

$$\begin{aligned}
\sin^2\left(\frac{sk_z-\chi_A}{2}\right)\Delta\left(k_z\right)&=-\frac{1}{4}\sum_n\Delta_ne^{-ik_zsn}\left[e^{i(k_zs-\chi_A)}+e^{-i(k_zs-\chi_A)}-2\right]\\
&=-\frac{1}{4}\sum_n\left[\Delta_{n+1}e^{-i\chi_A}+\Delta_{n-1}e^{i\chi_A}-2\Delta_n\right]e^{-ik_zsn}.
\end{aligned}$$

Equation (7.18) in the coordinate representation becomes

$$\begin{aligned}
&\alpha\Delta_n+\beta\left|\Delta_n\right|^2\Delta_n-\gamma_{ab}\left(\nabla-\frac{2ie}{c}\mathbf{A}\right)^2\Delta_n\\
&-\frac{\gamma_c}{s^2}\left(\Delta_{n+1}e^{-i\chi_A}+\Delta_{n-1}e^{+i\chi_A}-2\Delta_n\right)=0 \qquad (7.20)
\end{aligned}$$

where

$$\alpha=\nu^{(2)}\frac{T-T_c}{T_c},\beta=\frac{7\zeta\left(3\right)\nu^{(2)}}{8\pi^2T_c^2},$$

$$\gamma_{ab} = \frac{\pi \nu^{(2)} v_F^2 \tau \Lambda}{16 T_c}, \gamma_c = \frac{\pi \nu^{(2)} J^2 s^2 \tau \Lambda}{16 T_c}.$$

Equation (7.20) is the Lawrence–Doniach equation for layered superconductors. It has been derived microscopically by Klemm *et al.* (1975).

As in the three-dimensional case, we put

$$\xi_0 = \frac{v_F}{2\pi T_c}.$$

The temperature-dependent coherence length in the *ab* plane is

$$\xi_{ab}^2 (T) = \frac{\gamma_{ab}}{|\alpha|} = (\xi_0 \ell) \frac{\pi^2 \Lambda}{8} \frac{T_c}{T_c - T}.$$

The last line of eqn (7.20) is the discrete analogue of the second derivative. In the continuous limit, $s \to 0$, we can write

$$\Delta_{n+1} e^{-i\chi_A} + \Delta_{n-1} e^{+i\chi_A} - 2\Delta_n = s^2 \left(\frac{\partial}{\partial z} - \frac{2ie}{c} A_z \right)^2 \Delta (z).$$

Comparing this with the first term in the square brackets, we can define the coherence length ξ_c through the ratio

$$\frac{\xi_c^2}{\xi_{ab}^2} = \frac{\gamma_c}{\gamma_{ab}} = \frac{J^2 s^2}{v_F^2}.$$

In the continuous limit, $s \ll \xi_c$, eqn (7.20) transforms into the Ginzburg–Landau equation for anisotropic uniaxial superconductor

$$\alpha \Delta + \beta |\Delta|^2 \Delta - \gamma_{ab} \left(\nabla - \frac{2ie}{c} \mathbf{A} \right)^2 \Delta - \gamma_c \left(\frac{\partial}{\partial z} - \frac{2ie}{c} A_z \right)^2 \Delta = 0. \quad (7.21)$$

It has exactly the same form as eqn (6.68) for an anisotropic *d*-wave superconductor.

7.3.2 *Free energy and the supercurrent*

Equations (7.13) and (7.16) allow us to reconstruct the Lawrence–Doniach free energy. Using eqn (3.71) or (5.30) with the definition eqn (7.7) we find

$$\mathcal{F}_{sn} = s \sum_n \int d^2 r \left[\alpha |\Delta_n|^2 + \frac{\beta}{2} |\Delta_n|^4 \right.$$
$$\left. + \gamma_{ab} \left| \left(\nabla - \frac{2ie}{c} \mathbf{A} \right) \Delta_n \right|^2 + \frac{\gamma_c}{s^2} \left| \Delta_{n+1} e^{-i\chi_A} - \Delta_n \right|^2 \right]. \quad (7.22)$$

Equation (7.20) is reproduced when one calculates the variation

$$\frac{\delta \mathcal{F}_{sn}}{\delta \Delta_n^*} = 0.$$

The supercurrent can be found using the definition

$$\mathbf{j}_s = -c\frac{\delta \mathcal{F}_{sn}}{\delta \mathbf{A}}.$$

The supercurrent density in the ab direction at the position $z_n = ns$ of the n-th layer has the usual form

$$\mathbf{j}_s = -2ie\gamma_{ab}\left[\Delta_n^*\left(\nabla - \frac{2ie}{c}\mathbf{A}\right)\Delta_n - \Delta_n\left(\nabla + \frac{2ie}{c}\mathbf{A}\right)\Delta_n^*\right]. \tag{7.23}$$

The current density in the c direction at the n-th layer becomes

$$j_s^{(c)} = -\frac{ie\gamma_c}{s}\left[\left(\Delta_{n+1}\Delta_n^* + \Delta_n\Delta_{n-1}^*\right)e^{-i\chi_A} - \left(\Delta_{n+1}^*\Delta_n + \Delta_n^*\Delta_{n-1}\right)e^{i\chi_A}\right] \tag{7.24}$$

or

$$j_s^{(c)} = \frac{2e\gamma_c}{s}\left[|\Delta_{n+1}\Delta_n|\sin\left(\chi_{n+1} - \chi_n - \frac{2es}{c}A_z\right) + |\Delta_n\Delta_{n-1}|\sin\left(\chi_n - \chi_{n-1} - \frac{2es}{c}A_z\right)\right]. \tag{7.25}$$

This expression corresponds to the Josephson current which is produced by Cooper pair tunneling between the layers.

For a highly layered case, when $s \sim \xi_c$, the maximum Josephson current in eqn (7.25) is

$$j_{max} = 4e\gamma_c|\Delta|^2/s.$$

It enters the definition of the penetration length λ_c for currents flowing in the c direction. Indeed, for small fields, one can expand eqn (7.25) in a small argument to get

$$\begin{aligned} j_s^{(c)} &= -\frac{8e^2\gamma_c|\Delta|^2}{c}\left[A_z - \frac{c}{2es}\left(\chi_{n+1} - \chi_n\right)\right] \\ &= -\frac{2ej_{max}s}{c}\left[A_z - \frac{c}{2es}\left(\chi_{n+1} - \chi_n\right)\right]. \end{aligned}$$

Together with the Maxwell equation

$$j_s^{(c)} = \frac{c}{4\pi}\operatorname{curl}\operatorname{curl} A_z$$

this determines

$$\lambda_c^{-2} = \frac{8\pi^2 s j_{max}}{c\Phi_0} \tag{7.26}$$

which is called the Josephson length. If the Josephson coupling is small the penetration length λ_c^2 diverges.

When the order parameter is normalized in such a way that it is the wave function Ψ of superconducting electrons, the Lawrence–Doniach free energy equation (7.22) becomes

$$\mathcal{F}_{sn} = s \int d^2r \sum_n F(n) \tag{7.27}$$

where the energy density at a layer n is

$$F(n) = a\left|\Psi(n, \mathbf{r})\right|^2 + \frac{b}{2}\left|\Psi(n, \mathbf{r})\right|^4 + \frac{1}{2m_{ab}}\left|\left(\nabla - \frac{2ie}{c}\mathbf{A}\right)\Psi(n, \mathbf{r})\right|^2$$
$$+\frac{1}{2m_c s^2}\left|\Psi(n+1, \mathbf{r})e^{-i\chi_A} - \Psi(n, \mathbf{r})\right|^2. \tag{7.28}$$

This is similar to eqn (1.27). The constants a and b satisfy

$$\frac{|a|}{b} = \frac{m_{ab}c^2}{16\pi e^2 \lambda_{ab}^2} = |\Psi_{GL}|^2$$

where λ_{ab} is the London penetration length for currents flowing in the ab plane. The order parameter magnitude is

$$|\Psi_{GL}|^2 = 2m_{ab}\gamma_{ab}\Delta_\infty^2. \tag{7.29}$$

The coherence lengths are expressed through the "effective masses"

$$\xi_{ab}^2 = \frac{1}{2m_{ab}|a|}\,,\ \xi_c^2 = \frac{1}{2m_c|a|}. \tag{7.30}$$

The effective mass in the ab plane for clean superconductors $\ell \gg \xi_{ab}$ has the usual meaning

$$m_{ab}^{-1} = \partial^2 E_n^{(2)}/\partial p^2.$$

The "effective mass" m_c, however, can be written as $\left(\partial^2\psi/\partial p_z^2\right)^{-1}$ only in the continuous limit $\xi_c \gg s$.

The Lawrence–Doniach equations can also be derived using the model by Graf *al.* (1993) mentioned in the beginning of this section. In this case, the effective mass m_c is expressed through the interlayer diffusion.

7.3.3 *Microscopic derivation of the supercurrent*

The expression for the supercurrent can also be obtained microscopically using our results for the Green functions eqn (7.13). The current along the ab plane has a standard form and can be obtained exactly in the same way as it was done for the usual Ginzburg–Landau equation in Section 6.1.2. The current along c direction can be calculated from the corresponding terms in the Green function, namely,

$$g_1^{R(A)}(p_+, p_-) = \pm\left[\psi\left(p_+ - \frac{e}{c}A_z\right) - \psi\left(p_- + k_z' + \frac{e}{c}A_z\right)\right]$$

$$\times \frac{\Delta^*(k_z')\,\Delta(k_z - k_z')}{2\epsilon^2(2\epsilon \pm i/\tau)}$$
$$\mp \left[\psi\left(p_+ - k_z' + \frac{e}{c}A_z\right) - \psi\left(p_- - \frac{e}{c}A_z\right)\right]$$
$$\times \frac{\Delta^*(k_z - k_z')\,\Delta(k_z')}{2\epsilon^2(2\epsilon \pm i/\tau)}.$$

We now insert this expression into the definition of current, eqn (7.6), and make the shift of integration variable $p_z - (e/c)\,A_z \to p_z$. The terms which are odd in $p_z \to -p_z$ drop out after integration over dp_z and we get

$$j_s^{(c)}(k_z) = e\nu^{(2)} \int \frac{d\epsilon}{8} \tanh\left(\frac{\epsilon}{2T}\right)\left[\frac{1}{\epsilon^2(2\epsilon + i/\tau)} - \frac{1}{\epsilon^2(2\epsilon - i/\tau)}\right]$$
$$\times \int_{-\pi/s}^{\pi/s} \frac{s\,dp_z}{2\pi} \int \frac{d\alpha}{2\pi}\left[\frac{\partial\psi(p_+)}{\partial p_z} + \frac{\partial\psi(p_-)}{\partial p_z}\right]$$
$$\times \left[\psi\left(p_+ - k_z' + \frac{2eA_z}{c}\right)\Delta(k_z')\,\Delta^*(k_z - k_z')\right.$$
$$\left. + \psi\left(p_- + k_z' + \frac{2eA_z}{c}\right)\Delta^*(k_z')\,\Delta(k_z - k_z')\right].$$

Using eqn (7.2) we obtain

$$j_s^{(c)}(k_z) = -e\nu^{(2)} J^2 s \int \frac{d\epsilon}{8} \tanh\left(\frac{\epsilon}{2T}\right)\left[\frac{1}{\epsilon^2(2\epsilon + i/\tau)} - \frac{1}{\epsilon^2(2\epsilon - i/\tau)}\right]$$
$$\times [\sin(k_z's - \chi_A) - \sin(k_z s - k_z's + \chi_A)]\,\Delta(k_z')\,\Delta^*(k_z - k_z').$$

In the Fourier representation of eqn (7.19),

$$\sin(k_z's - \chi_A)\,\Delta(k_z')\,\Delta^*(k_z - k_z') \to \frac{\Delta_n^*}{2i}\left[\Delta_{n+1}e^{-i\chi_A} - \Delta_{n-1}e^{i\chi_A}\right],$$
$$\sin(k_z s - k_z's + \chi_A)\,\Delta^*(k_z - k_z')\,\Delta(k_z') \to \frac{\Delta_n}{2i}\left[\Delta_{n+1}^*e^{i\chi_A} - \Delta_{n-1}^*e^{-i\chi_A}\right].$$

Finally, we return to eqn (7.24).

7.4 Applications of the Lawrence–Doniach model

The Lawrence–Doniach model is widely used for layered superconductors; it is simple and quite general at the same time. It covers the whole class of anisotropic systems from three-dimensional weakly anisotropic to highly anisotropic layered and even two-dimensional superconductors. It is able to provide a reasonable description for many properties of the mixed state of layered superconductors. For example, layered superconductors can have two-dimensional analogues of the usual Abrikosov vortices, the so-called pancake vortices connected with each other by strings of Josephson vortices. Pancake vortices have been first considered by Artemenko and Kruglov (1990) and by Clem (1991). Pancakes turned out to be very important for high temperature superconductors most of which are highly anisotropic or layered compounds. Another interesting feature of layered superconductors is the unusual behavior of the upper critical magnetic field: For orientation parallel to the layers, it grows rapidly as temperature decreases and $\xi_c(T)$ approaches the interlayer distance s. Yet another important property is the so-called intrinsic pinning considered first by Tachiki and Takahashi (1989) and elaborated further by Ivlev and Kopnin (1989) and by Barone *et al.* (1990). Many problems specific for layered superconductors are discussed in the review by Blatter *et al.* (1994). Here we briefly outline two mentioned examples: the upper critical field and intrinsic pinning which will be used later when we consider the vortex dynamics in anisotropic and layered superconductors.

7.4.1 *Upper critical field*

Let the magnetic field be parallel to the layers. As in Section 1.1.2 we look for a second-order phase transition from the normal into the superconducting state which occurs with lowering the field. We choose the y-axis along the direction of the field with the gauge $\mathbf{A} = (0, 0, -Hx)$ and assume a solution in the form

$$\Delta_n(\mathbf{r}) = e^{iqns}Y(x).$$

The linearized equation (7.20) for $Y(x)$ becomes the Mathieu equation:

$$\xi_{ab}^2\frac{\partial^2 Y}{\partial x^2} - \frac{2\xi_c^2}{s^2}\left[1-\cos\left(qs+\frac{2eHsx}{c}\right)\right]Y + Y = 0. \tag{7.31}$$

There are two limiting cases: (i) Weakly layered (nearly continuous) limit, and (ii) highly layered case. Consider them separately.

7.4.1.1 *Weakly layered case*

In the continuous anisotropic case the upper critical field is known to be (Kats 1969, 1970, Lawrence and Doniach 1971):

$$H_{c2} = \frac{c}{2e\xi_{ab}\xi_c}. \tag{7.32}$$

This equation can be reproduced also from eqn (7.31). Since $x \sim \xi_{ab}$ we find that the argument of the cosine function in eqn (7.31) is

$$qs + \frac{sx}{\xi_{ab}\xi_c} \sim \frac{s}{\xi_c}.$$

In the continuous limit, $s \ll \xi_c$, the cosine function in eqn (7.31) can be expanded in a small argument. We obtain

$$\xi_{ab}^2\frac{\partial^2 Y}{\partial x^2} - \xi_c^2\left(q+\frac{2eHx}{c}\right)^2 Y + Y = 0. \tag{7.33}$$

This is the oscillator equation (1.33). Again, the lowest-level solution is

$$Y = \exp\left[-\frac{1}{2\xi_{ab}^2}\left(x+\frac{cq}{2eH_{c2}}\right)^2\right] \tag{7.34}$$

where H_{c2} is determined by eqn (7.32).

7.4.1.2 *Highly layered case* As we shall see, this limit corresponds to a high upper critical field and is realized for a certain relation between the interlayer distance s and the coherence length ξ_c. We introduce the parameter

$$h = \frac{2eHs^2\xi_{ab}}{c\xi_c}. \tag{7.35}$$

For a weakly layered superconductor, $h = s^2/\xi_c^2 \ll 1$. On the contrary, a highly layered case is realized for $h \gg 1$ when $H \to \infty$. For this limit, a solution of eqn (7.31) can be obtained as an expansion in powers of a small $1/h$. The lowest-level periodic solution corresponds to an even Mathieu function

$$Y = 1 + \frac{2}{h^2}\cos\left(\frac{2eHsx}{c}\right) + \frac{1}{2h^4}\cos\left(\frac{4eHsx}{c}\right) + \ldots$$

provided

$$h^2 = \frac{1}{1-(s^2/2\xi_c^2)}.$$

This relation determines the upper critical field

$$H_{c2} = \frac{\Phi_0}{2\pi s^2}\frac{\xi_c}{\xi_{ab}}\frac{1}{\sqrt{1-(s^2/2\xi_c^2)}}. \tag{7.36}$$

This solution is valid as long as $h \gg 1$, or $\xi_c(T) \to s/\sqrt{2}$.

We observe that the upper critical field diverges as the temperature-dependent coherence length $\xi_c(T)$ approaches $s/\sqrt{2}$ from above with lowering the temperature (Klemm *et al.* 1975). In this limit, vortex cores fit in-between the superconducting layers, and the supercurrents do not destroy superconductivity. Divergence of the upper critical field is similar to the Little and Parks effect for a non-singly-connected superconductor where the critical temperature is a periodic function of the magnetic field instead of vanishing with an increasing field. Therefore, an anisotropic superconductor that is essentially a three-dimensional system with $\xi_c(T) \gg s$ at a higher temperature, can undergo a transition into effectively a two-dimensional superconductor with a lowering of the temperature if the zero-temperature coherence length $\xi_c(0)$ is short enough,. This occurs when $\xi_c(T)$ decreases and reaches $s/\sqrt{2}$.

7.4.2 *Intrinsic pinning: Single vortex in a weak potential*

In this section we consider an important effect specific for layered systems, namely, an interaction of the order parameter with the underlying crystalline structure. More particularly, we discuss an example of vortices which are aligned parallel to the superconducting planes. The interaction with the inhomogeneous layered environment gives rise to the "intrinsic pinning": The vortex energy becomes dependent on the vortex position with respect to the planes and produces a pinning force which tries to keep vortices in between the superconducting layers (Tachiki and Takahashi 1989). We consider a nearly continuous limit $\xi_c \gg s$.

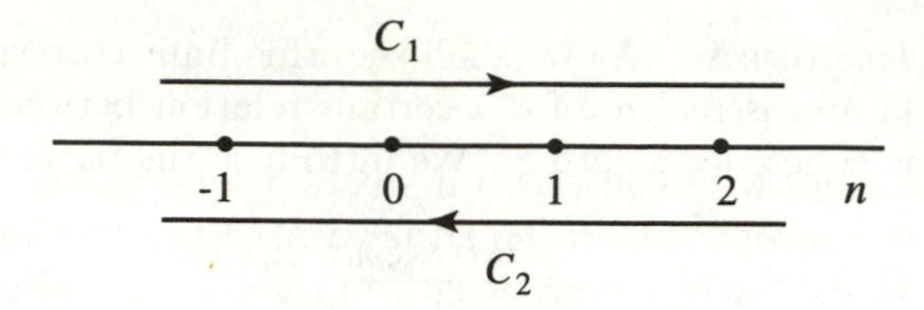

FIG. 7.1. Integration contours in the summation formula. The poles of $\cot(\pi n)$ are shown by dots.

Assume that the magnetic field is parallel to the layers and find the energy of a single vortex in a layered structure. We use the approach developed by Barone *et al.* (1990). To calculate the sum in eqn (7.27) we employ the summation formula

$$\sum_n F(n) = -\frac{1}{2i}\int_C \cot(\pi n)F(n)\,dn = \int_{-\infty}^{\infty} F(n)\,dn$$
$$+\frac{i}{2}\left(\int_{C_1}[\cot(\pi n)+i]\,F(n)\,dn + \int_{C_2}[\cot(\pi n)-i]\,F(n)\,dn\right).$$

The integration contour C goes above and below the poles $n = 0, \pm 1, \ldots$ of $\cot(\pi n)$ encircling them clockwise, while the contours C_1 and C_2 go above and below the poles, respectively, so that $C = C_1 + C_2$ (see Fig. 7.1). The total free energy becomes

$$\mathcal{F} = \int d^2r\, dz\left[\frac{H^2}{8\pi} + F(z)\right] + U_{pin} \tag{7.37}$$

where

$$U_{pin} = \frac{i}{2}\int d^2r\left(\int_{C_1}\left[\cot\frac{\pi z}{s} + i\right]F(z)\,dz + \int_{C_2}\left[\cot\frac{\pi z}{s} - i\right]F(z)\,dz\right). \tag{7.38}$$

The first term in eqn (7.37) is independent of the vortex position. However, the energy U_{pin} is a function of the vortex coordinate along the z axis.

For a weakly layered case, to the first approximation in $s \ll \xi_c$, a vortex solution has the same form as for an anisotropic superconductor. Indeed, let us take the y axis along the magnetic field and put

$$\Psi = \Psi_{GL} f(\tilde{\rho})e^{i\tilde{\chi}}$$

where

$$\tilde{\rho} = \sqrt{x^2/\xi_{ab}^2 + z^2/\xi_c^2}$$

and

$$\tan\tilde{\chi} = \frac{x/\xi_{ab}}{z/\xi_c}.$$

Assume that $\kappa = \lambda_{ab}/\xi_{ab}$ is large, $\kappa \gg 1$. We can neglect the magnetic field in the equation for f and obtain from eqns (7.21), (7.29), and (7.30)

$$\frac{d^2 f}{d\tilde{\rho}^2} + \frac{1}{\tilde{\rho}}\frac{df}{d\tilde{\rho}} - \frac{f}{\tilde{\rho}^2} + f - f^3 = 0. \tag{7.39}$$

This equation coincides with eqn (1.48) in Section 1.1.2 which we derived for a single vortex in an isotropic superconductor. We see that the anisotropy of the vortex core scales with the coherence length in the corresponding directions.

To calculate the pinning energy, we insert the exact vortex solution of eqn (7.20) into the free energy expression. In the small position-dependent part of the energy in eqn (7.38) one can use the approximate solution eqn (7.39) for an anisotropic case. To calculate the integrals in eqn (7.38), we shift the integration contours C_1 and C_2 to the upper and lower complex half planes of the integration variable z, respectively. The integral is then determined by a singularity of the function $f(z)$ which is the closest to the real axis. The function $f(z)$ has poles on the imaginary $\tilde{\rho}$ axis:

$$f(\rho) = \pm\frac{\sqrt{2}}{\tilde{\rho} \mp i\rho_0}. \tag{7.40}$$

This fact can easily be understood from a model expression for the vortex-core function $f(\tilde{\rho}) = \tanh(\tilde{\rho})$. It has poles at $i\rho_0 = i(\pi/2 + \pi k)$. The numerical solution (Barone *et al.* 1990) shows that the closest poles to the real axis have $\rho_0 = 2.51$ in eqn (7.40).

Let the vortex be located at the point $x = 0$, $z = z_0$. The vortex amplitude f is a function of the shifted variable

$$\tilde{\rho} = \sqrt{x^2/\xi_{ab}^2 + (z - z_0)^2/\xi_c^2}, \tag{7.41}$$

and its phase is now

$$\tan\tilde{\chi} = \frac{x/\xi_{ab}}{(z - z_0)/\xi_c}.$$

To the leading approximation in $\tilde{\rho} - \rho_0$, the energy density is determined by the highest singularity which comes from the gradient term:

$$F(z) = \frac{4\,|a|\,\Psi_{GL}^2}{(\tilde{\rho} \mp i\rho_0)^4}.$$

To calculate the integrals in eqn (7.38) we write

$$z = z_0 + \xi_c\sqrt{\tilde{\rho}^2 - (x/\xi_{ab})^2}\ ,\ dz = \xi_c\frac{\tilde{\rho}\,d\tilde{\rho}}{\sqrt{\tilde{\rho}^2 - (x/\xi_{ab})^2}}.$$

Consider first the integral over the contour C_1. At $\tilde{\rho} = \rho_0$, the imaginary part of $z = z' + iz''$ is large: $z''/s \gg 1$, therefore,

$$\cot\frac{\pi z}{s} = -i - 2i\exp\left(\frac{2i\pi z'}{s} - \frac{2\pi z''}{s}\right).$$

The integral over the contour C_1 gives

$$\frac{i}{2}\int dy\,dx\int_{C_1}\left[\cot\frac{\pi z}{s} + i\right]F(z)\,dz$$

$$
\begin{aligned}
&= 2i|a|\Psi_{GL}^2\xi_c L \int dx \int_{C_1} \left(\cot\left[\frac{\pi}{s}\left(z_0 + \xi_c\sqrt{\tilde{\rho}^2 - (x/\xi_{ab})^2}\right)\right] + i\right) \\
&\quad \times \frac{\tilde{\rho}\, d\tilde{\rho}}{\sqrt{\tilde{\rho}^2 - (x/\xi_{ab})^2}} \frac{1}{(\tilde{\rho} - i\rho_0)^4} \\
&= \frac{32\pi^4|a|\Psi_{GL}^2\xi_c^4 L}{3s^3} \exp\left(\frac{2i\pi z_0}{s}\right) \int \exp\left(-\frac{2\pi\xi_c}{s}\sqrt{\rho_0^2 + (x/\xi_{ab})^2}\right) dx \\
&= \frac{32\pi^4|a|\Psi_{GL}^2\xi_{ab}\xi_c\sqrt{\rho_0}L}{3}\left(\frac{\xi_c}{s}\right)^{\frac{5}{2}} \exp\left(-\frac{2\pi\xi_c\rho_0}{s}\right)\exp\left(\frac{2\pi i z_0}{s}\right).
\end{aligned}
$$

Here L is the vortex length along the y axis. During the integration we use that $x/\xi_{ab} \sim \sqrt{s/\xi_c} \ll 1$ and calculate the derivative only of the most rapidly varying exponential part of the function under the integral. Together with the integral over the contour C_2, this equation gives the energy per unit vortex length

$$
U_{pin} = U_0 \cos\left(\frac{2\pi z_0}{s}\right) \tag{7.42}
$$

where

$$
U_0 = \frac{4\pi\Phi_0^2\sqrt{\rho_0}}{3\lambda_{ab}^2\Gamma}\left(\frac{\xi_c}{s}\right)^{\frac{5}{2}} \exp\left(-\frac{2\pi\xi_c\sqrt{\rho_0}}{s}\right). \tag{7.43}
$$

We see that the vortex prefers to sit in-between the layers $z_0 = s/2$ where its energy is minimal. If the vortex is not exactly in the middle, $z_0 = s/2 + \delta z$, a restoring (pinning) force appears. The pinning force per unit vortex length is

$$
-\frac{\partial U_{pin}}{\partial \delta z} = -\frac{2\pi U_0}{s}\sin\left(\frac{2\pi\delta z}{s}\right). \tag{7.44}
$$

The pinning energy is exponentially small because the layered structure is weakly pronounced: the distance between the layers is small compared to the coherence length ξ_c.

Part III

Nonequilibrium superconductivity

8

NONSTATIONARY THEORY

We consider both the method of analytical continuation worked out by Eliashberg and the Keldysh diagram technique, which are designed for calculating the real-time Green functions of nonstationary superconductors. The Eliashberg phonon model of superconductivity is discussed. We derive equations for the Keldysh Green function for particles interacting with impurities, phonons and with each other.

8.1 The method of analytical continuation

This part of the book is devoted to introduction of general principles and tools of the microscopic nonstationary theory of superconductivity and is based on the general concepts we learned in the previous chapters. A physical system which evolves in time may be out of equilibrium with respect to the heat bath; moreover, various parts of the system may be not in equilibrium with each other. To treat a nonequilibrium situation, one needs to know the distribution of nonequilibrium excitations in addition to their spectrum. Moreover, the spectrum itself can be distorted as the system gets far from equilibrium. The specifics of superconductors are that the excitations which affect the response of a superconductor to an external field can accumulate near the energy gap when their relaxation is relatively slow. The major problem of the nonstationary theory is thus to find the distribution function of these excitations. In a nonlinear case, one needs this distribution also to calculate the modified spectrum in a self-consistent way.

Within the Green function formalism of the microscopic theory of nonstationary superconductivity, all these different tasks are combined into a single basic problem of finding the time-dependent Green function of the system. However, we encounter a major complication in this way: there is no simple recipe for calculating the real-time Green functions because the Dyson equations for them cannot be constructed by simply repeating the previous argumentation. There are two general methods developed so far for calculating the time-dependent Green functions. The first is due to Keldysh (1964). It deals directly with the set of real-time Green functions defined on the time axis with special rules of time ordering of the particle-field operators. The second method has been developed by Gor'kov and Eliashberg (1968) and by Eliashberg (1971). It operates with the Matsubara Green functions which, as functions of frequencies, are analytically continued from imaginary onto the real-frequency axis. We shall start with the analytical continuation technique. We demonstrate in Section 8.5 that both these methods are completely equivalent. In the present chapter, we derive the equations for those real-time Green functions which are appropriate for the superconducting response. The following chapters describe how to define and calculate the distribution function. We derive the kinetic equations for the distribution function; we solve them for various cases and show how to use the obtained distribution for further calculations.

As already mentioned, there is no simple way to calculate the real-time Green function. Instead, comparatively simple equations and the corresponding diagram technique exist for the Green functions which are defined for an imaginary time $t = -i\tau$ (Matsubara functions). This is because, in contrast to its imaginary counterpart, a real time cannot be directly time-ordered with the inverse temperature $-i/T$, and thus the Wick theorem cannot be directly applied for the real-time field operators. However, solving the imaginary-time equations or, equivalently, calculating all the diagrams, one can obtain the Matsubara functions. Since the field operators in both real and imaginary time representations obey the same equations with the substitution $t \to -i\tau$, one expects that the real-time Green function can be obtained from the Matsubara function by an analytical continuation from imaginary time onto the real-time axis provided the necessary initial conditions are satisfied. We assume that all the interactions between particles are switched on at the time $t = -\infty$. Therefore, the Green functions at $t = -\infty$ coincide with those for free particles which are in equilibrium with each other and with the heat bath with a temperature T.

For practical purposes, the Matsubara Green functions are usually calculated as the Fourier transforms for discrete imaginary (Matsubara) frequencies $\epsilon_n = 2\pi iT(n + 1/2)$ for Fermi particles, and $\omega_k = 2\pi iTk$ for Bose particles and external fields. The analytical continuation in time is equivalent to the analytical continuation in frequencies ϵ and ω from imaginary onto the real frequency axis. To satisfy the initial conditions, one has to make the continuation in such a way that all the external field frequencies have an infinitesimal positive imaginary part. In this case, all diagrams containing the field operators vanish for $t \to -\infty$ according to the causality principle. In other words, one has to continue all the field frequencies from the upper half-plane of complex ω.

In principle, one can first solve the particular physical problem and find the Matsubara Green function for it. After that, one can continue the Matsubara function analytically and obtain the required Green function in the real-time representation. This method, however, is not convenient. It operates with quantities which do not have a simple physical meaning and thus can cause confusion, which sometimes leads to incorrect results. The modern microscopic theory, instead, works with the real-time Green functions from the very beginning. In this representation, we can always keep track of the real physical phenomena involved in the particular problem. The price for that is more complicated equations needed to find the real-time functions. This chapter describes how to derive and solve equations for the real-time Green functions of superconductors. We shall follow the approach suggested by Eliashberg (1971).

All physical quantities like the current, the particle number, or the order

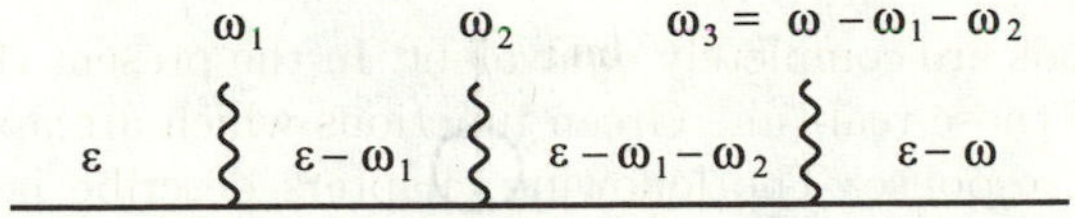

ω₁ ω₂ ω₃ = ω − ω₁ − ω₂

ε ε − ω₁ ε − ω₁ − ω₂ ε − ω

FIG. 8.1. Diagrammatic presentation of the Green function expansion in powers of the external field H (shown by wavy lines) up to the third-order term.

parameter are expressed through sums of certain matrix elements of the Green function $\check{G}$ over the Matsubara frequencies:

$$T \sum_n \check{G}_{\epsilon_n, \epsilon_n - \omega_k}. \tag{8.1}$$

In our case, $\epsilon_n = 2\pi i T(n + 1/2)$ are Matsubara frequencies for Fermionic operators. The frequencies $\omega_k = 2\pi i T k$ refer to the "external" Bose fields and to the time-dependent order parameter. Our goal is to find an analytical continuation of eqn (8.1) from the upper half plane of the external frequency ω. To make the analytical continuation in a general case, we consider the full function $\check{G}_{\epsilon_n, \epsilon_n - \omega_k}$ as a perturbation series expansion in terms of a generalized "field". We start with the Dyson equation for the Matsubara functions

$$\sum_{\epsilon_m} \int \frac{d^3 k_1}{(2\pi)^3} \left[\check{G}^{-1}_{\epsilon_n}(\mathbf{p} - \mathbf{k}_1, \omega_1) - \check{\Sigma}_{\epsilon_n, \epsilon_m}(\mathbf{p}, \mathbf{k}_1) \right] \check{G}_{\epsilon_m, \epsilon_k}(\mathbf{p} - \mathbf{k}_1, \mathbf{p} - \mathbf{k})$$
$$= \check{1} \cdot (2\pi)^3 \delta(\mathbf{k}) \delta_{\epsilon_n - \epsilon_k}. \tag{8.2}$$

Here $\omega_1 = \epsilon_n - \epsilon_m$ and

$$\check{G}^{-1}_{\epsilon}(\mathbf{p} - \mathbf{k}_1, \omega_1) = \begin{pmatrix} \xi_{\mathbf{p}} - \epsilon & 0 \\ 0 & \xi_{\mathbf{p}} + \epsilon \end{pmatrix} (2\pi)^3 \delta(\mathbf{k}_1) \delta_{\omega_1} + \check{H}_{\omega_1}(\mathbf{k}_1) \tag{8.3}$$

and $\check{H}$ is an "external field" matrix as defined by eqn (5.57). At the beginning, we do not consider the interaction with impurities and phonons.

8.1.1 *Clean superconductors*

Let us consider the N-th order term in the expansion of eqn (8.2) which we write as

$$\check{G}^{(N)}_{\epsilon_n, \epsilon_n - \omega} = (-1)^N G^{(0)}_{\epsilon_n} \check{H}_{\omega_1} G^{(0)}_{\epsilon_n - \omega_1} \check{H}_{\omega_2} G^{(0)}_{\epsilon_n - \omega_1 - \omega_2} \cdots \check{H}_{\omega_N} G^{(0)}_{\epsilon_n - \omega} \tag{8.4}$$

where $\omega = \omega_1 + \ldots + \omega_N$ and $G^{(0)}_{\epsilon_n}$ is the non-perturbed Green function of the normal state,

$$G^{(0)}_{\epsilon_n} = \frac{1}{\xi_p - \epsilon_n}.$$

The diagrammatic representation of eqn (8.4) is shown in Fig. 8.1.The summations of the Matsubara frequencies ω_i and integration over the corresponding

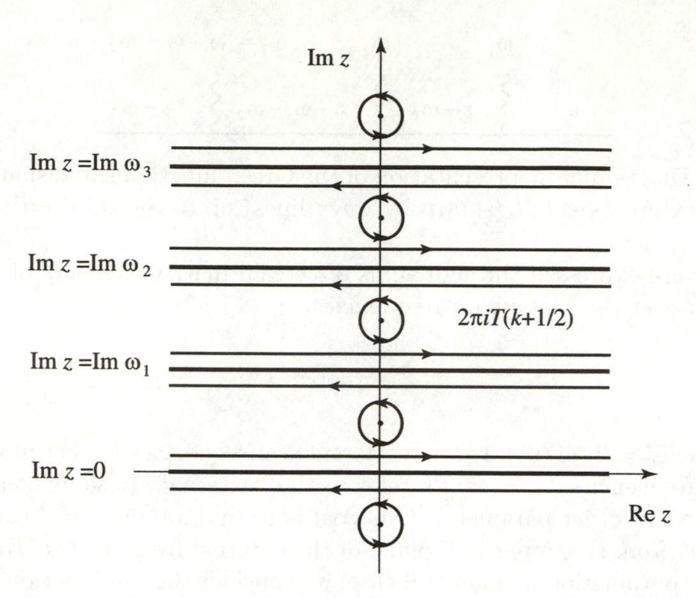

FIG. 8.2. The cuts (shown by thick solid lines) along the singular lines of the function $\check{G}^{(3)}_{z,z-\omega}$ on the plane of complex z. Contours around the points $z = 2\pi i T\,(k+1/2)$ (circles) are unrolled into contours going above and below the cuts.

momenta $\mathbf{k}_i$ are implicitly assumed. The sum of the type of eqn (8.1) can be written as the contour integral

$$T\sum_n \check{G}^{(N)}_{\epsilon_n,\epsilon_n-\omega_k} = \oint \check{G}^{(N)}_{z,z-\omega_k} \tanh\left(\frac{z}{2T}\right)\frac{dz}{4\pi i} \tag{8.5}$$

taken around each of the poles of the function tanh $(z/2T)$, i.e., around all $z = 2\pi i T\,(n+1/2)$. Consider the poles of $\check{G}^{(N)}_{z,z-\omega_k}$ as functions of z. The poles are at $z = \xi_p$, $z - \omega_1 = \xi_p$, $\ldots$, $z - \omega = \xi_p$. Since ξ_p is real, they are determined by equations $\mathrm{Im}\left(z - \sum_{j=0}^{l}\omega_j\right) = 0$, where $l = 0, \ldots, N$ and $\omega_0 = 0$.

Let us make $N+1$ horizontal cuts on the complex plane z passing through all $\mathrm{Im}\left(z - \sum_{j=0}^{l}\omega_j\right) = 0$ as shown in Fig. 8.2. Since $\omega_j = 2\pi i T k_j$, these cuts do not go through the poles of tanh $(z/2T)$. We shall continue from positive imaginary values of all ω_j, therefore we can consider all k_j to be positive, and numerate the cuts from $\mathrm{Im}z = 0$ to $\mathrm{Im}z = \mathrm{Im}\omega$ upwards in order of increasing k_j. Now we can unroll the contours going around each $z = 2\pi i T\,(n+1/2)$ into the contours going along the lower and upper coasts of each of $N+1$ cuts. At each cut labeled by l, the variable $z = \epsilon + i\mathrm{Im}\left(\sum_{j=0}^{l}\omega_j\right)$; the lower contour goes from right to left while the upper goes from left to right (see Fig. 8.2). As a result, we have

$$\oint \check{G}^{(N)}_{z,z-\omega} \tanh\left(\frac{z}{2T}\right) \frac{dz}{4\pi i} = \int \tanh\left(\frac{\epsilon}{2T}\right) \frac{d\epsilon}{4\pi i} \times \left[\delta_0 \check{G}^{(N)} + \delta_1 \check{G}^{(N)} + \ldots + \delta_N \check{G}^{(N)}\right].$$

Here

$$\delta_l \check{G}^{(N)} = \left[\check{G}^{(N)}_{z,z-\omega}\right]_{z=z_l+0} - \left[\check{G}^{(N)}_{z,z-\omega}\right]_{z=z_l-0}$$

is the jump of the Green function at the l-th cut where $z_l = \epsilon + i\mathrm{Im}\sum_{j=0}^{l}\omega_j$. Because of periodicity of tanh $(z/2T)$ along the imaginary axis with the period $2\pi iT$, the argument of the hyperbolic tangent is just ϵ at each cut. Let us write one of the jumps explicitly. For example,

$$\delta_1 \check{G}^{(N)} = (-1)^N G^{(0)}_{\epsilon+\omega_1} \check{H}_{\omega_1} \left[G^{(0)}_{\epsilon+i0} - G^{(0)}_{\epsilon-i0}\right] \check{H}_{\omega_2} G^{(0)}_{\epsilon-\omega_2} \ldots \check{H}_{\omega_N} G^{(0)}_{\epsilon+\omega_1-\omega}.$$

Here all ω_j are still discrete imaginary. When we unroll the contours, we make the analytical continuation in the variable z. At the cut $l = 1$ for example, we have, for a real ϵ, $\mathrm{Im}(\epsilon+\omega_1) > \mathrm{Im}(\epsilon + i0) > 0$ while $\mathrm{Im}(\epsilon + \omega_1 - \omega) < \ldots < \mathrm{Im}(\epsilon - \omega_2) < \mathrm{Im}(\epsilon - i0) < 0$. Therefore, we obtain

$$\delta_1 \check{G}^{(N)} = (-1)^N G^{(0)R}_{\epsilon+\omega_1} \check{H}_{\omega_1} \left[G^{(0)R}_{\epsilon} - G^{(0)A}_{\epsilon}\right] \check{H}_{\omega_2} G^{(0)A}_{\epsilon-\omega_2} \ldots \check{H}_{\omega_N} G^{(0)A}_{\epsilon+\omega_1-\omega}. \quad (8.6)$$

Now we can consider all ω_j as real continuous values. The sums over discrete Matsubara frequencies $\omega_j = 2\pi i T k_j$ in equations of the type of eqn (8.6) transform into the integrals according to the rule

$$T\sum_{k_j} = \int \frac{d\omega_j}{2\pi}.$$

We make the substitution of integration variable $\epsilon + \sum_{j=0}^{l}\omega_j \Rightarrow \epsilon$ at the l-th cut to restore the previous notation of incoming and outgoing frequencies and obtain

$$\oint \check{G}^{(N)}_{z,z-\omega_k} \tanh\left(\frac{z}{2T}\right) \frac{dz}{4\pi i}$$
$$= \int \left[\tanh\left(\frac{\epsilon}{2T}\right) [\delta_0 \check{G}^{(N)}]_{\epsilon,\epsilon-\omega} + \tanh\left(\frac{\epsilon-\omega_1}{2T}\right) [\delta_1 \check{G}^{(N)}]_{\epsilon,\epsilon-\omega} + \right.$$
$$+ \tanh\left(\frac{\epsilon-\omega_1-\omega_2}{2T}\right) [\delta_2 \check{G}^{(N)}]_{\epsilon,\epsilon-\omega} + \ldots$$
$$\left. + \tanh\left(\frac{\epsilon-\omega}{2T}\right) [\delta_N \check{G}^{(N)}]_{\epsilon,\epsilon-\omega}\right] \frac{d\epsilon}{4\pi i}.$$

Here $[\delta_j \check{G}^{(N)}]_{\epsilon,\epsilon-\omega}$ are constructed according to the scheme shown in eqn (8.6) but with shifted frequencies:

$$[\delta_1 \check{G}^{(N)}]_{\epsilon,\epsilon-\omega}$$

$$= (-1)^N G_\epsilon^{(0)R} \check{H}_{\omega_1} \left[G_{\epsilon-\omega_1}^{(0)R} - G_{\epsilon-\omega_1}^{(0)A} \right] \check{H}_{\omega_2} G_{\epsilon-\omega_1-\omega_2}^{(0)A} \ldots \check{H}_{\omega_N} G_{\epsilon-\omega}^{(0)A}. \quad (8.7)$$

Therefore, the sum in eqn (8.5) transforms into

$$\int \frac{d\epsilon}{4\pi i} \check{G}_{\epsilon,\epsilon-\omega}^{(N)K}$$

where

$$\check{G}_{\epsilon,\epsilon-\omega}^{(N)K} = \sum_{l=0}^{N} [\delta_l \check{G}^{(N)}]_{\epsilon,\epsilon-\omega} \tanh\left(\frac{\epsilon - \sum_{j=0}^{l} \omega_j}{2T} \right). \quad (8.8)$$

Writing all the jumps explicitly, we obtain

$$\begin{aligned} \check{G}_{\epsilon,\epsilon-\omega}^{(N)K} = & (-1)^N G_\epsilon^{(0)R} \check{H}_{\omega_1} G_{\epsilon-\omega_1}^{(0)R} \ldots \check{H}_{\omega_N} G_{\epsilon-\omega}^{(0)R} f^{(0)}(\epsilon - \omega) \\ & + (-1)^{N-1} G_\epsilon^{(0)R} \check{H}_{\omega_1} G_{\epsilon-\omega_1}^{(0)R} \ldots \check{h}_{\epsilon-\omega+\omega_N,\epsilon-\omega} G_{\epsilon-\omega}^{(0)A} \\ & + \ldots + (-1)^{N-1} G_\epsilon^{(0)R} \check{h}_{\epsilon,\epsilon-\omega_1} G_{\epsilon-\omega_1}^{(0)A} \ldots \check{H}_{\omega_N} G_{\epsilon-\omega}^{(0)A} \\ & - (-1)^N f^{(0)}(\epsilon) G_\epsilon^{(0)A} \check{H}_{\omega_1} G_{\epsilon-\omega_1}^{(0)A} \ldots \check{H}_{\omega_N} G_{\epsilon-\omega}^{(0)A} \end{aligned}$$

where

$$f^{(0)}(\epsilon) = \tanh\left(\frac{\epsilon}{2T} \right) \quad (8.9)$$

and

$$\check{h}_{\epsilon,\epsilon-\omega} = -\check{H}_\omega \left[f^{(0)}(\epsilon) - f^{(0)}(\epsilon - \omega) \right]. \quad (8.10)$$

Collecting now all orders of the perturbation expansion, we find that the analytical continuation gives

$$T \sum_n \check{G}_{\epsilon_n,\epsilon_n-\omega_k} \Rightarrow \int \frac{d\epsilon}{4\pi i} \check{G}_{\epsilon,\epsilon-\omega}^K \quad (8.11)$$

where we define the total time-dependent Green function

$$\check{G}_{\epsilon,\epsilon-\omega}^K = \check{G}_{\epsilon,\epsilon-\omega}^R f^{(0)}(\epsilon - \omega) - f^{(0)}(\epsilon) \check{G}_{\epsilon,\epsilon-\omega}^A + \check{G}_{\epsilon,\epsilon-\omega}^{(a)}. \quad (8.12)$$

It is this function which determines all the physical properties of a nonstationary system in a real time. We assign the superscript K to the total function to indicate that it coincides with the so-called Keldysh Green function introduced by Keldysh (1964). We shall prove this later.

The retarded and advanced functions are defined in the usual way as

$$\check{G}_{\epsilon,\epsilon-\omega}^R = \sum_{N=0}^{\infty} (-1)^N G_\epsilon^{(0)R} \check{H}_{\omega_1} G_{\epsilon-\omega_1}^{(0)R} \check{H}_{\omega_2} G_{\epsilon-\omega_1-\omega_2}^{(0)R} \ldots \check{H}_{\omega_N} G_{\epsilon-\omega}^{(0)R},$$

$$\check{G}_{\epsilon,\epsilon-\omega}^A = \sum_{N=0}^{\infty} (-1)^N G_\epsilon^{(0)A} \check{H}_{\omega_1} G_{\epsilon-\omega_1}^{(0)A} \check{H}_{\omega_2} G_{\epsilon-\omega_1-\omega_2}^{(0)A} \ldots \check{H}_{\omega_N} G_{\epsilon-\omega}^{(0)A}.$$

We assume integration over all ω_l and $\mathbf{k}_l$. These functions only contain all retarded or all advanced zero-order functions $G^{(0)}$, respectively, and are obtained by summation of all the diagrams exactly as for the Matsubara functions. They satisfy the Gor'kov equations with real frequencies ϵ.

The *anomalous function* is

$$\check{G}^{(a)}_{\epsilon,\epsilon-\omega} = \sum_{N=0}^{\infty} \check{G}^{(N)(a)}_{\epsilon,\epsilon-\omega}$$

where

$$\check{G}^{(N)(a)}_{\epsilon,\epsilon-\omega} = (-1)^{N-1} G^{(0)R}_{\epsilon} \check{h}_{\epsilon,\epsilon-\omega_1} G^{(0)A}_{\epsilon-\omega_1} \dots \check{H}_{\omega_N} G^{(0)A}_{\epsilon-\omega} + \dots$$
$$+ (-1)^{N-1} G^{(0)R}_{\epsilon} \check{H}_{\omega_1} \dots G^{(0)R}_{\epsilon-\omega+\omega_N} \check{h}_{\epsilon-\omega+\omega_N,\epsilon-\omega} G^{(0)A}_{\epsilon-\omega}. \quad (8.13)$$

It necessarily contains at least one incoming retarded together with one outgoing advanced function $G^{(0)}$.

The regular functions enter eqn (8.12) in combinations with the equilibrium distribution, eqn (8.9), which is

$$f^{(0)}(\epsilon) = 1 - 2n(\epsilon) \quad \text{where} \quad n(\epsilon) = \frac{1}{e^{\epsilon/T}+1} \quad (8.14)$$

is the Fermi function. It is the anomalous function which describes the deviation of the system from equilibrium. The summation over all N gives

$$\check{G}^{(a)}_{\epsilon,\epsilon-\omega} = \int \frac{d\epsilon' d\omega'}{(2\pi)^2} \frac{d^3k' d^3k''}{(2\pi)^6} \check{G}^{R}_{\epsilon,\epsilon'}(\mathbf{p},\mathbf{p}-\mathbf{k}') \check{h}_{\epsilon',\epsilon'-\omega'}(\mathbf{k}'')$$
$$\times \check{G}^{A}_{\epsilon'-\omega',\epsilon-\omega}(\mathbf{p}-\mathbf{k}'-\mathbf{k}''). \quad (8.15)$$

This simple expression for $\check{G}^{(a)}_{\epsilon,\epsilon-\omega}$ is only valid if there are no interactions with impurities, phonons, and particle–particle interactions. We now consider these interactions in turn.

8.1.2 *Impurities*

Consider first impurity scattering. We shall not write down all possible diagrams, instead, we concentrate on one typical example which demonstrates the general structure of the result of the analytical continuation. Consider the term which is of the third order in the field $\check{H}_\omega$ and of the second order in the impurity potential u. We have, similarly to the previous case,

$$\oint \check{G}^{(3,2)}_{z,z-\omega} \tanh\left(\frac{z}{2T}\right) \frac{dz}{4\pi i} = \int \tanh\left(\frac{\epsilon}{2T}\right) \frac{d\epsilon}{4\pi i}$$
$$\times \left[\delta_0 \check{G}^{(3)} + \delta_1 \check{G}^{(3)} + \delta_2 \check{G}^{(3)} + \delta_3 \check{G}^{(3)}\right]$$

where, for example,

$$\delta_1 \check{G}^{(3)} = (-1)^3 G^{(0)}_{\epsilon+\omega_1} \check{H}_{\omega_1} \left[G^{(0)}_{\epsilon+i0} u G^{(0)}_{\epsilon+i0} - G^{(0)}_{\epsilon-i0} u G^{(0)}_{\epsilon-i0}\right]$$
$$\times \check{H}_{\omega_2} G^{(0)}_{\epsilon-\omega_2} u G^{(0)}_{\epsilon-\omega_2} \check{H}_{\omega_3} G^{(0)}_{z-\omega_2-\omega_3}.$$

Note that the Green functions on both sides of the impurity potential U have the same frequencies because the impurity scattering is elastic and does not change the particle energy. Proceeding in the same way as before, we obtain for this particular term of the third order in $\check{H}_\omega$ and of the second order in U, after summation over all cuts:

$$\begin{aligned}
&\check{G}^{(3,2)K}_{\epsilon,\epsilon-\omega} \\
&= -G^{(0)R}_{\epsilon}\check{H}_{\omega_1}G^{(0)R}_{\epsilon-\omega_1}uG^{(0)R}_{\epsilon-\omega_1}\check{H}_{\omega_2}G^{(0)R}_{\epsilon-\omega_1-\omega_2}uG^{(0)R}_{\epsilon-\omega_1-\omega_2}\check{H}_{\omega_3}G^{(0)R}_{\epsilon-\omega}f^{(0)}(\epsilon-\omega) \\
&\quad +G^{(0)R}_{\epsilon}\check{H}_{\omega_1}G^{(0)R}_{\epsilon-\omega_1}uG^{(0)R}_{\epsilon-\omega_1}\check{H}_{\omega_2}G^{(0)R}_{\epsilon-\omega_1-\omega_2}uG^{(0)R}_{\epsilon-\omega_1-\omega_2}\check{h}_{\epsilon-\omega_1-\omega_2,\epsilon-\omega}G^{(0)A}_{\epsilon-\omega} \\
&\quad +G^{(0)R}_{\epsilon}\check{H}_{\omega_1}G^{(0)R}_{\epsilon-\omega_1}uG^{(0)R}_{\epsilon-\omega_1}\check{h}_{\epsilon-\omega_1,\epsilon-\omega_1-\omega_2}G^{(0)A}_{\epsilon-\omega_1-\omega_2}uG^{(0)A}_{\epsilon-\omega_1-\omega_2}\check{H}_{\omega_3}G^{(0)A}_{\epsilon-\omega} \\
&\quad +G^{(0)R}_{\epsilon}\check{h}_{\epsilon,\epsilon-\omega_1}G^{(0)A}_{\epsilon-\omega_1}uG^{(0)A}_{\epsilon-\omega_1}\check{H}_{\omega_2}G^{(0)A}_{\epsilon-\omega_1-\omega_2}uG^{(0)A}_{\epsilon-\omega_1-\omega_2}\check{H}_{\omega_3}G^{(0)A}_{\epsilon-\omega} \\
&\quad +f^{(0)}(\epsilon)G^{(0)A}_{\epsilon}\check{H}_{\omega_1}G^{(0)A}_{\epsilon-\omega_1}uG^{(0)A}_{\epsilon-\omega_1}\check{H}_{\omega_2}G^{(0)A}_{\epsilon-\omega_1-\omega_2}uG^{(0)A}_{\epsilon-\omega_1-\omega_2}\check{H}_{\omega_3}G^{(0)A}_{\epsilon-\omega}.
\end{aligned}$$

After averaging over positions of impurity atoms, we can present this term as

$$\begin{aligned}
\check{G}^{(3,2)K}_{\epsilon,\epsilon-\omega} &= \check{G}^{(3,2)R}_{\epsilon,\epsilon-\omega}f^{(0)}(\epsilon-\omega) - f^{(0)}(\epsilon)\check{G}^{(3,2)A}_{\epsilon,\epsilon-\omega} \\
&+\check{G}^{(2,2)R}_{\epsilon,\epsilon-\omega_1-\omega_2}\check{h}_{\epsilon-\omega_1-\omega_2,\epsilon-\omega}G^{(0)A}_{\epsilon-\omega} + G^{(0)R}_{\epsilon}\check{h}_{\epsilon,\epsilon-\omega_1}\check{G}^{(2,2)A}_{\epsilon-\omega_1,\epsilon-\omega} \\
&+\check{G}^{(1,0)R}_{\epsilon,\epsilon-\omega_1}\left(n_{imp}\int\frac{d^3p}{(2\pi)^3}|u(\theta)|^2 G^{(0)R}_{\epsilon-\omega_1}\check{h}_{\epsilon-\omega_1,\epsilon-\omega_1-\omega_2}G^{(0)A}_{\epsilon-\omega_1-\omega_2}\right) \\
&\times\check{G}^{(1,0)A}_{\epsilon,\epsilon-\omega}.
\end{aligned}$$

It is clear that, collecting all orders in $\check{H}_\omega$ and in the impurity potential, we again obtain eqn (8.12) where the anomalous function now is

$$\begin{aligned}
\check{G}^{(a)}_{\epsilon,\epsilon-\omega}(\mathbf{p},\mathbf{p}-\mathbf{k}) = \int\frac{d\epsilon' d\omega'}{(2\pi)^2}\frac{d^3k'\,d^3k''}{(2\pi)^6}\check{G}^{R}_{\epsilon,\epsilon'}(\mathbf{p},\mathbf{p}-\mathbf{k}') \\
\times\left[\check{h}_{\epsilon',\epsilon'-\omega'}(\mathbf{k}'') + \check{\Sigma}^{(a)}_{\epsilon',\epsilon'-\omega'}(\mathbf{p},\mathbf{k}'')\right] \\
\times\check{G}^{A}_{\epsilon'-\omega',\epsilon-\omega}(\mathbf{p}-\mathbf{k}'-\mathbf{k}'',\mathbf{p}-\mathbf{k}).
\end{aligned} \tag{8.16}$$

Sometimes, we shall use a symbolic form

$$\check{G}^{(a)} = \check{G}^{R}\left(\check{h} + \check{\Sigma}^{(a)}\right)\check{G}^{A} \tag{8.17}$$

where the frequencies and momenta are not shown, and the integration over all internal variables is assumed. Equation (8.16) differs from eqn (8.15) by the presence of the anomalous impurity self-energy $\Sigma^{(a)}$ which necessarily contains both retarder and advanced functions

$$\check{\Sigma}^{(a)}_{\epsilon,\epsilon-\omega}(\mathbf{p},\mathbf{k}) = n_{\text{imp}}\int\frac{d^3p'}{(2\pi)^3}|u(\theta)|^2\check{G}^{(a)}_{\epsilon,\epsilon-\omega}(\mathbf{p}',\mathbf{p}'-\mathbf{k}). \tag{8.18}$$

Here θ is the angle between $\mathbf{p}$ and $\mathbf{p}'$. The self-energy is obtained from $\check{G}^{(a)}_{\epsilon,\epsilon-\omega}$ in the same way as the self-energy $\check{\Sigma}^{R(A)}_{\epsilon,\epsilon-\omega}$ is obtained from the regular Green functions $\check{G}^{R(A)}_{\epsilon,\epsilon-\omega}$:

$$\check{\Sigma}^{R(A)}_{\epsilon,\epsilon-\omega}(\mathbf{p},\mathbf{k}) = n_{\text{imp}} \int \frac{d^3p'}{(2\pi)^3} |u(\theta)|^2 \check{G}^{R(A)}_{\epsilon,\epsilon-\omega}(\mathbf{p}',\mathbf{p}'-\mathbf{k}).$$

The regular functions are determined by the sum of diagrams of all orders in $\check{H}$ and $\check{\Sigma}^{R(A)}$, therefore, they satisfy the usual equations

$$\int \frac{d\omega_1}{2\pi} \frac{d^3k_1}{(2\pi)^3} \left[\check{G}^{-1}_{\epsilon}(\mathbf{p}-\mathbf{k}_1,\omega_1) - \check{\Sigma}^{R(A)}_{\epsilon,\epsilon-\omega_1}(\mathbf{p},\mathbf{k}_1)\right] \times \check{G}^{R(A)}_{\epsilon-\omega_1,\epsilon-\omega}(\mathbf{p}-\mathbf{k}_1,\mathbf{p}-\mathbf{k}) = \check{1}\cdot(2\pi)^4\delta(\omega)\,\delta(\mathbf{k}) \tag{8.19}$$

where the inverse operator $\check{G}^{-1}_{\epsilon}(\mathbf{p},\omega_1)$ is determined by eqn (8.3). Equation (8.19) in the symbolic form is

$$\left(\check{G}^{-1} - \check{\Sigma}^{R(A)}\right)\check{G}^{R(A)} = \check{1}. \tag{8.20}$$

One can obtain an equation for the anomalous function $\check{G}^{(a)}_{\epsilon,\epsilon-\omega}$ as well. Let us apply the operator in eqn (8.3) to eqn (8.16) from the left. We obtain, in a symbolic form,

$$\left(\check{G}^{-1} - \check{\Sigma}^{R}\right)\check{G}^{(a)} = \left(\check{h} + \check{\Sigma}^{(a)}\right)\check{G}^{A}. \tag{8.21}$$

Using the definition of the total time-dependent function $\check{G}^{K}$ and equations for the retarded and advanced functions one can prove that the total function $\check{G}^{K}$ satisfies a homogeneous equation of the type of eqn (8.21). Indeed, let us take the combination

$$\check{G}^{(r)}_{\epsilon,\epsilon-\omega} = \check{G}^{R}_{\epsilon,\epsilon-\omega}f^{(0)}(\epsilon-\omega) - f^{(0)}(\epsilon)\,\check{G}^{A}_{\epsilon,\epsilon-\omega} \tag{8.22}$$

and apply to it the operator $\check{G}^{-1}_{\epsilon} - \check{\Sigma}^{R}_{\epsilon,\epsilon'}$ from the left. We get

$$\left(\check{G}^{-1} - \check{\Sigma}^{R}\right)\check{G}^{(r)} = f^{(0)}(\epsilon)\left[1 - \left(\check{G}^{-1} - \check{\Sigma}^{A}\right)\check{G}^{A}\right] - \check{h}\check{G}^{A} + \left[\check{\Sigma}^{R}f^{(0)}(\epsilon') - f^{(0)}(\epsilon)\,\check{\Sigma}^{A}\right]\check{G}^{A}. \tag{8.23}$$

Adding eqns (8.21) and (8.23) together, we find

$$\left(\check{G}^{-1} - \check{\Sigma}^{R}\right)\check{G}^{K} - \check{\Sigma}^{K}\check{G}^{A} = 0. \tag{8.24}$$

Here

$$\check{\Sigma}^{K}_{\epsilon,\epsilon'} = \check{\Sigma}^{R}_{\epsilon,\epsilon'}f^{(0)}(\epsilon') - f^{(0)}(\epsilon)\,\check{\Sigma}^{A}_{\epsilon,\epsilon'} + \check{\Sigma}^{(a)}_{\epsilon,\epsilon'} \tag{8.25}$$

is the total self-energy. It is clear that

$$\check{\Sigma}^{K}_{\epsilon,\epsilon-\omega}(\mathbf{p},\mathbf{k}) = n_{\text{imp}} \int \frac{d^3p'}{(2\pi)^3} |u(\theta)|^2 \check{G}^{K}_{\epsilon,\epsilon-\omega}(\mathbf{p}',\mathbf{p}'-\mathbf{k}).$$

Similarly, applying the operator $\check{G}^{-1} - \check{\Sigma}^{A}$ from the right, we find

$$\check{G}^{(a)}\left(\check{G}^{-1} - \check{\Sigma}^{A}\right) = \check{G}^{R}\left(\check{h} + \check{\Sigma}^{(a)}\right) \tag{8.26}$$

and

$$\check{G}^{K}\left(\check{G}^{-1} - \check{\Sigma}^{A}\right) - \check{G}^{R}\check{\Sigma}^{K} = 0. \tag{8.27}$$

8.1.3 *Order parameter, current, and particle density*

The self-consistency order parameter equation can easily be obtained using the general equation for the order parameter in terms of the Matsubara Green function and the rules we have derived here for calculating the sums over the Matsubara frequencies in terms of real-frequency integrations, eqn (8.11). We find for an s-wave pairing

$$\frac{\Delta(\mathbf{k},\omega)}{|g|} = \int \frac{d^3p}{(2\pi)^3}\frac{d\epsilon}{4\pi i} F^K_{\epsilon_+,\epsilon_-}(\mathbf{p}_+,\mathbf{p}_-),$$
$$\frac{\Delta^*(\mathbf{k},\omega)}{|g|} = \int \frac{d^3p}{(2\pi)^3}\frac{d\epsilon}{4\pi i} F^{\dagger K}_{\epsilon_+,\epsilon_-}(\mathbf{p}_+,\mathbf{p}_-). \quad (8.28)$$

In the same way, the current takes the form

$$\mathbf{j}(\omega,\mathbf{k}) = -\frac{2e}{m}\int \frac{d^3p}{(2\pi)^3}\frac{d\epsilon}{4\pi i}\mathbf{p}G^K_{\epsilon_+,\epsilon_-}(\mathbf{p}_+,\mathbf{p}_-) - \frac{Ne^2}{mc}\mathbf{A}$$
$$= \frac{2e}{m}\int \frac{d^3p}{(2\pi)^3}\frac{d\epsilon}{4\pi i}\mathbf{p}\bar{G}^K_{\epsilon_+,\epsilon_-}(\mathbf{p}_+,\mathbf{p}_-) - \frac{Ne^2}{mc}\mathbf{A}. \quad (8.29)$$

The electron density becomes

$$N(\omega,\mathbf{k}) = N_0 - 2\int \frac{d^3p}{(2\pi)^3}\frac{d\epsilon}{4\pi i}\delta G^K_{\epsilon_+,\epsilon_-}(\mathbf{p}_+,\mathbf{p}_-)$$
$$= N_0 - 2\int \frac{d^3p}{(2\pi)^3}\frac{d\epsilon}{4\pi i}\delta\bar{G}^K_{\epsilon_+,\epsilon_-}(\mathbf{p}_+,\mathbf{p}_-). \quad (8.30)$$

8.2 The phonon model

8.2.1 *Self-energy*

Consider how one can generalize this approach for the phonon model where the pairing between electrons occurs due to their interaction with phonons. In the phonon model by Eliashberg (1960), the order parameter itself is regarded not as a "field" but rather as a self-energy in the Gor'kov Green function F. The starting point is the equation for the Matsubara Green functions eqn (8.2) where $\epsilon_n = 2\pi i T(n+1/2)$. The inverse operator $\check{G}^{-1}_{\epsilon_n}$ is defined as in eqn (8.3), the difference being that the "field" $\check{H}$ contains only the electromagnetic potentials. The phonon self-energy has the same diagram structure as the impurity self-energy shown in Fig. 4.2(b) where the dashed line stands for the phonon Green function D while the crosses denote the matrix element g_{ph} of the electron–phonon interaction:

$$\check{\Sigma}^{(\mathrm{ph})}_{\epsilon,\epsilon-\omega}(\mathbf{p},\mathbf{p}-\mathbf{k}) = T\sum_{\epsilon'}\int \frac{d^3p'}{(2\pi)^3} g^2_{\mathrm{ph}} D_{\epsilon'-\epsilon}(\mathbf{p}'-\mathbf{p})\,\check{G}_{\epsilon',\epsilon'-\omega}(\mathbf{p}',\mathbf{p}'-\mathbf{k}). \quad (8.31)$$

Here $\epsilon' - \epsilon = 2\pi i T m$. For simplicity, we do not include impurities here. We consider phonons as "free" particles assuming that their interaction with electrons

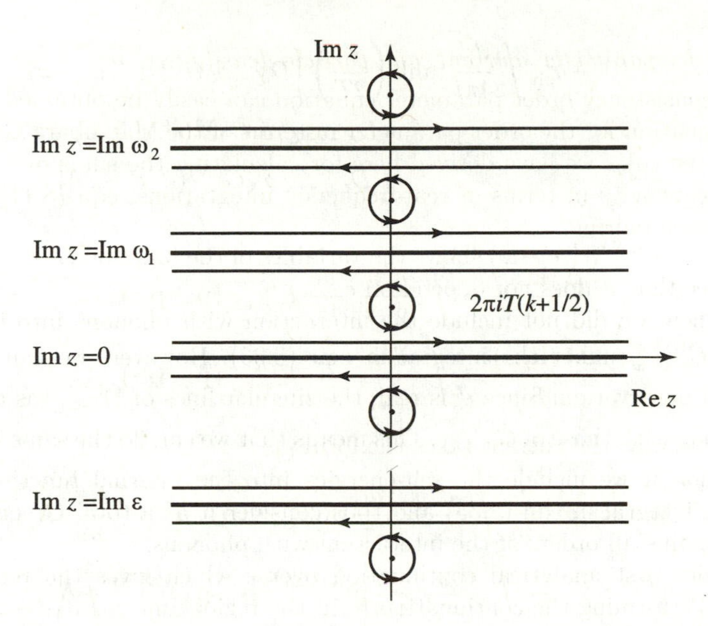

FIG. 8.3. The complex plane of frequencies with cuts along the singular lines of $\check{G}^{(2)}_{z,z-\omega}$ and of $D_{z-\epsilon}$.

does not change their properties considerably (Migdal 1958). According to eqn (2.57), the phonon Green function is

$$D_{\epsilon'-\epsilon}(\mathbf{p}'-\mathbf{p}) = \frac{\omega^2_{\mathbf{p}'-\mathbf{p}}}{\omega^2_{\mathbf{p}'-\mathbf{p}} - (\epsilon'-\epsilon)^2}. \tag{8.32}$$

Consider a contribution of the N-th order in field to the self-energy in eqn (8.31) and transform the sum over the Matsubara frequency ϵ' into the contour integral. We have

$$\check{\Sigma}^{(N)}_{\epsilon,\epsilon-\omega} = g^2_{\text{ph}} \oint \frac{dz}{4\pi i} \tanh\left(\frac{z}{2T}\right) \left[D_{z-\epsilon}\check{G}^{(N)}_{z,z-\omega}\right].$$

We do not show integration over d^3p', for brevity. Singularities of the function in the square brackets are at the lines $\text{Im}\,(z-\omega_j) = 0$ where $\check{G}^{(N)}_{z,z-\omega}$ has its poles, and at the line $\text{Im}\,(z-\epsilon) = 0$ where the poles of D are located (see Fig. 8.3). We make cuts along these singular lines, and then unroll the contours along the cuts. Doing this, we put $z = \epsilon' + i\text{Im}\omega_j$ at those cuts which correspond to the singular lines of $\check{G}^{(N)}_{z,z-\omega}$ and $z = \epsilon' + i\text{Im}\epsilon_n$ at the cut which corresponds to the singular line of the function D. Note that both $\epsilon = 2\pi iT(n+1/2)$ and $\omega_j = 2\pi iTk_j$ are still imaginary, but ϵ' is already a real variable. We obtain

$$\check{\Sigma}^{(N)}_{\epsilon,\epsilon-\omega} = g^2_{\text{ph}} \int \frac{d\epsilon'}{4\pi i} \left[\coth\left(\frac{\epsilon'}{2T}\right)\left(D^R_{\epsilon'} - D^A_{\epsilon'}\right)\check{G}^{(N)}_{\epsilon'+\epsilon,\epsilon'+\epsilon-\omega}\right]$$

$$+g_{\text{ph}}^2 \int \frac{d\epsilon'}{4\pi i} \tanh\left(\frac{\epsilon'}{2T}\right) \left[D_{\epsilon'-\epsilon}\delta_0 \check{G}^{(N)}(\epsilon', \epsilon' - \omega) \right.$$
$$+ D_{\epsilon'-\epsilon+\omega_1}\delta_1 \check{G}^{(N)}(\epsilon', \epsilon' - \omega) + \ldots$$
$$\left. + D_{\epsilon'-\epsilon+\omega}\delta_N \check{G}^{(N)}(\epsilon', \epsilon' - \omega) \right]. \quad (8.33)$$

The notation $\check{G}^{(N)}(\epsilon', \epsilon' - \omega)$ shows the variables of the function $\check{G}$; in particular, one can see that it does not depend on ϵ.

Until now we did not include the interaction with phonons into the Green function $\check{G}^{(N)}_{\epsilon,\epsilon-\omega}$ under the integral in eqn (8.33). However, we can make an important observation. Since ϵ' is real, the singular lines of $\check{\Sigma}^{(N)}_{\epsilon,\epsilon-\omega}$ as a function of ϵ coincide with those of $\check{G}^{(N)}_{\epsilon,\epsilon-\omega}$. This means that we can do the same analytical continuation if we include the self-energies into the internal function $\check{G}^{(N)}_{z,z-\omega}$ under the integral in eqn (8.33) and thus consider it as a total Green function which contains all orders of the interaction with phonons.

Consider first analytical continuation over ϵ which gives the regular self-energies. Performing the continuation from the region $\text{Im}\epsilon > \text{Im}\,(\epsilon - \omega) > 0$ we get the retarded self-energy. We now shift the integration variable $\epsilon' \Rightarrow \epsilon' - \sum_{l=0}^{j} \omega_l$ at each cut belonging to $\check{G}^{(N)}$ and also the shift $\epsilon' \Rightarrow \epsilon' - \epsilon$ at the cut for D. We have

$$\check{\Sigma}^{(N)R}_{\epsilon,\epsilon-\omega} = g_{\text{ph}}^2 \int \frac{d\epsilon'}{4\pi i} \left[\coth\left(\frac{\epsilon' - \epsilon}{2T}\right) \left(D^R_{\epsilon'-\epsilon} - D^A_{\epsilon'-\epsilon}\right) \check{G}^{(N)R}_{\epsilon',\epsilon'-\omega} \right.$$
$$\left. + D^A_{\epsilon'-\epsilon} \check{G}^{(N)K}_{\epsilon',\epsilon'-\omega} \right]$$

where $\check{G}^{(N)K}_{\epsilon',\epsilon'-\omega}$ is defined by eqn (8.8). Collecting all orders in N, we obtain

$$\check{\Sigma}^{(\text{ph})R}_{\epsilon,\epsilon-\omega}(\mathbf{p}, \mathbf{p} - \mathbf{k}) = g_{\text{ph}}^2 \int \frac{d\epsilon'}{4\pi i} \frac{d^3p'}{(2\pi)^3}$$
$$\times \left[\coth\left(\frac{\epsilon' - \epsilon}{2T}\right) \left(D^R_{\epsilon'-\epsilon} - D^A_{\epsilon'-\epsilon}\right) \check{G}^R_{\epsilon',\epsilon'-\omega} + D^A_{\epsilon'-\epsilon} \check{G}^K_{\epsilon',\epsilon'-\omega} \right]. \quad (8.34)$$

Similarly, continuing eqn (8.33) over ϵ from the region $\text{Im}\,(\epsilon - \omega) < \text{Im}\epsilon < 0$ and collecting all orders in N, we obtain the advanced self-energy

$$\check{\Sigma}^{(\text{ph})A}_{\epsilon,\epsilon-\omega}(\mathbf{p}, \mathbf{p} - \mathbf{k}) = g_{\text{ph}}^2 \int \frac{d\epsilon'}{4\pi i} \frac{d^3p'}{(2\pi)^3}$$
$$\times \left[\coth\left(\frac{\epsilon' - \epsilon}{2T}\right) \left(D^R_{\epsilon'-\epsilon} - D^A_{\epsilon'-\epsilon}\right) \check{G}^A_{\epsilon',\epsilon'-\omega} + D^R_{\epsilon'-\epsilon} \check{G}^K_{\epsilon',\epsilon'-\omega} \right]. \quad (8.35)$$

The phonon functions $D^{R(A)}(\mathbf{p} - \mathbf{p}')$ depend on the momentum transfer $\mathbf{p} - \mathbf{p}'$, while the electron functions have the variables $\check{G}_{\epsilon',\epsilon'-\omega}(\mathbf{p}', \mathbf{p}' - \mathbf{k})$. The total function $\check{G}^K_{\epsilon',\epsilon'-\omega}$ is determined by the same equations (8.12) and (8.16) as for the case with impurities

$$\check{G}^K_{\epsilon,\epsilon-\omega} = \sum_{N=0}^{\infty}\sum_{l=0}^{N}[\delta_l \check{G}^{(N)}]_{\epsilon,\epsilon-\omega} \tanh\left(\frac{\epsilon - \sum_{j=0}^{l}\omega_j}{2T}\right)$$
$$= \check{G}^R_{\epsilon,\epsilon-\omega} f^{(0)}(\epsilon-\omega) - f^{(0)}(\epsilon)\, \check{G}^A_{\epsilon,\epsilon-\omega} + \check{G}^{(a)}_{\epsilon,\epsilon-\omega}.$$

The retarded and advanced functions are defined as sums of all the contributions which contain only the retarded G^R and $\check{\Sigma}^R$ or only the advanced ones, G^A and $\check{\Sigma}^A$, respectively. The equation for retarded and advanced functions can thus be derived by the analytical continuation of the corresponding equation for the Matsubara functions from the region $\mathrm{Im}\epsilon > \mathrm{Im}(\epsilon-\omega) > 0$ and $\mathrm{Im}(\epsilon-\omega) < \mathrm{Im}\epsilon < 0$, respectively:

$$\left(\check{G}^{-1} - \check{\Sigma}^{(\mathrm{ph})R}\right)\check{G}^R = \check{1}, \quad \left(\check{G}^{-1} - \check{\Sigma}^{(\mathrm{ph})A}\right)\check{G}^A = \check{1}. \tag{8.36}$$

The total self-energy is defined exactly as in eqn (8.25):

$$\check{\Sigma}^K_{\epsilon,\epsilon-\omega} = \sum_{N=0}^{\infty}\sum_{l=0}^{N}[\delta_l \check{\Sigma}^{(N)}]_{\epsilon,\epsilon-\omega} \tanh\left(\frac{\epsilon - \sum_{j=0}^{l}\omega_j}{2T}\right)$$
$$= \check{\Sigma}^R_{\epsilon,\epsilon-\omega} f^{(0)}(\epsilon-\omega) - f^{(0)}(\epsilon)\, \check{\Sigma}^A_{\epsilon,\epsilon-\omega} + \check{\Sigma}^{(a)}_{\epsilon,\epsilon-\omega}. \tag{8.37}$$

Here $\check{\Sigma}^{(a)}$ necessarily contains both retarded and advanced functions.

We now need to express the total self-energy in eqn (8.37) through the total Green functions. Consider one of the jumps in the N-th order self-energy defined by eqn (8.33); it is

$$[\delta_l \check{\Sigma}^{(N)}]_{z,z-\omega} \tanh\left(\frac{z_l}{2T}\right)$$
$$= g_{\mathrm{ph}}^2 \int \frac{d\epsilon'}{4\pi i}\left[\coth\left(\frac{\epsilon'}{2T}\right)\left(D^R_{\epsilon'} - D^A_{\epsilon'}\right)[\delta_l \check{G}^{(N)}]_{\epsilon'+z,\epsilon'+z-\omega} \tanh\left(\frac{z_l}{2T}\right)\right]$$
$$+ g_{\mathrm{ph}}^2 \int \frac{d\epsilon'}{4\pi i} \tanh\left(\frac{\epsilon'}{2T}\right)\left[\tanh\left(\frac{z_l}{2T}\right)\delta_l D_{\epsilon'-z_l}\,\delta_l \check{G}^{(N)}(\epsilon',\epsilon'-\omega)\right]. \tag{8.38}$$

Here $z_l = z - \sum_{j=0}^{l}\omega_j$. We took into account that the jump of the phonon Green function at the cut l is nonzero only for the term with $\mathrm{Im}(\epsilon' - z + \sum_{j=0}^{l}\omega_j) = 0$.

The jump in the second line is taken at the cut where $\mathrm{Im}(\epsilon' + z - \sum_{j=0}^{l}\omega_j) = 0$. Putting $z = \epsilon + \sum_{j=0}^{l}\omega_j$ where ϵ is real and all ω_j are still imaginary, we obtain

$$[\delta_l \check{G}^{(N)}]_{\epsilon'+z,\epsilon'+z-\omega} \tanh\left(\frac{z_l}{2T}\right)$$
$$= \tanh\left(\frac{\epsilon}{2T}\right) G_{\epsilon'+\epsilon+\sum_j^l \omega_j} \ldots G_{\epsilon'+\epsilon+\omega_l} \check{H}_{\omega_l}$$

$$\times \left[G_{\epsilon'+\epsilon+i0} - G_{\epsilon'+\epsilon-i0}\right] \check{H}_{\omega_{l+1}} \ldots G_{\epsilon'+\epsilon-\omega+\sum_j^l \omega_j} \,. \tag{8.39}$$

Performing the shift of the integration variable $\epsilon + \epsilon' = \epsilon_1$, we have under the integral

$$\tanh\left(\frac{\epsilon}{2T}\right) \coth\left(\frac{\epsilon_1 - \epsilon}{2T}\right) \left(D^R_{\epsilon_1-\epsilon} - D^A_{\epsilon_1-\epsilon}\right) \delta_l \check{G}(\epsilon_1, \epsilon_1 - \omega).$$

The jump in the third line of eqn (8.38) is taken at the cut where $\mathrm{Im}(\epsilon' - z + \sum_{j=0}^l \omega_j) = 0$. Putting $z = \epsilon + \sum_{j=0}^l \omega_j$ in the third line of eqn (8.38) we get

$$\begin{aligned}
\tanh\left(\frac{z_l}{2T}\right) \delta_l D_{\epsilon'-z+\sum_j^l \omega_j} &= \tanh\left(\frac{\epsilon}{2T}\right) \left(D_{\epsilon'-\epsilon-i0} - D_{\epsilon'-\epsilon+i0}\right) \\
&= \tanh\left(\frac{\epsilon}{2T}\right) \left(D^A_{\epsilon'-\epsilon} - D^R_{\epsilon'-\epsilon}\right).
\end{aligned} \tag{8.40}$$

The full expression for the jump becomes

$$\begin{aligned}
&[\delta_l \check{\Sigma}^{(N)}]_{z,z-\omega} \tanh\left(\frac{z_l}{2T}\right) \\
&= g^2_{\mathrm{ph}} \int \frac{d\epsilon'}{4\pi i} \left[\coth\left(\frac{\epsilon' - \epsilon}{2T}\right) \tanh\left(\frac{\epsilon}{2T}\right) - \tanh\left(\frac{\epsilon'}{2T}\right) \tanh\left(\frac{\epsilon}{2T}\right)\right] \\
&\quad \times \delta_l \check{G}^{(N)}(\epsilon', \epsilon' - \omega) \left(D^R_{\epsilon'-\epsilon} - D^A_{\epsilon'-\epsilon}\right) \\
&= g^2_{\mathrm{ph}} \int \frac{d\epsilon'}{4\pi i} \left[\coth\left(\frac{\epsilon' - \epsilon}{2T}\right) \tanh\left(\frac{\epsilon'}{2T}\right) - 1\right] \\
&\quad \times \delta_l \check{G}^{(N)}(\epsilon', \epsilon' - \omega) \left(D^R_{\epsilon'-\epsilon} - D^A_{\epsilon'-\epsilon}\right).
\end{aligned} \tag{8.41}$$

Here we use the identity

$$\tanh\left(\frac{\epsilon'}{2T}\right) \tanh\left(\frac{\epsilon}{2T}\right) = 1 - \coth\left(\frac{\epsilon' - \epsilon}{2T}\right) \left[\tanh\left(\frac{\epsilon'}{2T}\right) - \tanh\left(\frac{\epsilon}{2T}\right)\right].$$

Let us sum up all the jumps and collect all orders in N. We note that

$$\int \frac{d\epsilon'}{4\pi i} \sum_{N=0}^{\infty} \sum_l \delta_l \check{G}^{(N)}(\epsilon', \epsilon' - \omega) = \int \frac{d\epsilon'}{4\pi i} \left[\check{G}^R_{\epsilon',\epsilon'-\omega} - \check{G}^A_{\epsilon',\epsilon'-\omega}\right]$$

and

$$\begin{aligned}
&\sum_{N=0}^{\infty} \tanh\left(\frac{\epsilon'}{2T}\right) \sum_l \delta_l \check{G}^{(N)}(\epsilon', \epsilon' - \omega) \\
&\qquad = \sum_{N=0}^{\infty} \sum_l [\delta_l \check{G}^{(N)}]_{\epsilon',\epsilon'-\omega} \tanh\left(\frac{\epsilon' - \sum_{j=0}^l \omega_j}{2T}\right) \\
&\qquad = \check{G}^K(\epsilon', \epsilon' - \omega).
\end{aligned}$$

Finally, we arrive at the expression for the total self-energy

$$\check{\Sigma}^K_{\epsilon,\epsilon-\omega}(\mathbf{p}, \mathbf{p} - \mathbf{k}) = g^2_{\mathrm{ph}} \int \frac{d\epsilon'}{4\pi i} \frac{d^3 p'}{(2\pi)^3} \left(D^R_{\epsilon'-\epsilon} - D^A_{\epsilon'-\epsilon}\right)$$

$$\times \left[\coth\left(\frac{\epsilon' - \epsilon}{2T}\right) \check{G}^K_{\epsilon',\epsilon'-\omega} - (\check{G}^R_{\epsilon',\epsilon'-\omega} - \check{G}^A_{\epsilon',\epsilon'-\omega}) \right]. \quad (8.42)$$

Here the phonon functions $D(\mathbf{p}'-\mathbf{p})$ depend on the momentum transfer while the electron functions are of the form of $\check{G}(\mathbf{p}', \mathbf{p}'-\mathbf{k})$.

8.2.2 *Order parameter*

In the phonon model, the order parameter is defined as a part of the self-energy

$$\Delta_{\epsilon,\epsilon-\omega}(\mathbf{p}, \mathbf{p}-\mathbf{k}) = \frac{g^2_{\rm ph}}{2} \int \frac{d\epsilon'}{4\pi i} \frac{d^3p'}{(2\pi)^3} \left(D^R_{\epsilon'-\epsilon} + D^A_{\epsilon'-\epsilon}\right) F^K_{\epsilon',\epsilon'-\omega}, \quad (8.43)$$

$$\Delta^*_{\epsilon,\epsilon-\omega}(\mathbf{p}, \mathbf{p}-\mathbf{k}) = \frac{g^2_{\rm ph}}{2} \int \frac{d\epsilon'}{4\pi i} \frac{d^3p'}{(2\pi)^3} \left(D^R_{\epsilon'-\epsilon} + D^A_{\epsilon'-\epsilon}\right) F^{\dagger K}_{\epsilon',\epsilon'-\omega}. \quad (8.44)$$

The frequencies of the order of Δ, i.e., in the range $\epsilon \ll \Omega_D$ where Ω_D is the Debye frequency, are only important because the function F^K vanishes for $\epsilon \gg \Delta$. For such frequencies, one can put $(D^R + D^A)/2 = 1$. As a result, Δ is actually independent of ϵ and of the momentum $\mathbf{p}$. It can be written simply in the form of a BCS-type s-wave self-consistency equation

$$\Delta_\omega(\mathbf{k}) = g^2_{\rm ph} \int \frac{d\epsilon}{4\pi i} \frac{d^3p}{(2\pi)^3} F^K_{\epsilon,\epsilon-\omega}. \quad (8.45)$$

which coincides with eqn (8.28) if the square of the matrix element of the electron–phonon interaction $g^2_{\rm ph}$ is replaced with $|g|$. We see that the phonon model in the form of eqn (8.32) favors an s-wave pairing.

Separating Δ from the self-energy we obtain

$$\Sigma^R_2 = \tilde{\Sigma}^R_2 + \Delta;\ \Sigma^A_2 = \tilde{\Sigma}^A_2 + \Delta$$

where

$$\tilde{\Sigma}^R_2 = g^2_{\rm ph} \int \frac{d\epsilon'}{4\pi i} \frac{d^3p'}{(2\pi)^3} \left(D^R_{\epsilon'-\epsilon} - D^A_{\epsilon'-\epsilon}\right) \times \left[\coth\left(\frac{\epsilon' - \epsilon}{2T}\right) F^R_{\epsilon',\epsilon'-\omega} - \frac{1}{2} F^K_{\epsilon',\epsilon'-\omega} \right],$$

and

$$\tilde{\Sigma}^A_2 = g^2_{\rm ph} \int \frac{d\epsilon'}{4\pi i} \frac{d^3p'}{(2\pi)^3} \left(D^R_{\epsilon'-\epsilon} - D^A_{\epsilon'-\epsilon}\right) \times \left[\coth\left(\frac{\epsilon' - \epsilon}{2T}\right) F^A_{\epsilon',\epsilon'-\omega} + \frac{1}{2} F^K_{\epsilon',\epsilon'-\omega} \right],$$

and similarly for $\tilde{\Sigma}^{\dagger R}_2$ and $\tilde{\Sigma}^{\dagger A}_2$.

The diagonal components of the self-energy determine the renormalization of the chemical potential

$$\delta\mu = \frac{g_{ph}^2}{2}\int \frac{d\epsilon'}{4\pi i}\frac{d^3p'}{(2\pi)^3}\left(D^R_{\epsilon'-\epsilon} + D^A_{\epsilon'-\epsilon}\right) G^K_{\epsilon',\epsilon'-\omega}. \tag{8.46}$$

We can separate the contribution from the normal state by writing $G^K = G^K_n + \left(G^K - G^K_n\right)$. For the difference $G^K - G^K_n$, the frequencies $\epsilon \sim \Delta$ are important. Therefore, we can neglect ϵ compared with Ω_D in the phonon Green functions for this term. For the first term, the frequencies $\epsilon \sim \Omega_D$ only participate. For such frequencies, the Green functions coincide with their values for the normal state with the accuracy $(\Delta/\Omega_D)^2$; the difference can thus be neglected in the weak coupling approximation. We have

$$\delta\mu = \delta\mu_{\rm norm} + g_{\rm ph}^2\int \frac{d\epsilon}{4\pi i}\frac{d^3p}{(2\pi)^3}\left(G^K_{\epsilon,\epsilon-\omega} - G^K_{\epsilon,\epsilon-\omega,\,{\rm norm}}\right) \tag{8.47}$$

where $\delta\mu$ is the renormalization of the chemical potential in the normal state due to the interaction with phonons. It only depends on the normal state properties and can be disregarded. The second term in the r.h.s. is $\frac{1}{2}g_{\rm ph}^2\delta N$ where δN is the difference between the densities of electrons in the superconducting and in the normal state according to eqn (8.30). In a good metal with a strong Coulomb interaction between electrons and ions of the crystalline lattice, the electronic density is equal to the ion density, and does not change at a transition to the superconducting state. If the Coulomb interaction is weak (or in case of electrically neutral systems), the density change is $\delta N \approx \nu(0)\delta\mu_{sn}$ where $\delta\mu_{sn}$ is the change in the chemical potential at the transition. In the weak coupling approximation, where the electron–phonon interaction constant $\lambda = \nu(0)\, g_{\rm ph}^2 \ll 1$, this term can also be neglected as compared to $\delta\mu_{sn}$ itself. As a result, the diagonal term eqn (8.46) can be included into the normal-state chemical potential and excluded from the consideration.

From now on we denote by $\check\Sigma^{R(A)}$ the new self-energy matrix which does not contain the order parameter or the renormalization of the chemical potential

$$\check\Sigma^{({\rm ph})R(A)}_{\epsilon,\epsilon-\omega}(\mathbf{p},\mathbf{p}-\mathbf{k}) = g_{\rm ph}^2\int \frac{d\epsilon'}{4\pi i}\frac{d^3p'}{(2\pi)^3}\left(D^R_{\epsilon'-\epsilon} - D^A_{\epsilon'-\epsilon}\right) \times\left[\coth\left(\frac{\epsilon'-\epsilon}{2T}\right)\check G^{R(A)}_{\epsilon',\epsilon'-\omega} \mp \frac{1}{2}\check G^K_{\epsilon',\epsilon'-\omega}\right] \tag{8.48}$$

where $\check G_{\epsilon',\epsilon'-\omega} \equiv \check G_{\epsilon',\epsilon'-\omega}(\mathbf{p}',\mathbf{p}'-\mathbf{k})$ and $D^{R(A)}_{\epsilon'-\epsilon} \equiv D^{R(A)}_{\epsilon'-\epsilon}(\mathbf{p}'-\mathbf{p})$. The self-energies defined by eqn (8.48) describe inelastic relaxation processes and may be important for dynamical problems. In the presence of impurities, the corresponding impurity self-energies should be added to the phonon self-energies.

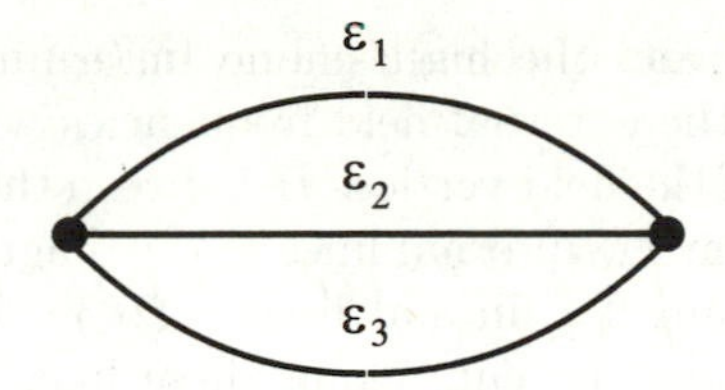

FIG. 8.4. Diagram representation for the particle–particle self-energy.

As a result, the equations for the retarded and advanced Green functions now have the form of eqns (8.19) with the operator [compare with eqn (8.3)]

$$\check{G}_{\epsilon}^{-1}(\mathbf{p}-\mathbf{k}_1,\omega_1) = \begin{pmatrix} \xi_{\mathbf{p}}-\epsilon & 0 \\ 0 & \xi_{\mathbf{p}}+\epsilon \end{pmatrix} (2\pi)^4\delta(\omega_1)\delta(\mathbf{k}_1) + \check{H}_{\omega_1} \tag{8.49}$$

with the effective Hamiltonian determined by eqn (3.70). It now includes both the electromagnetic potentials and the order parameter.

The equations for the anomalous and total Green functions are exactly the same as eqns (8.55) and (8.61). The only difference is that the self-energy contains the phonon contribution in the form of eqns (8.42), (8.48) in addition to the impurity self-energies.

8.3 Particle–particle collisions

Electron–electron collisions in usual superconductors are not very important for dynamic processes. This is in contrast with another Fermi superfluid, ^{3}He, where the particle–particle collisions are not only responsible for kinetics of excitations but also determine the pairing itself. In our discussion of superconductors, however, we consider the pairing within the BCS model; particle–particle collisions are only taken into account as long as they provide a relaxation mechanism for excitations.

For degenerate Fermi systems such as metals and ^{3}He, we can only consider pairwise interactions of excitations which result in a self-energy shown in Fig. 8.4. The self-energy diagrams contain two vertexes connected by three particle Green functions. In this section we do not consider the particular expressions for the particle–particle self-energy. Instead, we shall only discuss the procedure of analytical continuation for the self-energy of particle–particle interaction. The particular expression for the relaxation parts of the self-energies are given in Section 10.4.

We are interested in a sum in the form

$$\check{\Sigma}_{\epsilon,\epsilon-\omega} \propto T^2 \sum_{\epsilon_1,\epsilon_2} \check{G}_{\epsilon_1,\epsilon_1-\omega_1}\check{G}_{\epsilon_2,\epsilon_2-\omega_2}\check{G}_{\epsilon-\epsilon_1-\epsilon_2,\epsilon-\epsilon_1-\epsilon_2-\omega-\omega_1-\omega_2}. \tag{8.50}$$

Consider the contribution of the order N in the external field as a function of the complex variable ϵ for fixed imaginary frequencies of the field operator $\check{H}$. It has the singular lines determined by $\mathrm{Im}\,(\epsilon-\Omega_k)=0$ which are located between the lowest line $\mathrm{Im}\epsilon = 0$ and the highest line $\mathrm{Im}(\epsilon - \omega) = 0$. The frequencies Ω_k are combinations of the external field frequencies which now depend on the particular distribution of the field vertices $\check{H}$ between the functions $\check{G}$ in the sum in eqn (8.50), i.e., between the internal lines of the diagram in Fig. 8.4. Consider the term which has the singular lines at $\mathrm{Im}(\epsilon - \Omega_{1i}) = 0$, $\mathrm{Im}(\epsilon - \Omega_{2k}) = 0$, and $\mathrm{Im}(\epsilon - \Omega_{3l}) = 0$ and make the cuts along these lines. We now transform the double sum over frequencies in eqn (8.50) into the double contour integral. We shall make this transformation in two stages. On the first stage, the sum over one frequency, say ϵ_2, is transformed into the contour integral while keeping ϵ_1 as a fixed imaginary frequency. On the second stage, the sum over ϵ_1 is transformed. During these transformations, the external frequency ϵ and all field frequencies are fixed at their imaginary values. The transformation of the particular term in eqn (8.50) gives

$$\begin{aligned}
T^2 \sum_{\epsilon_1,\epsilon_2} \check{G}_{\epsilon_1} \check{G}_{\epsilon_2} \check{G}_{\epsilon-\epsilon_1-\epsilon_2} \Rightarrow & \int\!\!\int \frac{dz_1\, dz_2}{(4\pi i)^2} \tanh\left(\frac{z_2}{2T}\right) \\
\times \Bigg[& \sum_{i,k} \left(\delta_i \check{G}_{z_1+\Omega_{1i}}\right) \left(\delta_k \check{G}_{z_2+\Omega_{2k}}\right) \check{G}_{\epsilon-\Omega_{1i}-\Omega_{2k}-z_1-z_2} \tanh\left(\frac{z_1}{2T}\right) \\
& - \sum_{k,l} \check{G}_{\epsilon+z_1-\Omega_{2k}-\Omega_{3l}} \left(\delta_k \check{G}_{z_2+\Omega_{2k}}\right) \left(\delta_l \check{G}_{-z_1-z_2+\Omega_{3l}}\right) \coth\left(\frac{z_1}{2T}\right) \\
& + \sum_{k,l} \check{G}_{\epsilon+z_1-\Omega_{2k}-\Omega_{3l}} \left(\delta_k \check{G}_{z_2-z_1+\Omega_{2k}}\right) \left(\delta_l \check{G}_{-z_2+\Omega_{3l}}\right) \coth\left(\frac{z_1}{2T}\right) \\
& - \sum_{i,l} \left(\delta_i \check{G}_{z_1+\Omega_{1i}}\right) \check{G}_{\epsilon-z_1+z_2-\Omega_{1i}-\Omega_{3l}} \left(\delta_l \check{G}_{-z_2+\Omega_{3l}}\right) \tanh\left(\frac{z_1}{2T}\right) \Bigg] . \qquad (8.51)
\end{aligned}$$

Here we omitted the second index at each $\check{G}$ for brevity.

Making now the analytical continuation of eqn (8.51) in ϵ from the region $\mathrm{Im}(\epsilon - \omega) > 0$ and, after that, in all field frequencies ω_i from the upper half-plane, we obtain Σ^R. Making the continuation in ϵ from the region $\mathrm{Im}\epsilon < 0$ and, after that, in all field frequencies ω_i from the upper half-plane, we obtain Σ^A. We are interested in the relaxation part $\Sigma^R - \Sigma^A$. It is

$$\begin{aligned}
\Sigma^R - \Sigma^A \Rightarrow \int\!\!\int \frac{d\epsilon_1\, d\epsilon_2}{(4\pi i)^2} & \left[\check{G}_1^K \check{G}_2^K \left(\check{G}_3^R - \check{G}_3^A\right)\right. \\
& + \check{G}_1^K \left(\check{G}_2^R - \check{G}_2^A\right) \check{G}_3^K + \left(\check{G}_1^R - \check{G}_1^A\right) \check{G}_2^K \check{G}_3^K \\
& \left. + \left(\check{G}_1^R - \check{G}_1^A\right)\left(\check{G}_2^R - \check{G}_2^A\right)\left(\check{G}_3^R - \check{G}_3^A\right)\right] . \qquad (8.52)
\end{aligned}$$

Here $\check{G}_1$ has the frequencies $\epsilon_1, \epsilon_1 - \omega_1$, etc., with $\epsilon_3 = \epsilon - \epsilon_1 - \epsilon_2$, and $\omega_3 = \omega - \omega_1 - \omega_2$.

Define the total self-energy according to eqn (8.37) as

$$\begin{aligned}\check{\Sigma}^K_{\epsilon,\epsilon-\omega} &= \sum_{N=0}^{\infty}\sum_j [\delta_j \check{\Sigma}^{(N)}]_{\epsilon,\epsilon-\omega} \tanh\left(\frac{\epsilon-\Omega_j}{2T}\right) \\ &= \check{\Sigma}^R_{\epsilon,\epsilon-\omega} f^{(0)}(\epsilon-\omega) - f^{(0)}(\epsilon)\,\check{\Sigma}^A_{\epsilon,\epsilon-\omega} + \check{\Sigma}^{(a)}_{\epsilon,\epsilon-\omega}. \end{aligned} \tag{8.53}$$

We obtain

$$\begin{aligned}\check{\Sigma}^K \Rightarrow \int\!\!\int \frac{d\epsilon_1\, d\epsilon_2}{(4\pi i)^2} &\left[\check{G}^K_1\left(\check{G}^R_2 - \check{G}^A_2\right)\left(\check{G}^R_3 - \check{G}^A_3\right)\right. \\ &+ \left(\check{G}^R_1 - \check{G}^A_1\right)\check{G}^K_2\left(\check{G}^R_3 - \check{G}^A_3\right) \\ &+ \left.\left(\check{G}^R_1 - \check{G}^A_1\right)\left(\check{G}^R_2 - \check{G}^A_2\right)\check{G}^K_3 + \check{G}^K_1\check{G}^K_2\check{G}^K_3\right]. \end{aligned} \tag{8.54}$$

8.4 Transport-like equations and the conservation laws

One can derive kinetic-type equations for the anomalous and the total Green functions. To do this, let us subtract eqn (8.26) from eqn (8.21). We obtain

$$\begin{aligned}&\left[\check{G}^{-1}\check{G}^{(a)}\right]_{\epsilon,\epsilon-\omega} - \left[\check{G}^{(a)}\check{G}^{-1}\right]_{\epsilon,\epsilon-\omega} \\ &\quad - \left[\check{\Sigma}^R\check{G}^{(a)}\right]_{\epsilon,\epsilon-\omega} + \left[\check{G}^{(a)}\check{\Sigma}^A\right]_{\epsilon,\epsilon-\omega} - \left[\check{\Sigma}^{(a)}\check{G}^A\right]_{\epsilon,\epsilon-\omega} + \left[\check{G}^R\check{\Sigma}^{(a)}\right]_{\epsilon,\epsilon-\omega} \\ &\quad = \left[\check{h}_{\epsilon,\epsilon-\omega'}\check{G}^A_{\epsilon-\omega',\epsilon-\omega}\right]_{\epsilon,\epsilon-\omega} - \left[\check{G}^R_{\epsilon,\epsilon-\omega+\omega'}\check{h}_{\epsilon-\omega+\omega',\epsilon-\omega}\right]_{\epsilon,\epsilon-\omega}. \end{aligned} \tag{8.55}$$

The integration over internal variables is assumed for the expressions in the square brackets

$$[AB]_{\epsilon,\epsilon_1;\mathbf{p},\mathbf{p}_1} = \int A_{\epsilon,\epsilon'}(\mathbf{p},\mathbf{p}')\,B_{\epsilon',\epsilon_1}(\mathbf{p}',\mathbf{p}_1)\,\frac{d\epsilon'}{2\pi}\,\frac{d^3p'}{(2\pi)^3}. \tag{8.56}$$

The internal frequency variables are shown explicitly in the last line of eqn (8.55). The momentum variables are not shown.

Equation (8.55) can also be written as

$$\begin{aligned}&\left(\xi_{\mathbf{p}+\mathbf{k}/2} - \check{\tau}_3\epsilon_+\right)\check{G}^{(a)}_{\epsilon_+,\epsilon_-} - \check{G}^{(a)}_{\epsilon_+,\epsilon_-}\left(\xi_{\mathbf{p}-\mathbf{k}/2} - \check{\tau}_3\epsilon_-\right) \\ &\quad - \left[\check{\Sigma}^R\check{G}^{(a)}\right]_{\epsilon_+,\epsilon_-} + \left[\check{G}^{(a)}\check{\Sigma}^A\right]_{\epsilon_+,\epsilon_-} - \left[\check{\Sigma}^{(a)}\check{G}^A\right]_{\epsilon_+,\epsilon_-} + \left[\check{G}^R\check{\Sigma}^{(a)}\right]_{\epsilon_+,\epsilon_-} \\ &\quad + \left[\check{H}\check{G}^{(a)}\right]_{\epsilon_+,\epsilon_-} - \left[\check{G}^{(a)}\check{H}\right]_{\epsilon_+,\epsilon_-} = \left[\check{h}\check{G}^A\right]_{\epsilon_+,\epsilon_-} - \left[\check{G}^R\check{h}\right]_{\epsilon_+,\epsilon_-} \end{aligned} \tag{8.57}$$

where $\epsilon_\pm = \epsilon \pm \omega/2$.

Subtracting eqn (8.27) from eqn (8.24) we find the equation for the total function

$$\left(\xi_{\mathbf{p}+\mathbf{k}/2} - \check{\tau}_3\epsilon_+\right)\check{G}^K_{\epsilon_+,\epsilon_-} - \check{G}^K_{\epsilon_+,\epsilon_-}\left(\xi_{\mathbf{p}-\mathbf{k}/2} - \check{\tau}_3\epsilon_-\right)$$

$$+\left[\check{H}\check{G}^K\right]_{\epsilon_+,\epsilon_-} - \left[\check{G}^K\check{H}\right]_{\epsilon_+,\epsilon_-} = \check{\mathcal{I}}^K_{\epsilon_+,\epsilon_-} . \tag{8.58}$$

The "collision integral" is

$$\check{\mathcal{I}}^K_{\epsilon_+,\epsilon_-} = \left[\check{\Sigma}^R\check{G}^K - \check{G}^K\check{\Sigma}^A - \check{G}^R\check{\Sigma}^K + \check{\Sigma}^K\check{G}^A\right]_{\epsilon_+,\epsilon_-} . \tag{8.59}$$

Equation (8.58) can be written as

$$\begin{aligned}\mathbf{v}_F\mathbf{k}\check{G}^K_{\epsilon_+,\epsilon_-} - \epsilon_+\check{\tau}_3\check{G}^K_{\epsilon_+,\epsilon_-} + \check{G}^K_{\epsilon_+,\epsilon_-}\epsilon_-\check{\tau}_3 \\ + \left[\check{H}\check{G}^K\right]_{\epsilon_+,\epsilon_-} - \left[\check{G}^K\check{H}\right]_{\epsilon_+,\epsilon_-} = \check{\mathcal{I}}^K_{\epsilon_+,\epsilon_-}\end{aligned} \tag{8.60}$$

because $\xi_{\mathbf{p}+\mathbf{k}/2} - \xi_{\mathbf{p}-\mathbf{k}/2} = \mathbf{v}_F\mathbf{k}$. It can also be represented in a more general form

$$\left[\check{G}_0^{-1}\check{G}^K\right] - \left[\check{G}^K\check{G}_0^{-1}\right] = \check{\mathcal{I}}^K . \tag{8.61}$$

Here

$$\check{G}_0^{-1}(\mathbf{p}) = -i\frac{\partial}{\partial t}\check{\tau}_3 + \begin{pmatrix} E_n\left(\mathbf{p}-\frac{e}{c}\mathbf{A}\right) - E_F & -\Delta \\ \Delta^* & E_n\left(\mathbf{p}+\frac{e}{c}\mathbf{A}\right) - E_F \end{pmatrix} + e\varphi\check{1}$$

is shown in the coordinate representation. The momentum operator is $\mathbf{p} = -i\hbar\nabla$ and $E_n(\mathbf{p})$ is the normal-state electronic spectrum. Equations (8.61, 8.25) or eqns (8.55, 8.18) together with eqn (8.19) for the regular functions make the basis for describing nonstationary properties of superconductors.

Equation (8.61) contains the conservation of particle number and the energy conservation. To get the particle conservation

$$e\frac{\partial N}{\partial t} + \operatorname{div}\mathbf{j} = 0 \tag{8.62}$$

we multiply eqn (8.61) with $\check{\tau}_3$ then take the trace and integrate it over frequency $d\epsilon/4\pi i$ and momentum $d^3p/(2\pi)^3$. The terms of the form of Δ^*F^K and $\Delta F^{\dagger K}$ cancel out due to the self-consistency equation. One can also check that all the collision integrals vanish under this operation (for example, the impurity collision integral already vanishes after integration over the momentum directions).

To obtain the energy conservation, we multiply eqn (8.61) by ϵ, take the trace and integrate it over frequency $d\epsilon/4\pi i$ and momentum $d^3p/(2\pi)^3$. We find

$$\frac{\partial\mathcal{E}}{\partial t} + \operatorname{div}\mathbf{j}_{\mathcal{E}} = \mathbf{j}\cdot\mathbf{E} \tag{8.63}$$

The internal energy is

$$\mathcal{E} = -\int\epsilon\left[G^K_{\epsilon_+,\epsilon_-}(\mathbf{p}_+,\mathbf{p}_-) - \bar{G}^K_{\epsilon_+,\epsilon_-}(\mathbf{p}_+,\mathbf{p}_-)\right]\frac{d\epsilon}{4\pi i}\frac{d^3p}{(2\pi)^3} + \frac{|\Delta|^2}{|g|} - Ne\varphi \tag{8.64}$$

while the internal energy current is

$$\mathbf{j}_{\mathcal{E}} = -\frac{1}{m}\int\epsilon\left[\left(\mathbf{p} - \frac{e}{c}\mathbf{A}\right)G^K_{\epsilon_+,\epsilon_-}(\mathbf{p}_+,\mathbf{p}_-)\right.$$

$$+\left(\mathbf{p}+\frac{e}{c}\mathbf{A}\right)\bar{G}^{K}_{\epsilon_+,\epsilon_-}(\mathbf{p}_+,\mathbf{p}_-)\Big]\frac{d\epsilon}{4\pi i}\frac{d^3p}{(2\pi)^3}-\mathbf{j}\varphi. \tag{8.65}$$

We assume that the electronic system is in equilibrium with phonons. Otherwise, there will be the additional energy source in the r.h.s. of eqn (8.63)

$$\left(\frac{\partial\mathcal{E}}{\partial t}\right)_{\text{ph}}=-i\int\epsilon\,\text{Tr}\check{\mathcal{I}}^{K(\text{ph})}_{\epsilon_+,\epsilon_-}\frac{d\epsilon}{4\pi i}\frac{d^3p}{(2\pi)^3}$$

due to the interaction with phonons. The electron–electron collision integral conserves the energy of interacting particles and thus drops out after integration over $d\epsilon$ with ϵ. The elastic impurity integral also conserves the energy; it already vanishes after integration over the momentum directions.

The total energy

$$\mathcal{E}_{\text{tot}}=\mathcal{E}+\frac{H^2}{8\pi}$$

obeys the conservation equation

$$\frac{\partial\mathcal{E}_{\text{tot}}}{\partial t}+\text{div}\,\mathbf{j}_{\mathcal{E}_{\text{tot}}}=0$$

where the total-energy current is

$$\mathbf{j}_{\mathcal{E}_{\text{tot}}}=\mathbf{j}_{\mathcal{E}}+\frac{c}{4\pi}\left[\mathbf{E}\times\mathbf{H}\right].$$

8.5 The Keldysh diagram technique

8.5.1 *Definitions of the Keldysh functions*

We have already mentioned that it is the absence of the proper time-ordering procedure in the definitions of the real-time Green functions that produces practical difficulties with calculations of the real-time functions. However, one can establish the necessary time-ordering for some auxiliary Green functions in such a way that the Wick theorem works for them. After the Wick theorem is restored, one can derive the Dyson equation (or to formulate the diagram technique) thus providing the algorithm of calculating these auxiliary Green functions. The next step is to relate the auxiliary functions to the true real-time Green functions or to express the physical observables through some of the auxiliary Green functions (Keldysh 1964). In this section we discuss the correspondence between the method of the analytical continuation and the Keldysh technique. We shall see that the both methods lead to the same results: The equations for the Green functions which determine the physical observables are identical in the both cases. One can use any of these two methods. Since we have already derived the basic equations using the method of analytical calculation, we discuss the Keldysh approach only briefly for the most simple case. A more detailed description of the Keldysh technique can be found in the review by Rammer and Smith (1986).

Let us consider the statistical average of a single-particle-observable operator which is linear in the particle-field operators $\tilde{\psi}\tilde{\psi}^{\dagger}$ in the Heisenberg representation as introduced in Section 2.1. It contains the average

$$\sum^{(d)} \left[\exp\left(\frac{\Omega + \mu\hat{\mathcal{N}} - \hat{\mathcal{H}}(t_0)}{T} \right) \left(\tilde{\psi}_\alpha(\mathbf{r}_1, t_1) \tilde{\psi}^{\dagger}_\beta(\mathbf{r}_2, t_2) \right) \right]$$
$$= \left\langle \tilde{\psi}_\alpha(\mathbf{r}_1, t_1) \tilde{\psi}^{\dagger}_\beta(\mathbf{r}_2, t_2) \right\rangle_{st}$$

where $\tilde{\psi}_\alpha(1)$ is the Heisenberg operator

$$\tilde{\psi}_\alpha(\mathbf{r}, t) = \hat{S}^{-1}(t, t_0)\, \psi_\alpha(\mathbf{r}, t_0) \hat{S}(t, t_0)$$

defined through the S-matrix introduced by eqn (2.63):

$$\hat{S}(t, t_0) = T_t \exp\left[-i \int_{t_0}^{t} (\hat{\mathcal{H}} - \mu\hat{\mathcal{N}}) dt' \right].$$

It is assumed that the system was in equilibrium at $t = t_0$ and had a temperature T. We put $t_0 = -\infty$ assuming that the nonequilibrium interaction is turned on at $t_0 = -\infty$. We see that the operators depend on times which are confined to the interval ranging from $-\infty$ up to the largest time $\max\{t_1, t_2\}$. Moreover, the time instants belonging to the operator $\tilde{\psi}$ are not ordered with respect to the instants which belong to the operator $\tilde{\psi}^{\dagger}$. We want to relate this average to the average of a sequence of operators whose times are fully ordered with respect to each other. The proper ordering is achieved along the time contour depicted in Fig. 8.5. It starts at $-\infty$ and runs up to the largest time $\max\{t_1, t_2\}$ and then returns back to the time $-\infty$. We denote the ordering along this contour by T_c and define the auxiliary Green function

$$G_{\alpha\beta}(1, 2) = i \left\langle T_c \tilde{\psi}_\alpha(\mathbf{r}_1, t_1) \tilde{\psi}^{\dagger}_\beta(\mathbf{r}_2, t_2) \right\rangle_{st}. \tag{8.66}$$

The time-ordering T_c orders the operators as follows

$$T_c \tilde{\psi}(\mathbf{r}_1, t_1) \tilde{\psi}^{\dagger}(\mathbf{r}_2, t_2) = \begin{cases} \tilde{\psi}(\mathbf{r}_1, t_1) \tilde{\psi}^{\dagger}(\mathbf{r}_2, t_2), & t_1 >_c t_2, \\ \mp \tilde{\psi}^{\dagger}(\mathbf{r}_2, t_2) \tilde{\psi}(\mathbf{r}_1, t_1), & t_1 <_c t_2. \end{cases}$$

The relation $t_1 >_c t_2$ means that t_1 is later than t_2 along the contour c. The contour c is shown in Fig. 8.5 (a) for $t_1 >_c t_2$. The upper sign refers to Fermi particles while the lower sign is for the Bose particles. The brackets denote the statistical average.

We introduce also the other auxiliary functions

$$G^{>}_{\alpha\beta}(1, 2) = i \left\langle \tilde{\psi}_\alpha(1) \tilde{\psi}^{\dagger}_\beta(2) \right\rangle_{st},$$

(a)

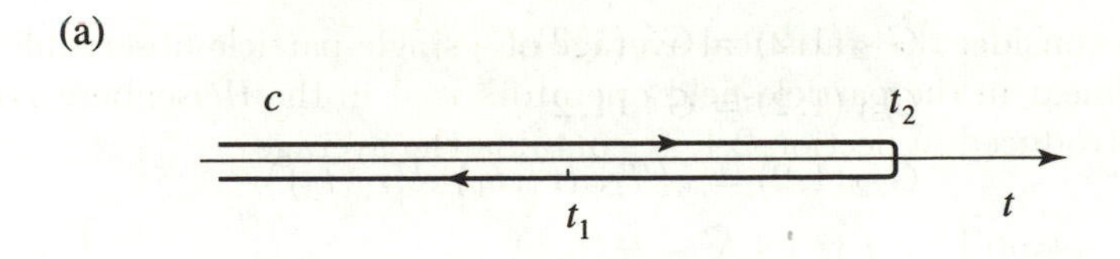

(b)

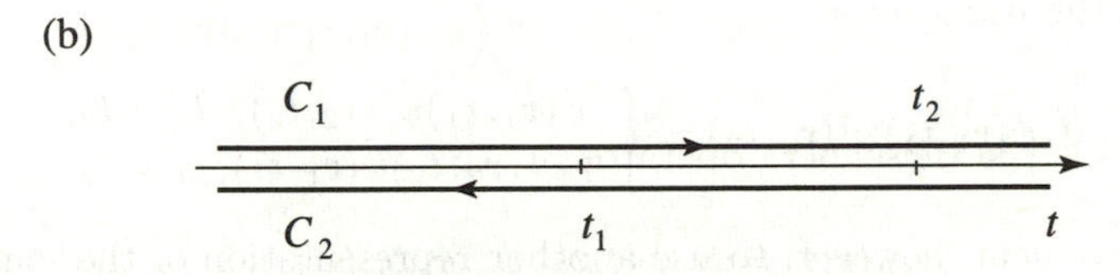

FIG. 8.5. The Keldysh time contours for real-time ordered Green functions.

$$G^{<}_{\alpha\beta}(1,2) = \mp i \left\langle \tilde{\psi}^{\dagger}_{\beta}(2)\tilde{\psi}_{\alpha}(1) \right\rangle_{st} .$$

One can see that

$$G\,(1,2) = \begin{cases} G^{>}_{\alpha\beta}(1,2),\, t_1 >_c t_2 \\ G^{<}_{\alpha\beta}(1,2),\, t_1 <_c t_2 \end{cases}$$

We can extend the contour c up to $+\infty$ by introducing the identical unity

$$1 \equiv S\,(\max\{t_1,t_2\}\,, +\infty)\, S\,(+\infty, \max\{t_1,t_2\})$$

under the averaging in eqn (8.66). The contour c transforms into the contour C which consists of two parts: the contour C_1 runs from $-\infty$ to $+\infty$ and the contour C_2 which runs back from $+\infty$ to $-\infty$. The contour C is shown in Fig. 8.5 (b).

The physical observables are expressed through the function $G^{<}_{\alpha\beta}(1;2)$. For example, the Fermionic density is

$$N = i \sum_{\alpha} \left[G^{<}_{\alpha\alpha}(1;1)\right] \tag{8.67}$$

where the sum is over the spin indices. At the same time, it is the contour-ordered Green function G in eqn (8.66) that possesses a simple perturbative expansion. Indeed, we can prove the Wick theorem for the average of the contour-ordered particle operators exactly in the same way as it was done for averages of the imaginary-time ordered operators. We want now to establish a connection between $G^{<}$ and the time-ordered function G.

Let us now construct the matrix in the so-called Keldysh space

$$\underline{\breve{G}} = \begin{pmatrix} G_{11} & G_{12} \\ G_{21} & G_{22} \end{pmatrix} \tag{8.68}$$

whose components are defined as follows

$$G_{11}\,(1,2) = i \left\langle T_t \tilde{\psi}(\mathbf{r}_1,t_1)\tilde{\psi}^{\dagger}(\mathbf{r}_2,t_2) \right\rangle_{st} ,$$

$$G_{12}(1,2) = G^{<}(1,2),$$
$$G_{21}(1,2) = G^{>}(1,2),$$
$$G_{22}(1,2) = i\left\langle \bar{T}_t \tilde{\psi}(\mathbf{r}_1,t_1)\psi^{\dagger}(\mathbf{r}_2,t_2)\right\rangle_{st}. \tag{8.69}$$

The time-ordering operator T_t is the usual time-ordering along the contour C_1 and the operator $\bar{T}_t$ orders along the contour C_2, i.e., it is the "inverse-time" ordering in the usual sense

$$\bar{T}_t \tilde{\psi}(\mathbf{r}_1,t_1)\tilde{\psi}^{\dagger}(\mathbf{r}_2,t_2) = \begin{cases} \tilde{\psi}(\mathbf{r}_1,t_1)\tilde{\psi}^{\dagger}(\mathbf{r}_2,t_2), & t_1 < t_2, \\ \mp\tilde{\psi}^{\dagger}(\mathbf{r}_2,t_2)\tilde{\psi}(\mathbf{r}_1,t_1), & t_1 > t_2. \end{cases}$$

It is convenient, however, to use another representation of the matrix Green function which is obtained from eqn (8.68) by a linear transformation. The point is that the four components in eqn (8.69) are not linearly independent. Indeed, we observe that

$$G_{11}(1,2) - G_{12}(1,2) = G_{21}(1,2) - G_{22}(1,2) \equiv G^{R}(1,2),$$
$$G_{11}(1,2) - G_{21}(1,2) = G_{12}(1,2) - G_{22}(1,2) \equiv G^{A}(1,2),$$
$$G_{21}(1,2) + G_{12}(1,2) = G_{11}(1,2) + G_{22}(1,2) \equiv G^{K}(1,2).$$

We introduce here the retarded and advanced Green functions G^R and G^A, respectively, defined in the usual way

$$G^{R}_{\alpha\beta}(1;2) = \begin{cases} i\left\langle\left(\tilde{\psi}_{\alpha}(1)\tilde{\psi}^{\dagger}_{\beta}(2) \pm \tilde{\psi}^{\dagger}_{\beta}(2)\tilde{\psi}_{\alpha}(1)\right)\right\rangle_{st}, & t_1 > t_2, \\ 0, & t_1 < t_2, \end{cases}$$

and

$$G^{A}_{\alpha\beta}(1;2) = \begin{cases} -i\left\langle\left(\tilde{\psi}_{\alpha}(1)\tilde{\psi}^{\dagger}_{\beta}(2) \pm \tilde{\psi}^{\dagger}_{\beta}(2)\tilde{\psi}_{\alpha}(1)\right)\right\rangle_{st}, & t_1 < t_2, \\ 0, & t_1 > t_2. \end{cases}$$

The function G^K is defined as

$$G^{K}(1,2) = G^{>}(1,2) + G^{<}(1,2). \tag{8.70}$$

It is called the Keldysh function. Performing the transformation

$$\breve{G} = L\,\breve{\tau}_3\,\underline{\breve{G}}\,L^{+}$$

where

$$L = \left(\breve{1} - i\breve{\tau}_2\right)/\sqrt{2},$$

we arrive at the matrix Green function

$$\breve{G} = \begin{pmatrix} G^{R} & G^{K} \\ 0 & G^{A} \end{pmatrix}. \tag{8.71}$$

Equation (8.71) is the standard representation accepted in the literature.

The physical observables can be expressed through the Keldysh function. Indeed, using the identity

$$\lim_{t_1 \to t_2} \left[G^{>}_{\alpha\beta}(1,2) - G^{<}_{\alpha\beta}(1,2) \right] = i\delta\left(\mathbf{r}_1 - \mathbf{r}_2\right)\delta_{\alpha\beta}$$

which follows from the general commutation rules eqn (2.15) we find from eqn (8.70)

$$\lim_{t_1 \to t_2} G^{<}_{\alpha\beta}(1,2) = \frac{1}{2} \lim_{t_1 \to t_2} G^{K}_{\alpha\beta}\left(1,2\right) - i\delta\left(\mathbf{r}_1 - \mathbf{r}_2\right)\delta_{\alpha\beta}. \tag{8.72}$$

Let us calculate the particle density using eqn (8.67). We write $N = N_0 + \delta N$ where N_0 is the density for noninteracting particles. We have

$$N = N_0 + \frac{i}{2} \sum_{\alpha} \left[\lim_{t_1 \to t_2} \delta G^{K}_{\alpha\alpha}\left(1,2\right) \right]_{\mathbf{r}_1 = \mathbf{r}_2}.$$

In the frequency representation

$$G^{K}\left(1,2\right) = \int e^{-i\epsilon(t_1 - t_2) - i\omega(t_1 + t_2)/2} e^{i\mathbf{p}(\mathbf{r}_1 - \mathbf{r}_2) + i\mathbf{k}(\mathbf{r}_1 + \mathbf{r}_2)/2}$$
$$\times G^{K}_{\epsilon_+,\epsilon_-}\left(\mathbf{p}_+, \mathbf{p}_-\right) \frac{d\epsilon\, d\omega}{(2\pi)^2} \frac{d^3p\, d^3k}{(2\pi)^6}$$

we get

$$N\left(\omega, \mathbf{k}\right) = N_0 - \sum_{\alpha} \int \delta G^{K}_{\epsilon_+,\epsilon_-}\left(\mathbf{p}_+, \mathbf{p}_-\right) \frac{d\epsilon}{4\pi i} \frac{d^3p}{(2\pi)^3}.$$

This expression coincides with eqn (8.30) obtained by the analytical continuation of the Matsubara Green function.

8.5.2 *Dyson equation*

Consider a series expansion for the simplest example of the interaction with the external potential U. The first-order diagram of the type shown in Fig. 8.1 gives for the contour-ordered Green function

$$G^{(1)}\left(1,1'\right) = -\int d^3r_2 \int_C dt_2 G^{(0)}\left(1,2\right) U\left(2\right) G^{(0)}\left(2,1'\right).$$

The integral along the contour C can be separated into the integrals along C_1 and C_2. Interchanging the limits of integration on the lower contour C_2 we present $G^{(1)}$ as the sum of two terms which can be written as components in the Keldysh space

$$G^{(1)}_{ik}\left(1,1'\right) = -\int d^3r_2 \int_{-\infty}^{+\infty} dt_2 G^{(0)}_{ij}\left(1,2\right) U_{jl}\left(2\right) G^{(0)}_{lk}\left(2,1'\right) \tag{8.73}$$

where we introduce the matrix

$$\check{U}(2) = U(2)\check{\tau}_3 = \begin{pmatrix} U(2) & 0 \\ 0 & -U(2) \end{pmatrix}$$

and assume summation over repeated indices. Equation (8.73) can be written in the matrix form

$$\underline{\check{G}}^{(1)} = -\underline{\check{G}}^{(0)}\check{U}\underline{\check{G}}^{(0)}$$

where we assume integration over internal coordinates and time. It is now clear that collecting all orders of the perturbation expansion in U we arrive at the Dyson equation for the full Green function $\underline{\check{G}}$

$$\underline{\check{G}} = \underline{\check{G}}^{(0)} - \underline{\check{G}}^{(0)}\check{U}\underline{\check{G}}. \tag{8.74}$$

One can easily check that eqn (8.74) transforms into

$$\check{G} = \check{G}^{(0)} - \check{G}^{(0)}U\check{G} \tag{8.75}$$

for the matrices in the representation of eqn (8.71). The potential U is proportional to the unit matrix in the Keldysh space.

We introduce now the inverse operator for the Green function of noninteracting particles

$$G_0^{-1} = -i\frac{\partial}{\partial t} + E_n(\mathbf{p}) - E_F.$$

Applying the inverse operator to eqn (8.75) we obtain the Dyson equation in the differential form

$$\left(G_0^{-1} + U\right)\check{G} = \check{1}.$$

The interaction with impurities, phonons, and electron–electron interactions can also be included into the same scheme. These interactions introduce the corresponding self-energies which also are Keldysh matrices:

$$\check{\Sigma} = \begin{pmatrix} \Sigma^R & \Sigma^K \\ 0 & \Sigma^A \end{pmatrix}.$$

The resulting Dyson equation takes the form

$$\left(G_0^{-1} - \check{\Sigma}\right)\check{G} = \check{1}. \tag{8.76}$$

We do not specify the rules as to how to construct the elements of these matrices within the Keldysh formalism: they are the same as those obtained using the method of analytical continuation. The detailed description of the Keldysh formulation of various interactions can be found in the review by Rammer and Smith (1986).

8.5.3 *Keldysh functions in the BCS theory*

We have four different Green functions in the BCS theory. Accordingly, the components of the matrix in the Keldysh space used in the BSC theory are matrices in the Nambu space. We define the Keldysh matrices

$$\check{G} = \begin{pmatrix} \check{G}^R & \check{G}^K \\ 0 & \check{G}^A \end{pmatrix}, \quad \check{\Sigma} = \begin{pmatrix} \check{\Sigma}^R & \check{\Sigma}^K \\ 0 & \check{\Sigma}^A \end{pmatrix}, \quad \check{1} = \begin{pmatrix} \check{1} & 0 \\ 0 & \check{1} \end{pmatrix}.$$

From eqns (8.20) and (8.24) or eqn (8.27) one can directly see that the total function $\check{G}^K$ introduced as a result of analytical continuation of the Matsubara Green functions is nothing but the Keldysh function because it satisfies the same equations. Indeed, instead of two equations (8.20) and one equation (8.24) we obtain one matrix equation

$$\left(\check{1}\check{G}^{-1} - \check{\Sigma}\right) \check{G} = \check{1}$$

which is a generalization of eqn (8.76).

The expressions for the order parameter and for the current density in terms of the Keldysh functions are the same as obtained by the analytical continuation. Indeed, the δ-term in eqn (8.72) is absent for the F-component of the Keldysh function $\check{G}^K$. This directly results in eqn (8.28) for the order parameter. The δ-term also disappears if one applies the momentum operator to the function G^K when calculating the current. As a result, eqn (8.29) is reproduced as well. We thus have confirmed that the method of analytical continuation and the Keldysh method are completely equivalent.

9

QUASICLASSICAL METHOD FOR NONSTATIONARY PHENOMENA

We apply the quasiclassical approximation to nonstationary problems in the theory of superconductivity and derive the Eliashberg equations for the quasiclassical Green functions. We express the Keldysh function in terms of a generalized two-component distribution function. As an example, we consider a nonlinear response of a superconductor to microwave irradiation: the order parameter becomes enhanced as a result of a nonequilibrium distribution of excitations under irradiation.

9.1 Eliashberg equations

The quasiclassical method based on the fundamental assumption that the Fermi momentum is larger than the inverse coherence length can be applied to nonstationary problems, as well. As in Chapter 5, this assumption again allows us to consider the Green functions only in a vicinity of the Fermi surface. As in the static case, we introduce the quasiclassical matrix Green function integrated over the energy variable $d\xi_{\mathbf{p}}$. In addition to $\check{g}^{R(A)}$ we define

$$\check{g}^K_{\epsilon_+,\epsilon_-}(\mathbf{p}_F,\mathbf{k}) = \oint \frac{d\xi_{\mathbf{p}}}{\pi i}\check{G}^K_{\epsilon_+,\epsilon_-}(\mathbf{p}_+,\mathbf{p}_-) = \begin{pmatrix} g^K & f^K \\ -f^{\dagger K} & \bar{g}^K \end{pmatrix}$$

and similarly for $\check{g}^{(a)}$. The momentum $\mathbf{p}_F$ lies at the Fermi surface; its magnitude is fixed, thus the Green function only depends on the direction of the momentum. One can obtain equations for the quasiclassical functions using the corresponding transport-like equations derived in the previous section in a full analogy to the static case.

We start the derivation of the quasiclassical equations with an important observation that the transport-like equation (8.60) does not contain $\xi_{\mathbf{p}}$ explicitly. Within the quasiclassical approximation, the Hamiltonian $\check{H}$ contains $\mathbf{v}_F$ taken at the Fermi surface, therefore it also depends only on the direction of the momentum. This is completely similar to the stationary case where we were able to make a transformation to the quasiclassical Green functions by performing an integration of the corresponding equation over the energy variable $d\xi_{\mathbf{p}}$. Integrating now the nonstationary equation (8.60) over $d\xi_{\mathbf{p}}$ we find

$$\mathbf{v}_F\mathbf{k}\check{g}^K_{\epsilon_+,\epsilon_-} - \epsilon_+\check{\tau}_3\check{g}^K_{\epsilon_+,\epsilon_-} + \check{g}^K_{\epsilon_+,\epsilon_-}\epsilon_-\check{\tau}_3 + \left[\check{H}\check{g}^K\right]_{\epsilon_+,\epsilon_-} - \left[\check{g}^K\check{H}\right]_{\epsilon_+,\epsilon_-} = \check{I}^K_{\epsilon_+,\epsilon_-}. \quad (9.1)$$

Here we introduce the quasiclassical collision integral

$$\check{I}^K_{\epsilon_+,\epsilon_-}(\mathbf{p}_F,\mathbf{k}) = \int \frac{d\xi_{\mathbf{p}}}{\pi i}\left[\check{\Sigma}^R\check{G}^K - \check{G}^K\check{\Sigma}^A - \check{G}^R\check{\Sigma}^K + \check{\Sigma}^K\check{G}^A\right]_{\epsilon_+,\epsilon_-}. \tag{9.2}$$

We again use the shortcut notation eqn (8.56) which now reads, for example,

$$\left[\check{H}\check{g}\right]_{\epsilon_+,\epsilon_-} = \int \frac{d^3k_1}{(2\pi)^3}\frac{d\omega_1}{2\pi}\check{H}_{\omega_1}(\mathbf{k}_1)\check{g}_{\epsilon_+-\omega_1,\epsilon_-}(\mathbf{v}_F,\mathbf{k}-\mathbf{k}_1). \tag{9.3}$$

Equation (9.2) was first obtained by Eliashberg (1971).

The integration over $d\xi_{\mathbf{p}}$ performed in eqn (9.2) for the coordinate representation means, as in Section 5.1, that we take the limit $|\mathbf{r}_1 - \mathbf{r}_2| \to 0$. Doing so we average out the coordinate dependence of the order parameter and the vector potential over distances of order of p_F^{-1}, i.e., over distances of the atomic scale. In the momentum representation, the limit $\mathbf{r}_1 - \mathbf{r}_2 \to 0$ implies that we neglect terms of the first order in $k/p_F \sim 1/p_F\xi(T)$, i.e., terms like

$$\frac{\partial \check{G}}{\partial \mathbf{p}}\nabla \check{H} \tag{9.4}$$

in the expansion of the convolution $[\check{G}\check{H}]$ in powers of k/p_F and keep only zero–order terms like the simple product $\check{G}(\mathbf{p}_+,\mathbf{p}_-)\check{H}(\mathbf{k})$. We shall discuss this procedure in more detail later in Section 10.1. The same approximation was used when we derived the Eilenberger equations for the stationary Green functions. These terms can indeed be neglected in the Eilenberger equations for the regular functions. However, they may sometimes become important in the equations for the Keldysh functions. This is due to the fact that some leading expansion terms vanish. We shall see later in Chapter 10 that one cannot neglect the momentum derivatives for clean superconductors when the collision integrals are small such that the mean free path ℓ approaches $\xi(T)\,[E_F/\Delta]$. The accuracy of eqn (9.2) becomes poor already for clean superconductors with $\ell \gg \xi(T)$: some phenomena are lost in this description. The quasiclassical equation (9.2) is thus applicable for superconductors if the mean free path of quasiparticles is not very long. The limit of clean superconductors is discussed in more detail in Section 10.2.

The corresponding equation for the anomalous function can be obtained from eqn (8.55) in exactly the same way:

$$\begin{aligned}\mathbf{v}_F\mathbf{k}\check{g}^{(a)}_{\epsilon_+,\epsilon_-} - \epsilon_+\check{\tau}_3\check{g}^{(a)}_{\epsilon_+,\epsilon_-} + \check{g}^{(a)}_{\epsilon_+,\epsilon_-}\check{\tau}_3\epsilon_- + \left[\check{H}\check{g}^{(a)} - \check{g}^{(a)}\check{H}\right]_{\epsilon_+,\epsilon_-}\\ = \check{I}^{(a)}_{\epsilon_+,\epsilon_-} + \left[\check{h}\check{g}^A - \check{g}^R\check{h}\right]_{\epsilon_+,\epsilon_-},\end{aligned}$$

where

$$\check{I}^{(a)}_{\epsilon_+,\epsilon_-} = \int \frac{d\xi_{\mathbf{p}}}{\pi i}\left[\check{\Sigma}^R\check{G}^{(a)} - \check{G}^{(a)}\check{\Sigma}^A - \check{G}^R\check{\Sigma}^{(a)} + \check{\Sigma}^{(a)}\check{G}^A\right]_{\epsilon_+,\epsilon_-}, \tag{9.5}$$

and $\check{h}_{\epsilon,\epsilon-\omega}$ are defined by eqn (8.10) on page 148.

Of course, one can also obtain similar equations for the regular functions, as well. Subtracting the left-handed and right-handed equations

$$\left(\check{G}^{-1} - \check{\Sigma}^{R(A)}\right)\check{G}^{R(A)} = \check{1}, \quad \check{G}^{R(A)}\left(\check{G}^{-1} - \check{\Sigma}^{R(A)}\right) = \check{1} \tag{9.6}$$

from each other, and integrating the result over $d\xi_{\mathbf{p}}$, we obtain

$$\mathbf{v}_F\mathbf{k}\check{g}^{R(A)}_{\epsilon_+,\epsilon_-} - \epsilon_+\check{\tau}_3\check{g}^{R(A)}_{\epsilon_+,\epsilon_-} + \check{g}^{R(A)}_{\epsilon_+,\epsilon_-}\check{\tau}_3\epsilon_- + \left[\check{H}\check{g}^{R(A)} - \check{g}^{R(A)}\check{H}\right]_{\epsilon_+,\epsilon_-} = \check{I}^{R(A)}_{\epsilon_+,\epsilon_-} \tag{9.7}$$

where the corresponding collision integral is

$$\check{I}^{R(A)}_{\epsilon_+,\epsilon_-} = \int \frac{d\xi_{\mathbf{p}}}{\pi i}\left[\check{\Sigma}^{R(A)}\check{G}^{R(A)} - \check{G}^{R(A)}\check{\Sigma}^{R(A)}\right]_{\epsilon_+,\epsilon_-}. \tag{9.8}$$

9.1.1 *Self-energies*

The self-energies in eqns (9.2), (9.5), and (9.8) also contain only the $\xi_{\mathbf{p}}$-integrated Green functions; they depend only on the direction of the momentum whose magnitude is fixed by the Fermi surface. Indeed, for interaction with nonmagnetic impurities, we have

$$\begin{aligned}\check{\Sigma}^K_{\epsilon,\epsilon-\omega}(\mathbf{p}_F,\mathbf{k}) &= n_{\text{imp}}\int \frac{d^3p'}{(2\pi)^3}\,|u(\theta)|^2\,\check{G}^K_{\epsilon,\epsilon-\omega}(\mathbf{p}',\mathbf{p}'-\mathbf{k})\\ &= \pi i\nu(0)\,n_{\text{imp}}\int\frac{d\Omega_{\mathbf{p}'}}{4\pi}\,|u(\theta)|^2\,\check{g}^K_{\epsilon,\epsilon-\omega}(\mathbf{p}'_F,\mathbf{k}),\end{aligned} \tag{9.9}$$

and

$$\check{\Sigma}^{R(A)}_{\epsilon,\epsilon-\omega}(\mathbf{p}_F,\mathbf{k}) = \pi i\nu(0)\,n_{\text{imp}}\int\frac{d\Omega_{\mathbf{p}'}}{4\pi}\,|u(\theta)|^2\,\check{g}^{R(A)}_{\epsilon,\epsilon-\omega}(\mathbf{p}'_F,\mathbf{k}) \tag{9.10}$$

in a complete analogy to eqn (5.69). The impurity self-energy is written in the Born approximation. The Fourier transformed impurity potential squared is $|u(\theta)|^2 = 2v_F\sigma_{\mathbf{pp}'}/\nu(0)$ where $\sigma_{\mathbf{pp}'}$ is the scattering cross section. One can thus present the self-energy as

$$\begin{aligned}\check{\Sigma}^K_{\epsilon,\epsilon-\omega}(\mathbf{p}_F,\mathbf{k}) &= \frac{i}{2}v_F n_{\text{imp}}\int d\Omega_{\mathbf{p}'}\sigma_{\mathbf{pp}'}\check{g}^K_{\epsilon,\epsilon-\omega}(\mathbf{p}'_F,\mathbf{k}),\\ \check{\Sigma}^{R(A)}_{\epsilon,\epsilon-\omega}(\mathbf{p}_F,\mathbf{k}) &= \frac{i}{2}v_F n_{\text{imp}}\int d\Omega_{\mathbf{p}'}\sigma_{\mathbf{pp}'}\check{g}^{R(A)}_{\epsilon,\epsilon-\omega}(\mathbf{p}'_F,\mathbf{k}).\end{aligned}$$

Remind that the scattering rate in the normal sate is

$$\frac{1}{\tau_{\text{imp}}} = 4\pi\left\langle\sigma_{\mathbf{pp}'}\right\rangle v_F n_{\text{imp}}.$$

The phonon self-energy can also be expressed through the quasiclassical functions. It is

$$\check{\Sigma}^{(\text{ph})K}_{\epsilon,\epsilon-\omega}(\mathbf{p}_F,\mathbf{k}) = \nu(0)\,g^2_{ph}\int\frac{d\epsilon'}{4}\frac{d\Omega_{\mathbf{p}'}}{4\pi}\left(D^R_{\epsilon'-\epsilon} - D^A_{\epsilon'-\epsilon}\right)$$

$$\times \left[\coth\left(\frac{\epsilon' - \epsilon}{2T}\right) \check{g}^K_{\epsilon', \epsilon'-\omega} - \left(\check{g}^R_{\epsilon', \epsilon'-\omega} - \check{g}^A_{\epsilon', \epsilon'-\omega}\right)\right].$$

The relaxation part of the regular phonon self-energy takes the form

$$\check{\Sigma}^{(\mathrm{ph})R(A)}_{\epsilon, \epsilon-\omega}(\mathbf{p}_F, \mathbf{k}) = \nu(0)\, g^2_{\mathrm{ph}} \int \frac{d\epsilon'}{4} \frac{d\Omega_{\mathbf{p}'}}{4\pi} \left(D^R_{\epsilon'-\epsilon} - D^A_{\epsilon'-\epsilon}\right) \times \left[\coth\left(\frac{\epsilon' - \epsilon}{2T}\right) \check{g}^{R(A)}_{\epsilon', \epsilon'-\omega} \mp \frac{1}{2} \check{g}^K_{\epsilon', \epsilon'-\omega}\right].$$

The function $\check{g}^R_{\epsilon', \epsilon'-\omega}$ in these equations carries the momentum $\mathbf{p}'$, i.e., $\check{g}^R_{\epsilon', \epsilon'-\omega} \equiv \check{g}^R_{\epsilon', \epsilon'-\omega}(\mathbf{p}', \mathbf{k})$. Since the phonon energy is $\omega \sim T$ and satisfies the inequality $\omega \ll \Omega_D$ where $\Omega_D \sim s p_F$ is the Debye frequency, the phonon energy is $\omega_{\mathbf{pp}'} = 2 s p_F |\sin(\theta/2)|$ where s is the velocity of sound. Because the relative change in the momentum during the emission of a phonon is small

$$|\mathbf{p} - \mathbf{p}'| \sim \Delta/s \ll p_F,$$

the the scattering angle of a phonon $\theta \ll 1$. We have

$$\int \frac{d\Omega_{\mathbf{p}'}}{4\pi} \left[D^R_\omega(\theta) - D^A_\omega(\theta)\right] = \frac{\pi i}{2(s p_F)^2} \omega^2 \mathrm{sign}\omega. \tag{9.11}$$

As a result,

$$\check{\Sigma}^{(\mathrm{ph})K}_{\epsilon, \epsilon-\omega}(\mathbf{p}_F, \mathbf{k}) = \frac{\pi i \nu(0) g^2_{\mathrm{ph}}}{2(s p_F)^2} \int \frac{d\epsilon'}{4} (\epsilon' - \epsilon)^2 \,\mathrm{sign}\,(\epsilon' - \epsilon) \times \left[\coth\left(\frac{\epsilon' - \epsilon}{2T}\right) \check{g}^K_{\epsilon', \epsilon'-\omega}(\mathbf{p}, \mathbf{k}) - \left(\check{g}^R_{\epsilon', \epsilon'-\omega}(\mathbf{p}, \mathbf{k}) - \check{g}^A_{\epsilon', \epsilon'-\omega}(\mathbf{p}, \mathbf{k})\right)\right], \tag{9.12}$$

$$\check{\Sigma}^{(\mathrm{ph})R(A)}_{\epsilon, \epsilon-\omega}(\mathbf{p}_F, \mathbf{k}) = \frac{\pi i \nu(0) g^2_{\mathrm{ph}}}{2(s p_F)^2} \int \frac{d\epsilon'}{4} (\epsilon' - \epsilon)^2 \,\mathrm{sign}\,(\epsilon' - \epsilon) \times \left[\coth\left(\frac{\epsilon' - \epsilon}{2T}\right) \check{g}^{R(A)}_{\epsilon', \epsilon'-\omega}(\mathbf{p}, \mathbf{k}) \mp \frac{1}{2} \check{g}^K_{\epsilon', \epsilon'-\omega}(\mathbf{p}, \mathbf{k})\right]. \tag{9.13}$$

Note that the functions $\check{g}_{\epsilon'}(\mathbf{p}, \mathbf{r})$ in eqns (9.12, 9.13) have the same momentum $\mathbf{p}$ as the self-energy itself since the relative change in the momentum during interaction with the acoustic phonons is small.

The collision integral eqn (9.2) takes the form

$$\check{I}^K_{\epsilon_+, \epsilon_-}(\mathbf{p}_F, \mathbf{k}) = \left[\check{\Sigma}^R \check{g}^K - \check{g}^K \check{\Sigma}^A - \check{g}^R \check{\Sigma}^K + \check{\Sigma}^K \check{g}^A\right]_{\epsilon_+, \epsilon_-} \tag{9.14}$$

where all terms are expressed through quasiclassical functions.

The collision integral eqn (9.5) for the anomalous function is

$$\check{I}^{(a)}_{\epsilon_+,\epsilon_-} = \left[\check{\Sigma}^R \check{g}^{(a)} - \check{g}^{(a)}\check{\Sigma}^A - \check{g}^R\check{\Sigma}^{(a)} + \check{\Sigma}^{(a)}\check{g}^A\right]_{\epsilon_+,\epsilon_-} . \tag{9.15}$$

For regular functions in eqn (9.7) we obtain

$$\check{I}^{R(A)}_{\epsilon_+,\epsilon_-} = \left[\check{\Sigma}^{R(A)}\check{g}^{R(A)} - \check{g}^{R(A)}\check{\Sigma}^{R(A)}\right]_{\epsilon_+,\epsilon_-} . \tag{9.16}$$

9.1.2 *Order parameter, current, and particle density*

The order parameter, current, and the particle density can be expressed through the quasiclassical functions in the same way as for the static case. The order parameter equation becomes for a general pairing

$$\begin{aligned}
\frac{\Delta_{\mathbf{p}}(\mathbf{k},\omega)}{\lambda} &= \int V(\hat{\mathbf{p}},\hat{\mathbf{p}}')\, f^K_{\epsilon_+,\epsilon_-}(\hat{\mathbf{p}}'_F,\mathbf{k})\,\frac{d\Omega_{\mathbf{p}'}}{4\pi}\,\frac{d\epsilon}{4},\\
\frac{\Delta^*_{\mathbf{p}}(\mathbf{k},\omega)}{\lambda} &= \int V(\hat{\mathbf{p}},\hat{\mathbf{p}}')\, f^{\dagger K}_{\epsilon_+,\epsilon_-}(\hat{\mathbf{p}}'_F,\mathbf{k})\,\frac{d\Omega_{\mathbf{p}'}}{4\pi}\,\frac{d\epsilon}{4}.
\end{aligned} \tag{9.17}$$

The expression for current is

$$\begin{aligned}
\mathbf{j}(\omega,\mathbf{k}) &= -2e\nu(0)\int \frac{d\Omega_{\mathbf{p}}}{4\pi}\,\frac{d\epsilon}{4}\,\mathbf{v}_F g^K_{\epsilon_+,\epsilon_-}(\mathbf{p}_F,\mathbf{k})\\
&= 2e\nu(0)\int \frac{d\Omega_{\mathbf{p}}}{4\pi}\,\frac{d\epsilon}{4}\,\mathbf{v}_F \bar{g}^K_{\epsilon_+,\epsilon_-}(\mathbf{p}_F,\mathbf{k}),
\end{aligned} \tag{9.18}$$

and the electron density has the form

$$N(\omega,\mathbf{k}) = N_0(2\pi)^4\,\delta(\mathbf{k})\,\delta(\omega) + \delta N(\omega,\mathbf{k})$$

where

$$\begin{aligned}
\delta N(\omega,\mathbf{k}) &= -2\nu(0)\int \frac{d\Omega_{\mathbf{p}}}{4\pi}\,\frac{d\epsilon}{4}\,g^K_{\epsilon_+,\epsilon_-}(\mathbf{p}_F,\mathbf{k})\\
&= -2\nu(0)\int \frac{d\Omega_{\mathbf{p}}}{4\pi}\,\frac{d\epsilon}{4}\,\bar{g}^K_{\epsilon_+,\epsilon_-}(\mathbf{p}_F,\mathbf{k}).
\end{aligned} \tag{9.19}$$

They are easily obtained from the corresponding equations in terms of the total Green functions and the quasiclassical expressions derived earlier for the static case. The diamagnetic term in the current, eqn (9.18), disappears in the quasiclassical representation since it is canceled by the normal-state contribution subtracted from the ξ-integrated function g as it has been shown earlier.

9.1.3 *Normalization of the quasiclassical functions*

Here we derive useful identities for the total quasiclassical Green functions. First of all, we demonstrate that the normalization condition

$$\int \frac{d\epsilon'}{2\pi} \frac{d^3k'}{(2\pi)^3} \check{g}^{R(A)}_{\epsilon,\epsilon'} (\mathbf{p}_F, \mathbf{k})\, \check{g}^{R(A)}_{\epsilon',\epsilon-\omega} (\mathbf{p}_F, \mathbf{k} - \mathbf{k}') = \check{1}\, (2\pi)^4\, \delta(\omega)\, \delta(\mathbf{k}) \tag{9.20}$$

holds in the nonstationary case, as well. To see this, we multiply eqn (9.7) by $\check{g}^{R(A)}_{\epsilon,\epsilon'}$ first from the left and then from the right and add the two equations. We obtain

$$\begin{aligned} \mathbf{v}_F\mathbf{k}\, [\check{g}\check{g}]^{R(A)}_{\epsilon_+,\epsilon_-} - \epsilon_+\check{\tau}_3\, [\check{g}\check{g}]^{R(A)}_{\epsilon_+,\epsilon_-} + [\check{g}\check{g}]^{R(A)}_{\epsilon_+,\epsilon_-}\, \epsilon_-\check{\tau}_3 \\ + \check{H}_{\omega_1}\, [\check{g}\check{g}]^{R(A)}_{\epsilon_+-\omega_1,\epsilon_-} - [\check{g}\check{g}]^{R(A)}_{\epsilon_+,\epsilon_-+\omega_1}\, \check{H}_{\omega_1} \\ = \left[\check{\Sigma}^{R(A)}\, [\check{g}\check{g}]^{R(A)} - [\check{g}\check{g}]^{R(A)}\, \check{\Sigma}^{R(A)}\right]_{\epsilon_+,\epsilon_-} \end{aligned} \tag{9.21}$$

where the shortcut $[\check{g}\check{g}]$ again stands for the expression in the l.h.s. of eqn (9.20). We observe that $[\check{g}\check{g}]^{R(A)} = \check{1} \cdot const$ is a solution of eqn (9.21). We can use the same argumentation as for the static case. We assume that, at large distances the system is in equilibrium and is homogeneous. Therefore the condition $[\check{g}\check{g}]^{R(A)} = \check{1}$, being fulfilled in equilibrium, should also hold in a general situation. We thus arrive at eqn (9.20).

Let us now multiply eqn (9.2) first by $\check{g}^R$ from the left and then by $\check{g}^A$ from the right. Next, we multiply eqn (9.7) for $\check{g}^R$ by $\check{g}^K$ from the right, and then eqn (9.7) for $\check{g}^A$ by $\check{g}^K$ from the left. We now add the four equations together and obtain with the help of eqn (9.20)

$$\begin{aligned} \mathbf{v}_F\mathbf{k}\check{C}_{\epsilon_+,\epsilon_-} - \epsilon_+\check{\tau}_3\check{C}_{\epsilon_+,\epsilon_-} + \check{C}_{\epsilon_+,\epsilon_-}\epsilon_-\check{\tau}_3 + \left[\check{H}\check{C}\right]_{\epsilon_+,\epsilon_-} - \left[\check{C}\check{H}\right]_{\epsilon_+,\epsilon_-} \\ = \left[\check{\Sigma}^R\check{C} - \check{C}\check{\Sigma}^A\right]_{\epsilon_+,\epsilon_-} \end{aligned} \tag{9.22}$$

where $\check{C} = \left[\check{g}^R\check{g}^K + \check{g}^K\check{g}^A\right]$. We prove now that $\check{C} = 0$. Indeed, the solution $\check{C} = 0$ satisfies eqn (9.22) and the boundary conditions for an equilibrium system where

$$\check{g}^K_{\epsilon_+,\epsilon_-} = \left(\check{g}^R_\epsilon - \check{g}^A_\epsilon\right) \tanh\left(\frac{\epsilon}{2T}\right) (2\pi)\, \delta(\omega)\,. \tag{9.23}$$

Therefore, in a general situation

$$\begin{aligned} \int \frac{d\epsilon'}{2\pi} \frac{d^3k'}{(2\pi)^3} \left[\check{g}^R_{\epsilon,\epsilon'} (\mathbf{p}_F, \mathbf{k}')\, \check{g}^K_{\epsilon',\epsilon-\omega} (\mathbf{p}_F, \mathbf{k} - \mathbf{k}')\right. \\ \left. + \check{g}^K_{\epsilon,\epsilon'} (\mathbf{p}_F, \mathbf{k}')\, \check{g}^A_{\epsilon',\epsilon-\omega} (\mathbf{p}_F, \mathbf{k} - \mathbf{k}')\right] = 0. \end{aligned} \tag{9.24}$$

This result was first obtained by Larkin and Ovchinnikov (1975). Using eqn (9.20) we can see that eqn (9.24) also holds for the anomalous function $\check{g}^{(a)}$.

9.2 Generalized distribution function

Equation (9.24) allows us to present the Keldysh function in the form

$$\check{g}^K_{\epsilon_+,\epsilon_-} (\mathbf{p}_F, \mathbf{k}) = \int \frac{d\epsilon'}{2\pi} \frac{d^3k'}{(2\pi)^3} \left[\check{g}^R_{\epsilon_+,\epsilon'} (\mathbf{p}_F, \mathbf{k}')\, \check{f}_{\epsilon',\epsilon_-} (\mathbf{p}_F, \mathbf{k} - \mathbf{k}')\right.$$

$$-\check{f}_{\epsilon_+,\epsilon'}(\mathbf{p}_F,\mathbf{k}+\mathbf{k}')\,\check{g}^A_{\epsilon',\epsilon_-}(\mathbf{p}_F,\mathbf{k}')\Big] \tag{9.25}$$

where $\check{f}_{\epsilon,\epsilon'}$ is another unknown matrix. Comparing this equation with eqn (9.23) we can see that $\check{f}_{\epsilon,\epsilon'}$ has the meaning of a generalized "distribution function" which describes the system in a nonequilibrium situation. Recall that the hyperbolic tangent in eqn (9.23) comes from the equilibrium distribution eqn (8.14).

We shall show that the matrix $\check{f}_{\epsilon,\epsilon'}$ has only two independent components. Consider the operators $\mathcal{L}_\pm\{\check{A}\}$ acting on a matrix $\check{A}$ which are defined by

$$\mathcal{L}_+\{\check{A}\} = [\check{g}^R\check{A}] + [\check{A}\check{g}^A],\ \mathcal{L}_-\{\check{A}\} = [\check{g}^R\check{A}] - [\check{A}\check{g}^A].$$

Here we again imply integration over internal frequencies and momenta. Using the definition of the $\mathcal{L}$ operators and the normalization condition eqn (9.20) for the regular functions we find

$$\mathcal{L}_+\mathcal{L}_- = \mathcal{L}_-\mathcal{L}_+ = 0. \tag{9.26}$$

Consider the eigen-matrices of these operators

$$\mathcal{L}_+\{\check{A}_+\} = \lambda_+\check{A}_+;\mathcal{L}_-\{\check{A}_-\} = \lambda_-\check{A}_-. \tag{9.27}$$

Each of these operators has four eigen-matrices belonging to four eigenvalues. It is easy to check that the eigen-matrix $\check{A}_+$ belonging to a nonzero eigenvalue λ_+ of the operator $\mathcal{L}_+$ is simultaneously an eigen-matrix of the operator $\mathcal{L}_-$ with eigenvalue zero, and vice versa: the eigen-matrix $\check{A}_-$ belonging to a nonzero eigenvalue λ_- of the operator $\mathcal{L}_-$ is simultaneously an eigen-matrix of the operator $\mathcal{L}_+$ with eigenvalue zero. To see this, we apply the operator $\mathcal{L}_-$ to the first equation (9.27) using eqn (9.26) and obtain $\mathcal{L}_-\{\check{A}_+\} = 0$. We can also apply $\mathcal{L}_+$ to the second equation (9.27) to get $\mathcal{L}_+\{\check{A}_-\} = 0$. Applying the operator $\mathcal{L}_-$ twice to the eigen-matrix $\check{A}_-$ we have

$$2\left(\check{A}_- - [\check{g}^R\check{A}_-\check{g}^A]\right) = \lambda_-^2\check{A}_-.$$

Applying the operator $\mathcal{L}_-$ again we find $\lambda_-\left(\lambda_-^2 - 4\right) = 0$. In the same way we get $\lambda_+\left(\lambda_+^2 - 4\right) = 0$. Therefore, the four eigenvalues of the operator $\mathcal{L}_-$ are $\lambda_-^{(1)} = +2$, $\lambda_-^{(2)} = -2$, and twice degenerate $\lambda_-^{(3,4)} = 0$. The corresponding eigen-matrices $\check{A}_1$, $\check{A}_2$, $\check{A}_3$, $\check{A}_4$ are simultaneously the eigen-matrices of the operator $\mathcal{L}_+$ with the eigenvalues 0, 0, +2, and −2, respectively.

Let us now expand the matrix $\check{f}_{\epsilon,\epsilon'}$ in the matrices $\check{A}_1$, $\check{A}_2$, $\check{A}_3$, $\check{A}_4$. We have

$$\check{f}_{\epsilon,\epsilon'} = \check{A}_1\tilde{f}^{(1)} + \check{A}_2\tilde{f}^{(2)} + \check{A}_3\tilde{f}^{(3)} + \check{A}_4\tilde{f}^{(4)}.$$

It is clear that the coefficients $\tilde{f}^{(3)}$ and $\tilde{f}^{(4)}$ will not appear in the total Green function $\check{g}^K = \mathcal{L}_-\{\check{f}\}$ because the matrices $\check{A}_3$, $\check{A}_4$ correspond to zero eigenvalues of the operator $\mathcal{L}_-$. Therefore, the matrix $\check{f}_{\epsilon,\epsilon'}$ in eqn (9.25) has only two independent components $\tilde{f}^{(1)}$ and $\tilde{f}^{(2)}$. The two remaining coefficients $\tilde{f}^{(3)}$ and $\tilde{f}^{(4)}$ can be chosen arbitrarily. For, example, one can choose them in such a way as to make the function $\check{f}_{\epsilon,\epsilon'}$ diagonal

$$\check{f}_{\epsilon,\epsilon'} = \check{1} f^{(1)}(\epsilon,\epsilon') + \check{\tau}_3 f^{(2)}(\epsilon,\epsilon'). \tag{9.28}$$

This form of presentation for the distribution function was introduced by Schmid and Schön (1975) and by Larkin and Ovchinnikov (1977) We note that, in equilibrium,

$$f^{(1)}(\epsilon,\epsilon') = 2\pi\delta(\epsilon-\epsilon') f^{(0)}(\epsilon), \quad f^{(2)} = 0,$$

where, as in eqn (8.9),

$$f^{(0)}(\epsilon) = \tanh\frac{\epsilon}{2T}.$$

We use this form of presentation of the distribution function throughout the book. Another choice for the free parameters $\tilde{f}^{(1)}$ and $\tilde{f}^{(2)}$ has been suggested by Shelankov (1985). It has an advantage in some cases when the relation to the Boltzmann kinetic equation is considered.

In Chapter 10 we discuss the generalized two-component distribution function in detail and derive the kinetic equations which determine $f^{(1)}$ and $f^{(2)}$. The obtained solutions will be used for various applications later in this book. In the rest of the present chapter, we consider several examples of a direct implementation of the Eliashberg equations in their original form.

9.3 s-wave superconductors with a short mean free path

Here we discuss an approach which can be used for superconductors with an impurity mean free path $\ell \ll \xi(T)$. This limit is realized for some s-wave superconducting alloys. For d-wave superconductors, where always $\ell \gg \xi_0$, this approach can also be used in a close vicinity of the critical temperature such that $\xi(T)$ already exceeds ℓ. We consider the d-wave case later in Section 11.3 while concentrating here on s-wave superconductors. The method is similar to the one used in Section 5.6 to derive the Usadel equations.

The gradients of the order parameter and of other physical quantities are generally of the order of $\xi^{-1}(T)$ and are thus small compared with ℓ^{-1}. We can use a "hydrodynamic" approximation, assuming that the Green functions are, to the first approximation, independent of the momentum direction, and restrict ourselves to the first correction in small gradients. Since these corrections are vectors, we write

$$\check{g} = \langle\check{g}\rangle + \hat{\mathbf{p}}\check{\mathbf{g}}$$

for all (retarded, advanced, and Keldysh) Green functions. Here $\hat{\mathbf{p}}$ is the unit vector in the momentum direction, and $\langle\ldots\rangle$ denotes an average over the momentum directions in accordance with eqns (5.72) or (5.73). Both $\langle\check{g}\rangle$ and $\check{\mathbf{g}}$ are independent of the momentum direction. The vector part of the Green function is small compared to its averaged value $\langle\check{g}\rangle$.

Averaging of the normalization condition eqn (9.20) gives for the regular functions

$$\left[\left\langle \check{g}^{R(A)}\right\rangle \left\langle \check{g}^{R(A)}\right\rangle\right]_{\epsilon_+,\epsilon_-} = \check{1}\,(2\pi)^4\,\delta\left(\mathbf{k}\right)\delta\left(\omega\right). \tag{9.29}$$

Averaging it with the momentum direction $\hat{\mathbf{p}}$ results in

$$\left[\left\langle \check{g}^{R(A)}\right\rangle \check{\mathbf{g}}^{R(A)} + \check{\mathbf{g}}^{R(A)}\left\langle \check{g}^{R(A)}\right\rangle\right]_{\epsilon_+,\epsilon_-} = 0. \tag{9.30}$$

For the Keldysh function we have from eqn (9.24)

$$\left[\left\langle \check{g}^{R}\right\rangle \left\langle \check{g}^{K}\right\rangle + \left\langle \check{g}^{K}\right\rangle \left\langle \check{g}^{A}\right\rangle\right]_{\epsilon_+,\epsilon_-} = 0, \tag{9.31}$$

and

$$\left[\left\langle \check{g}^{R}\right\rangle \check{\mathbf{g}}^{K} + \check{\mathbf{g}}^{K}\left\langle \check{g}^{A}\right\rangle + \check{\mathbf{g}}^{R}\left\langle \check{g}^{K}\right\rangle + \left\langle \check{g}^{K}\right\rangle \check{\mathbf{g}}^{A}\right]_{\epsilon_+,\epsilon_-} = 0. \tag{9.32}$$

The impurity collision integrals for nonmagnetic scattering eqn (9.14) or (9.16) are expressed through nonmagnetic self-energies defined by eqn (5.76). The nonmagnetic collision integrals vanish after averaging over the momentum directions:

$$\left\langle \check{I}^{R(A)}_{\text{imp}}\left(\mathbf{p}\right)\right\rangle = \left\langle \check{I}^{K}_{\text{imp}}\left(\mathbf{p}\right)\right\rangle = 0.$$

Using this fact, we can simplify the quasiclassical equations in the same way as it has been done in Section 5.6. On the first stage, we average the kinetic equation over directions of the momentum. For the Keldysh function we get from eqn (9.2) in coordinate representation

$$\begin{aligned}
&\frac{1}{3}v_F\left(-i\nabla\check{\mathbf{g}}^{K}_{\epsilon_+,\epsilon_-} - \frac{e}{c}\left(\check{\tau}_3\left[\mathbf{A}\check{\mathbf{g}}^{K}\right]_{\epsilon_+,\epsilon_-} - \left[\check{\mathbf{g}}^{K}\mathbf{A}\right]_{\epsilon_+,\epsilon_-}\check{\tau}_3\right)\right)\\
&\qquad -\epsilon_+\check{\tau}_3\left\langle \check{g}^{K}_{\epsilon_+,\epsilon_-}\right\rangle + \left\langle \check{g}^{K}_{\epsilon_+,\epsilon_-}\right\rangle \epsilon_-\check{\tau}_3 + \left[\check{H}_{\Delta}\left\langle \check{g}^{K}\right\rangle\right]_{\epsilon_+,\epsilon_-} - \left[\left\langle \check{g}^{K}\right\rangle \check{H}_{\Delta}\right]_{\epsilon_+,\epsilon_-}\\
&\quad = \left\langle \check{I}^{(\text{ph})K}_{\epsilon_+,\epsilon_-}\right\rangle + \left\langle \check{I}^{(s)K}_{\epsilon_+,\epsilon_-}\right\rangle.
\end{aligned} \tag{9.33}$$

Here

$$\check{H}_{\Delta} = \begin{pmatrix} 0 & -\Delta \\ \Delta^* & 0 \end{pmatrix}.$$

We note here an important difference between s-wave and d-wave superconductors. For the s-wave case, the average of $\check{H}_{\Delta}$ does not disappear because Δ is independent of the momentum directions. On the contrary, for a d-wave case, this average vanishes.

Since the largest part of the total collision integral associated with non-magnetic impurities disappears after averaging we have to keep other terms. In the phonon and spin-flip collision integrals, one can replace the full Green functions by their averaged parts using smallness of the vector parts. We have, for example

$$\left\langle \check{I}^{(s)K}_{\epsilon_+,\epsilon_-} \right\rangle = \frac{i}{2\tau_s} \left[\check{\tau}_3 \left\langle \check{g}^R \right\rangle \check{\tau}_3 \left\langle \check{g}^K \right\rangle - \left\langle \check{g}^R \right\rangle \check{\tau}_3 \left\langle \check{g}^K \right\rangle \check{\tau}_3 \right.$$
$$\left. - \left\langle \check{g}^K \right\rangle \check{\tau}_3 \left\langle \check{g}^A \right\rangle \check{\tau}_3 + \check{\tau}_3 \left\langle \check{g}^K \right\rangle \check{\tau}_3 \left\langle \check{g}^A \right\rangle \right]_{\epsilon_+,\epsilon_-} . \qquad (9.34)$$

Let us now multiply eqn (9.2) by the momentum unit vector $\hat{\mathbf{p}}$ and average over its directions. We obtain

$$v_F \left(-i\nabla \left\langle \check{g}^K_{\epsilon_+,\epsilon_-} \right\rangle - \frac{e}{c} \left(\check{\tau}_3 \left[\mathbf{A} \left\langle \check{g}^K \right\rangle \right]_{\epsilon_+,\epsilon_-} - \left[\left\langle \check{g}^K \right\rangle \mathbf{A} \right]_{\epsilon_+,\epsilon_-} \check{\tau}_3 \right) \right)$$
$$-\epsilon_+ \check{\tau}_3 \check{\mathbf{g}}^K_{\epsilon_+,\epsilon_-} + \check{\mathbf{g}}^K_{\epsilon_+,\epsilon_-} \epsilon_- \check{\tau}_3 + \left[\check{H}_\Delta \check{\mathbf{g}}^K \right]_{\epsilon_+,\epsilon_-} - \left[\check{\mathbf{g}}^K \check{H}_\Delta \right]_{\epsilon_+,\epsilon_-}$$
$$= \frac{i}{2\tau_{\rm tr}} \left[\left\langle \check{g}^R \right\rangle \check{\mathbf{g}}^K - \check{\mathbf{g}}^R \left\langle \check{g}^K \right\rangle + \left\langle \check{g}^K \right\rangle \check{\mathbf{g}}^A - \check{\mathbf{g}}^K \left\langle \check{g}^A \right\rangle \right]_{\epsilon_+,\epsilon_-}$$
$$+3 \left\langle \hat{\mathbf{p}} \check{I}^{(\rm ph)K}_{\epsilon_+,\epsilon_-} \right\rangle + 3 \left\langle \hat{\mathbf{p}} \check{I}^{(s)K}_{\epsilon_+,\epsilon_-} \right\rangle .$$

We use eqn (5.80) here and the definition eqn (5.81) of the transport mean free time. The relaxation rate $1/\tau_{\rm tr}$ is usually large compared with $1/\tau_s$. In the dirty limit it is also large compared to T and Δ, to say nothing about the electron-phonon relaxation rate. Therefore, we can neglect the corresponding terms here and obtain

$$v_F \left(-i\nabla \left\langle \check{g}^K_{\epsilon_+,\epsilon_-} \right\rangle - \frac{e}{c} \left(\check{\tau}_3 \left[\mathbf{A} \left\langle \check{g}^K \right\rangle \right]_{\epsilon_+,\epsilon_-} - \left[\left\langle \check{g}^K \right\rangle \mathbf{A} \right]_{\epsilon_+,\epsilon_-} \check{\tau}_3 \right) \right)$$
$$= \frac{i}{2\tau_{\rm tr}} \left[\left\langle \check{g}^R \right\rangle \check{\mathbf{g}}^K - \check{\mathbf{g}}^R \left\langle \check{g}^K \right\rangle + \left\langle \check{g}^K \right\rangle \check{\mathbf{g}}^A - \check{\mathbf{g}}^K \left\langle \check{g}^A \right\rangle \right]_{\epsilon_+,\epsilon_-} . \qquad (9.35)$$

This equation allows one to find the vector part of the Keldysh function. Let us multiply eqn (9.35) by $\left\langle \check{g}^R \right\rangle$ from the left and integrate over internal frequencies and momenta, then multiply eqn (9.35) again by $\left\langle \check{g}^A \right\rangle$ from the right and subtract the two equations. Using eqns (9.29)-(9.32) we obtain

$$-iv_F \left[\left\langle \check{g}^R \right\rangle \left(\nabla \left\langle \check{g}^K \right\rangle \right) - \left(\nabla \left\langle \check{g}^K \right\rangle \right) \left\langle \check{g}^A \right\rangle \right]_{\epsilon_+,\epsilon_-}$$
$$-\frac{e}{c} v_F \left[\left(\left\langle \check{g}^R \right\rangle \check{\tau}_3 + \check{\tau}_3 \left\langle \check{g}^R \right\rangle \right) \mathbf{A} \left\langle \check{g}^K \right\rangle + \left\langle \check{g}^K \right\rangle \mathbf{A} \left(\left\langle \check{g}^A \right\rangle \check{\tau}_3 + \check{\tau}_3 \left\langle \check{g}^A \right\rangle \right) \right]_{\epsilon_+,\epsilon_-}$$
$$= \frac{i}{\tau_{\rm tr}} \left[\mathbf{g}^K - \left\langle \check{g}^R \right\rangle \mathbf{g}^K \left\langle \check{g}^A \right\rangle \right]_{\epsilon_+,\epsilon_-} .$$

The r.h.s. can also be written as

$$\frac{i}{\tau_{\rm tr}} \left[2\mathbf{g}^K - \mathbf{g}^R \left\langle \check{g}^R \right\rangle \left\langle \check{g}^K \right\rangle - \left\langle \check{g}^K \right\rangle \left\langle \check{g}^A \right\rangle \mathbf{g}^A \right]_{\epsilon_+,\epsilon_-} .$$

Having found the vector part from this equation, we can then use it to determine the averaged functions from eqn (9.33). However, further transformations of these equations cannot easily be performed in a general case. We shall consider some examples later.

The equations for regular functions can be obtained in the same way. The averaged equation is

$$\frac{1}{3}v_F\left(-i\nabla\check{\mathbf{g}}^{R(A)}_{\epsilon_+,\epsilon_-}-\frac{e}{c}\left(\check{\tau}_3\left[\mathbf{A}\check{\mathbf{g}}^{R(A)}\right]_{\epsilon_+,\epsilon_-}-\left[\check{\mathbf{g}}^{R(A)}\mathbf{A}\right]_{\epsilon_+,\epsilon_-}\check{\tau}_3\right)\right)$$
$$-\epsilon_+\check{\tau}_3\left\langle\check{g}^{R(A)}_{\epsilon_+,\epsilon_-}\right\rangle+\left\langle\check{g}^{R(A)}_{\epsilon_+,\epsilon_-}\right\rangle\epsilon_-\check{\tau}_3$$
$$+\left[\check{H}_\Delta\left\langle\check{g}^{R(A)}\right\rangle\right]_{\epsilon_+,\epsilon_-}-\left[\left\langle\check{g}^{R(A)}\right\rangle\check{H}_\Delta\right]_{\epsilon_+,\epsilon_-}$$
$$=\left\langle\check{I}^{(\mathrm{ph})R(A)}_{\epsilon_+,\epsilon_-}\right\rangle+\left\langle\check{I}^{(s)R(A)}_{\epsilon_+,\epsilon_-}\right\rangle \tag{9.36}$$

where

$$\left\langle\check{I}^{(s)R(A)}_{\epsilon_+,\epsilon_-}\right\rangle=\frac{i}{2\tau_s}\left[\check{\tau}_3\left\langle\check{g}^{R(A)}\right\rangle\check{\tau}_3\left\langle\check{g}^{R(A)}\right\rangle-\left\langle\check{g}^{R(A)}\right\rangle\check{\tau}_3\left\langle\check{g}^{R(A)}\right\rangle\check{\tau}_3\right]_{\epsilon_+,\epsilon_-}. \tag{9.37}$$

The vector part is

$$\ell_{\mathrm{tr}}\left(-i\nabla\left\langle\check{g}^{R(A)}_{\epsilon_+,\epsilon_-}\right\rangle-\frac{e}{c}\left(\check{\tau}_3\left[\mathbf{A}\left\langle\check{g}^{R(A)}\right\rangle\right]_{\epsilon_+,\epsilon_-}-\left[\left\langle\check{g}^{R(A)}\right\rangle\mathbf{A}\right]_{\epsilon_+,\epsilon_-}\check{\tau}_3\right)\right)$$
$$=\frac{i}{2}\left[\left\langle\check{g}^{R(A)}\right\rangle\check{\mathbf{g}}^{R(A)}-\check{\mathbf{g}}^{R(A)}\left\langle\check{g}^{R(A)}\right\rangle\right]_{\epsilon_+,\epsilon_-} \tag{9.38}$$

where $\ell_{\mathrm{tr}}=v_F\tau_{\mathrm{tr}}$. The vector part of the regular functions can be found using the normalization of eqns (9.29)–(9.30). Multiplying eqn (9.38) by $\langle\check{g}^{R(A)}\rangle$ from the left, we get

$$\check{\mathbf{g}}^{R(A)}_{\epsilon_+,\epsilon_-}=-\ell_{\mathrm{tr}}\left[\left\langle\check{g}^{R(A)}\right\rangle\nabla\left\langle\check{g}^{R(A)}\right\rangle\right]_{\epsilon_+,\epsilon_-}$$
$$-i\ell_{\mathrm{tr}}\frac{e}{c}\left[\mathbf{A}\check{\tau}_3-\left\langle\check{g}^{R(A)}\right\rangle\check{\tau}_3\mathbf{A}\left\langle\check{g}^{R(A)}\right\rangle\right]_{\epsilon_+,\epsilon_-}. \tag{9.39}$$

Sometimes, one uses the equations for anomalous functions, as well. We obtain in exactly the same way

$$\frac{1}{3}v_F\left(-i\nabla\check{\mathbf{g}}^{(a)}_{\epsilon_+,\epsilon_-}-\frac{e}{c}\left(\check{\tau}_3\left[\mathbf{A}\check{\mathbf{g}}^{(a)}\right]_{\epsilon_+,\epsilon_-}-\left[\check{\mathbf{g}}^{(a)}\mathbf{A}\right]_{\epsilon_+,\epsilon_-}\check{\tau}_3\right)\right)$$
$$-\epsilon_+\check{\tau}_3\left\langle\check{g}^{(a)}_{\epsilon_+,\epsilon_-}\right\rangle+\left\langle\check{g}^{(a)}_{\epsilon_+,\epsilon_-}\right\rangle\epsilon_-\check{\tau}_3+\left[\check{H}_\Delta\left\langle\check{g}^{(a)}\right\rangle\right]_{\epsilon_+,\epsilon_-}-\left[\left\langle\check{g}^{(a)}\right\rangle\check{H}_\Delta\right]_{\epsilon_+,\epsilon_-}$$
$$=\left[\check{h}_\Delta\left\langle\check{g}^A\right\rangle\right]_{\epsilon_+,\epsilon_-}-\left[\left\langle\check{g}^R\right\rangle\check{h}_\Delta\right]_{\epsilon_+,\epsilon_-}+\frac{1}{3}\left(\left[\check{\mathbf{h}}_A\check{\mathbf{g}}^A\right]_{\epsilon_+,\epsilon_-}-\left[\check{\mathbf{g}}^R\check{\mathbf{h}}_A\right]_{\epsilon_+,\epsilon_-}\right)$$
$$+\left\langle\check{I}^{(\mathrm{ph})(a)}_{\epsilon_+,\epsilon_-}\right\rangle+\left\langle\check{I}^{(s)(a)}_{\epsilon_+,\epsilon_-}\right\rangle. \tag{9.40}$$

Here

$$\check{h}_\Delta=-\left[\check{H}_\Delta\right]_{\epsilon-\epsilon'}\left[f^{(0)}(\epsilon)-f^{(0)}(\epsilon')\right],$$

$$\check{\mathbf{h}}_A = \check{\tau}_3 \frac{e}{c} v_F \mathbf{A}_{\epsilon-\epsilon'} \left[f^{(0)}(\epsilon) - f^{(0)}(\epsilon') \right].$$

The vector part of the equation is

$$v_F \left(-i\nabla \left\langle \check{g}^{(a)}_{\epsilon_+,\epsilon_-} \right\rangle - \frac{e}{c} \left(\check{\tau}_3 \left[\mathbf{A} \left\langle \check{g}^{(a)} \right\rangle \right]_{\epsilon_+,\epsilon_-} - \left[\left\langle \check{g}^{(a)} \right\rangle \mathbf{A} \right]_{\epsilon_+,\epsilon_-} \check{\tau}_3 \right) \right)$$
$$= \left[\check{\mathbf{h}}_A \langle \check{g}^A \rangle \right]_{\epsilon_+,\epsilon_-} - \left[\langle \check{g}^R \rangle \check{\mathbf{h}}_A \right]_{\epsilon_+,\epsilon_-}$$
$$+ \frac{i}{2\tau_{\rm tr}} \left[\langle \check{g}^R \rangle \check{\mathbf{g}}^{(a)} - \check{\mathbf{g}}^{(a)} \langle \check{g}^A \rangle - \check{\mathbf{g}}^R \left\langle \check{g}^{(a)} \right\rangle + \left\langle \check{g}^{(a)} \right\rangle \check{\mathbf{g}}^A \right]_{\epsilon_+,\epsilon_-}. \quad (9.41)$$

9.4 Stimulated superconductivity

We consider here an example that demonstrates how the quasiparticle distribution affects the superconducting order parameter. This example is closely related with experiment and is instructive in that it treats a nonlinear response of a superconductor to an applied time-dependent perturbation. Consider a superconducting film placed in a microwave electromagnetic field. A part of electromagnetic energy is absorbed in the film. The intuitive picture is that, the superconducting being heated, its temperature grows and the order parameter decreases. However, this is not always the case. The real behavior is more tricky. The temperature, indeed, can increase. However, the overall rise in temperature depends on the cooling rate provided by the thermal contact of the film with the heat bath. If the heat contact is good, the temperature does not increase considerably. What happens is as follows. The microwave irradiation with a frequency of the order of the energy gap produces redistribution of energy between excitations and shifts the quasiparticle distribution to higher energies. The states with $\epsilon \sim \Delta$ become depleted. This causes an increase in the energy gap as compared to its value without irradiation. This effect is called stimulated superconductivity. It was first theoretically predicted by Eliashberg (1970) and then observed in many experiments. The detailed discussion of stimulated superconductivity can be found in the review by Eliashberg and Ivlev (1986).

We consider the simplest case of temperatures close to T_c, and restrict ourselves to dirty superconductors with the order parameter homogeneous in space. We first find the (averaged over time) distribution function taking into account absorption of energy from an external electromagnetic field with a frequency $\omega \sim \Delta$. For such frequencies, the order parameter performs only small oscillations near its average value (see Section 11.2), therefore we can take $\Delta = const$.

To find the time-averaged distribution function we use the kinetic equation for the anomalous function taken at zero frequency. Within the linear approximation in A^2 we have from eqn (9.40)

$$\mathrm{Tr} \left\langle \check{I}^{(\mathrm{ph})(a)}_{\epsilon,\epsilon} \right\rangle = \frac{e v_F}{3c} \mathrm{Tr} \left(\check{\tau}_3 \left[\mathbf{A} \check{\mathbf{g}}^{(a)} \right]_{\epsilon,\epsilon} - \left[\check{\mathbf{g}}^{(a)} \mathbf{A} \right]_{\epsilon,\epsilon} \check{\tau}_3 \right)$$
$$- \frac{1}{3} \mathrm{Tr} \left(\left[\check{\mathbf{h}}_A \check{\mathbf{g}}^A \right]_{\epsilon,\epsilon} - \left[\check{\mathbf{g}}^R \check{\mathbf{h}}_A \right]_{\epsilon,\epsilon} \right). \quad (9.42)$$

We assume that magnetic impurities are absent. The vector parts of both the regular functions $\check{\mathbf{g}}^R$ and $\check{\mathbf{g}}^A$ and of the anomalous function $\check{\mathbf{g}}^{(a)}$ are proportional to the vector potential $\mathbf{A}$ and have Fourier components at the frequency of the external field. Therefore, we obtain from eqn (9.41)

$$\operatorname{Tr}\left\langle \check{I}^{(\mathrm{ph})(a)}_{\epsilon,\epsilon}\right\rangle = \frac{ev_F}{3c}\operatorname{Tr}\left(\check{\tau}_3\mathbf{A}_{-\omega}\check{\mathbf{g}}^{(a)}_{\epsilon+\omega,\epsilon} - \check{\mathbf{g}}^{(a)}_{\epsilon,\epsilon-\omega}\mathbf{A}_{-\omega}\check{\tau}_3\right)$$
$$-\frac{1}{3}\operatorname{Tr}\left(\check{\mathbf{h}}_{A\,\epsilon,\epsilon-\omega}\check{\mathbf{g}}^A_{\epsilon-\omega,\epsilon} - \check{\mathbf{g}}^R_{\epsilon,\epsilon+\omega}\check{\mathbf{h}}_{A\,\epsilon+\omega,\epsilon}\right),$$

where

$$\check{\mathbf{h}}_{A\,\epsilon,\epsilon-\omega} = \check{\tau}_3\frac{e}{c}v_F\mathbf{A}_\omega\left(f^{(0)}(\epsilon) - f^{(0)}(\epsilon-\omega)\right),$$
$$\check{\mathbf{h}}_{A\,\epsilon+\omega,\epsilon} = \check{\tau}_3\frac{e}{c}v_F\mathbf{A}_\omega\left(f^{(0)}(\epsilon+\omega) - f^{(0)}(\epsilon)\right).$$

The vector components of regular Green functions are found from eqn (9.39). Since the averaged Green functions do not depend on time

$$\left\langle \check{g}^{R(A)}_{\epsilon,\epsilon'}\right\rangle = 2\pi\delta\left(\epsilon-\epsilon'\right)\left\langle \check{g}^{R(A)}_{\epsilon}\right\rangle$$

we find

$$\check{\mathbf{g}}^{R(A)}_{\epsilon_+,\epsilon_-} = i\tau_{\mathrm{tr}}v_F\frac{e}{c}\mathbf{A}_\omega\left[\left\langle \check{g}^{R(A)}_{\epsilon_+}\right\rangle\check{\tau}_3\left\langle \check{g}^{R(A)}_{\epsilon_-}\right\rangle - \check{\tau}_3\right].$$

Making the corresponding shifts in the frequencies, we have

$$\check{\mathbf{g}}^A_{\epsilon-\omega,\epsilon} = i\ell_{\mathrm{tr}}\frac{e}{c}\mathbf{A}_{-\omega}\left[\langle \check{g}^A_{\epsilon-\omega}\rangle\check{\tau}_3\langle \check{g}^A_{\epsilon}\rangle - \check{\tau}_3\right],$$
$$\check{\mathbf{g}}^R_{\epsilon,\epsilon+\omega} = i\ell_{\mathrm{tr}}\frac{e}{c}\mathbf{A}_{-\omega}\left[\langle \check{g}^R_{\epsilon}\rangle\check{\tau}_3\langle \check{g}^R_{\epsilon+\omega}\rangle - \check{\tau}_3\right].$$

The vector part of the anomalous function is to be found from eqn (9.41) which gives

$$\left\langle \check{g}^R_{\epsilon_+}\right\rangle\check{\mathbf{h}}_{A\,\epsilon_+,\epsilon_-} - \check{\mathbf{h}}_{A\,\epsilon_+,\epsilon_-}\left\langle \check{g}^A_{\epsilon_-}\right\rangle = \frac{i}{2\tau_{\mathrm{tr}}}\left[\left\langle \check{g}^R_{\epsilon_+}\right\rangle\check{\mathbf{g}}^{(a)}_{\epsilon_+,\epsilon_-} - \check{\mathbf{g}}^{(a)}_{\epsilon_+,\epsilon_-}\left\langle \check{g}^A_{\epsilon_-}\right\rangle\right]$$

within a linear approximation in $\mathbf{A}^2$. Consider the relation

$$\left\langle \check{g}^R_{\epsilon_+}\right\rangle\check{\mathbf{g}}^{(a)}_{\epsilon_+,\epsilon_-} + \check{\mathbf{g}}^{(a)}_{\epsilon_+,\epsilon_-}\left\langle \check{g}^A_{\epsilon_-}\right\rangle + \left\langle \check{g}^{(a)}_{\epsilon_+}\right\rangle\check{\mathbf{g}}^A_{\epsilon_+,\epsilon_-} + \check{\mathbf{g}}^R_{\epsilon_+,\epsilon_-}\left\langle \check{g}^{(a)}_{\epsilon_-}\right\rangle = 0$$

similar to eqn (9.32). We can neglect the last two terms with the vector components of regular functions which are small in $\mathbf{A}^2$. We find from eqn (9.4)

$$\frac{i}{\tau_{\mathrm{tr}}}\check{\mathbf{g}}^{(a)}_{\epsilon_+,\epsilon_-} = \check{\mathbf{h}}_{A\,\epsilon_+,\epsilon_-} - \left\langle \check{g}^R_{\epsilon_+}\right\rangle\check{\mathbf{h}}_{A\,\epsilon_+,\epsilon_-}\left\langle \check{g}^A_{\epsilon_-}\right\rangle.$$

Making the corresponding shifts in frequencies, we have

$$\check{\mathbf{g}}^{(a)}_{\epsilon+\omega,\epsilon} = -i\tau_{\mathrm{tr}}\left(\check{\mathbf{h}}_{A\,\epsilon+\omega,\epsilon} - \langle \check{g}^R_{\epsilon+\omega}\rangle\check{\mathbf{h}}_{A\,\epsilon+\omega,\epsilon}\langle \check{g}^A_{\epsilon}\rangle\right),$$

$$\check{\mathbf{g}}^{(a)}_{\epsilon,\epsilon-\omega} = -i\tau_{\rm tr}\left(\check{\mathbf{h}}_{A\,\epsilon,\epsilon-\omega} - \langle\check{g}^R_\epsilon\rangle\,\check{\mathbf{h}}_{A\,\epsilon,\epsilon-\omega}\,\langle\check{g}^A_{\epsilon-\omega}\rangle\right).$$

Finally, inserting $\check{\mathbf{g}}^{(a)}$ together with $\check{\mathbf{g}}^{R(A)}$ into eqn (9.42) we obtain the equation for the distribution function

$$\begin{aligned}\mathrm{Tr}\left\langle \check{I}^{(\mathrm{ph})(a)}_{\epsilon,\epsilon}\right\rangle &= iD\frac{e^2}{c^2}\mathbf{A}_\omega\mathbf{A}_{-\omega}\\ &\times\Big[\left(f^{(0)}(\epsilon)-f^{(0)}(\epsilon-\omega)\right)\mathrm{Tr}\left[\check{\tau}_3\langle\check{g}^A_{\epsilon-\omega}\rangle\,\check{\tau}_3\left(\langle\check{g}^R_\epsilon\rangle-\langle\check{g}^A_\epsilon\rangle\right)\right]\\ &+\left(f^{(0)}(\epsilon+\omega)-f^{(0)}(\epsilon)\right)\mathrm{Tr}\left[\check{\tau}_3\langle\check{g}^R_{\epsilon+\omega}\rangle\,\check{\tau}_3\left(\langle\check{g}^R_\epsilon\rangle-\langle\check{g}^A_\epsilon\rangle\right)\right]\Big].\end{aligned}$$

For temperatures close to T_c, the phonon collision integral at zero frequency can be taken in a τ-approximation (see Section 10.4 for more detail)

$$\mathrm{Tr}\left\langle \check{I}^{(ph)(a)}_{\epsilon,\epsilon}\right\rangle = \frac{2i}{\tau_{\rm ph}}\left(g^R_\epsilon - g^A_\epsilon\right)f_1$$

where $f_1 = f^{(1)} - f^{(0)}$ is the correction to the distribution function $f^{(1)}$ defined in eqn (9.28). As a result,

$$\begin{aligned}\left(g^R_\epsilon - g^A_\epsilon\right)f_1 &= \tau_{\rm ph}D\frac{e^2}{c^2}\mathbf{A}_\omega\mathbf{A}_{-\omega}\\ &\times\Big[\left(f^{(0)}(\epsilon)-f^{(0)}(\epsilon-\omega)\right)\left[g^A_{\epsilon-\omega}\left(g^R_\epsilon-g^A_\epsilon\right)+f^A_{\epsilon-\omega}\left(f^R_\epsilon-f^A_\epsilon\right)\right]\\ &+\left(f^{(0)}(\epsilon+\omega)-f^{(0)}(\epsilon)\right)\left[\left(g^R_\epsilon-g^A_\epsilon\right)g^R_{\epsilon+\omega}+\left(f^R_\epsilon-f^A_\epsilon\right)f^R_{\epsilon+\omega}\right]\Big].\end{aligned}$$

We omit the averaging brackets for brevity.

The regular functions are given by eqns (5.46), (5.47), and (5.50), (5.51). Therefore, we find for $\epsilon > 0$

$$\begin{aligned}f_1 = -\alpha\frac{df^{(0)}}{d\epsilon}&\left[\frac{\epsilon(\epsilon-\omega)+\Delta^2}{\epsilon\sqrt{(\epsilon-\omega)^2-\Delta^2}}\Theta(\epsilon-\Delta-\omega)\right.\\ &-\frac{\epsilon(\epsilon+\omega)+\Delta^2}{\epsilon\sqrt{(\epsilon+\omega)^2-\Delta^2}}\Theta(\epsilon-\Delta)\\ &\left.+\frac{\epsilon(\omega-\epsilon)-\Delta^2}{\epsilon\sqrt{(\omega-\epsilon)^2-\Delta^2}}\Theta(\epsilon-\Delta)\,\Theta(\omega-\Delta-\epsilon)\right].\end{aligned}\tag{9.43}$$

The term

$$\alpha = \tau_{\rm ph}D\frac{e^2}{c^2}\omega\mathbf{A}_\omega\mathbf{A}_{-\omega}$$

in eqn (9.43) is proportional to the absorbed power. Indeed, since $2\mathbf{A}_\omega\mathbf{A}_{-\omega} = \mathbf{A}^2(t)$ the dissipated energy per volume is

$$\sigma_n \overline{E^2} = 2\nu(0) De^2\omega^2 \overline{\mathbf{A}^2(t)}/c^2$$

multiplied by the inelastic relaxation time τ_{ph}. This energy is distributed among $\delta N \sim \nu(0)\Delta$ particles. Therefore, the quantity $(\omega/\Delta)\alpha$ is the absorbed energy per particle. The last term in eqn (9.43) only exists if $\omega > 2\Delta$. It corresponds to creation of new excitations over the energy gap Δ.

The nonequilibrium correction to the distribution function is negative. This means that the number of excitations with energies $\epsilon \sim \Delta$ decreases. It is equivalent to a decrease in an "effective temperature" of excitations. Let us study how this affects the order parameter. We define the function

$$W(\Delta) = \frac{1}{2}\int_\Delta^\infty \frac{d\epsilon}{\sqrt{\epsilon^2 - \Delta^2}} f_1.$$

We have for $\Delta \ll T$

$$W(\Delta) = \frac{\alpha}{4T} P\left(\frac{\omega}{\Delta}\right)$$

where the function $P(\omega/\Delta)$ is

$$P(w) = w\int_1^\infty \frac{x(x+w)+1}{x(x+w)\sqrt{x^2-1}\sqrt{(x+w)^2-1}}\,dx$$

$$-\Theta(w-2)\int_1^{w-1} \frac{(w-x)x-1}{x\sqrt{x^2-1}\sqrt{(w-x)^2-1}}\,dx$$

with $w = \omega/\Delta$. The last term is nonzero for $\omega > \Delta/2$. As we mentioned, it corresponds to creation of quasiparticle excitations by the external irradiation. The function

$$P(w) = \begin{cases} w\ln(8/ew), & w \ll 1, \\ \pi/w, & w \gg 1. \end{cases}$$

It has a maximum at $w = 2$ where its first derivative is discontinuous.

Let us consider the equation for the order parameter. We obtain the Keldysh function from eqns (9.25) and (9.28) in the form

$$\langle f^K_{\epsilon,\epsilon}\rangle = \left(\langle f^R_\epsilon\rangle - \langle f^A_\epsilon\rangle\right)\left(f^{(0)} + f_1\right).$$

Let us insert this into the order-parameter equation (9.17)

$$\frac{\Delta(\mathbf{k},\omega)}{\lambda} = \int \frac{d\epsilon}{4}\left(\langle f^R_\epsilon\rangle - \langle f^A_\epsilon\rangle\right)\left(f^{(0)} + f_1\right).$$

The equilibrium part of the distribution function gives the usual stationary Ginzburg–Landau equation (6.23). The resulting equation has the form

$$-\frac{\pi D}{8T_c}\frac{4e^2\mathbf{A}_\omega\mathbf{A}_{-\omega}}{c^2} + \frac{T_c - T}{T_c} + W'(\Delta) = 0 \tag{9.44}$$

where

$$W'(\Delta) = W(\Delta) - \frac{7\zeta(3)}{8\pi^2 T_c^2}|\Delta|^2.$$

One can generally neglect the first term in eqn (9.44) when $\omega\tau_{\text{ph}} \gg 1$.

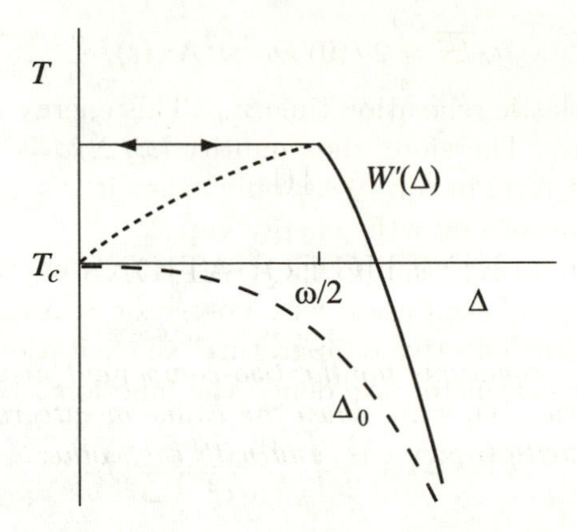

FIG. 9.1. The order parameter as a function of temperature. The long-dashed line $\Delta_0(T)$ shows the order parameter in the absence of irradiation. The increasing part of $W'(\Delta)$ shown by short-dashed line is unstable. The arrows indicate the order parameter jumps with varying the temperature.

One can see that a nonzero order parameter exists even if $T > T_c$ because $W' > 0$ for not very high Δ. For the irradiation intensity such that $\omega/T \ll \alpha \ll 1$ the temperature dependence of the order parameter is determined mainly by the first term in W'. The maximum temperature $T_{\max}$ in this limit is determined by

$$\frac{T_{\max} - T_{c0}}{T_{c0}} = \frac{\alpha\omega}{2T} P(2) = \frac{\alpha\omega}{2T} \frac{\pi}{\sqrt{3}}.$$

At this temperature, $\Delta = \omega/2$. If the temperature is increased further the order parameter jumps to zero. The increasing branch of $W(\Delta)$ is unstable. The order parameter approaches the equilibrium value with decreasing temperature. The temperature dependence $\Delta(T)$ is shown schematically in Fig. 9.1.

10

KINETIC EQUATIONS

We derive the kinetic equations for the two-component distribution function in a gauge-invariant form. We write down the collision integrals for interaction of excitations with impurities, phonons, and with each other.

10.1 Gauge-invariant Green functions

In this section we introduce a more convenient description of a time-dependent state of a superconductor in terms of the gauge-invariant Green functions. We know that the gauge invariance is the basic property of the superconducting state. The Eilenberger equations are gauge-invariant and so are the regular (retarded and advanced) quasiclassical Green functions for a *time-independent state* of a superconductor which are determined by these equations. Of course, the gauge-invariance holds for a nonstationary state, as well. However, the generalized *distribution function* introduced in Chapter 9 and the regular functions for a *nonstationary state*, treated separately, can become non-gauge-invariant if special precautions are not taken.

Remind that according to the general gauge-invariance (see also page 19), the electromagnetic potentials can be changed according to

$$\mathbf{A} \to \mathbf{A} + \frac{\hbar c}{2e}\nabla f; \ \varphi \to \varphi - \frac{\hbar}{2e}\frac{\partial f}{\partial t} \tag{10.1}$$

where f is an arbitrary function, without changing the electromagnetic fields

$$\mathbf{E} = -\frac{1}{c}\frac{\partial \mathbf{A}}{\partial t} - \nabla\varphi; \ \mathbf{H} = \operatorname{curl}\mathbf{A}. \tag{10.2}$$

In superconductors, the transformation eqn (10.1) is accompanied by the transformation of the order parameter phase

$$\chi \to \chi + f. \tag{10.3}$$

Therefore the combinations

$$\mathbf{Q} = \mathbf{A} - \frac{c}{2e}\nabla\chi, \ \Phi = \varphi + \frac{1}{2e}\frac{\partial \chi}{\partial t} \tag{10.4}$$

are also gauge-invariant; they are called the gauge-invariant potentials. Of course, the order parameter magnitude is gauge-invariant, too. Therefore, the equations for the Green functions and for the distribution function should contain, in addition to the electric field $\mathbf{E}$ and magnetic field $\mathbf{H}$, the vector and scalar potentials in proper combinations with the spatial and time derivatives of the order parameter phase such that they appear as the gauge-invariant potentials eqn (10.4). Note that the time-dependent Ginzburg–Landau model discussed earlier in Section 1.2.1 is an example of such gauge invariant theory because it only contain the potentials (10.4) and the order parameter magnitude.

In this section we define the Green functions and the generalized distribution functions in a gauge-invariant way. We start with the transport-like equations obtained in Section 8.4 and derive the kinetic equations for the distribution function in a state which slowly varies in time. More particularly, we assume that the time variations of the superconducting parameters are such that the characteristic frequency is $\omega \ll \Delta$. These equations will make the basis for further discussions of nonequilibrium properties of superconductors.

Let us turn back to the definitions of non-quasiclassical Green functions in Chapter 8. We recall the operator eqn (8.3) acting on the matrix Green function in the Nambu space. It can be written as

$$\check{G}^{-1} = \check{G}_0^{-1} + \check{H}. \tag{10.5}$$

where

$$\check{G}_0^{-1}(\mathbf{p}) = \begin{pmatrix} -i\partial/\partial t + E_n(\mathbf{p}) - E_F & 0 \\ 0 & i\partial/\partial t + E_n(\mathbf{p}) - E_F \end{pmatrix}, \tag{10.6}$$

and $\mathbf{p} = -i\nabla$ is the particle momentum operator. The Hamiltonian is

$$\check{H} = \begin{pmatrix} H(\mathbf{p}) & -\Delta_{\mathbf{p}} \\ \Delta^*_{\mathbf{p}} & H(-\mathbf{p}) \end{pmatrix} \tag{10.7}$$

where

$$H(\mathbf{p}) = -\frac{e}{2c}(\mathbf{v}\mathbf{A} + \mathbf{A}\mathbf{v}) + \frac{e^2A^2}{2m^*c^2} + e\varphi, \tag{10.8}$$

see eqn (5.57). The velocity is $\mathbf{v} = \partial\epsilon(\mathbf{p})/\partial\mathbf{p}$, and m^* is the effective mass of an electron. As distinct from eqn (3.70), the effective Hamiltonian contains the scalar potential φ which should be separated from a spatially independent part of the chemical potential, E_F.

The equation for the regular Green function can be written as

$$\left[\left(\check{G}^{-1}(\mathbf{p}_1) - \check{\Sigma}^{R(A)}\right) \otimes \check{G}^{R(A)}\right](\mathbf{r}_1, t_1; \mathbf{r}_2, t_2) = \delta(\mathbf{r}_1 - \mathbf{r}_2)\delta(t_1 - t_2). \tag{10.9}$$

The full convolution in the coordinate space is

$$\left[\check{A} \otimes \check{B}\right](\mathbf{r}_1, t_1; \mathbf{r}_2, t_2) = \int \check{A}(\mathbf{r}_1, t_1; \mathbf{r}', t')\check{B}(\mathbf{r}', t'; \mathbf{r}_2, t_2)\, d^3r'\, dt'. \tag{10.10}$$

It is equivalent to the shortcut $[AB]$ used in the momentum–frequency representation. The equation of motion for the total Green function is obtained from eqn (8.61):

$$\left[\check{G}^{-1}(\mathbf{p}_1)\otimes\check{G}^K-\check{G}^K\otimes\check{G}^{-1}(\mathbf{p}_2)\right]=\check{\mathcal{I}}. \tag{10.11}$$

The collision-integral matrix is

$$\check{\mathcal{I}}=\begin{pmatrix}\mathcal{I}_1 & \mathcal{I}_2\\ -\mathcal{I}_2^\dagger & \bar{\mathcal{I}}_1\end{pmatrix}=\check{\Sigma}^R\otimes\check{G}^K-\check{G}^K\otimes\check{\Sigma}^A+\check{\Sigma}^K\otimes\check{G}^A-\check{G}^R\otimes\check{\Sigma}^K. \tag{10.12}$$

The regular Green functions satisfy eqn (10.11) with the collision integral

$$\check{\mathcal{I}}^{R(A)}=\check{\Sigma}^{R(A)}\otimes\check{G}^{R(A)}-\check{G}^{R(A)}\otimes\check{\Sigma}^{R(A)}\ . \tag{10.13}$$

As before, we use the mixed coordinate–momentum representation

$$\check{G}_\epsilon\left(\mathbf{p};\mathbf{r},t\right)=\int\check{G}_\epsilon\left(\mathbf{p};\mathbf{k},\omega\right)e^{-i\omega t+\mathbf{kr}}\frac{d^3k\,d\omega}{(2\pi)^4} \tag{10.14}$$

in the center-of-mass momentum and frequency for the Green function defined through

$$\check{G}_{\epsilon_+,\epsilon_-}\left(\mathbf{p}_+,\mathbf{p}_-\right)\equiv\check{G}_\epsilon\left(\mathbf{p};\mathbf{k},\omega\right), \tag{10.15}$$

where $\mathbf{p}_\pm=\mathbf{p}\pm\mathbf{k}/2$ and $\epsilon_\pm=\epsilon\pm\omega/2$. We are going to derive equations for the function $\check{G}_\epsilon\left(\mathbf{p};\mathbf{r},t\right)$.

10.1.1 *Equations of motion for the invariant functions*

We denote the center-of-mass coordinate and time $\mathbf{r}=(\mathbf{r}_1+\mathbf{r}_2)/2$ and $t=(t_1+t_2)/2$ and introduce the modified Green functions

$$\begin{aligned}G(\mathbf{r}_1,t_1;\mathbf{r}_2,t_2)&=\tilde{G}(\mathbf{r}_1,t_1;\mathbf{r}_2,t_2)\exp\left[i\psi(\mathbf{r}_1,t_1)-i\psi(\mathbf{r}_2,t_2)\right],\\ F(\mathbf{r}_1,t_1;\mathbf{r}_2,t_2)&=\tilde{F}(\mathbf{r}_1,t_1;\mathbf{r}_2,t_2)\exp\left[i\psi(\mathbf{r}_1,t_1)+i\psi(\mathbf{r}_2,t_2)\right],\end{aligned} \tag{10.16}$$

and same for $\bar{G}$ and $F^\dagger$ with $\psi\to-\psi$ in the corresponding exponents. Here

$$\psi(\mathbf{r}_1,t_1)=\frac{e}{c}\int_{\mathbf{r}_0}^{\mathbf{r}_1}\mathbf{A}(\mathbf{r}',t_1)\,d\mathbf{r}'-e\int_{t_0}^{t_1}\varphi(\mathbf{r}_1,t')dt'. \tag{10.17}$$

Equations (10.16) refer both to regular and Keldysh Green functions. The point $(\mathbf{r}_0,t_0)$ is taken close, i.e., on a scale of p_F^{-1}, to the points $(\mathbf{r}_1,t_1)$ and $(\mathbf{r}_2,t_2)$. Integration in the first term is carried out along the shortest path connecting the points $\mathbf{r}_0$ and $\mathbf{r}_1$. We shall write the equation in the differential form at the point $(\mathbf{r},t)$, therefore, in the final equations, we can put $(\mathbf{r}_0,t_0)\to(\mathbf{r},t)$.

Since the relative distances are always small, $\mid\mathbf{r}_1-\mathbf{r}_2\mid\sim p_F^{-1}\ll\xi$, we can expand all the quantities in $\mid\mathbf{r}_1-\mathbf{r}_2\mid/\xi$. For further derivation, we adopt one more assumption, namely, that the electric field $\mathbf{E}=-\nabla\varphi-(1/c)\partial\mathbf{A}/\partial t$ is small: $eE\ll\Delta/\xi$ or, equivalently, that the time variations are slow: $\omega\ll\Delta$. This means that we restrict ourselves to a linear response to the electric field. In return, we are able to expand in the relative time, $(t_1-t_2)\omega\sim\omega/\Delta$, as well.

The equation of motion for the modified regular Green function can be obtained (Kopnin 1994) after substitution of eqn (10.16) into eqn (10.9) and expansion of the operators $\check{G}^{-1}$ and Σ in powers of $\mid\mathbf{r}_1-\mathbf{r}_2\mid/\xi$ and $(t_1-t_2)\omega$

which is performed in a standard way (compare with Serene and Rainer 1983, Aronov *et al.* 1981, Eckern and Schmid 1981). Simultaneously, we expand in φ and $\mathbf{E}$.

On the next step we go into the mixed momentum–coordinate representation performing the Fourier transformation with respect to the relative coordinates $\mathbf{r}_1 - \mathbf{r}_2$ and $t_1 - t_2$ to the momentum $\mathbf{p}$ and frequency ϵ, respectively. Equations for thus obtained modified functions $\tilde{\check{G}}_\epsilon(\mathbf{p}; \mathbf{r}, t)$ become

$$\tilde{\check{G}}_R^{-1} \tilde{\check{G}}^{R(A)} - \tilde{\Sigma}^{R(A)} \otimes \tilde{\check{G}}^{R(A)} = \check{1}, \tag{10.18}$$

or

$$\tilde{\check{G}}^{R(A)} \tilde{\check{G}}_L^{-1} - \tilde{\check{G}}^{R(A)} \otimes \tilde{\Sigma}^{R(A)} = \check{1}. \tag{10.19}$$

Here we defined the gauge-invariant operators

$$\begin{aligned} \tilde{\check{G}}_R^{-1} &= \tilde{\check{G}}_0^{-1} - \frac{i}{2}\check{\tau}_3 \frac{\partial}{\partial t} - \frac{i}{2}\mathbf{v} \cdot \nabla + \frac{1}{2}\tilde{\check{H}}_d, \\ \tilde{\check{G}}_L^{-1} &= \tilde{\check{G}}_0^{-1} + \frac{i}{2}\check{\tau}_3 \frac{\partial}{\partial t} + \frac{i}{2}\mathbf{v} \cdot \nabla - \frac{1}{2}\tilde{\check{H}}_d, \end{aligned} \tag{10.20}$$

where

$$\tilde{\check{G}}_0^{-1} = \begin{pmatrix} -\epsilon + \xi_\mathbf{p} & -\Delta_\mathbf{p} \\ \Delta_\mathbf{p}^* & \epsilon + \xi_\mathbf{p} \end{pmatrix}.$$

The differential operator $\tilde{\check{H}}_d = \tilde{\check{H}}_\epsilon + \tilde{\check{H}}_p$ contains

$$\tilde{\check{H}}_\epsilon = \begin{pmatrix} e\mathbf{v}_F\mathbf{E} & -\hat{\partial}\Delta/\partial t \\ \hat{\partial}\Delta^*/\partial t & -e\mathbf{v}_F\mathbf{E} \end{pmatrix} \frac{\partial}{\partial \epsilon}, \tag{10.21}$$

and

$$\begin{aligned} \tilde{\check{H}}_p = &\begin{pmatrix} e\mathbf{E} + (e/c)\,[\mathbf{v}_F \times \mathbf{H}] & -\hat{\nabla}\Delta \\ \hat{\nabla}\Delta^* & e\mathbf{E} - (e/c)\,[\mathbf{v}_F \times \mathbf{H}] \end{pmatrix} \frac{\partial}{\partial \mathbf{p}} \\ &+ \begin{pmatrix} 0 & \partial\Delta/\partial\mathbf{p} \\ -\partial\Delta/\partial\mathbf{p} & 0 \end{pmatrix} \nabla. \end{aligned} \tag{10.22}$$

The gauge-invariant derivatives $\hat{\partial}/\partial t$ and $\hat{\nabla}$ are

$$\frac{\hat{\partial}\Delta}{\partial t} = (\frac{\partial}{\partial t} + 2ie\varphi)\Delta\ , \quad \hat{\nabla}\Delta = (\nabla - \frac{2ie}{c}\mathbf{A})\Delta, \tag{10.23}$$

$$\frac{\hat{\partial}\Delta^*}{\partial t} = (\frac{\partial}{\partial t} - 2ie\varphi)\Delta^*\ , \quad \hat{\nabla}\Delta^* = (\nabla + \frac{2ie}{c}\mathbf{A})\Delta^*. \tag{10.24}$$

The operator $\partial/\partial t$ in $\tilde{\check{G}}^{-1}$ acts on the center-of-mass time t.

The convolution for the modified Green functions in our approximation is replaced with

$$\tilde{\check{G}}_1 \otimes \tilde{\check{G}}_2 = \tilde{\check{G}}_1 \tilde{\check{G}}_2 + \frac{i}{2}\left(\frac{\partial \tilde{\check{G}}_1}{\partial \epsilon}\frac{\partial \tilde{\check{G}}_2}{\partial t} - \frac{\partial \tilde{\check{G}}_1}{\partial t}\frac{\partial \tilde{\check{G}}_2}{\partial \epsilon}\right)$$
$$-\frac{i}{2}\left(\frac{\partial \tilde{\check{G}}_1}{\partial \mathbf{p}}\nabla \tilde{\check{G}}_2 - \nabla \tilde{\check{G}}_1 \frac{\partial \tilde{\check{G}}_2}{\partial \mathbf{p}}\right). \qquad (10.25)$$

The differential operator $\tilde{\check{H}}_d$ in eqn (10.21), (10.22) and the derivatives $\partial/\partial\epsilon$ and $\partial/\partial\mathbf{p}$ in eqn (10.25) appear as a result of expansion of the convolution near the point $(\mathbf{r}_1 + \mathbf{r}_2)/2$ and $(t_1 + t_2)/2$ in small $\mathbf{r}_1 - \mathbf{r}_2$ and $t_1 - t_2$. To demonstrate this, let us consider the time-dependent part in the expansion of eqn (10.25). On can check that, in the frequency representation,

$$\check{G}_{\epsilon_+,\epsilon'}\check{G}_{\epsilon',\epsilon_-} \equiv \check{G}_{\epsilon+(\omega-\omega')/2}\left(\omega'\right)\check{G}_{\epsilon-\omega'/2}\left(\omega-\omega'\right)$$
$$= \check{G}_{\epsilon}\left(\omega'\right)\check{G}_{\epsilon}\left(\omega-\omega'\right) - \frac{\omega'}{2}\check{G}_{\epsilon}\left(\omega'\right)\frac{\partial \check{G}_{\epsilon}\left(\omega-\omega'\right)}{\partial\epsilon}$$
$$+\frac{\partial \check{G}_{\epsilon}\left(\omega'\right)}{\partial \epsilon}\frac{\omega-\omega'}{2}\check{G}_{\epsilon}\left(\omega-\omega'\right) \qquad (10.26)$$

where we put $\epsilon' = \epsilon + \omega/2 - \omega'$ and use eqn (10.15). Performing the Fourier transformation over the center-of-mass frequencies ω' and $\omega - \omega'$ we obtain the first line of eqn (10.25). The momentum part can be expanded similarly.

A transport-like equation for the modified regular Green function is obtained after subtracting eqn (10.19) from eqn (10.18):

$$\tilde{\check{G}}_R^{-1}\tilde{\check{G}}^{R(A)} - \tilde{\check{G}}^{R(A)}\tilde{\check{G}}_L^{-1} = \tilde{\check{\mathcal{I}}}^{R(A)}. \qquad (10.27)$$

Equation (10.11) for the modified Keldysh function can be transformed in the same way. We have

$$\tilde{\check{G}}_R^{-1}\tilde{\check{G}}^{K} - \tilde{\check{G}}^{K}\check{G}_L^{-1} = \tilde{\check{\mathcal{I}}}^{K}. \qquad (10.28)$$

The gauge-invariant equation (10.28) is the basis for our further derivation. As an example, we write down two elements of the matrix eqn (10.28):

$$\frac{\partial}{\partial t}\tilde{G}^K + \mathbf{v}\cdot\nabla\tilde{G}^K + e\mathbf{v}\cdot\mathbf{E}\frac{\partial \tilde{G}^K}{\partial\epsilon} + e\mathbf{E}\cdot\frac{\partial \tilde{G}^K}{\partial \mathbf{p}} + \frac{e}{c}[\mathbf{v}\times\mathbf{H}]\cdot\frac{\partial \tilde{G}^K}{\partial \mathbf{p}}$$
$$+i(\Delta\tilde{F}^{\dagger K} - \tilde{F}^K\Delta^*) + \frac{1}{2}\left[\frac{\partial\Delta}{\partial t}\frac{\partial\tilde{F}^{\dagger K}}{\partial\epsilon} + \frac{\partial\tilde{F}^K}{\partial\epsilon}\frac{\partial\Delta^*}{\partial t}\right]$$
$$-\frac{1}{2}\left[\left(\hat{\nabla}\Delta\right)\cdot\frac{\partial\tilde{F}^{\dagger K}}{\partial\mathbf{p}} + \frac{\partial\tilde{F}^K}{\partial\mathbf{p}}\cdot\left(\hat{\nabla}\Delta^*\right)\right]$$

$$+\frac{1}{2}\left[\frac{\partial\Delta}{\partial\mathbf{p}}\cdot\nabla\tilde{F}^{\dagger K}+\frac{\partial\Delta^*}{\partial\mathbf{p}}\cdot\nabla\tilde{F}^{K}\right]=i\tilde{\mathcal{I}}_1, \tag{10.29}$$

and

$$\begin{aligned}&-2i\epsilon\tilde{F}^K+\mathbf{v}\cdot\nabla\tilde{F}^K+e\mathbf{E}\cdot\frac{\partial\tilde{F}^K}{\partial\mathbf{p}}-i\Delta\left(\tilde{\bar{G}}^K-\tilde{G}^K\right)\\&-\frac{1}{2}\frac{\hat{\partial}\Delta}{\partial t}\left(\frac{\partial\tilde{\bar{G}}^K}{\partial\epsilon}+\frac{\partial\tilde{G}^K}{\partial\epsilon}\right)+\frac{1}{2}\left(\frac{\partial\tilde{\bar{G}}^K}{\partial\mathbf{p}}+\frac{\partial\tilde{G}^K}{\partial\mathbf{p}}\right)\cdot\hat{\nabla}\Delta\\&-\frac{1}{2}\left(\nabla\tilde{G}^K+\nabla\tilde{\bar{G}}^K\right)\cdot\frac{\partial\Delta}{\partial\mathbf{p}}=i\tilde{\mathcal{I}}_2\,.\end{aligned} \tag{10.30}$$

In eqns (10.29), (10.30) the arguments ϵ, $\mathbf{p}$, $\mathbf{r}$, and t of the Green functions $\tilde{G}_\epsilon(\mathbf{p};\mathbf{r},t)$, $\tilde{F}_\epsilon(\mathbf{p};\mathbf{r},t)$, etc., are omitted for brevity. Equations for $\bar{G}$ and $\tilde{F}^\dagger$ can be obtained by the transformation $\tilde{G}\to\bar{G}$, and $\tilde{F}\to\tilde{F}^\dagger$ under complex conjugation and the change of signs $\epsilon\to-\epsilon$, and $\mathbf{p}\to-\mathbf{p}$ in eqns (10.29) and (10.30). The same equations hold for the regular functions with the corresponding collision integrals in the right-hand side.

From eqns (10.27) and (10.28) it is clear that the functions $\tilde{\check{G}}^K$ and $\tilde{\check{G}}^{R(A)}$ are gauge-invariant while the usual functions $\check{G}^K$ are "gauge-covariant". Note that one can put locally $\tilde{F}_\epsilon(\mathbf{p};\mathbf{r},t)=F_\epsilon(\mathbf{p};\mathbf{r},t)$ and $\tilde{F}^\dagger_\epsilon(\mathbf{p};\mathbf{r},t)=F^\dagger_\epsilon(\mathbf{p};\mathbf{r},t)$ since we take the limit $\mathbf{r}_0\to\mathbf{r}$, $t_0\to t$ in eqn (10.16). However, the derivatives should be transformed according to eqn (10.16)

$$\begin{aligned}\nabla\tilde{F}_\epsilon(\mathbf{p};\mathbf{r},t)&=\left(\nabla-\frac{2ie}{c}\mathbf{A}\right)F_\epsilon(\mathbf{p};\mathbf{r},t),\\\frac{\partial}{\partial t}\tilde{F}_\epsilon(\mathbf{p};\mathbf{r},t)&=\left(\frac{\partial}{\partial t}+2ie\varphi\right)F_\epsilon(\mathbf{p};\mathbf{r},t),\end{aligned} \tag{10.31}$$

and similarly for $\tilde{F}^\dagger_\epsilon$. At the same time, eqn (10.16) implies that $\tilde{G}$ is the function of the "particle" kinetic momentum $\tilde{\mathbf{p}}=\mathbf{p}-(e/c)\,\mathbf{A}$ and an energy $\epsilon-e\varphi$, while $\tilde{\bar{G}}$ is the function of the "hole" kinetic momentum $\tilde{\mathbf{p}}=\mathbf{p}+(e/c)\,\mathbf{A}$ and an energy $\epsilon+e\varphi$. We shall use the matrix gauge-invariant time-derivative acting on a gauge-covariant functions defined through

$$\frac{\hat{\partial}\check{G}}{\partial t}=\begin{pmatrix}\frac{\partial}{\partial t}G & \left(\frac{\partial}{\partial t}+2ie\varphi\right)F\\-\left(\frac{\partial}{\partial t}-2ie\varphi\right)F^\dagger & \frac{\partial}{\partial t}\bar{G}\end{pmatrix}, \tag{10.32}$$

and similarly for $\hat{\nabla}\check{G}$. Therefore,

$$\frac{\partial\tilde{\check{G}}}{\partial t}=\frac{\hat{\partial}\check{G}}{\partial t};\quad\nabla\tilde{\check{G}}=\hat{\nabla}\check{G}.$$

10.2 Quasiclassical kinetic equations

In the previous chapter, we have neglected the terms with the momentum derivatives of the type of eqn (9.4) when deriving the Eliashberg equations from the transport-like equations. These terms appear now in eqns (10.29) and (10.30) which are of a higher accuracy in the quasiclassical parameter $1/p_F\xi$. Such terms have also been neglected during derivation of the Eilenberger equations in Chapter 5. With this approximation, the Eliashberg and Eilenberger equations are local in the momentum $\mathbf{p}$ which reduces immensely the complexity of the full Green function formalism. However, as we have already mentioned on page 171, this approximation can be insufficient for the Keldysh Green function. We shall see later in this section that, within this approximation, the distribution functions $f^{(1)}$ and $f^{(2)}$ enter the kinetic equations in such a way that the left-hand side of one equation, for example, contains the time-derivative of $f^{(1)}$ and the coordinate derivative of $f^{(2)}$ while the function $f^{(1)}$ itself is contained only in the corresponding collision integral. On the other hand, some of the neglected contributions with the momentum derivatives do contain the function $f^{(1)}$ in the form of $\partial f^{(1)}/\partial \mathbf{p}$. If the collision term and/or the time-derivative is larger than the momentum derivative, the latter can be safely neglected. The problem arises when the collision terms are small. If they are of the order of momentum-derivatives, the latter can no longer be neglected. Therefore, in general, we have to keep the momentum-derivatives in our equations. A complication which we immediately obtain is that there are many other terms of the order of $1/p_F\xi$ in the expansion of the full Green function transport equations in the quasiclassical parameter. Fortunately, not all of them are equally important.

To separate the important contributions of the first order in the quasiclassical parameter $1/p_F\xi$ we note that various terms in the transport equations of the type of eqn (10.29) for the full Green functions, which result from the expansion of the convolution $\check{H} \otimes \check{G}$ in the small parameter $1/p_F\xi$ are of two different types. First, there are corrections to already existing terms in the quasiclassical kinetic equations derived in Chapter 9 within the zero order in $1/p_F\xi$. Second, there are new contributions which appear only as the first approximation in $1/p_F\xi$. As we already mentioned, the new terms (though they are formally of the relative order of $1/p_F\xi$) enter the kinetic equations in the same way as the collision integrals, and, therefore, they have to be compared with the terms of the order of $1/\tau$. In clean superconductors with $1/\tau \ll \Delta$, the terms of the new structure may happen to be not negligible compared to the (small) collision integrals. Thus one can neglect the corrections of the first type and, at the same time, keep the corrections of the new structure. The possibility to separate the most important contributions from all the numerous terms of the expansion in $1/p_F\xi$ exists for clean superconductors with the mean free path $\ell \gg \xi(T)$. For dirty superconductors with $\ell \sim \xi(T)$, one has two options. If the quasiclassical accuracy is sufficient for the particular problem, the first-order terms in $1/p_F\xi$ including the momentum-derivatives should be all neglected. If the quasiclassical approximation is not sufficient, one has to take into account other expansion terms in $1/p_F\xi$. This approach has been developed by Larkin and Ovchinnikov (1995) for the problems of the Hall effect in vortex dynamics (see Chapter 14). This approach is much more complicated, and we do not describe it here.

In what follows we shall work with the quasiclassical functions integrated over $d\xi_{\mathbf{p}}/\pi i$. The Keldysh Green function can be generally parametrized in terms of four linearly independent matrices. We can write

$$\tilde{\check{g}}^K = \tilde{\check{g}}^R \otimes \check{f}^{(-)} - \check{f}^{(-)} \otimes \tilde{\check{g}}^A + \tilde{\check{g}}^R \otimes \check{f}^{(+)} + \check{f}^{(+)} \otimes \tilde{\check{g}}^A \ , \tag{10.33}$$

where $\check{f}$ is the matrix distribution function; it consists of the diagonal matrix $\check{f}^{(-)}$

$$\check{f}_\epsilon^{(-)}(\mathbf{p};\mathbf{r},t) = \check{1} f^{(1)} + \check{\tau}_3 f^{(2)} \tag{10.34}$$

and another two-component matrix $\check{f}^{(+)}$. In the zero-order quasiclassical approximation (see Chapter 9), $\check{f}^{(+)} = 0$ due to the normalization of the quasiclassical Keldysh function. If we include non-quasiclassical corrections, the matrix $\check{f}^{(+)}$ becomes nonzero. The components of $\check{f}^{(+)}$ are thus small in the quasiclassical parameter $(p_F\xi)^{-1}$. We shall write $f^{(1)} = f^{(0)} + f_1$ and $f^{(2)} \equiv f_2$ where f_1 and f_2 are the nonequilibrium corrections. The equilibrium distribution function $f^{(0)}(\epsilon)$ depends only on energy; $f_1(\epsilon, \mathbf{p}; \mathbf{r}, t)$ and $f_2(\epsilon, \mathbf{p}; \mathbf{r}, t)$ are corrections proportional to time-derivatives of the order parameter and to the electric field. The functions f_1 and f_2 are respectively the odd and even in ϵ and $\mathbf{p}$ parts of the total distribution function. We have

$$\begin{aligned}\tilde{\check{g}}^K &= \left[\tilde{\check{g}}^R \check{f}^{(-)} - \check{f}^{(-)}\tilde{\check{g}}^A\right] + \left[\tilde{\check{g}}^R \check{f}^{(+)} + \check{f}^{(+)}\tilde{\check{g}}^A\right] - \frac{i}{2}\frac{df^{(0)}}{d\epsilon}\left[\frac{\partial \tilde{\check{g}}^R}{\partial t} + \frac{\partial \tilde{\check{g}}^A}{\partial t}\right] \\ &- \frac{i}{2}\left[\frac{\partial \tilde{\check{g}}^R}{\partial \mathbf{p}}\nabla \check{f}^{(-)} + \nabla \check{f}^{(-)}\frac{\partial \tilde{\check{g}}^A}{\partial \mathbf{p}}\right] + \frac{i}{2}\left[\nabla \tilde{\check{g}}^R \frac{\partial \check{f}^{(-)}}{\partial \mathbf{p}} + \frac{\partial \check{f}^{(-)}}{\partial \mathbf{p}}\nabla \tilde{\check{g}}^A\right]. \end{aligned} \tag{10.35}$$

In the second line we neglect the function $\check{f}^{(+)}$ since it is already small in the quasiclassical parameter.

Now we make the next step in derivation: we integrate eqns (10.27) and (10.28) over $d\xi_{\mathbf{p}}/\pi i$. As we explained earlier, we omit the corrections of the order of $1/p_F\xi$ to the terms which do already exist in the zero-order approximation. Within this accuracy, we neglect the dependence of all the factors on $\xi_{\mathbf{p}}$, and put $\mathbf{v} = \mathbf{v}_F$. Moreover, we replace the derivative $\partial/\partial\mathbf{p} = \mathbf{v}_F(\partial/\partial\xi_{\mathbf{p}}) + \partial/\partial\mathbf{p}_F$ with just the second term. Here $\delta\mathbf{p}_F$ is the increment of the momentum $\mathbf{p}_F$ belonging to the Fermi surface $\xi_{\mathbf{p}}(\mathbf{p}) = 0$, i.e., $\delta\mathbf{p}_F$ is perpendicular to $\mathbf{v}_F$. We obtain

$$\begin{aligned} i\check{\tau}_3\frac{\partial \tilde{\check{g}}^{R(A)}}{\partial t} + i\mathbf{v}_F\nabla\tilde{\check{g}}^{R(A)} &= \check{H}_0\tilde{\check{g}}^{R(A)} - \tilde{\check{g}}^{R(A)}\check{H}_0 \\ &\quad - \frac{i}{2}\left[\tilde{\check{H}}_d\tilde{\check{g}}^{R(A)} + \tilde{\check{g}}^{R(A)}\tilde{\check{H}}_d\right] - \check{I}^{R(A)}, \end{aligned} \tag{10.36}$$

and

$$i\check{\tau}_3\frac{\partial \tilde{\check{g}}^K}{\partial t} + i\mathbf{v}_F\nabla\tilde{\check{g}}^K = \check{H}_0\tilde{\check{g}}^K - \tilde{\check{g}}^K\check{H}_0$$

$$-\frac{i}{2}\left[\check{\tilde{H}}_d\check{\tilde{g}}^K + \check{\tilde{g}}^K\check{\tilde{H}}_d\right] - \check{I}^K. \quad (10.37)$$

Here

$$\check{H}_0 = \begin{pmatrix} -\epsilon & -\Delta \\ \Delta^* & \epsilon \end{pmatrix}$$

and $\check{\tilde{H}}_d$ is defined by eqns (10.21) and (10.22).

Using eqn (10.33) we find

$$\begin{aligned}
&-i\check{\tau}_3\frac{\partial\check{\tilde{g}}^K}{\partial t} - i\mathbf{v}_F\nabla\check{\tilde{g}}^K + \check{H}_0\check{\tilde{g}}^K - \check{\tilde{g}}^K\check{H}_0 \\
&-\frac{i}{2}\left[\check{\tilde{H}}_d\left(\underline{\check{\tilde{g}}^R}\check{f}^{(-)} - \check{f}^{(-)}\underline{\check{\tilde{g}}^A}\right) + \left(\underline{\check{\tilde{g}}^R}\check{f}^{(-)} - \check{f}^{(-)}\underline{\check{\tilde{g}}^A}\right)\check{\tilde{H}}_d\right] \\
&-\frac{i}{2}\left[\check{\tilde{H}}_d\left(\check{\tilde{g}}^R\underline{\check{f}}^{(-)} - \underline{\check{f}}^{(-)}\check{\tilde{g}}^A\right) + \left(\check{\tilde{g}}^R\underline{\check{f}}^{(-)} - \underline{\check{f}}^{(-)}\check{\tilde{g}}^A\right)\check{\tilde{H}}_d\right] = \check{I}^K. \quad (10.38)
\end{aligned}$$

The operators $\partial/\partial\epsilon$, $\partial/\partial\mathbf{p}$, and ∇ in $\check{\tilde{H}}_d$ act on the underlined functions. The components of $\check{f}^{(+)}$ can again be neglected if they appear together with the operator $\check{\tilde{H}}_d$.

Equations for the distribution functions $\check{f}^{(-)}$ and $\check{f}^{(+)}$ can be obtained from eqn (10.38) by projecting it onto the eigenfunctions of the operators

$$\mathcal{L}_-\left\{\check{A}\right\} = \check{g}^R\check{A} - \check{A}\check{g}^A \text{ and } \mathcal{L}_+\left\{\check{A}\right\} = \check{g}^R\check{A} + \check{A}\check{g}^A,$$

respectively. Here $\check{A}$ is the matrix equation (10.38).

10.2.1 *Superconductors in electromagnetic fields*

In many situations, the kinetic equations can be simplified considerably. The wide class of such phenomena includes, in particular, responses of superconductors to applied electromagnetic fields. These are the cases when the non-quasiclassical corrections in the equation for the function f_2 can be neglected. We specify the corresponding conditions later.

Using eqn (10.36) and its time-derivative, we find in this case

$$\begin{aligned}
&-i\check{\tau}_3\left(\check{\tilde{g}}^R - \check{\tilde{g}}^A\right)\frac{\partial f_1}{\partial t} - i\left(\check{\tilde{g}}^R - \check{\tilde{g}}^A\right)\mathbf{v}_F\nabla f_1 - i\check{\tau}_3\frac{\partial\delta\check{\tilde{g}}^K}{\partial t} - i\mathbf{v}_F\nabla\delta\check{\tilde{g}}^K \\
&+\left[\check{H}_0\delta\check{\tilde{g}}^K - \delta\check{\tilde{g}}^K\check{H}_0\right] - \frac{i}{2}\left[\check{\tilde{H}}_d\underline{\delta\check{\tilde{g}}^K} + \underline{\delta\check{\tilde{g}}^K}\check{\tilde{H}}_d\right] \\
&-\frac{i}{2}\left[\check{\tilde{H}}_d\left(\check{\tilde{g}}^R - \check{\tilde{g}}^A\right) + \left(\check{\tilde{g}}^R - \check{\tilde{g}}^A\right)\check{\tilde{H}}_d\right]\underline{f_1} \\
&+\frac{i}{2}\left[\check{\tilde{H}}_\epsilon\left(\check{\tilde{g}}^R + \check{\tilde{g}}^A\right) - \left(\check{\tilde{g}}^R + \check{\tilde{g}}^A\right)\check{\tilde{H}}_\epsilon\right]\underline{f_1} = \check{K} \quad (10.39)
\end{aligned}$$

where

$$\check{K} = \check{I} - \left(\check{I}^R - \check{I}^A\right)\left(f^{(0)} + f_1\right) + \frac{i}{2}\frac{\hat{\partial}}{\partial t}\left(\check{I}^R + \check{I}^A\right)\frac{\partial f^{(0)}}{\partial\epsilon}. \quad (10.40)$$

Within our approximation, we can omit the terms of order of $1/p_F\xi_0$ in the collision integral. We also define

$$\delta\widetilde{\check{g}}^K = \widetilde{\check{g}}^K - \left[\widetilde{\check{g}}_\epsilon^R - \widetilde{\check{g}}_\epsilon^A\right] f_1 + \frac{i}{2}\frac{df^{(0)}}{d\epsilon}\left[\frac{\partial\widetilde{\check{g}}^R}{\partial t} + \frac{\partial\widetilde{\check{g}}^A}{\partial t}\right].$$

We now take the trace of eqn (10.39). The first line gives

$$-2i\left(g^R - g^A\right)\frac{\partial f_1}{\partial t} - 2i\left(g^R - g^A\right)\mathbf{v}_F\nabla f_1 - i\mathbf{v}_F\nabla\,\mathrm{Tr}\left[\delta\widetilde{\check{g}}^K\right].$$

We neglect the non-quasiclassical corrections in the first two terms for the reason which we discuss later. Consider the trace

$$\begin{aligned}\mathrm{Tr}\delta\check{g}^K &= 2\left(g^R - g^A\right)f_2 + \mathrm{Tr}\left(\check{g}^R\check{f}^{(+)} + \check{f}^{(+)}\check{g}^A\right)\\ &\quad -i\left[\frac{\partial g^R}{\partial\mathbf{p}} + \frac{\partial g^A}{\partial\mathbf{p}}\right]\nabla f_2 + i\frac{\partial f_2}{\partial\mathbf{p}}\nabla\left(g^R + g^A\right).\end{aligned}$$

All terms except for the first one contain small non-quasiclassical correction and can be neglected.

Putting this into eqn (10.39), we write the final equation for the gauge covariant functions using the transformation rules of eqn (10.23) and (10.24):

$$\begin{aligned}&\left(e\mathbf{v}_F\cdot\mathbf{E}g_- + \frac{1}{2}\left[f_-\frac{\hat\partial\Delta^*}{\partial t} + f_-^\dagger\frac{\hat\partial\Delta}{\partial t}\right]\right)\frac{\partial f^{(0)}}{\partial\epsilon} + \mathbf{v}_F\cdot\nabla(f_2 g_-) + g_-\frac{\partial f_1}{\partial t}\\ &-\left(\frac{1}{2}\left[(\hat\nabla\Delta)f_-^\dagger + (\hat\nabla\Delta^*)f_-\right] - \frac{e}{c}[\mathbf{v}_F\times\mathbf{H}]g_-\right)\frac{\partial f_1}{\partial\mathbf{p}}\\ &+\frac{1}{2}\left(\frac{\partial\Delta}{\partial\mathbf{p}}f_-^\dagger + \frac{\partial\Delta^*}{\partial\mathbf{p}}f_-\right)\nabla f_1 = J_1.\end{aligned} \tag{10.41}$$

Here

$$\check{g}_\pm = \frac{1}{2}(\check{g}_\epsilon^R \pm \check{g}_\epsilon^A) \tag{10.42}$$

and

$$J_1 = \frac{i}{4}\mathrm{Tr}\check{K}. \tag{10.43}$$

An equation for f_2 is obtained from eqn (10.39) after multiplying it by $\check{\tau}_3$ and taking the trace. Neglecting the non-quasiclassical corrections in this equation we get

$$g_-\mathbf{v}_F\nabla f_1 + \frac{1}{2}\left[f_+\frac{\hat\partial\Delta^*}{\partial t} - f_+^\dagger\frac{\hat\partial\Delta}{\partial t}\right]\frac{\partial f^{(0)}}{\partial\epsilon} + i\left[\Delta^* f_+ + \Delta f_+^\dagger\right]f_2 = J_2\,. \tag{10.44}$$

Here

$$J_2 = \frac{i}{4}\mathrm{Tr}\left[\check{\tau}_3\check{K}\right]. \tag{10.45}$$

Equation (10.44) shows that the term $\mathrm{Tr}\left(\widetilde{\check{g}}^R - \widetilde{\check{g}}^A\right)\mathbf{v}_F\nabla f_1$ omitted in eqn (10.39) during the derivation of eqn (10.41) introduces corrections proportional to the quasiclassical parameter $(p_F\xi)^{-1}$ to the terms which are already present in eqn (10.41); it thus should be neglected.

Equation (10.41) can be further modified using the equation for the static Green functions

$$\mathbf{v}_F\nabla g_\epsilon^{R(A)} + i\left(\Delta f_\epsilon^{\dagger R(A)} - f_\epsilon^{R(A)}\Delta^*\right) = \int \left[I_1^{R(A)} - \bar{I}_1^{R(A)}\right]\frac{d\xi_\mathbf{p}}{2\pi}. \qquad (10.46)$$

The non-quasiclassical correction can be omitted here. Multiplying this equation by f_2 and subtracting it from eqn (10.41), we obtain

$$\begin{aligned}&\left(e\mathbf{v}_F\cdot\mathbf{E}g_- + \frac{1}{2}\left[f_-\frac{\partial\Delta^*}{\partial t} + f_-^\dagger\frac{\partial\Delta}{\partial t}\right]\right)\frac{\partial f^{(0)}}{\partial\epsilon} + g_-\mathbf{v}_F\cdot\nabla f_2\\ &\quad + i\left[\Delta^* f_- - \Delta f_-^\dagger\right]f_2 - \left(\frac{1}{2}\left[(\hat\nabla\Delta)f_-^\dagger + (\hat\nabla\Delta^*)f_-\right] - \frac{e}{c}[\mathbf{v}_F\times\mathbf{H}]g_-\right)\frac{\partial f_1}{\partial\mathbf{p}}\\ &\quad + \frac{1}{2}\left(\frac{\partial\Delta}{\partial\mathbf{p}}f_-^\dagger + \frac{\partial\Delta^*}{\partial\mathbf{p}}f_-\right)\nabla f_1 = J_1', \qquad (10.47)\end{aligned}$$

where

$$J' = \frac{i}{4}\mathrm{Tr}\check{K}' \qquad (10.48)$$

and

$$\check{K}' = \check{I} - \left(\check{I}^R - \check{I}^A\right)\left(f^{(0)} + f_1 + \check\tau_3 f_2\right) + \frac{i}{2}\frac{\hat\partial}{\partial t}\left(\check{I}^R + \check{I}^A\right)\frac{\partial f^{(0)}}{\partial\epsilon}. \qquad (10.49)$$

In the literature, the function f_2' is used sometimes instead of the function f_2, it is defined according to

$$f_2 = f_2' + \frac{1}{2}\left(\frac{\partial\chi}{\partial t} + 2e\varphi\right)\frac{df^{(0)}}{d\epsilon} \qquad (10.50)$$

where χ is the phase of the order parameter. Using the fact that the collision integral $J_1'^{(2)}$ vanishes for any function independent of the momentum-direction [see eqn (10.76)] we find instead of eqns (10.47)

$$\begin{aligned}&\left(-\frac{e}{c}\mathbf{v}_F\frac{\partial\mathbf{Q}}{\partial t}g_- + \frac{1}{2}\left[f_-\frac{\tilde\partial\Delta^*}{\partial t} + f_-^\dagger\frac{\tilde\partial\Delta}{\partial t}\right]\right)\frac{\partial f^{(0)}}{\partial\epsilon}\\ &\quad + g_-\mathbf{v}_F\cdot\nabla f_2' + i\left[\Delta^* f_- - \Delta f_-^\dagger\right]f_2'\\ &\quad - \left(\frac{1}{2}\left[(\hat\nabla\Delta)f_-^\dagger + (\hat\nabla\Delta^*)f_-\right] - \frac{e}{c}[\mathbf{v}_F\times\mathbf{H}]g_-\right)\frac{\partial f_1}{\partial\mathbf{p}}\\ &\quad + \frac{1}{2}\left(\frac{\partial\Delta}{\partial\mathbf{p}}f_-^\dagger + \frac{\partial\Delta^*}{\partial\mathbf{p}}f_-\right)\nabla f_1\end{aligned}$$

$$= J_1'\{f_1, f_2'\}\,, \tag{10.51}$$

where the operator $\tilde{\partial}/\partial t$ acts only on the modulus of the order parameter. Equation (10.51) can be further transformed into

$$g_-\frac{\partial f_1}{\partial t}+\left(-\frac{e}{c}\mathbf{v}_F\cdot\frac{\partial \mathbf{Q}}{\partial t}g_-+\frac{1}{2}\left[f_-\frac{\tilde{\partial}\Delta^*}{\partial t}+f_-^\dagger\frac{\tilde{\partial}\Delta}{\partial t}\right]\right)\frac{\partial f^{(0)}}{\partial\epsilon}$$
$$+\mathbf{v}_F\cdot\nabla(f_2'g_-)-\left(\frac{1}{2}\left[(\hat{\nabla}\Delta)f_-^\dagger+(\hat{\nabla}\Delta^*)f_-\right]-\frac{e}{c}[\mathbf{v}_F\times\mathbf{H}]g_-\right)\frac{\partial f_1}{\partial\mathbf{p}}$$
$$+\frac{1}{2}\left(\frac{\partial\Delta}{\partial\mathbf{p}}f_-^\dagger+\frac{\partial\Delta^*}{\partial\mathbf{p}}f_-\right)\nabla f_1 = J_1\{f_1, f_2'\}\,. \tag{10.52}$$

Equation (10.44) becomes

$$g_-\mathbf{v}_F\nabla f_1+\frac{1}{2}\left[f_+\frac{\tilde{\partial}\Delta^*}{\partial t}-f_+^\dagger\frac{\tilde{\partial}\Delta}{\partial t}\right]\frac{\partial f^{(0)}}{\partial\epsilon}$$
$$+i\left[\Delta^* f_+ + \Delta f_+^\dagger\right]f_2' = J_2\{f_1, f_2'\}\,. \tag{10.53}$$

These are the equations for the new function f_2'.

10.2.2 *Discussion*

The kinetic equations eqn (10.41) [or eqn (10.47)], and eqn (10.44) for the distribution functions f_1 and f_2 are the central result of this chapter. The kinetic equations contain the momentum derivatives of $\partial f/\partial \mathbf{p}$ multiplied by "forces" which include the Lorentz force $\mathbf{f}_L = (e/c)[\mathbf{v}_F\times\mathbf{H}]$ and the forces arising due to spatial variations of the order parameter. Note that the electric field is included in the terms which describe the energy variation in time, since $\mathbf{E}$ is of the first order in time-derivative. For this reason, it does not enter the force terms.

As we saw already on page 195, the function f_1 appears in the l.h.s. of eqn (10.41) only as $\partial f_1/\partial t$ and in combinations with the momentum derivatives $\partial f_1/\partial \mathbf{p}$ or as $(\partial\Delta/\partial\mathbf{p})\nabla f_1$. For a slow variation in time, $\omega \ll 1/\tau$, the only terms with f_1 are those with the momentum derivatives of the distribution function and of the order parameter. Such terms only exist in the first-order approximation in $1/p_F\xi$. All other non-quasiclassical contributions are omitted: they turn out to be corrections to the first line of eqn (10.41) which already exists in the zero approximation.

From eqn (10.41) we note that the momentum derivatives are of the same order of magnitude as the collision integrals when

$$\frac{\Delta}{p_F\xi}\sim\frac{1}{\tau}. \tag{10.54}$$

The relative importance of the terms with the momentum derivatives decreases as the mean free path gets shorter. For $\tau \ll E_F/\Delta^2$, their contribution to the distribution function is of the order of $\Delta^2\tau/E_F$ as compared to that determined by the collision integral. This correction becomes of the order of $1/p_F\xi$ when $\Delta\tau$ decreases down to values of the order of unity. For such short τ, the contributions from $\partial f_1/\partial \mathbf{p}$ and $\partial\Delta/\partial \mathbf{p}$ become comparable or less than the terms which we have neglected during the derivation of the quasiclassical kinetic equations. Thus, it becomes illegal to keep the momentum derivatives $\partial f_1/\partial \mathbf{p}$ and $\partial\Delta/\partial \mathbf{p}$ in equation (10.41) for $\Delta\tau \sim 1$ and they should be neglected. Therefore, for dirty superconductors, eqns (10.41, 10.44) should be used with the momentum derivatives neglected. In the dirty case we thus obtain from eqn (10.41)

$$\left(e\mathbf{v}_F \cdot \mathbf{E} g_- + \frac{1}{2}\left[f_- \frac{\hat{\partial}\Delta^*}{\partial t} + f_-^\dagger \frac{\hat{\partial}\Delta}{\partial t}\right]\right)\frac{\partial f^{(0)}}{\partial \epsilon} + \mathbf{v}_F \cdot \nabla(f_2 g_-) + g_- \frac{\partial f_1}{\partial t} = J_1. \tag{10.55}$$

Equation (10.44) can be used as it stands. These equations can be also obtained directly from the Eliashberg equations (9.2) using the Keldysh function in the form of eqn (9.25) with the distribution function from eqn (9.28). The derivation of the quasiclassical set of equations (10.55) and (10.44) for dirty superconductors is described in the review by Larkin and Ovchinnikov (1986).

For clean superconductors with $\Delta\tau \gg 1$, on the contrary, the momentum derivatives should be included into the kinetic equations. They give small contributions in the so-called moderately clean regime when $\Delta^{-1} \ll \tau \ll E_F/\Delta^2$. As τ increases further, their role becomes more important. Finally, in the collisionless limit (superclean regime) $\tau \gg E_F/\Delta^2$, the momentum derivatives $\partial f/\partial \mathbf{p}$ and $\partial\Delta/\partial \mathbf{p}$ dominate.

One more observation is that, for clean superconductors interacting with electromagnetic fields, $f_2 \ll f_1$. Indeed, eqn (10.41) gives

$$f_1 \sim \tau\omega\Delta \frac{\partial f^{(0)}}{\partial \epsilon}.$$

On the other hand, eqn (10.44) shows that the variation of f_1 is $\delta f_1 \sim \omega \partial f^{(0)}/\partial\epsilon$ along a trajectory part with a length of the order of ξ. This variation is much smaller than f_1 itself. Therefore, eqn (10.44) reduces simply to

$$g_- \mathbf{v}_F \nabla f_1 = 0 \tag{10.56}$$

i.e., the function f_1 is constant along the quasiparticle trajectory. As a result, eqn (10.41) implies that $f_2 \sim \omega\partial f^{(0)}/\partial\epsilon$, i.e., $f_2 \ll f_1$. It is this condition that allows one to neglect non-quasiclassical corrections to f_2 and to use eqn (10.56) in a clean superconductor.

Equation (10.56) holds for a trajectory which is not very long. The maximum length L is determined by the shortest of the three length scales: $L \ll \min\{\ell, (E_F/\Delta)l_\Delta, r_H\}$ where l_Δ is the characteristic scale of variations of the order parameter, and $r_H = v_F/\omega_c$ is the Larmor radius. If this condition is violated, one cannot neglect the other terms in eqn (10.39) any more. This condition, in particular, determines the necessary requirement of validity of eqn (10.56); it is $L \gg \xi$ which is equivalent, of course, to the clean-limit condition $\ell \gg \xi$.

The non-quasiclassical corrections to eqn (10.44) which were omitted during the derivation may become important for problems associated with the heat transport, when there exists a temperature gradient in a superclean regime with $\tau \sim E_F/\Delta^2$ and longer. In this case, on the contrary, the function f_2 is larger than f_1 and eqn (10.41) gives $g_- f_2 = const$. The second equation (10.44) should then be derived including the non-quasiclassical corrections. We shall not consider such problems here.

10.3 Observables in the gauge-invariant representation

Using the modified gauge-invariant Green functions, the current can be written as

$$\mathbf{j} = -e \int \tilde{\mathbf{v}}_F \left[\tilde{G}^K_\epsilon(\tilde{\mathbf{p}};\mathbf{r},t) - \tilde{\bar{G}}^K_\epsilon(\tilde{\mathbf{p}};\mathbf{r},t) \right] \frac{d^3p}{(2\pi)^3}\frac{d\epsilon}{4\pi i}. \tag{10.57}$$

Here $\tilde{\mathbf{p}} = \mathbf{p} - (e/c)\,\mathbf{A}$ for $\tilde{G}$ and $\tilde{\mathbf{p}} = \mathbf{p} + (e/c)\,\mathbf{A}$ for $\tilde{\bar{G}}$ according to the definition of the gauge-invariant functions. We see that the diamagnetic term does not appear explicitly for the gauge-invariant representation in the quasiclassical limit. Performing shifts in the momentum under the integrals we can write the current through the gauge-invariant quasiclassical Keldysh functions in the usual form

$$\mathbf{j} = -e\nu(0) \int \mathbf{v}_F \left[\tilde{g}^K_\epsilon(\mathbf{p};\mathbf{r},t) - \tilde{\bar{g}}^K_\epsilon(\mathbf{p};\mathbf{r},t) \right] \frac{d\Omega_\mathbf{p}}{4\pi}\frac{d\epsilon}{4}. \tag{10.58}$$

The order parameter is

$$\frac{\Delta(\mathbf{r},t)}{|\,g\,|} = \int \tilde{F}^K_\epsilon(\mathbf{p};\mathbf{r},t)\,\frac{d^3p}{(2\pi)^3}\frac{d\epsilon}{4\pi i}, \tag{10.59}$$

where $|\,g\,|$ is the pairing interaction. The quasiclassical version of this expression is

$$\frac{\Delta(\mathbf{r},t)}{\lambda} = \int \tilde{f}^K_\epsilon(\mathbf{p};\mathbf{r},t)\,\frac{d\Omega_\mathbf{p}}{4\pi}\frac{d\epsilon}{4}, \tag{10.60}$$

where $\lambda = \nu(0)\,|\,g\,|$ is the pairing constant.

The electron density is (Eliashberg 1971)

$$\begin{aligned} N - N_0 &= -\int \left[G^K_\epsilon(\mathbf{p};\mathbf{r},t) + \bar{G}^K_\epsilon(\mathbf{p};\mathbf{r},t) \right] \frac{d^3p}{(2\pi)^3}\frac{d\epsilon}{4\pi i} \\ &= -2\nu(0)e\varphi - \nu(0)\int \left[g^K_\epsilon(\mathbf{p}_F;\mathbf{r},t) + \bar{g}^K_\epsilon(\mathbf{p}_F;\mathbf{r},t) \right] \frac{d\Omega_\mathbf{p}}{4\pi}\frac{d\epsilon}{4} \end{aligned}$$

in terms of the usual Green functions. The scalar potential appears in the r.h.s. due to a change in the normal-state density caused by a $e\varphi$ variation of the chemical potential. The electron density becomes

$$N - N_0 = -2\nu(0)e\varphi - \nu(0)\int \left[\tilde{g}^K_{\epsilon - e\varphi}(\mathbf{p}_F;\mathbf{r},t) + \tilde{\bar{g}}^K_{\epsilon + e\varphi}(\mathbf{p}_F;\mathbf{r},t) \right] \frac{d\Omega_\mathbf{p}}{4\pi}\frac{d\epsilon}{4}$$

$$= -\nu(0) \int \left[\tilde{g}_\epsilon^K(\mathbf{p}_F;\mathbf{r},t) + \tilde{\bar{g}}_\epsilon^K(\mathbf{p}_F;\mathbf{r},t) \right] \frac{d\Omega_{\mathbf{p}}}{4\pi} \frac{d\epsilon}{4} \tag{10.61}$$

in terms of the gauge-invariant Green functions. Here we use that

$$\tilde{g}^K_{\epsilon-e\varphi}(\mathbf{p}_F;\mathbf{r},t) = \tilde{g}^K_\epsilon(\mathbf{p}_F;\mathbf{r},t) - e\varphi \frac{\partial \tilde{g}^K_\epsilon}{\partial \epsilon}$$

and

$$\int \frac{\partial \tilde{g}^K_\epsilon}{\partial \epsilon} \frac{d\epsilon}{4} = 1.$$

The latter follows from the observation that for $|\epsilon| \to \infty$

$$\check{g}_\epsilon^{R(A)} \to \pm \check{1},\ f^{(1)} \to f^{(0)},\ \text{and}\ f^{(2)} \to 0.$$

We can now replace the gauge-invariant functions in equations (10.58), (10.60), and (10.61) with the usual (gauge-covariant) functions. The only condition is that they have to be determined by the gauge-invariant kinetic equations (10.41) and (10.44) together with the gauge-invariant parameterization for the Keldysh function through f_1 and f_2.

The Keldysh function takes the form

$$\begin{aligned} \check{g}^K_\epsilon &= \left[\check{g}^R_\epsilon - \check{g}^A_\epsilon\right] f^{(0)} - \frac{i}{2}\frac{df^{(0)}}{d\epsilon}\left[\frac{\partial \hat{g}^R_\epsilon}{\partial t} + \frac{\partial \hat{g}^A_\epsilon}{\partial t}\right] \\ &+ \left[\check{g}^R_\epsilon \delta\check{f}^{(-)} - \delta\check{f}^{(-)}\check{g}^A_\epsilon\right] + \left[\check{g}^R_\epsilon \check{f}^{(+)} + \check{f}^{(+)}\check{g}^A_\epsilon\right] \\ &- \frac{i}{2}\left[\frac{\partial \check{g}^R_\epsilon}{\partial \mathbf{p}}\hat{\nabla}\check{f}^{(-)} + \hat{\nabla}\check{f}^{(-)}\frac{\partial \check{g}^R_\epsilon}{\partial \mathbf{p}}\right] + \frac{i}{2}\left[\hat{\nabla}\check{g}^R_\epsilon\frac{\partial \check{f}^{(-)}}{\partial \mathbf{p}} + \frac{\partial \check{f}^{(-)}}{\partial \mathbf{p}}\hat{\nabla}\check{g}^A_\epsilon\right] \end{aligned}$$

where $\delta\check{f}^{(-)} = \check{f}^{(-)} - \check{1}f^{(0)}$. We shall neglect the small non-quasiclassical corrections and write

$$\begin{aligned} \check{g}^K_{\epsilon_+,\epsilon_-} &= \left[\check{g}^R_\epsilon - \check{g}^A_\epsilon\right] f^{(0)}_\epsilon - \frac{i}{2}\frac{\hat{\partial}}{\partial t}\left(\check{g}^R_\epsilon + \check{g}^A_\epsilon\right)\frac{\partial f^{(0)}}{\partial \epsilon} \\ &+ \left(\check{g}^R_\epsilon - \check{g}^A_\epsilon\right) f_1 + \left(\check{g}^R_\epsilon \check{\tau}_3 - \check{\tau}_3 \check{g}^A_\epsilon\right) f_2. \end{aligned} \tag{10.62}$$

In this approximation, the Keldysh function is parametrized by two distribution functions f_1 and f_2.

With eqn (10.62), the current and the order parameter have their usual form

$$\mathbf{j} = -e\nu(0) \int \mathbf{v}_F \left[g^K_\epsilon(\mathbf{p};\mathbf{r},t) - \bar{g}^K_\epsilon(\mathbf{p};\mathbf{r},t)\right] \frac{d\Omega_{\mathbf{p}}}{4\pi}\frac{d\epsilon}{4}, \tag{10.63}$$

and

$$\frac{\Delta(\mathbf{r},t)}{\lambda} = \int f^K_\epsilon(\mathbf{p};\mathbf{r},t)\frac{d\Omega_{\mathbf{p}}}{4\pi}\frac{d\epsilon}{4}. \tag{10.64}$$

The expressions for the internal energy and the internal energy current can also be written in terms of the gauge-invariant functions. We have

$$\mathcal{E} = -\int \left[(\epsilon - e\varphi) G^K_\epsilon(\mathbf{p};\mathbf{r},t) - (\epsilon + e\varphi)\bar{G}^K_\epsilon(\mathbf{p};\mathbf{r},t)\right]\frac{d\epsilon}{4\pi i}\frac{d^3p}{(2\pi)^3} + \frac{|\Delta|^2}{|g|}$$

$$= -\int \left[(\epsilon - e\varphi) \tilde{G}^K_{\epsilon - e\varphi} (\mathbf{p}; \mathbf{r}, t) - (\epsilon + e\varphi) \tilde{\tilde{G}}^K_{\epsilon + e\varphi} (\mathbf{p}; \mathbf{r}, t) \right] \frac{d\epsilon}{4\pi i} \frac{d^3p}{(2\pi)^3} + \frac{|\Delta|^2}{|g|}$$

$$= -\int \epsilon \left[\tilde{G}^K_\epsilon (\mathbf{p}; \mathbf{r}, t) - \tilde{\tilde{G}}^K_\epsilon (\mathbf{p}; \mathbf{r}, t) \right] \frac{d\epsilon}{4\pi i} \frac{d^3p}{(2\pi)^3} + \frac{|\Delta|^2}{|g|}.$$

In the quasiclassical representation, it becomes

$$\mathcal{E} = \mathcal{E}_n - \nu(0) \int \epsilon \left[\delta g^K_\epsilon (\mathbf{p}; \mathbf{r}, t) - \delta \bar{g}^K_\epsilon (\mathbf{p}; \mathbf{r}, t) \right] \frac{d\epsilon}{4} \frac{d\Omega_\mathbf{p}}{4\pi} + \frac{|\Delta|^2}{|g|}. \tag{10.65}$$

Here $\mathcal{E}_n$ is the energy in the normal state, and δg is the deviation of the Green function from the normal state. We have separated the normal state contribution because the normal state energy cannot be calculated within the quasiclassical approximation: the momentum integral is determined by $\xi_\mathbf{p}$ far from the Fermi surface. The energy current takes the form

$$\mathbf{j}_\mathcal{E} = -\int \epsilon \mathbf{v} \left[\tilde{G}^K_\epsilon (\mathbf{p}; \mathbf{r}, t) + \tilde{\tilde{G}}^K_\epsilon (\mathbf{p}; \mathbf{r}, t) \right] \frac{d\epsilon}{4\pi i} \frac{d^3p}{(2\pi)^3}$$

$$= -\nu(0) \int \epsilon \mathbf{v}_F \left[g^K_\epsilon (\mathbf{p}; \mathbf{r}, t) + \bar{g}^K_\epsilon (\mathbf{p}; \mathbf{r}, t) \right] \frac{d\epsilon}{4} \frac{d\Omega_\mathbf{p}}{4\pi}. \tag{10.66}$$

10.3.1 *The electron density and charge neutrality*

The electron density eqn (10.61) takes the form

$$N - N_0 = -\nu(0) \int \left[g^K_\epsilon (\mathbf{p}; \mathbf{r}, t) + \bar{g}^K_\epsilon (\mathbf{p}; \mathbf{r}, t) \right] \frac{d\Omega_\mathbf{p}}{4\pi} \frac{d\epsilon}{4}$$

$$= -\nu(0) \int \left(\mathrm{Tr} \left[\check{g}^R_\epsilon - \check{g}^A_\epsilon \right] f^{(0)} + 4 g_- f_2 \right) \frac{d\Omega_\mathbf{p}}{4\pi} \frac{d\epsilon}{4}.$$

The combination $\mathrm{Tr} \left[\check{g}^R_\epsilon (\mathbf{p}; \mathbf{r}, t) - \check{g}^A_\epsilon (\mathbf{p}; \mathbf{r}, t) \right]$ is nonzero in a nonstationary situation. To find it, we use the normalization condition

$$\check{g}^R \otimes \check{g}^R \equiv \check{g}^R_\epsilon \check{g}^R_\epsilon + \frac{i}{2} \left(\frac{\partial \check{g}^R_\epsilon}{\partial \epsilon} \frac{\hat{\partial} \check{g}^R_\epsilon}{\partial t} - \frac{\hat{\partial} \check{g}^R_\epsilon}{\partial t} \frac{\partial \check{g}^R_\epsilon}{\partial \epsilon} \right) = \check{1}.$$

We put $\check{g}^R = \check{g}^R_0 + \check{g}^R_1$ where $\check{g}^R_0$ is the stationary Green function and $\check{g}^R_1$ is a nonstationary correction and find

$$\check{g}^R_0 \check{g}^R_1 + \check{g}^R_1 \check{g}^R_0 = -\frac{i}{2} \left(\frac{\partial \check{g}^R_0}{\partial \epsilon} \frac{\hat{\partial} \check{g}^R_0}{\partial t} - \frac{\hat{\partial} \check{g}^R_0}{\partial t} \frac{\partial \check{g}^R_0}{\partial \epsilon} \right).$$

We multiply this equation by $\check{g}^R_0$ and take the trace

$$2\mathrm{Tr} \check{g}^R_1 = \frac{i}{2} \left(\check{g}^R_0 \frac{\hat{\partial} \check{g}^R_0}{\partial t} - \frac{\hat{\partial} \check{g}^R_0}{\partial t} \check{g}^R_0 \right) \frac{\partial \check{g}^R_0}{\partial \epsilon}.$$

With help of the identity $g^2 - f f^\dagger = 1$ we obtain

$$\mathrm{Tr} \check{g}^R_1 = -\frac{i}{2} \left[\frac{1}{f^R} \left(\frac{\partial}{\partial t} + 2ie\phi \right) f^R - \frac{1}{f^{\dagger R}} \left(\frac{\partial}{\partial t} - 2ie\phi \right) f^{\dagger R} \right] \frac{\partial g^R}{\partial \epsilon}$$

$$+\frac{1}{2g^R}\left[\frac{1}{f^R}\frac{\partial f^R}{\partial t}+\frac{1}{f^{\dagger R}}\frac{\partial f^{\dagger R}}{\partial t}\right] \tag{10.67}$$

and the same for the advanced function. Next, we note that according to the particle–hole symmetry of the Eilenberger equations (5.89)–(5.91), the static Green functions transform as

$$g^R \to -g^A,\ f^R \to f^A,\ f^{\dagger R} \to f^{\dagger A} \tag{10.68}$$

under the transformation $\epsilon \to -\epsilon$, $\mathbf{p} \to -\mathbf{p}$. Therefore, the first line of eqn (10.67) transforms into

$$-\frac{i}{2}\left[\frac{1}{f^A}\left(\frac{\partial}{\partial t}+2ie\varphi\right)f^A-\frac{1}{f^{\dagger A}}\left(\frac{\partial}{\partial t}-2ie\phi\right)f^{\dagger A}\right]\frac{\partial g^A}{\partial \epsilon},$$

while the second line transforms into

$$-\frac{1}{2g^A}\left[\frac{1}{f^A}\frac{\partial f^A}{\partial t}+\frac{1}{f^{\dagger A}}\frac{\partial f^{\dagger A}}{\partial t}\right].$$

As a result, we can write

$$\mathrm{Tr}\left(\check{g}^R_{\epsilon_+,\epsilon_-}-\check{g}^A_{\epsilon_+,\epsilon_-}\right)=\left(\frac{\partial \chi}{\partial t}+2e\varphi\right)\left(\frac{\partial g^R_\epsilon}{\partial \epsilon}-\frac{\partial g^A_\epsilon}{\partial \epsilon}\right)+A(\epsilon,\mathbf{p})$$

where $A(\epsilon,\mathbf{p})$ is an even function of $(\epsilon,\mathbf{p})$ while the first term is an odd function of these variables. The term with $A(\epsilon,\mathbf{p})$ vanishes after integration with the odd function $f^{(0)}=\tanh(\epsilon/2T)$. The electron density becomes

$$N=N_0-\nu(0)\,e\Phi\int_{-\infty}^{\infty}\left\langle\frac{\partial g_-}{\partial \epsilon}\right\rangle f^{(0)}\,d\epsilon-\nu(0)\int_{-\infty}^{\infty}\langle g_- f_2\rangle\,d\epsilon \tag{10.69}$$

where Φ is defined by eqn (10.4), and the averaging over momentum directions eqn (5.72) is implied. Integrating by parts in the first line and using that

$$g_-|_{\epsilon=\pm\infty}=\pm 1;\ f^{(0)}(\epsilon=\pm\infty)=\pm 1$$

we find

$$N=N_0-\nu(0)\,e\Phi+\nu(0)\,e\Phi\int_{-\infty}^{\infty}\langle g_-\rangle\frac{\partial f^{(0)}}{\partial \epsilon}\,d\epsilon-\nu(0)\int_{-\infty}^{\infty}\langle g_- f_2\rangle\,d\epsilon. \tag{10.70}$$

With the definition equation (10.50) of the function f_2', the electron density can be written as

$$N=N_0-\nu(0)\,e\Phi-\nu(0)\int_{-\infty}^{\infty}\langle g_- f_2'\rangle\,d\epsilon. \tag{10.71}$$

In metals, all bulk charges are screened. It can be seen from the Poisson equation for the normal state

$$\operatorname{div}\mathbf{E} = -\nabla^2\varphi = 4\pi e(N_e - N_i)$$

where N_e and N_i are densities of electrons and ions, respectively. In equilibrium, $N_e = N_i$. Introduction of an additional electronic charge results in a shift of the chemical potential $\delta\mu = -e\delta\varphi$ such that

$$\delta N = N_e - N_i = \frac{\partial N_e}{\partial\mu}\delta\mu = -2\nu e\delta\varphi.$$

The Poisson equation becomes

$$\nabla^2\delta\varphi = 8\pi e^2\nu\delta\varphi.$$

It demonstrates that the potential φ together with δN decay at distances of the order of the Debye screening length

$$\lambda_D^2 = \frac{1}{8\pi e^2\nu}$$

which is of the order of interatomic distance in good metals. Therefore, variations in the electronic charge density are practically zero at distances of the order of the coherence length, and we have to put the constraint $\delta N_e = 0$ or

$$\operatorname{div}\mathbf{j} = 0. \tag{10.72}$$

The latter follows from the continuity equation

$$\frac{\partial(eN_e)}{\partial t} + \operatorname{div}\mathbf{j} = 0.$$

Since $N_i = N_0$ where N_0 is the normal-state electron density, the charge neutrality condition requires

$$N = N_0. \tag{10.73}$$

10.4 Collision integrals

Collision integrals describe interaction of excitations with impurities, phonons and with each other. These interactions are responsible for establishing equilibrium in an electronic subsystem in a superconductor assuming that the crystal lattice (phonons) together with impurities are themselves in equilibrium and form a heat bath. In practical superconducting compounds, impurity concentrations are usually such that the most effective relaxation is brought about through scattering of electrons by impurities; the electron–phonon relaxation rate is usually much smaller, to say nothing about electron–electron collisions. If not specified otherwise, we assume in what follows that the impurity collision integral is the largest source of relaxation.

In this section we consider various collision integrals and derive expressions for them in terms of the distribution functions $f^{(1)}$ and $f^{(2)}$ which will be used later in our discussions. We note that all the collision integrals vanish for the equilibrium distribution $f^{(1)} = f^{(0)}$ and $f^{(2)} = 0$. They are thus proportional to deviations from equilibrium. Therefore, one can use the stationary Green functions $\check{g}^{R(A)}$ in the collision integrals.

10.4.1 *Impurities*

10.4.1.1 *Nonmagnetic scattering* We shall assume for simplicity that the impurity scattering is isotropic. Remind that the self-energies for nonmagnetic impurities are

$$\check{\Sigma}^{K}_{\epsilon,\epsilon-\omega}(\mathbf{k}) = \frac{i}{2\tau}\left\langle \check{g}^{K}_{\epsilon,\epsilon-\omega}(\mathbf{p}_F,\mathbf{k})\right\rangle ,$$
$$\check{\Sigma}^{R(A)}_{\epsilon,\epsilon-\omega}(\mathbf{k}) = \frac{i}{2\tau}\left\langle \check{g}^{R(A)}_{\epsilon,\epsilon-\omega}(\mathbf{p}_F,\mathbf{k})\right\rangle .$$

Here we use notation τ instead of $\tau_{\rm imp}$ for brevity. The collision integrals $J^{(\rm imp)}_{1,2}$ are found from eqns (9.14) and (9.16) using the definitions eqns (10.43), (10.45), and (10.48):

$$J_1^{(\rm imp)} = J_1^{(1)}\{f^{(1)}\} + J_1^{(2)}\{f^{(2)}\}, \quad J_1^{\prime(\rm imp)} = J_1^{(1)}\{f^{(1)}\} + J_1^{\prime(2)}\{f^{(2)}\},$$
$$J_2^{(\rm imp)} = J_2^{(1)}\{f^{(1)}\} + J_2^{(2)}\{f^{(2)}\},$$

where

$$J_1^{(1)} = -\frac{1}{\tau}\left[\left(f^{(1)}\langle g_-\rangle - \langle f^{(1)} g_-\rangle\right) g_- - \frac{1}{2}\left(f^{(1)}\langle f_-^\dagger\rangle - \langle f^{(1)} f_-^\dagger\rangle\right) f_- - \frac{1}{2}\left(f^{(1)}\langle f_-\rangle - \langle f^{(1)} f_-\rangle\right) f_-^\dagger\right], \tag{10.74}$$

$$J_2^{(2)} = -\frac{1}{\tau}\left[\left(f^{(2)}\langle g_-\rangle - \langle f^{(2)} g_-\rangle\right) g_- - \frac{1}{2}\left(f^{(2)}\langle f_+^\dagger\rangle - \langle f^{(2)} f_+^\dagger\rangle\right) f_+ - \frac{1}{2}\left(f^{(2)}\langle f_+\rangle - \langle f^{(2)} f_+\rangle\right) f_+^\dagger\right], \tag{10.75}$$

and

$$J_1^{(2)} = -\frac{1}{2\tau}\left[\left(f^{(2)}\langle f_-^\dagger\rangle f_+ - \langle f^{(2)} f_+\rangle f_-^\dagger\right) - \left(f^{(2)}\langle f_-\rangle f_+^\dagger - \langle f^{(2)} f_+^\dagger\rangle f_-\right)\right],$$
$$J_2^{(1)} = -\frac{1}{2\tau}\left[\left(f^{(1)}\langle f_-\rangle - \langle f^{(1)} f_-\rangle\right) f_+^\dagger - \left(f^{(1)}\langle f_-^\dagger\rangle - \langle f^{(1)} f_-^\dagger\rangle\right) f_+\right].$$

In addition

$$J_1^{\prime(2)} = -\frac{1}{2\tau}\left[\left(f^{\prime(2)}\langle f_+\rangle - \left\langle f^{\prime(2)} f_+\right\rangle\right) f_-^\dagger - \left(f^{\prime(2)}\left\langle f_+^\dagger\right\rangle - \langle f^{\prime(2)} f_+^\dagger\rangle\right) f_-\right]. \tag{10.76}$$

Note that the collision integrals J_1' and J_1 differ in two respects. First, the collision integral J_1 vanishes after integration over the momentum directions while the integral J_1' does not:

$$\left\langle J_1^{\prime(2)}\right\rangle \neq 0$$

At the same time $J_1^{\prime(2)}$ vanishes for any function $f^{\prime(2)}$ independent of the momentum directions while $J_1^{(2)}$ does not.

10.4.1.2 *Scattering by magnetic impurities* Keep in mind that the self-energy in eqn (5.76) for magnetic impurities is

$$\Sigma_s(\mathbf{k}) = \frac{i}{2\tau_s}\langle\check{\tau}_3\check{g}(\mathbf{p}_F,\mathbf{k})\check{\tau}_3\rangle \tag{10.77}$$

The spin-flip collision integrals become

$$\begin{aligned}J_1^{(s)(1)} &= -\frac{1}{\tau_s}\left[\left(f^{(1)}\langle g_-\rangle - \langle f^{(1)}g_-\rangle\right)g_- + \frac{1}{2}\left(f^{(1)}\langle f_-^\dagger\rangle - \langle f^{(1)}f_-^\dagger\rangle\right)f_-\right.\\ &\quad\left.+\frac{1}{2}\left(f^{(1)}\langle f_-\rangle - \langle f^{(1)}f_-\rangle\right)f_-^\dagger\right],\end{aligned} \tag{10.78}$$

$$\begin{aligned}J_2^{(s)(2)} &= -\frac{1}{\tau_s}\left[\left(f^{(2)}\langle g_-\rangle - \langle f^{(2)}g_-\rangle\right)g_- + \frac{1}{2}\left(f^{(2)}\langle f_+^\dagger\rangle - \langle f^{(2)}f_+^\dagger\rangle\right)f_+\right.\\ &\quad\left.+\frac{1}{2}\left(f^{(2)}\langle f_+\rangle - \langle f^{(2)}f_+\rangle\right)f_+^\dagger\right],\end{aligned} \tag{10.79}$$

and

$$\begin{aligned}J_1^{(s)(2)} &= -\frac{1}{2\tau_s}\left[\left(f^{(2)}\langle f_-\rangle f_+^\dagger - \langle f^{(2)}f_+^\dagger\rangle f_-\right) - \left(f^{(2)}\langle f_-^\dagger\rangle f_+ - \langle f^{(2)}f_+\rangle f_-^\dagger\right)\right],\\ J_2^{(s)(1)} &= -\frac{1}{2\tau_s}\left[\left(f^{(1)}\langle f_-^\dagger\rangle - \langle f^{(1)}f_-^\dagger\rangle\right)f_+ - \left(f^{(1)}\langle f_-\rangle - \langle f^{(1)}f_-\rangle\right)f_+^\dagger\right].\end{aligned}$$

10.4.2 *Electron–phonon collision integral*

For small deviations from the equilibrium distribution function and for $\omega \ll \Delta$, we can consider only the terms linear in $f^{(2)}$. The electron–phonon self-energies have the form of eqns (9.12) and (9.13). One obtains for the collision integrals:

$$\begin{aligned}J_1^{(\mathrm{ph})} &= -\frac{\pi\lambda}{16(sp_F)^2}\int\frac{d\epsilon'}{2}(\epsilon'-\epsilon)^2\mathrm{sign}(\epsilon'-\epsilon)\left[2(g_{\epsilon'}^R - g_{\epsilon'}^A)(g_\epsilon^R - g_\epsilon^A)\right.\\ &\quad\left.-(f_{\epsilon'}^R - f_{\epsilon'}^A)(f_\epsilon^{\dagger R} - f_\epsilon^{\dagger A}) - (f_\epsilon^R - f_\epsilon^A)(f_{\epsilon'}^{\dagger R} - f_{\epsilon'}^{\dagger A})\right]\\ &\quad\times\left[\coth\left(\frac{\epsilon'-\epsilon}{2T}\right)(f_\epsilon^{(1)} - f_{\epsilon'}^{(1)}) - f_\epsilon^{(1)}f_{\epsilon'}^{(1)} + 1\right].\end{aligned} \tag{10.80}$$

Here $\lambda = \nu(0)\,g_{\mathrm{ph}}^2$ is the electron–phonon interaction constant. Terms linear in $f^{(2)}$ do not appear in $J_1^{(\mathrm{ph})}$. The second collision integral contains both $f^{(1)}$ and $f^{(2)}$. It has a form $J_2^{(\mathrm{ph})} = J_2^{(\mathrm{ph})(1)} + J_2^{(\mathrm{ph})(2)}$. The first part is

$$\begin{aligned}J_2^{(\mathrm{ph})(2)} &= -\frac{\pi\lambda}{16(sp_F)^2}\int\frac{d\epsilon'}{2}(\epsilon'-\epsilon)^2\mathrm{sign}(\epsilon'-\epsilon)\left[2(g_{\epsilon'}^R - g_{\epsilon'}^A)(g_\epsilon^R - g_\epsilon^A)\right.\\ &\quad\left.-(f_{\epsilon'}^R +_{\epsilon'}^A)(f_\epsilon^{\dagger R} + f_\epsilon^{\dagger A}) - (f_\epsilon^R + f_\epsilon^A)(f_{\epsilon'}^{\dagger R} + f_{\epsilon'}^{\dagger A})\right]\end{aligned}$$

$$\times \left[\coth\left(\frac{\epsilon' - \epsilon}{2T}\right)(f_\epsilon^{(2)} - f_{\epsilon'}^{(2)}) - f_\epsilon^{(1)} f_{\epsilon'}^{(2)} - f_\epsilon^{(2)} f_{\epsilon'}^{(1)}\right]. \quad (10.81)$$

The function $f^{(1)}$ here should be considered equilibrium: $f^{(1)} = \tanh(\epsilon/2T)$. The part $J_2^{(\mathrm{ph})(1)}$ only depends on $f^{(1)}$

$$\begin{aligned} J_2^{(\mathrm{ph})(1)} &= -\frac{\pi\lambda}{16(sp_F)^2}\int \frac{d\epsilon'}{2}(\epsilon' - \epsilon)^2 \mathrm{sign}(\epsilon' - \epsilon) \\ &\times \left[(f_{\epsilon'}^R - f_{\epsilon'}^A)(f_\epsilon^{\dagger R} + f_\epsilon^{\dagger A}) - (f_\epsilon^R + f_\epsilon^A)(f_{\epsilon'}^{\dagger R} - f_{\epsilon'}^{\dagger A})\right] \\ &\times \left[\coth\left(\frac{\epsilon' - \epsilon}{2T}\right)(f_\epsilon^{(1)} - f_{\epsilon'}^{(1)}) - f_\epsilon^{(1)} f_{\epsilon'}^{(1)} + 1\right]. \end{aligned} \quad (10.82)$$

The functions $\check{g}_\epsilon(\mathbf{p}, \mathbf{r})$ and $\check{g}_{\epsilon'}(\mathbf{p}, \mathbf{r})$ in eqns (10.80)–(10.82) have the same momentum since the relative change in $\mathbf{p}$ during interaction with the acoustic phonons is small.

We note that the collision integrals $J_1^{(\mathrm{ph})}$ and $J_2^{(\mathrm{ph})(1)}$ vanish for the equilibrium $f^{(1)} = \tanh(\epsilon/2T)$. In turn, the collision integral $J_2^{(\mathrm{ph})(2)}$ vanishes for any function of the form of $f^{(2)} = const\, \left(df^{(0)}/d\epsilon\right)$; the latter corresponds to the invariance with respect to variations in the chemical potential.

10.4.2.1 *Simplifications for $T \to T_c$* Near the critical temperature when $\Delta \ll T$, one usually needs the distribution functions for energies $\epsilon \sim \Delta$, i.e., for $\epsilon \ll T$, while the deviation from equilibrium $f_1 \equiv f^{(1)} - f^{(0)}$ is small for $\epsilon' \sim T$. The phonon collision integral can be simplified considerably in this case.

For $\Delta \ll T$, and for $\epsilon \ll T$, $\epsilon' \sim T$ we have $g_{\epsilon'}^{R(A)} \approx \pm 1$ and $f_{\epsilon'}^{R(A)} \ll 1$. As a result

$$\begin{aligned} J_1^{(\mathrm{ph})} &= -\frac{\pi\lambda}{4(sp_F)^2}\int \frac{d\epsilon'}{2}(\epsilon' - \epsilon)^2 \mathrm{sign}(\epsilon' - \epsilon) \\ &\times (g_\epsilon^R - g_\epsilon^A)\left[\coth\left(\frac{\epsilon' - \epsilon}{2T}\right)(f_\epsilon^{(1)} - f_{\epsilon'}^{(1)}) - f_\epsilon^{(1)} f_{\epsilon'}^{(1)} + 1\right] \\ &= -\frac{1}{\tau_{\mathrm{ph}}} g_- f_1. \end{aligned} \quad (10.83)$$

where the electron–phonon mean free time is

$$\frac{1}{\tau_{\mathrm{ph}}} = \frac{\pi\lambda}{(sp_F)^2}\int_0^\infty \frac{\epsilon^2\, d\epsilon}{\sinh(\epsilon/T)} = \frac{7\zeta(3)\,\pi\lambda T^3}{2(sp_F)^2}. \quad (10.84)$$

The second collision integral becomes

$$\begin{aligned} J_2^{(\mathrm{ph})} &= -\frac{\pi\lambda}{4(sp_F)^2}\int \frac{d\epsilon'}{2}(\epsilon' - \epsilon)^2 \mathrm{sign}(\epsilon' - \epsilon) \\ &\times (g_\epsilon^R - g_\epsilon^A)\left[\coth\left(\frac{\epsilon' - \epsilon}{2T}\right)\left(f_\epsilon^{(2)} - f_{\epsilon'}^{(2)}\right) - f_\epsilon^{(2)} f_{\epsilon'}^{(1)} - f_\epsilon^{(1)} f_{\epsilon'}^{(2)}\right] \end{aligned}$$

$$= -\frac{1}{\tau_{\rm ph}} g_- \frac{df^{(0)}}{d\epsilon} \left[\chi_\epsilon^{(2)} - \chi_{\epsilon'}^{(2)} \right]. \quad (10.85)$$

Here

$$f_\epsilon^{(2)} = \chi_\epsilon^{(2)} \frac{df^{(0)}}{d\epsilon}. \quad (10.86)$$

The electron–phonon relaxation rate, eqn (10.84), is $\tau_{\rm ph}^{-1} \sim \lambda T^3/\Omega_D^2$. Usually the Debye frequency is higher than the critical temperature, $T_c \ll \Omega_D$ and also $\lambda \lesssim 1$. The electron–phonon relaxation rate is thus small compared to the order parameter for temperatures which are not very close to T_c: $\tau_{\rm ph}^{-1} \ll \Delta(T)$. High-temperature superconductors can be exceptions because the critical temperature is roughly of the same order as Ω_D. Nevertheless, experimentally, the inequality $\tau_{\rm ph}^{-1} \ll \Delta(T)$ is believed to hold. This should be especially correct for *d*-wave superconductors where it is crucial for the very existence of *d*-wave superconductivity as we have pointed out in Section 6.3.

10.4.3 *Electron–electron collision integral*

Self-energies for the electron–electron collisions can be obtained from the expressions established (Eliashberg 1971) for the electron–electron interaction in Section 8.3. The transformation to the quasiclassical functions is made according to the general rules described earlier. The only difference is that the self-energy contains three G functions and only two integrations over the momenta. However, we can put

$$\check{G} = \pi i \check{g} \delta(\xi_{\mathbf{p}})$$

because we are only interested in energies close to the Fermi surface. We obtain

$$\Sigma_1^R - \Sigma_1^A = \frac{\nu^2 \pi i}{16 p_F v_F} \int d\epsilon_1\, d\epsilon_2 \int \frac{d\Omega_{\mathbf{p}_1}\, d\Omega_{\mathbf{p}_2}}{(4\pi)^2} \delta\left(\frac{p_3}{p_F} - 1\right) \times \left[A\{g_1 g_2 \bar{g}_3\}^{(r)} - B\{f_1 f_2^+ g_3\}^{(r)} \right], \quad (10.87)$$

$$\Sigma_2^R - \Sigma_2^A = \frac{\nu^2 \pi i}{16 p_F v_F} \int d\epsilon_1\, d\epsilon_2 \int \frac{d\Omega_{\mathbf{p}_1}\, d\Omega_{\mathbf{p}_2}}{(4\pi)^2} \delta\left(\frac{p_3}{p_F} - 1\right) \times \left[B\{g_1 \bar{g}_2 f_3\}^{(r)} - A\{f_1 f_2 f_3^+\}^{(r)} \right], \quad (10.88)$$

and

$$\Sigma_1 = \frac{\nu^2 \pi i}{16 p_F v_F} \int d\epsilon_1\, d\epsilon_2 \int \frac{d\Omega_{\mathbf{p}_1}\, d\Omega_{\mathbf{p}_2}}{(4\pi)^2} \delta\left(\frac{p_3}{p_F} - 1\right) \times \left[A\{g_1 g_2 \bar{g}_3\}^{(t)} - B\{f_1 f_2^+ g_3\}^{(t)} \right], \quad (10.89)$$

$$\Sigma_2 = \frac{\nu^2 \pi i}{16 p_F v_F} \int d\epsilon_1\, d\epsilon_2 \int \frac{d\Omega_{\mathbf{p}_1}\, d\Omega_{\mathbf{p}_2}}{(4\pi)^2} \delta\left(\frac{p_3}{p_F} - 1\right) \times \left[B\{g_1 \bar{g}_2 f_3\}^{(t)} - A\{f_1 f_2 f_3^+\}^{(t)} \right]. \quad (10.90)$$

Here A and B are the matrix elements of the interaction potential; different signs appear because of different spin structure of the matrix elements between the states $\psi^\dagger$ and ψ as compared to those between the states ψ and ψ. The curly brackets are defined as

$$\begin{aligned}\{g_1 g_2 \bar{g}_3\}^{(r)} &= g_1^K g_2^K (\bar{g}_3^R - \bar{g}_3^A) + g_1^K (g_2^R - g_2^A)\bar{g}_3^K + (g_1^R - g_1^A) g_2^K \bar{g}_3^K \\ &+ (g_1^R - g_1^A)(g_2^R - g_2^A)(\bar{g}_3^R - \bar{g}_3^A), \\ \{g_1 g_2 \bar{g}_3\}^{(t)} &= g_1^K g_2^K \bar{g}_3^K + g_1^K (g_2^R - g_2^A)(\bar{g}_3^R - \bar{g}_3^A) \\ &+ (g_1^R - g_1^A) g_2^K (\bar{g}_3^R - \bar{g}_3^A) \\ &+ (g_1^R - g_1^A)(g_2^R - g_2^A)\bar{g}_3^K .\end{aligned}$$

The energies and momenta satisfy the conservation laws:

$$\epsilon = \epsilon_1 + \epsilon_2 + \epsilon_3; \ \mathbf{p} = \mathbf{p}_1 + \mathbf{p}_2 + \mathbf{p}_3. \tag{10.91}$$

For a point interaction $V(\mathbf{r}) = V\delta(\mathbf{r})$ one has

$$A = -|V|^2; \ B = |V|^2.$$

The electron–electron interaction constant is defined as $\lambda_{ee} = \nu V$. We introduce the operator

$$\hat{L}[\ldots] = \frac{\lambda_{ee}^2 \pi}{2 p_F v_F} \int d\epsilon_1 \, d\epsilon_2 \int \frac{d\Omega_{\mathbf{p}_1} \, d\Omega_{\mathbf{p}_2}}{(4\pi)^2} \, \delta\left(\frac{p_3}{p_F} - 1\right)[\ldots]. \tag{10.92}$$

Self-energies become

$$\Sigma_1^R - \Sigma_1^A = -\frac{i}{8}\hat{L}\left[\{g_1 g_2 \bar{g}_3\}^{(r)} + \{f_1 f_2^+ g_3\}^{(r)}\right], \tag{10.93}$$

$$\Sigma_2^R - \Sigma_2^A = \frac{i}{8}\hat{L}\left[\{g_1 \bar{g}_2 f_3\}^{(r)} + \{f_1 f_2 f_3^+\}^{(r)}\right], \tag{10.94}$$

$$\Sigma_1 = -\frac{i}{8}\hat{L}\left[\{g_1 g_2 \bar{g}_3\}^{(t)} + \{f_1 f_2^+ g_3\}^{(t)}\right], \tag{10.95}$$

$$\Sigma_2 = \frac{i}{8}\hat{L}\left[\{g_1 \bar{g}_2 f_3\}^{(t)} + \{f_1 f_2 f_3^+\}^{(t)}\right]. \tag{10.96}$$

We write down only one electron–electron collision integral

$$J_1^{(\mathrm{e-e})} = J_1^{(1)}\{f^{(1)}\} + J_1^{(2)}\{f^{(2)}\}.$$

To calculate $J_1^{(1)}\{f^{(1)}\}$ we can put $f^{(2)} = 0$. Using that Tr $\hat{I}^{R(A)} = 0$ we obtain

$$J_1^{(1)}\{f^{(1)}\} = -\hat{L}\left[M_1 g_-(\epsilon, \mathbf{p}) g_-(\epsilon_1, \mathbf{p}_1) g_-(\epsilon_2, \mathbf{p}_2) g_-(\epsilon_3, \mathbf{p}_3) F(\epsilon, \epsilon_1, \epsilon_2, \epsilon_3)\right] \tag{10.97}$$

where

$$M_1 = 1 - \frac{f_-(1) f_-^+(2) + f_-^+(1) f_-(2)}{2 g_-(1) g_-(2)} + \frac{f_-(0) f_-^+(3) + f_-^+(0) f_-(3)}{2 g_-(0) g_-(3)}$$

$$-\frac{f_-(1)f_-(2)f_-^+(3)f_-^+(0)+f_-^+(1)f_-^+(2)f_-(3)f_-(0)}{2g_-(1)g_-(2)g_-(3)g_-(0)}$$

and

$$\begin{aligned}F(\epsilon,\epsilon_1,\epsilon_2,\epsilon_3) = {} & f^{(1)}(0)f^{(1)}(1)f^{(1)}(2)+f^{(1)}(0)f^{(1)}(1)f^{(1)}(3)\\ & +f^{(1)}(0)f^{(1)}(2)f^{(1)}(3)-f^{(1)}(1)f^{(1)}(2)f^{(1)}(3)\\ & +f^{(1)}(0)-f^{(1)}(1)-f^{(1)}(2)-f^{(1)}(3).\end{aligned}$$

Here the index 0 refers to ϵ, $\mathbf{p}$, while the index 1 is for ϵ_1, $\mathbf{p}_1$, etc.

The equilibrium distribution function $f^{(0)}=\tanh(\epsilon/2T)$ turns the function $F(\epsilon,\epsilon_1,\epsilon_2,\epsilon_3)$ to zero. If we put now

$$f^{(1)}=f^{(0)}+\frac{\partial f^{(0)}}{\partial\epsilon}\chi$$

where χ is a small correction, we obtain

$$\begin{aligned}F = {} & \left(1+f^{(0)}(1)f^{(0)}(2)+f^{(0)}(1)f^{(0)}(3)+f^{(0)}(2)f^{(0)}(3)\right)\\ & \times\frac{\partial f^{(0)}}{\partial\epsilon}\left(\chi(0)-\chi(1)-\chi(2)-\chi(3)\right)\\ = {} & \frac{1}{2T}\,\frac{[\chi(0)-\chi(1)-\chi(2)-\chi(3)]}{\cosh(\epsilon/2T)\cosh(\epsilon_1/2T)\cosh(\epsilon_2/2T)\cosh(\epsilon_3/2T)}.\end{aligned}$$

10.5 Kinetic equations for dirty s-wave superconductors

To make an example, we apply our kinetic equations to dirty s-wave superconductors such that the mean free time satisfies the condition

$$\tau\Delta(T)\ll 1.$$

The starting equations have the form of eqns (10.55) and (10.44). We write them again

$$\left(e\mathbf{v}_F\mathbf{E}g_-+\frac{1}{2}\left[f_-\frac{\partial\hat{\Delta}^*}{\partial t}+f_-^\dagger\frac{\partial\hat{\Delta}}{\partial t}\right]\right)\frac{\partial f^{(0)}}{\partial\epsilon}+\mathbf{v}_F\nabla(f_2g_-)=J_1. \tag{10.98}$$

$$g_-\mathbf{v}_F\nabla f_1+\frac{1}{2}\left[f_+\frac{\partial\hat{\Delta}^*}{\partial t}-f_+^\dagger\frac{\partial\hat{\Delta}}{\partial t}\right]\frac{\partial f^{(0)}}{\partial\epsilon}+i\left[\Delta^*f_++\Delta f_+^\dagger\right]f_2=J_2\,. \tag{10.99}$$

The collision integrals are given by eqns (10.74)–(10.75) for scattering by impurities, and by eqns (10.80)–(10.82) for electron–phonon relaxation.

We use the method employed earlier in Sections 5.6 and 9.3 for superconductors with a short mean free path. Let us first average eqn (10.98) over the directions of $\mathbf{v}_F$. The nonmagnetic impurity collision integral vanishes and we obtain

$$\left(\frac{1}{3}ev_F\left(\mathbf{E}\cdot\mathbf{g}_-\right)+\frac{1}{2}\left[\langle f_-\rangle\frac{\hat{\partial}\Delta^*}{\partial t}+\left\langle f_-^\dagger\right\rangle\frac{\hat{\partial}\Delta}{\partial t}\right]\right)\frac{\partial f^{(0)}}{\partial\epsilon}$$
$$+\nabla\cdot\langle\mathbf{v}_F\left(f_2 g_-\right)\rangle=\left\langle J_1^{(\mathrm{ph})}\right\rangle+\left\langle J_1^{(\mathrm{s})}\right\rangle. \tag{10.100}$$

This equation contains the vector component of $\mathbf{f}_2$ in the left-hand side and the average component of f_1 in the right-hand side. Now we multiply eqn (10.99) with $\mathbf{v}_F$ and average it:

$$v_F^2\langle g_-\rangle\nabla f_1=3\left\langle\mathbf{v}_F J_2^{(\mathrm{imp})}\right\rangle. \tag{10.101}$$

Here we keep the largest collision term due to scattering by nonmagnetic impurities. The vector components of regular functions in eqn (10.101) are neglected as compared to the averaged components. This equation also contains the averaged f_1 and the vector $\mathbf{f}_2$.

The second set of equations for vector $\mathbf{f}_1$ and averaged f_2 components is obtained as follows. First, we average eqn (10.99):

$$\langle g_-\rangle\frac{v_F}{3}\nabla\mathbf{f}_1+\frac{1}{2}\left[\langle f_+\rangle\frac{\hat{\partial}\Delta^*}{\partial t}-\left\langle f_+^\dagger\right\rangle\frac{\hat{\partial}\Delta}{\partial t}\right]\frac{\partial f^{(0)}}{\partial\epsilon}$$
$$+i\left[\Delta^*\langle f_+\rangle+\Delta\left\langle f_+^\dagger\right\rangle\right]f_2=\left\langle J_2^{(\mathrm{ph})}\right\rangle+\left\langle J_2^{(\mathrm{s})}\right\rangle. \tag{10.102}$$

Next, we multiply eqn (10.98) with $\mathbf{v}_F$ and average it:

$$ev_F^2\mathbf{E}\langle g_-\rangle\frac{\partial f^{(0)}}{\partial\epsilon}+v_F^2\nabla\langle f_2 g_-\rangle=3\left\langle\mathbf{v}_F J_1^{(\mathrm{imp})}\right\rangle. \tag{10.103}$$

The four equations (10.100), (10.101), (10.102), and (10.103) determine the four functions f_1, f_2, $\mathbf{f}_1$, and $\mathbf{f}_2$.

10.5.1 *Small gradients without magnetic impurities*

The regular functions are determined by the Usadel equations (5.98), (5.99). Assume that the order parameter gradients are small $Dk^2 \ll \Delta$. We also assume that the electron–phonon relaxation is slow such that $\tau_{\mathrm{ph}}^{-1} \ll \Delta$ which is usually the case for temperatures not extremely close to T_c. The Usadel equations give the expressions for the Green functions as in a homogeneous case

$$\langle g_-\rangle=\frac{\epsilon}{\sqrt{\epsilon^2-|\Delta|^2}}\Theta\left(\epsilon^2-|\Delta|^2\right),$$
$$\langle f_-\rangle=\left\langle f_-^\dagger\right\rangle^*=\frac{\Delta}{\sqrt{\epsilon^2-|\Delta|^2}}\Theta\left(\epsilon^2-|\Delta|^2\right), \tag{10.104}$$

and

$$\langle g_+ \rangle = \frac{-i\epsilon}{\sqrt{|\Delta|^2 - \epsilon^2}} \Theta\left(|\Delta|^2 - \epsilon^2\right),$$

$$\langle f_+ \rangle = \left\langle f_+^\dagger \right\rangle^* = \frac{-i\Delta}{\sqrt{|\Delta|^2 - \epsilon^2}} \Theta\left(|\Delta|^2 - \epsilon^2\right), \qquad (10.105)$$

in the leading approximation in small gradients. As a result, we have from eqn (10.75) for $\epsilon^2 > |\Delta|^2$

$$\left\langle \mathbf{v}_F J_2^{(\mathrm{imp})} \right\rangle = -\frac{1}{\tau_{\mathrm{imp}}} \langle \mathbf{v}_F (g_- f_2) \rangle \langle g_- \rangle + \frac{v_F}{3\tau_{\mathrm{imp}}} \mathbf{g}_- \langle g_- f_2 \rangle .$$

We can neglect the second term in the r.h.s. since the vector part of the Green function is small compared to the averaged one. Equation (10.101) gives

$$\langle \mathbf{v}_F (g_- f_2) \rangle = -D\nabla f_1. \qquad (10.106)$$

Using eqn (10.100) we find

$$D\nabla^2 f_1 + \left\langle J_1^{(\mathrm{ph})} \right\rangle = \frac{1}{2} \left[\langle f_- \rangle \frac{\hat{\partial}\Delta^*}{\partial t} + \left\langle f_-^\dagger \right\rangle \frac{\hat{\partial}\Delta}{\partial t} \right] \frac{\partial f^{(0)}}{\partial \epsilon}. \qquad (10.107)$$

The vector part of the Green function is small compared to the averaged component and can be neglected.

Equation (10.102) should be used together with eqn (10.103) which determines the vector part $\mathbf{f}_1$. The impurity collision integral $J_1^{(2)}$ vanishes because $f_+ f_- = 0$ for slow gradients. The part $J_1^{(1)}$ gives

$$\mathbf{f}_1 = -e\ell \langle g_- \rangle \mathbf{E} \frac{\partial f^{(0)}}{\partial \epsilon} - \ell \nabla \langle g_- f_2 \rangle . \qquad (10.108)$$

The function f_2 is obtained from eqn (10.102) where one inserts $\mathbf{f}_1$ from eqn (10.108). Note that the Green functions $g_\pm$ and $f_\pm$ in eqn (10.102) should be taken with the proper account of the phonon scattering rate since the last term in the l.h.s. is proportional to τ_{ph}^{-1}. The so-obtained function f_2 should then be used in the charge neutrality condition to find the equation which determines the scalar potential φ.

10.5.2 *Heat conduction*

Equation (10.106) allows one to calculate the thermal conductivity. Using the definition of the internal energy current, eqn (10.66), we have

$$\mathbf{j}_{\mathcal{E}} = -\nu(0) \int \epsilon \mathbf{v}_F \left(g^K + \bar{g}^K\right) \frac{d\epsilon}{4} \frac{d\Omega_{\mathbf{p}}}{4\pi} = -\nu(0) \int \epsilon \langle \mathbf{v}_F (g_- f_2) \rangle \, d\epsilon.$$

We find from eqn (10.106)

$$\mathbf{j}_{\mathcal{E}} = 2\nu\,(0)\,D\int_{\Delta}^{\infty} \epsilon\nabla f^{(1)}\,d\epsilon.$$

Let us assume that the distribution function is $f^{(1)} = f^{(0)} = \tanh(\epsilon/2T)$ where T is a slow function of coordinates. The gradient becomes

$$\nabla f_1 = -\frac{df^{(0)}}{d\epsilon}\frac{\epsilon}{T}\nabla T.$$

The heat current takes the form $\mathbf{j}_{\mathcal{E}} = -\kappa\nabla T$ where the thermal conductivity is

$$\kappa = 2\nu\,(0)\,D\int_{\Delta}^{\infty} \frac{df^{(0)}}{d\epsilon}\frac{\epsilon^2}{T}\,d\epsilon. \tag{10.109}$$

This expression for the thermal conductivity was first derived by Bardeen *et al.* (1959).

11

THE TIME-DEPENDENT GINZBURG–LANDAU THEORY

We find out the conditions when the time-dependent Ginzburg–Landau model can be justified microscopically. We demonstrate that the TDGL model is exact for gapless superconductors. It is not justified, however, for systems with a finite energy gap. The role of nonequilibrium excitations is elucidated in the dynamics of superconductors. The charge imbalance and decay of a d.c. electric field in a superconductor is discussed.

11.1 Gapless superconductors with magnetic impurities

We already discussed the time-dependent Ginzburg–Landau (TDGL) model which generalizes the usual Ginzburg–Landau (GL) theory and provides a simple description of nonstationary processes in superconductors (see Section 1.2.1). Consider now how and under which conditions, the microscopic theory can justify the phenomenological TDGL model.

We start with the simplest example that requires a minimum of calculations if we use the preparatory work done in the previous chapters. This is the case of dirty superconductors which have a large concentration of magnetic impurities such that $\tau_s T_c \ll 1$ (Gor'kov and Eliashberg 1968). Note that this is exactly the condition of gapless superconductivity. In this case also $\tau_s \Delta \ll 1$. The regular functions are found from eqn (6.45):

$$g^R = -g^A = 1,\ f^R = \frac{\Delta}{\epsilon + i/\tau_s},\ f^A = -\frac{\Delta}{\epsilon - i/\tau_s}, \tag{11.1}$$

within the first-order approximation in Δ and the zero-order approximation in spatial gradients. The characteristic energy scale in eqn (11.1) is $\epsilon \sim \tau_s^{-1}$. Therefore, the external frequency $\omega \sim \Delta$ is small compared to the energy scale of the regular Green functions. One can thus safely use the kinetic equations obtained by expansion in small ω. The collision integrals for magnetic scattering in eqns (10.100), (10.102) become

$$J_1^{(s)(1)} = -\frac{1}{\tau_s}\left(f_1 - \langle f_1 \rangle\right),\ J_1^{(s)(2)} = -\frac{1}{\tau_s}\left(f_2 - \langle f_2 \rangle\right).$$

All other terms are small in $\tau_s^2 \Delta^2$.

Kinetic equations (10.100), (10.102) become

$$f_- \frac{\partial \Delta}{\partial t}\frac{\partial f^{(0)}}{\partial \epsilon} = J_1^{(s)}, \tag{11.2}$$

and

$$-i\frac{\partial f^{(0)}}{\partial \epsilon}\left(\frac{\partial \chi}{\partial t}+2e\varphi\right)\Delta f_{+}+2i\Delta f_{+}f_{2}=J_{2}^{(s)}. \tag{11.3}$$

The phonon collision integrals are small as compared to the magnetic scattering and can be neglected. We see that the corrections f_1 and f_2 to the equilibrium $f^{(0)}$ are proportional to $\tau_s^2\Delta^2$ and are thus small as compared to the terms with $f^{(0)}$ in the Keldysh function eqn (10.62). Therefore

$$f^{K}_{\epsilon_+,\epsilon_-}=\left(f^{R}_{\epsilon_+,\epsilon_-}-f^{A}_{\epsilon_+,\epsilon_-}\right)f^{(0)}_{\epsilon}-\frac{i}{2}\left(\frac{\partial}{\partial t}+2ie\phi\right)\left(f^{R}_{\epsilon}+f^{A}_{\epsilon}\right)\frac{\partial f^{(0)}}{\partial \epsilon}. \tag{11.4}$$

As a result, the equation for the order parameter takes the form

$$\begin{aligned}\frac{\Delta(\mathbf{k},\omega)}{\lambda}=\int\frac{d\Omega_{\mathbf{p}}}{4\pi}\frac{d\epsilon}{4}&\left[\left(f^{R}_{\epsilon_+,\epsilon_-}-f^{A}_{\epsilon_+,\epsilon_-}\right)f^{(0)}_{\epsilon}\right.\\&\left.-\frac{i}{2}\left(\frac{\partial}{\partial t}+2ie\varphi\right)\left(f^{R}_{\epsilon}+f^{A}_{\epsilon}\right)\frac{\partial f^{(0)}}{\partial \epsilon}\right].\end{aligned} \tag{11.5}$$

It is instructive to compare this equation with eqn (1.85). These two equations coincide if one replaces the imaginary part i/τ_s in eqn (11.1) for the functions $f^{R(A)}$ with an infinitely small $i\delta$. However, the expression (11.4) for the Keldysh function which has resulted in eqn (11.5) is not exact; it requires that the deviations from equilibrium $f_{1,2}$ are small. It is the gapless condition $\Delta\tau_s \ll 1$ that makes $f_{1,2}$ small. We thus confirm the conclusion of Section 1.2.2 that the TDGL scheme works only for gapless superconductors when deviations from equilibrium are negligible.

Equation (11.5) yields

$$\begin{aligned}\frac{\Delta(\mathbf{k},\omega)}{\lambda}=&\int\frac{d\Omega_{\mathbf{p}}}{4\pi}\frac{d\epsilon}{4}\left(f^{R}_{\epsilon}-f^{A}_{\epsilon}\right)f^{(0)}_{\epsilon}\\&-\left(\frac{\partial}{\partial t}+2ie\varphi\right)\Delta\int\frac{d\epsilon}{4\tau_s\left(\epsilon^2+1/\tau_s^2\right)}\frac{\partial f^{(0)}}{\partial \epsilon}.\end{aligned} \tag{11.6}$$

Since $T\ll 1/\tau_s$, the derivative of the distribution function becomes $\partial f^{(0)}/\partial\epsilon = 2\delta(\epsilon)$. The last term in eqn (11.6) gives

$$\left(\frac{\partial}{\partial t}+2ie\varphi\right)\Delta\int\frac{d\epsilon}{4\tau_s\left(\epsilon^2+1/\tau_s^2\right)}\frac{\partial f^{(0)}}{\partial \epsilon}=\frac{\tau_s}{2}\left(\frac{\partial}{\partial t}+2ie\varphi\right)\Delta.$$

The first term in the r.h.s. of eqn (11.6) gives the stationary part of the Ginzburg–Landau equation (Abrikosov and Gor'kov 1960). We have from eqn (11.6)

$$\left(\frac{\partial}{\partial t}+2ie\varphi\right)\Delta=\frac{\tau_s}{3}\left[\pi^2\left(T_c^2-T^2\right)-\frac{|\Delta|^2}{2}\right]\Delta+D\left(\nabla-\frac{2ie}{c}\mathbf{A}\right)^2\Delta. \tag{11.7}$$

The expression for the current is obtained in exactly the same way as before. We have for the vector part of the distribution function from eqn (10.103)

$$\mathbf{f}_1 = -ev_F\tau\frac{\partial f^{(0)}}{\partial\epsilon}\mathbf{E}.$$

Thus the normal current is $\mathbf{j}_n = \sigma_n\mathbf{E}$. The total current becomes

$$\mathbf{j} = \sigma_n\mathbf{E} - \frac{2\sigma_n\tau_s|\Delta|^2}{c}\mathbf{Q}. \tag{11.8}$$

One can easily see that the system of equations (11.7) and (11.8) can be written in the form of general equations of relaxation dynamics, eqns (1.65), (1.67) used to derive the TDGL model in Section 1.2.1 with the superconducting free energy

$$\mathcal{F}_{sn} = \nu(0)\int\left(\frac{\tau_s^2}{6}\left[\pi^2\left(T^2 - T_c^2\right)|\Delta|^2 + \frac{|\Delta|^4}{4}\right]\right.$$
$$\left. + \frac{D\tau_s}{2}\left|\left(\nabla - \frac{2ie}{c}\mathbf{A}\right)\Delta\right|^2\right)dV.$$

The order parameter relaxation constant as defined in Section 1.2.1 is

$$\Gamma = \nu(0)\,\tau_s/2.$$

The relaxation times are

$$\tau_j = \frac{1}{4\pi^2\tau_s\left(T_c^2 - T^2\right)},\quad \tau_\Delta = \frac{3}{\pi^2\tau_s\left(T_c^2 - T^2\right)}.$$

Their ratio $u = 12$.

We conclude therefore that gapless superconductors with large concentration of magnetic impurities can be exactly described by the TDGL model. The reason is that the distribution of excitations is very close to equilibrium. It is the small parameter $\Delta\tau_s$ which makes the deviation from equilibrium negligible. We shall see in the next example, that the TDGL model starts to fail as the deviation from equilibrium increases.

11.2 Generalized TDGL equations

The example which we discuss here is more interesting and more realistic in a sense that it does not require strong limitations necessary for gapless superconductivity. Of course, we need some restrictions to simplify the general kinetic equations. Consider a dirty superconducting alloy with the mean free path $\ell \ll \xi_0$ or $T_c\tau_{\text{imp}} \ll 1$. We assume that the temperature is close to T_c. Moreover, we assume that variations in space and time are slow:

$$\omega, Dk^2 \ll \tau_{\text{ph}}^{-1}. \tag{11.9}$$

In our example, the inelastic scattering by phonons is of major importance. Since the largest wave vector is $k^2 \sim \xi^{-2}(T)$ and $D\xi^{-2}(T) \propto \Delta^2/T_c$, the condition of slow spatial variations is certainly satisfied if $T_c - T \ll \tau_{\rm ph}^{-1}$. The gradients can satisfy eqn (11.9) also because the characteristic scale of the particular problem may happen to be large compared to $\xi(T)$ for one or another reason; the condition $T_c - T \ll \tau_{\rm ph}^{-1}$ can be lifted in this case. As a reward to the restriction of eqn (11.9) we obtain a possibility to derive a closed set of time-dependent equations for the order parameter magnitude and phase which resemble the TDGL model. On the other hand, they are more general and allow us to follow the crossover from the simple TDGL scheme to a more realistic dynamics for systems with strong deviations from equilibrium. The TDGL-like equations in this case were derived by Watts-Tobin *et al.* (1981).

Consider first the regular Green functions. Within the leading approximation in spatial gradients and in the external frequency, one has from Eilenberger equation (5.84)

$$-\epsilon f^{R(A)} + \Delta g^{R(A)} = \Sigma_1^{R(A)} f^{R(A)} - g^{R(A)} \Sigma_2^{R(A)}. \tag{11.10}$$

The phonon self-energies are

$$\Sigma_1^{R(A)} = \frac{i\pi\lambda}{4\,(sp_F)^2} \int_{-\infty}^{\infty} (\epsilon' - \epsilon)^2 \,\text{sign}\,(\epsilon' - \epsilon)\, \frac{d\epsilon'}{2} \times \left[\coth\left(\frac{\epsilon' - \epsilon}{2T}\right) g_{\epsilon'}^{R(A)} \mp \frac{1}{2} f_{\epsilon'}^{(0)} \left(g_{\epsilon'}^{R} - g_{\epsilon'}^{A}\right)\right], \tag{11.11}$$

and

$$\Sigma_2^{R(A)} = \frac{i\pi\lambda}{4\,(sp_F)^2} \int_{-\infty}^{\infty} (\epsilon' - \epsilon)^2 \,\text{sign}\,(\epsilon' - \epsilon)\, \frac{d\epsilon'}{2} \times \left[\coth\left(\frac{\epsilon' - \epsilon}{2T}\right) f_{\epsilon'}^{R(A)} \mp \frac{1}{2} f_{\epsilon'}^{(0)} \left(f_{\epsilon'}^{R} - f_{\epsilon'}^{A}\right)\right]. \tag{11.12}$$

Energies $\epsilon' \sim T$ dominate in the integrals in eqns (11.11) and (11.12). At these energies, the functions $g^{R(A)} \approx \pm 1$ and $f^{R(A)} \sim \Delta/T \ll 1$, therefore,

$$\Sigma_1^{R(A)} = \pm \frac{i}{2\tau_{\rm ph}}$$

and $\Sigma_2 \sim (\Delta/T_c)\,\Sigma_1 \ll \Sigma_1$, where the electron–phonon mean free time $\tau_{\rm ph}$ is defined by eqn (10.84).

Equation (11.10) gives

$$\Delta g^{R(A)} - \left(\epsilon \pm \frac{i}{2\tau_{\rm ph}}\right) f^{R(A)} = 0.$$

Using the normalization $\left[g^{R(A)}\right]^2 - f^{R(A)} f^{\dagger R(A)} = 1$ we obtain

$$g^{R(A)} = \pm \frac{\epsilon \pm i/2\tau_{\rm ph}}{\sqrt{(\epsilon \pm i/2\tau_{\rm ph})^2 - \Delta^2}}, \quad f^{R(A)} = \pm \frac{\Delta}{\sqrt{(\epsilon \pm i/2\tau_{\rm ph})^2 - \Delta^2}}. \tag{11.13}$$

We choose Δ to be real for brevity at this point, and put $f^{R(A)} = f^{\dagger R(A)}$ within the leading approximation.

Kinetic equations (10.100) takes the form

$$f_- \frac{\partial \Delta}{\partial t} \frac{\partial f^{(0)}}{\partial \epsilon} = J_1^{(\mathrm{ph})}, \tag{11.14}$$

whence

$$f_1 = -\tau_{\mathrm{ph}} \frac{\partial \Delta}{\partial t} \frac{\partial f^{(0)}}{\partial \epsilon} \frac{f_-}{g_-}. \tag{11.15}$$

Equation (10.102) becomes

$$-i \frac{\partial f^{(0)}}{\partial \epsilon} \left(\frac{\partial \chi}{\partial t} + 2e\varphi \right) \Delta f_+ + 2i \Delta f_+ f_2 = J_2^{(\mathrm{ph})}. \tag{11.16}$$

We can neglect the spatial derivative in eqn (11.16) because it is small as compared to the other terms when $\epsilon \sim \Delta, \tau_{\mathrm{ph}}^{-1}$. Fortunately, it is this range of ϵ which gives the main contribution to the equation for the order parameter. Indeed, for higher energies $\epsilon \sim T$, the spatial gradient could no longer be ignored since the other terms in eqn (11.16) also become small. However, as we shall see in a moment, the function f_2 decreases rapidly as ϵ increases. It becomes small for $\epsilon \sim T$ so that the energy range $\epsilon \sim T$ does not contribute much to the integrals which contain f_2. Equation (11.16) yields, after a little algebra,

$$f_2 = \frac{\partial f^{(0)}}{\partial \epsilon} e\Phi + f_2', \; f_2' = C \frac{\partial f^{(0)}}{\partial \epsilon} \left(1 - \frac{\Delta}{\epsilon} \frac{f_-}{g_-} \right) \tag{11.17}$$

where C is independent of ϵ. In contrast to f_2, the function f_2' is not expected to decay for $\epsilon \sim T$. Equation (11.17) has been derived using the argumentation as follows. We note that the first term in f_2 satisfies eqn (11.16) exactly while f_2' transforms into $C\partial f^{(0)}/\partial \epsilon$ for energies $\epsilon \sim T$ and larger. Since energies $\epsilon' \sim T$ dominate in the collision integral the expression (10.85) for J_2 takes the form

$$J_2 = -\frac{g_-}{\tau_{\mathrm{ph}}} \left[f_2'(\epsilon) - C \frac{\partial f^{(0)}}{\partial \epsilon} \right].$$

Remind that it is assumed $\epsilon \ll T$. Inserting this into eqn (11.16) we arrive at eqn (11.17).

The constant C is to be found from the charge neutrality condition. The change in the electron density (10.71) due to nonequilibrium processes is

$$\delta N^{(\mathrm{nst})} = -\nu(0)\, e\Phi - \nu(0) \int_{-\infty}^{\infty} \langle g_- f_2' \rangle \, d\epsilon. \tag{11.18}$$

The change in the electron density should be zero. The second term in the r.h.s. of eqn (11.18) is determined by $\epsilon \sim T$. For these energies, $g_- \approx 1$. Using eqn (11.17) we find

$$C = -e\Phi.$$

Therefore

$$f_2 = e\Phi \frac{\partial f^{(0)}}{\partial \epsilon} \frac{\Delta}{\epsilon} \frac{f_-}{g_-}. \tag{11.19}$$

The function f_2 decreases with ϵ and becomes small for $\epsilon \gg \Delta$. This behavior is exactly what we expect for f_2.

Consider now the self-consistency equation for the order parameter. We have

$$\frac{\Delta(\mathbf{k}, \omega)}{\lambda} = \int \frac{d\Omega_{\mathbf{p}}}{4\pi} \frac{d\epsilon}{4} f^K_{\epsilon_+, \epsilon_-}.$$

The Keldysh function in eqn (10.62) is

$$f^K_{\epsilon_+, \epsilon_-} = \left(f^R_{\epsilon_+, \epsilon_-} - f^A_{\epsilon_+, \epsilon_-} \right) f^{(0)}_\epsilon + f^{(\mathrm{nst})}_{\epsilon_+, \epsilon_-} \tag{11.20}$$

where the non-stationary part is

$$\begin{aligned} f^{(\mathrm{nst})}_{\epsilon_+, \epsilon_-} &= -\frac{i}{2} \left(\frac{\partial}{\partial t} + 2ie\varphi \right) \left(f^R_\epsilon + f^A_\epsilon \right) \frac{\partial f^{(0)}}{\partial \epsilon} \\ &\quad + \left(f^R_\epsilon - f^A_\epsilon \right) f_1 - \left(f^R_\epsilon + f^A_\epsilon \right) f_2. \end{aligned} \tag{11.21}$$

For the energy integration, we put

$$x = \frac{1}{2} \left[\sqrt{\left(\epsilon + \frac{i}{2\tau_{\mathrm{ph}}} \right)^2 - \Delta^2} + \sqrt{\left(\epsilon - \frac{i}{2\tau_{\mathrm{ph}}} \right)^2 - \Delta^2} \right]. \tag{11.22}$$

The two radicals are determined as analytical functions on the complex plane of ϵ with the cuts connecting $-\Delta \mp i/2\tau_{\mathrm{ph}}$ and $\Delta \mp i/2\tau_{\mathrm{ph}}$, respectively (compare with Fig. 5.1). The variable x is real. It increases from $-\infty$ to $+\infty$ as ϵ varies from $-\infty$ to $+\infty$ along the real axis. It is convenient to use the identities

$$\frac{dx}{d\epsilon} = g_-, \; \epsilon = x\sqrt{\frac{x^2 + \Delta^2 + 1/4\tau^2_{\mathrm{ph}}}{x^2 + 1/4\tau^2_{\mathrm{ph}}}},$$

and

$$\left(\frac{f_-}{g_-} \right)^2 = \frac{x^2 \Delta^2}{\left(x^2 + 1/4\tau^2_{\mathrm{ph}} \right) \left(x^2 + \Delta^2 + 1/4\tau^2_{\mathrm{ph}} \right)},$$

$$\frac{f_+}{g_-} = -i \frac{\Delta/2\tau_{\mathrm{ph}}}{x^2 + 1/4\tau^2_{\mathrm{ph}}},$$

which follow from eqn (11.22). Now it is easy to perform the integration. For example,

$$\int_{-\infty}^{\infty} \frac{d\epsilon}{4} \left[-\frac{i}{2} \frac{\partial}{\partial t} \left(f^R_\epsilon + f^A_\epsilon \right) \frac{\partial f^{(0)}}{\partial \epsilon} + \left(f^R_\epsilon - f^A_\epsilon \right) f_1 \right]$$

$$= -\frac{1}{4}\frac{\partial \Delta}{\partial t}\int_{-\infty}^{\infty} 2\tau_{\rm ph}\frac{\partial f^{(0)}}{\partial \epsilon}\frac{\Delta^2 + 1/4\tau_{\rm ph}^2}{x^2 + \Delta^2 + 1/4\tau_{\rm ph}^2}\, dx$$
$$= -\frac{\pi}{8T_c}\sqrt{1 + 4\tau_{\rm ph}^2\Delta^2}\,\frac{\partial \Delta}{\partial t}.$$

Here we use that the characteristic $x \ll T_c$. Similarly,

$$\int_{-\infty}^{\infty}\frac{d\epsilon}{4}\left[\frac{1}{2}\left(\frac{\partial \chi}{\partial t} + 2e\varphi\right)\left(f_\epsilon^R + f_\epsilon^A\right)\frac{\partial f^{(0)}}{\partial \epsilon} - \left(f_\epsilon^R + f_\epsilon^A\right) f_2\right]$$
$$= -\frac{ie\Phi\Delta}{8\tau_{\rm ph}}\int_{-\infty}^{\infty}\frac{1}{x^2 + \Delta^2 + 1/4\tau_{\rm ph}^2}\, dx = -\frac{\pi}{8T_c}\frac{2ie\Phi\Delta}{\sqrt{1 + 4\tau_{\rm ph}^2\Delta^2}}.$$

The stationary part of the Keldysh function defined by eqn (11.20) gives the usual Ginzburg–Landau contribution. Here we note that

$$\int_{-\infty}^{\infty}\frac{d\Omega_{\mathbf{p}}}{4\pi}\frac{d\epsilon}{4}\left(f_{\epsilon_+,\epsilon_-}^R - f_{\epsilon_+,\epsilon_-}^A\right) f_\epsilon^{(0)}$$

is even in ω, therefore, one can use here only the zero-order terms in ω. Finally, we obtain

$$\frac{\pi}{8T_c}\left[\sqrt{1 + 4|\Delta|^2\tau_{\rm ph}^2}\,\frac{\partial |\Delta|}{\partial t} + \frac{2ie\Phi|\Delta|}{\sqrt{1 + 4|\Delta|^2\tau_{\rm ph}^2}}\right]$$
$$= \frac{\pi D}{8T_c}\left(\nabla - \frac{2ie\mathbf{Q}}{c}\right)^2 |\Delta| + \frac{T_c - T}{T_c}|\Delta| - \frac{7\zeta(3)}{8\pi^2 T_c^2}|\Delta|^3. \qquad (11.23)$$

Here we restore the complex–valuedness of the order parameter.

Expression for the current is

$$\mathbf{j}(\omega, \mathbf{k}) = -e\nu(0)\int \frac{d\Omega_{\mathbf{p}}}{4\pi}\frac{d\epsilon}{4}\mathbf{v}_F\left(g_{\epsilon_+,\epsilon_-}^K - \bar{g}_{\epsilon_+,\epsilon_-}^K\right).$$

In the leading approximation in Δ, eqn (10.62) gives

$$g_{\epsilon_+,\epsilon_-}^K - \bar{g}_{\epsilon_+,\epsilon_-}^K = 4g_- f^{(0)} + 4g_- f_1.$$

The stationary part of the Keldysh function eqn (11.20) gives the supercurrent. The vector part of the distribution function $f_1 = \langle f_1\rangle + \hat{\mathbf{v}}_F\mathbf{f}_1$ is found from eqn (10.108). The characteristic frequencies in the integral for the current are $\epsilon \sim T$, therefore, one can neglect the function f_2 and put $g_- = 1$. Therefore

$$\mathbf{f}_1 = -e\ell\frac{\partial f^{(0)}}{\partial \epsilon}\mathbf{E}$$

where $\ell = v_F\tau$. The total current becomes

$$\mathbf{j} = \mathbf{j}_n + \mathbf{j}_s.$$

Here the supercurrent has its usual form

$$\mathbf{j}_s = -\frac{\pi\sigma_n|\Delta|^2}{2cT}\mathbf{Q}. \tag{11.24}$$

The normal current is again $\mathbf{j}_n = \sigma_n\mathbf{E}$.

Note that the generalized TDGL equation (11.23) cannot be presented in the form of eqns (1.65), (1.67) because the relaxation of the order parameter magnitude is governed by a different process compared to the relaxation of the order parameter phase. Nevertheless, the real part of eqn (11.23) can be separately written as

$$\frac{\pi}{4T}\sqrt{1+4\left|\Delta\right|^2\tau_{\text{ph}}^2}\frac{\partial\left|\Delta\right|}{\partial t} = -\frac{\delta\mathcal{F}_{sn}}{\delta\left|\Delta\right|} \tag{11.25}$$

where we use expression (6.30) for the free energy of a dirty superconductor.

Relaxation of the order parameter phase determines conversion of normal current into a supercurrent and penetration of the electric field into a superconductor. Using eqns (11.23), (11.24), and the charge neutrality condition $\text{div}\mathbf{j} = 0$ we can obtain the equation for gauge-invariant potential Φ. The imaginary part of eqn (11.23) gives

$$\text{div}\mathbf{j}_s = \frac{\pi\nu\left(0\right)e^2}{T}\frac{\Phi\left|\Delta\right|^2}{\sqrt{1+4|\Delta|^2\tau_{\text{ph}}^2}}. \tag{11.26}$$

From the charge neutrality we find

$$-\sigma\left(\frac{1}{c}\text{div}\frac{\partial\mathbf{Q}}{\partial t} + \nabla^2\Phi\right) + \frac{\pi\nu\left(0\right)e^2}{T}\frac{\Phi\left|\Delta\right|^2}{\sqrt{1+4|\Delta|^2\tau_{\text{ph}}^2}} = 0. \tag{11.27}$$

This equation determines the relaxation length for the potential Φ, i.e., the electric field penetration length. For a constant Δ,

$$l_E^2 = \frac{2DT}{\pi\left|\Delta\right|^2}\sqrt{1+4|\Delta|^2\tau_{\text{ph}}^2}. \tag{11.28}$$

The penetration length is $l_E \sim \xi\left(T\right)\left(1+4|\Delta|^2\tau_{\text{ph}}^2\right)^{1/4}$; it can be much longer than $\xi\left(T\right)$ when $|\Delta|\tau_{\text{ph}} \gg 1$. Equation (11.28) was first derived by Tinkham (1972), Schmid and Schön (1975), and by Artemenko and Volkov (1975) for $\left|\Delta\right|\tau_{\text{ph}} \gg 1$. We discuss this phenomenon in more detail later in Section 11.4.

Equation (11.23) transforms into the usual TDGL equation when $\Delta\tau_{\text{ph}} \ll 1$. This limit corresponds to the gapless superconductivity where the pair breaking is due to the inelastic scattering by phonons. One can see this from comparing eqn (11.13) with eqn (11.1) obtained for magnetic impurities. The phenomenological relaxation parameter in this limit is defined by eqn (1.86). The relaxation time for current defined in Section 1.2.1 is

$$\tau_j = \frac{7\zeta(3)}{4\pi^3 T_c}\left(1 - \frac{T}{T_c}\right)^{-1}. \tag{11.29}$$

The order parameter relaxation time is

$$\tau_\Delta = \frac{\pi}{8T_c}\left(1 - \frac{T}{T_c}\right)^{-1}.$$

Ratio of the two times is

$$u = \frac{\pi^4}{14\zeta(3)} \approx 5.79.$$

In a more realistic limit one has, however, $\Delta\tau_{\rm ph} \gg 1$. Equation (11.23) no longer has a TDGL form. Indeed, the relaxation time for the order parameter becomes Δ-dependent:

$$\tau_\Delta = \frac{\pi\,|\Delta|\,\tau_{\rm ph}}{4\,(T_c - T)}.$$

Moreover, it is much longer than the TDGL time and has a tendency to approach $\tau_{\rm ph}$ as temperature decreases away from T_c. A long order parameter relaxation time means that the order parameter in a rapidly oscillating external field is unable to follow the instantaneous magnitude of the field. In fact, it oscillates slightly near an average value, the latter being determined by eqn (11.23) averaged over time. After averaging over fast oscillations, the time derivative drops out; on the other hand, one can replace $|\Delta|$ with its averaged value.

11.3 TDGL theory for *d*-wave superconductors

The unusual *d*-wave symmetry affects both thermodynamic and dynamic properties of superconductors. However, *d*-wave superconductors are expected to display a more "conventional" type of behavior in the region of temperatures close to the critical temperature. From a thermodynamical point of view, this is the temperature range where the Ginzburg–Landau theory is applicable for both *s*- and *d*-wave superconductors. As far as dynamics is concerned, *d*-wave superconductors appear to be even more simple than the usual *s*-wave superconductors. Indeed, for *s*-wave superconductors, in addition to solving the equation for the order parameter one has to take care of nonequilibrium excitations which become extremely important due to a singular density of states near the energy gap. As we already know, a comparatively simple time-dependent Ginzburg–Landau (TDGL) theory is only available for several particular situations. The most important example is the gapless superconductivity realized in materials with large enough concentration of magnetic impurities. The energy gap is suppressed due to a spin-dependent scattering which breaks spin-singlet Cooper pairs. *d*-wave superconductors with nonmagnetic impurities are in a sense similar to *s*-wave superconductors with magnetic impurities: the energy gap is destroyed due to breaking of the momentum coherence of a d-wave Cooper pair by scattering at impurity atoms (see Section 6.3). In the limit $\tau\Delta(T) \ll 1$, all memory of the energy gap disappears completely. This is where one can expect a comparatively simple nonequilibrium dynamics.

The "gapless" behavior is a natural high-temperature limit for d-wave superconductors with impurities in contrast to s-wave superconductors where it is quite an exotic situation. The d-wave superconductivity can only exist in clean compounds with the impurity mean free path ℓ much larger than the zero-temperature coherence length ξ_0; this is equivalent to the condition $\tau_{\rm imp}T_c \gg 1$. The "gapless" regime is thus realized in close vicinity of T_c where $1 - T/T_c \ll (T_c\tau_{\rm imp})^{-2}$. In a sense, d-wave superconductors in close vicinity of the critical temperature display a much simpler dynamics than s-wave superconductors near T_c. Nevertheless, it is richer than the conventional TDGL theory.

In the present section we demonstrate that the general nonstationary microscopic theory of superconductivity in this temperature range can be reduced to a set of equations resembling the usual TDGL equations with one extra term responsible for relaxation of nonequilibrium excitations (Kopnin 1998 a). In an inhomogeneous situation, relaxation is diffusion dominated. If the diffusion occurs at distances of the order of the coherence length $\xi(T)$, the relaxation is comparatively fast and results in a dissipation of the same order of magnitude as in the usual TDGL theory. The extra term becomes more important when the diffusion slows down, i.e., when the scale of spatial variations is large. In a homogeneous situation, the extra term is determined by yet slower inelastic relaxation and thus can reach rather high values. The overall behavior is reminiscent of an s-wave superconducting alloy with a small concentration of magnetic impurities considered by Eliashberg (1968). To compare the results, one needs to replace the spin-flip relaxation time τ_s with $2\tau_{\rm imp}$.

We start our derivation with the general self-consistency equation for the d-wave order parameter, eqn (5.29). For a time-dependent problem, it has the form

$$\frac{\Delta_{\mathbf{p}}(\mathbf{k},\omega)}{\lambda} = \int \frac{d\epsilon}{4\pi i}\left\langle V_d(\hat{\mathbf{p}},\hat{\mathbf{p}}_1) f^K_{\epsilon_+,\epsilon_-}(\mathbf{p}_{1+},\mathbf{p}_{1-})\right\rangle_{\mathbf{p}_1} . \tag{11.30}$$

The angle brackets, as usual, denote an average over the Fermi surface according to eqn (5.73) having in mind that, in d-wave superconductors, the Fermi surface may be nonspherical. The d-wave pairing potential $V_d(\hat{\mathbf{p}},\hat{\mathbf{p}}_1)$ is normalized in such a way that

$$\langle V_d(\hat{\mathbf{p}},\hat{\mathbf{p}}_1)\Delta_{\mathbf{p}_1}\rangle_{\mathbf{p}_1} = \Delta_{\mathbf{p}}. \tag{11.31}$$

To find the regular quasiclassical Green functions we use the Eilenberger equations

$$-i\mathbf{v}_F\left(\nabla - \frac{2ie}{c}\mathbf{A}\right) f_\epsilon - 2\epsilon f_\epsilon + 2\Delta_{\mathbf{p}} g_\epsilon - 2\Sigma_1 f_\epsilon + 2g_\epsilon\Sigma_2 = 0, \tag{11.32}$$

$$i\mathbf{v}_F\left(\nabla + \frac{2ie}{c}\mathbf{A}\right) f^\dagger_\epsilon - 2\epsilon f^\dagger_\epsilon + 2\Delta^*_{\mathbf{p}} g_\epsilon - 2\Sigma_1 f^\dagger_\epsilon + 2g_\epsilon\Sigma^\dagger_2 = 0, \tag{11.33}$$

for both retarded and advanced functions. The impurity self-energies are

$$\Sigma_1 = \frac{i}{2\tau_{\rm imp}} \langle g \rangle , \; \Sigma_2 = \frac{i}{2\tau_{\rm imp}} \langle f \rangle . \tag{11.34}$$

For the distribution function we use the kinetic equations (10.98) and (10.99) in the "dirty limit" since we assume that $\Delta\tau_{\rm imp} \ll 1$. The collision integrals may include both impurity and inelastic (electron–phonon) scattering $J = J^{(\rm imp)} + J^{(\rm ph)}$. We assume that the inelastic mean free time is much larger that the impurity scattering time, $\tau_{\rm ph} \gg \tau_{\rm imp}$.

We consider here the so-called "gapless" limit which is realized at temperatures close enough to the critical temperature. The corresponding condition is $\tau_{\rm imp} T_c (1 - T/T_c)^{1/2} \ll 1$; this is equivalent to $\ell \ll \xi(T)$. In this gapless limit, one can perform expansion in the small order parameter $\Delta\tau_{\rm imp} \ll 1$ as well as in small gradients and in the small vector potential. Since $\langle \Delta_{\mathbf{p}} \rangle = 0$, the Eilenberger equations in the first approximation give $g^{R(A)} = \pm 1$ and

$$f_\epsilon^{R(A)} = \pm \frac{\Delta_{\mathbf{p}}}{\epsilon \pm i/2\tau_{\rm imp}}, \; f_\epsilon^{\dagger\, R(A)} = \pm \frac{\Delta_{\mathbf{p}}^*}{\epsilon \pm i/2\tau_{\rm imp}}. \tag{11.35}$$

In case of a small admixture of an s-wave component into the order parameter, eqn (11.35) is valid as long as $|\langle \Delta_{\mathbf{p}} \rangle| \ll |\Delta_{\mathbf{p}}|$.

The impurity collision integrals become

$$J_1^{(\rm imp)} = -\frac{1}{\tau_{\rm imp}} (f_1 - \langle f_1 \rangle) , \; J_2^{(\rm imp)} = -\frac{1}{\tau_{\rm imp}} (f_2 - \langle f_2 \rangle) . \tag{11.36}$$

The inelastic collision integral in eqn (10.98) is $J_1^{(\rm ph)} = -f_1/\tau_{\rm ph}$.

We put

$$f_1 = \langle f_1 \rangle + \tilde{f}_1, \; f_2 = \langle f_2 \rangle + \tilde{f}_2 \tag{11.37}$$

where $\tilde{f}_1 \ll \langle f_1 \rangle$ and $\tilde{f}_2 \ll \langle f_2 \rangle$, and obtain from eqn (10.101) and (10.103)

$$\tilde{f}_1 = -\tau_{\rm imp} e \mathbf{v}_F \cdot \mathbf{E} \frac{df^{(0)}}{d\epsilon} - \tau_{\rm imp} \mathbf{v}_F \cdot \nabla \langle f_2 \rangle , \; \tilde{f}_2 = -\tau_{\rm imp} \mathbf{v}_F \cdot \nabla \langle f_1 \rangle . \tag{11.38}$$

Equation (10.100) gives

$$\frac{\epsilon}{2(\epsilon^2 + 1/4\tau_{\rm imp}^2)} \frac{\partial \langle |\Delta_{\mathbf{p}}|^2 \rangle}{\partial t} \frac{df^{(0)}}{d\epsilon} - D\nabla^2 \langle f_1 \rangle + \frac{1}{\tau_{\rm ph}} \langle f_1 \rangle = 0. \tag{11.39}$$

From now on we assume an isotropic Fermi surface for simplicity and put

$$\langle \tau_{\rm imp} v_{F\,i} v_{F\,k} \rangle = \delta_{ik} D,$$

where D is the diffusion constant.

We introduce a new function of coordinates $U(\mathbf{r})$ defined as

$$\langle f_1\rangle = -\frac{U(\mathbf{r})\epsilon}{2\tau(\epsilon^2+1/4\tau_{\rm imp}^2)}\frac{df^{(0)}}{d\epsilon}. \tag{11.40}$$

The function U satisfies the diffusion equation

$$D\nabla^2 U - \frac{1}{\tau_{\rm ph}}U = -\tau_{\rm imp}\frac{\partial\left\langle|\Delta_{\mathbf{p}}|^2\right\rangle}{\partial t}. \tag{11.41}$$

The second averaged kinetic equation (10.102) gives

$$-\frac{\left\langle|\Delta_{\mathbf{p}}|^2\right\rangle}{\tau_{\rm imp}(\epsilon^2+1/4\tau_{\rm imp}^2)}\frac{df^{(0)}}{d\epsilon}e\Phi + \frac{\left\langle|\Delta_{\mathbf{p}}|^2 f_2\right\rangle}{\tau_{\rm imp}(\epsilon^2+1/4\tau_{\rm imp}^2)}$$
$$-D\nabla^2\langle f_2\rangle - De\,\mathrm{div}\,\mathbf{E}\frac{df^{(0)}}{d\epsilon} = \left\langle J_2^{(\rm ph)}\right\rangle. \tag{11.42}$$

Here χ is the order parameter phase.

Using the charge neutrality $\delta N^{(\rm nst)}=0$ we can exclude f_2 from eqn (11.42). Indeed, since in eqn (10.69), $\partial g_-/\partial\epsilon \sim |\Delta|^2/T^3$, the distribution function should satisfy $f_2 \ll e\Phi/T$ such that $\int_{-\infty}^{\infty} f_2\,d\epsilon$ is zero within the leading approximation in Δ^2/T^2. We can thus omit f_2 from the first line of eqn (11.42) together with the inelastic collision integral and integrate this equation over $d\epsilon$. The term with $D\nabla^2\langle f_2\rangle$ in the second line of eqn (11.42) vanishes as compared to div $\mathbf{E}$. As a result, f_2 disappears completely, and we find

$$De\,\mathrm{div}\,\mathbf{E} + \frac{\pi}{2T}\left\langle|\Delta_{\mathbf{p}}|^2\right\rangle e\Phi = 0. \tag{11.43}$$

Equation (11.43) yields, in particular

$$\mathrm{div}\,\mathbf{j}_s = \frac{\pi\nu(0)}{T}\left\langle|\Delta_{\mathbf{p}}|^2\right\rangle e^2\Phi, \tag{11.44}$$

which describes conversion of supercurrent into a normal current. It is similar to eqn (11.26).

Equation (11.43) introduces the characteristic electric field penetration length

$$l_E = \sqrt{\frac{2DT}{\pi\left\langle|\Delta_{\mathbf{p}}|^2\right\rangle}} \tag{11.45}$$

which describes the relaxation of the scalar potential Φ in a d-wave superconductor. We have already encountered a similar length on page 220 in this section and also on page 21 in Section 1.2.1. This penetration length is of the order of $\sqrt{T_c\tau}\xi(T)$, i.e., is much longer than $\xi(T)$. We refer the reader to the following section for further discussion of the effects associated with an electric field.

Using eqn (11.21) together with eqn (11.40) we find for the nonstationary part of the Keldysh Green function

$$f^{(\mathrm{nst})}_{\epsilon_+,\epsilon_-} = -\frac{1}{2\tau_{\mathrm{imp}}(\epsilon^2 + 1/4\tau^2_{\mathrm{imp}})} \frac{df^{(0)}}{d\epsilon}\left[\frac{\hat{\partial}\Delta_{\mathbf{p}}}{\partial t} + \frac{2\epsilon^2\Delta_{\mathbf{p}}U}{\epsilon^2 + 1/4\tau^2_{\mathrm{imp}}}\right].$$

As the result, we obtain for the order parameter

$$-\frac{\delta\mathcal{F}_{sn}}{\delta\Delta^*_{\mathbf{p}}} = \frac{\pi\nu(0)}{8T_c}\left(\frac{\partial\Delta_{\mathbf{p}}}{\partial t} + 2ie\varphi\Delta_{\mathbf{p}}\right) + \frac{\pi\nu(0)}{8T_c}\Delta_{\mathbf{p}}U. \tag{11.46}$$

The free energy in the left-hand side of eqn (11.46) is given by the Ginzburg–Landau expression (6.69). Equations (11.46) together with eqn (11.41) coincide with the corresponding equations obtained by Eliashberg (1968) if one puts there $\tau_s = 2\tau_{\mathrm{imp}}$.

The total current $\mathbf{j} = \mathbf{j}_s + \mathbf{j}_n$ is determined by the variation of $\mathcal{F}$ with respect to the vector potential. We find, using eqn (11.38)

$$-\frac{\delta\mathcal{F}_{sn}}{\delta\mathbf{A}} = \frac{1}{c}\left(\mathbf{j} - \mathbf{j}_n\right), \tag{11.47}$$

where the normal current can again be written as $\mathbf{j}_n = \sigma_n\mathbf{E}$ (remind that we assume an isotropic Fermi surface). The complex equation (11.46) contains the equations for the order parameter magnitude and phase. Together with eqn (11.47) and the current conservation $\operatorname{div}\mathbf{j} = 0$, the latter is equivalent to eqn (11.43).

Equation (11.46) differs from a simple TDGL equation by the additional term with U; it is determined by a diffusive relaxation of excitations driven out of equilibrium by external perturbations. It becomes particularly important in situations where the diffusion is slow, i.e., where the characteristic scale of spatial variations is larger than ξ. For example, in a spatially homogeneous case,

$$U = \tau_{\mathrm{ph}}\tau_{\mathrm{imp}}\frac{\partial\left\langle|\Delta_{\mathbf{p}}|^2\right\rangle}{\partial t}.$$

The U-term is much larger than the TDGL term as long as the temperature is not very close to T_c , i.e., when the condition $\tau_{\mathrm{imp}}\tau_{\mathrm{ph}}\Delta^2 \gg 1$ holds.

To summarize our discussion of the microscopic justification of the TDGL model we conclude that the microscopic theory supports the TDGL model only in a very limited situation where the deviation of the excitations from equilibrium is small. The best result, i.e., when the TDGL model occurs to be exact is achieved for the gapless limit which can be associated with various mechanisms such as scattering by magnetic impurities, phonons, or by nonmagnetic impurities in unconventional superconductors. However, as soon as the distribution of excitations starts to deviate from equilibrium, the TDGL model breaks down. We have seen this both for the case when the relaxation of excitations is dominated by electron–phonon scattering and when the relaxation is by diffusion.

When the distribution is essentially nonequilibrium, one can no longer construct a simple model which involves a closed set of equations for only the order parameter magnitude and phase. The distribution function of excitations appears as a new important variable. In our examples, the function U in eqn (11.40) plays the role of such a variable. In general, one would need to invoke the full Green function technique described in the previous chapters. Of course, to get the results one has to work hard and solve the whole set of equations which are basically the Eilenberger equations and the kinetic equations for the distribution function. Next step would be to self-consistently determine the order parameter and electromagnetic fields through the order parameter and Maxwell equations. In the following part of the book we consider how one can carry out this program for a very important and interesting problem of the vortex dynamics. Before going to the vortex dynamics, however, we discuss one more aspect which can be elucidated using the TDGL model.

11.4 d.c. electric field in superconductors. Charge imbalance

Let us discuss the TDGL equations (1.80), (11.27), or (11.43) which all can be written in the form

$$\nabla^2\Phi - \eta\left(\frac{|\Delta|}{\Delta_\infty}\right) l_E^{-2}\Phi = -\frac{1}{c}\mathrm{div}\frac{\partial \mathbf{Q}}{\partial t} \tag{11.48}$$

where $\eta\left(|\Delta|/\Delta_\infty\right)$ is a function of the ratio of the order parameter magnitude $|\Delta|$ and its homogeneous value Δ_∞; η is defined in such a way that $\eta = 1$ for $|\Delta|/\Delta_\infty = 1$. The function $\eta = f^2$ for the usual TDGL model, eqn (1.80). The magnitude of l_E and the specific form of η depend on parameters of the superconductor.

We already mentioned that l_E determines the length over which a d.c. electric field decays into a bulk superconductor. According to the definition of the gauge invariant potentials in eqns (1.46) and (1.75), the electric field is

$$\mathbf{E} = -\frac{1}{c}\frac{\partial \mathbf{Q}}{\partial t} - \nabla\Phi.$$

A d.c. electric field is simply $\mathbf{E} = -\nabla\Phi$. Consider an interface between a normal metal and a superconductor, and assume that the superconductor occupies the space $x > 0$. If there is a current along the x axis flowing from the normal metal into the superconductor, the electric field $E = j/\sigma_n$ exists in the normal metal such that $d\varphi/dx = -j/\sigma_n$. However, in the bulk superconductor, a d.c. electric field cannot exist: It would produce continuous acceleration of superconducting electrons and destroy superconductivity. It is eqn (11.48) that describes relaxation of E. We have for a d.c. electric field

$$\nabla^2\Phi = l_E^{-2}\Phi$$

whence

$$\Phi = \frac{jl_E}{\sigma_n} \exp\left(-\frac{x}{l_E}\right).$$

The constant is chosen to satisfy the continuity of current through the interface. According to this equation, the electric field and the normal current decay into the superconductor, while supercurrent increases until $j_s = j - j_n$ takes all the current flowing into the superconductor: a normal current

$$j_n(x) = -\sigma_n d\Phi/dx$$

converges into supercurrent over the distance l_E.

Because a d.c. electric field exists in a superconductor within the length on order l_E near its boundary with the normal metal, the interface offers an extra resistance to the current. The voltage across the conversion layer is jl_E/σ_n and thus the extra resistance is $\delta R = l_E/\sigma_n$. Of course, such a simple picture takes place only as long as the generalized TDGL model holds. In general, conditions at the interface itself also cause changes in the electric field which contribute to the extra resistance. This problem was considered by many researchers and the results are summarized in the review by Artemenko and Volkov (1979). We are not going into these details here but rather concentrate on another aspect of the problem.

Equation (11.48) tells us that, in a bulk superconductor, we have $\Phi = 0$ for a stationary electric field. In other words,

$$\varphi + \frac{1}{2e}\frac{\partial \chi}{\partial t} = 0 \tag{11.49}$$

which is the celebrated Josephson equation. We thus conclude that the Josephson equation holds only in equilibrium.

Another way of looking at eqn (11.49) is as follows. The quantity $\delta\mu_n = -e\varphi$ can be considered as a change in the chemical potential of a (normal) electron due to the presence of electromagnetic potential φ. The quantity $\partial\chi/\partial t$ is then the chemical potential of a Cooper pair. Equation (11.49) now expresses that

$$\delta\mu_n - \delta\mu_s = 0 \tag{11.50}$$

where

$$\delta\mu_s = \frac{1}{2}\frac{\partial \chi}{\partial t}$$

is the chemical potential of a superconducting electron measured from the Fermi energy. Equation (11.50) implies that, under static conditions and in a bulk superconductor, normal and superconducting electrons are in equilibrium with each other.

Equilibrium $\Phi = 0$ may be violated if time-dependent processes take place in the superconductor and/or near its border. A nonequilibrium state with a nonzero Φ is known as a state with the *charge imbalance* (Tinkham 1972). This terminology comes from the following observation. In a static situation, $g_\epsilon^{R(A)} =$

$-\bar{g}_\epsilon^{R(A)}$ which reflects the particle–hole symmetry. This causes, in particular, that always

$$\mathrm{Tr}\check{g}^K = 0$$

in a static case. This is the condition of a constant particle density: there are exactly as many particle-like excitations as there are hole-like ones; moving an electron from under the Fermi level to a position above it we do not create a new particle. The situation changes if $\Phi \neq 0$: now $\mathrm{Tr}\check{g}^K$ is nonzero if a special condition on the chemical potential is not imposed. This means that the numbers of particle-like and hole-like excitations are no longer equal. To keep the overall particle density constant one needs to shift the chemical potential. This is called the charge imbalance: the particle-like and the hole-like charges are not equal. Of course, the overall charge density remains constant because we adjust the chemical potential exactly in such a way as to maintain zero value of $\mathrm{Tr}\check{g}^K$. This effect is sometimes also called "branch mixing" because the relaxation of Φ is associated with equalizing the populations of the particle-like and hole-like branches of the excitation spectrum.

The process of relaxation of the charge imbalance with $\delta\mu_n \neq \delta\mu_s$ is associated with an exchange of particles between the normal part and the condensate in the superconductor. Any process which facilitates the exchange contributes to the relaxation. Inelastic processes are definitely the ones which work for such an exchange. During an inelastic scattering event, the particle energy changes together with the state of a given particle. It causes redistribution of particles between the normal part and the condensate thus contributing to the charge imbalance relaxation. This is why the electric field penetration length contains the electron–phonon relaxation time in eqn (11.28). In the gapless regime, the energy gap disappears, the exchange is very effective, and the only characteristic time is τ_Δ. The electric field penetration length squared is thus the diffusion constant multiplied with this time which makes $l_E \sim \xi$. The same happens during the spin-flip scattering process because the energy of an electron in the order-parameter field depends on its spin. In an s-wave superconductor, the scattering by impurities will not cause the exchange because an elastic scattering does not change the state of a particle if the energy gap is independent of the direction of the particle momentum. However, if in the presence of a supercurrent, and/or for a d-wave superconductor, the energy does depend on the momentum direction, and impurity scattering does change the state of a particle. This is why eqn (11.45) determines l_E through τ_{imp} without characteristics of the inelastic relaxation. For a more detailed discussion, see the book by Tinkham (1996).

Part IV

Vortex dynamics

12

TIME-DEPENDENT GINZBURG–LANDAU ANALYSIS

Vortex dynamics is considered within the framework of the TDGL model. The balance of forces acting on a moving vortex is derived from the free energy considerations. The vortex viscosity and the flux flow conductivity are calculated. A modification of the TDGL model is considered which allows us to account for the flux flow Hall effect.

12.1 Introduction

A very simple question that can arise in connection with the title of the present part of the book is as follows. What is special in the vortex dynamics and why do we want to study it? To answer this we note that the majority of known superconductors are type II superconductors whose London penetration length λ_L is longer than the coherence length ξ. The exact condition which separates type II from type I superconductors is $\kappa \equiv \lambda_L/\xi > 1/\sqrt{2}$, see Section 1.1.2, page 12. Almost all alloys and films made of conventional (low temperature) superconducting materials are of type II. All known high-temperature and heavy-Fermionic superconductors are also of type II. We can also mention superfluid ^{3}He where $\lambda_L = \infty$ since particles which make Cooper pairs are uncharged. If we place a type II superconductor in a magnetic field (or rotate superfluid ^{3}He) vortices appear in the range of fields $H_{c1} < H < H_{c2}$, and the mixed state arises, see Section 1.1.2. Since the lower critical magnetic field $H_{c1} \sim H_{c2}/\kappa^2$ (see, for example, de Gennes 1966, Saint-James *et al.* 1969), the mixed state exists in a broad range of magnetic fields if the Ginzburg–Landau parameter κ is substantially larger than unity which is the case in high-temperature superconductors where κ can be as high as 100.

In superconductors, each vortex carries exactly one magnetic–flux quantum $\Phi_0 = \pi c/|e|$. Being magnetically active, vortices determine magnetic properties of superconductors. In addition, they are mobile if the material is homogeneous and there are no defects which can attract vortices and "pin" them somewhere in a superconductor. In fact, a superconductor in the mixed state is not fully superconducting any more. Indeed, there is no complete Meissner effect: some magnetic field penetrates into superconductor via vortices. In addition, regions with the normal phase appear. Indeed, since the order parameter turns to zero at the vortex axis (see Fig. 1.1) and is suppressed around each vortex axis within a vortex core region of the order of the coherence length, a finite low-energy density of states appears in the vortex cores. Moreover, mobile vortices move in the presence of an average (transport) current: the Lorentz force

$$\mathbf{F}_L = \frac{\Phi_0}{c}\left[\mathbf{j}_{\rm tr} \times \hat{\mathbf{H}}\right] \tag{12.1}$$

pushes vortices in a direction perpendicular to the current (here $\hat{\mathbf{H}}$ is the unit vector along the magnetic field). An electric field perpendicular to the vortex velocity is generated by a moving flux, and a voltage appears across the superconductor. Since vortices move at an angle to the transport current, there may be components of the electric field both parallel and perpendicular to the current. The longitudinal component produces dissipation in a superconductor while the transverse one is responsible for the Hall effect. We see that a finite resistivity appears (the so-called flux flow resistivity): a superconductor is no longer "superconducting"! This is certainly an important effect.

The magnitude and direction of the vortex velocity is determined by a balance of the Lorentz force and the forces acting on a moving vortex from the environment. In the absence of pinning these forces include friction (longitudinal with respect to the vortex velocity) and gyroscopic (transverse) forces (see Fig. 12.1). The friction force accounts for dissipation, i.e., for an effective longitudinal or Ohmic "flux flow" conductivity while the transverse force determines the Hall conductivity.

Experimental studies of flux flow effects began with the work by Kim *et al.* (1965). Since then enormous efforts have been undertaken to find out and understand the processes involved in the vortex dynamics and the vortex physics in general. One now uses a notion of "vortex matter" to comprise all features which vortices introduce to physics of superconductivity. We can mention the reviews by Gor'kov and Kopnin (1975) and by Larkin and Ovchinnikov (1986) which deal with the vortex dynamics, and a review by Blatter *et al.* (1994) which contains many basic concepts of the vortex physics especially those which are relevant to vortex lattices, vortex pinning, flux creep, etc.

In this part of the book, we concentrate on the theoretical description of the vortex dynamics based on the microscopic theory of nonstationary superconductivity. We consider the most representative examples among all numerous situations studied by many researchers during several decades of intensive work. We shall see that motion of vortices initiates almost all nonequilibrium processes and involves all relaxation mechanisms which work in superconductors. This is one more reason why it is important to understand the vortex dynamics. We start with the simplest theory which uses the TDGL model. The next chapter deals with more complicated physics in superconducting alloys when nonequilibrium excitations come seriously into play. The last two chapters describe the most interesting and intriguing phenomena in clean superconductors, i.e., in those classes of materials which include, in particular, new high-temperature superconductors. We do not consider effects associated with the vortex pinning by random defects in a superconducting material assuming that vortices are free to move in a homogeneous environment. An exception will be made for the intrin-

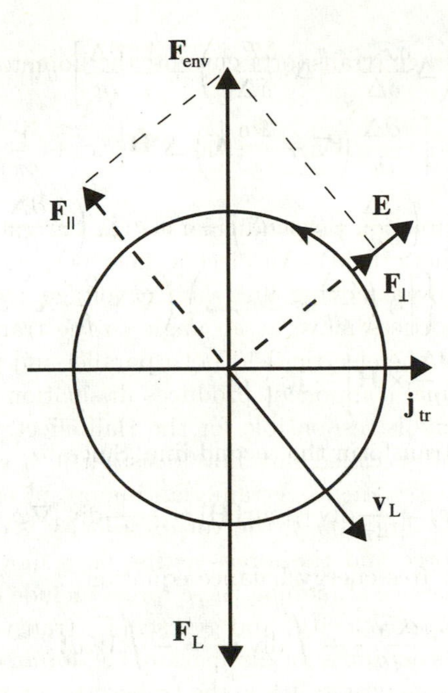

FIG. 12.1. Forces on a moving vortex: the Lorenz force from the supercurrent $\mathbf{j}_{\rm tr}$ is balanced by a force from environment $\mathbf{F}_{\rm env}$. The latter has the longitudinal (friction) $\mathbf{F}_{\parallel}$ and the transverse $\mathbf{F}_{\perp}$components. The moving vortex generates an electric field $\mathbf{E}$ perpendicular to the vortex velocity $\mathbf{v}_L$.

sic pinning when vortices interact with the regular crystal structure in layered superconductors.

12.2 Energy balance

The TDGL model provides the simplest basis for dealing with nonstationary processes in superconductors. Using it, one can establish quite general relationships which are particularly useful for our goal of understanding the vortex dynamics. The first is the energy balance (Schmid 1966). Let us calculate the time-derivative of the total GL free energy eqn (1.11). We have

$$\frac{\partial \mathcal{F}_{\rm tot}}{\partial t} = \int \left(\frac{\partial \Delta}{\partial t} \left[-|\alpha|\Delta^* + \beta|\Delta|^2\Delta^* - \gamma\left(\nabla + \frac{2ie}{c}\mathbf{A}\right)^2 \Delta^* \right] + c.c. \right.$$
$$\left. + \frac{1}{c}(\mathbf{j} - \mathbf{j}_s)\frac{\partial \mathbf{A}}{\partial t} \right) dV$$
$$+ \int {\rm div} \left[\gamma \frac{\partial \Delta}{\partial t}\left(\nabla + \frac{2ie}{c}\mathbf{A}\right)\Delta^* + c.c. + \frac{1}{4\pi}\frac{\partial \mathbf{A}}{\partial t} \times \mathbf{H}\right] dV.$$

With help of the TDGL equations (1.68) and (1.66) we obtain

$$\frac{\partial \mathcal{F}_{\rm tot}}{\partial t} = \int \left[\frac{\delta \mathcal{F}_{sn}}{\delta \Delta}\left(\frac{\partial \Delta}{\partial t} + 2ie\phi\Delta\right) + \frac{\delta \mathcal{F}_{sn}}{\delta \Delta^*}\left(\frac{\partial \Delta^*}{\partial t} - 2ie\phi\Delta^*\right)\right.$$

$$-2ie\phi\left(\Delta\frac{\delta\mathcal{F}_{sn}}{\delta\Delta}-\Delta^*\frac{\delta\mathcal{F}_{sn}}{\delta\Delta^*}\right)+\frac{\mathbf{j}_n}{c}\frac{\partial\mathbf{A}}{\partial t}\Bigg]\,dV$$
$$+\int\operatorname{div}\left[\gamma\frac{\partial\Delta}{\partial t}\left(\nabla+\frac{2ie}{c}\mathbf{A}\right)\Delta^*+c.c.+\frac{1}{4\pi}\left[\frac{\partial\mathbf{A}}{\partial t}\times\mathbf{H}\right]\right]dV$$
$$=\int\left[\frac{\delta\mathcal{F}_{sn}}{\delta\Delta}\left(\frac{\partial\Delta}{\partial t}+2ie\varphi\Delta\right)+c.c.+\mathbf{j}_n\left(\frac{1}{c}\frac{\partial\mathbf{A}}{\partial t}+\nabla\varphi\right)\right]dV$$
$$+\int\operatorname{div}\left\{\left[\gamma\left(\frac{\partial\Delta}{\partial t}+2ie\varphi\Delta\right)\left(\nabla+\frac{2ie}{\hbar c}\mathbf{A}\right)\Delta^*+c.c.\right]\right.$$
$$\left.+\frac{1}{4\pi}\left[\frac{\partial\mathbf{A}}{\partial t}\times\mathbf{H}\right]-\mathbf{j}\varphi\right\}dV.$$

We use eqn (1.76) to transform the second line. Since

$$\operatorname{div}(\mathbf{j}\varphi)=\frac{c}{4\pi}\operatorname{div}[\varphi\operatorname{curl}\mathbf{H}]=-\frac{c}{4\pi}\operatorname{div}[\nabla\varphi\times\mathbf{H}]$$

we finally arrive at the free-energy balance equation

$$\frac{\partial\mathcal{F}_{\mathrm{tot}}}{\partial t}+\int\operatorname{div}\mathbf{j}_{\mathcal{F}}=-\int W\,dV. \tag{12.2}$$

Here the free–energy current density is

$$\mathbf{j}_{\mathcal{F}}=-\gamma\left[\left(\frac{\partial\Delta}{\partial t}+2ie\varphi\Delta\right)\left(\nabla+\frac{2ie}{c}\mathbf{A}\right)\Delta^*+c.c.\right]+\frac{c}{4\pi}[\mathbf{E}\times\mathbf{H}], \tag{12.3}$$

and the dissipation function density has the form

$$W=2\Gamma\left|\left(\frac{\partial\Delta}{\partial t}+\frac{2ie\varphi\Delta}{\hbar}\right)\right|^2+\sigma_n\mathbf{E}^2. \tag{12.4}$$

One can identify two different terms in the dissipation function. The first contains the time-derivative of the order parameter. It describes dissipation produced by relaxation of the order parameter. The second is due to normal currents and is similar to that in usual normal metals.

12.3 Moving vortex

Vortex moves when there is a transport current $\mathbf{j}_{\mathrm{tr}}$ giving rise to the Lorentz force eqn (12.1). We can calculate the flux flow conductivity σ_f directly from the dissipation function, putting it to

$$W=\mathbf{j}_{\mathrm{tr}}\langle\mathbf{E}\rangle_L=\sigma_f\langle\mathbf{E}\rangle_L^2 \tag{12.5}$$

where $\langle\mathbf{E}\rangle_L$ is the electric field averaged over the vortex array. The average is defined as

$$\langle\ldots\rangle_L=S_0^{-1}\int_{S_0}(\ldots)\,dS \tag{12.6}$$

where $S_0=\Phi_0/B$ is the area occupied by each vortex.

All the values in eqn (12.4) can easily be calculated if the vortex velocity $\mathbf{v}_L$ is small. To calculate the time-derivative of the order parameter and of the vector potential we note that, for a small vortex velocity,

$$\Delta(\mathbf{r}) = \Delta_{st}(\mathbf{r} - \mathbf{v}_L t) + \Delta_1 \ , \ \mathbf{A}(\mathbf{r}) = \mathbf{A}_{st}(\mathbf{r} - \mathbf{v}_L t) + \mathbf{A}_1, \tag{12.7}$$

where Δ_{st} and $\mathbf{A}_{st}$ are the values for a static vortex, and Δ_1 and $\mathbf{A}_1$ are small corrections. Therefore, within the main terms in the vortex velocity $\mathbf{v}_L$, we can write

$$\frac{\partial \Delta}{\partial t} = -(\mathbf{v}_L \cdot \nabla)\Delta_{st} \ , \ \frac{\partial \mathbf{A}}{\partial t} = -(\mathbf{v}_L \cdot \nabla)\mathbf{A}_{st}. \tag{12.8}$$

The scalar potential φ is to be found from eqn (1.80). For small $\mathbf{v}_L$ it contains only the static vortex order parameter.

Let us calculate the average electric field. The local electric field is

$$\mathbf{E} = -\frac{1}{c}\frac{\partial \mathbf{A}}{\partial t} - \nabla\varphi = -\frac{1}{c}\frac{\partial \mathbf{Q}}{\partial t} - \nabla\Phi - \frac{1}{2e}\left(\frac{\partial}{\partial t}\nabla\chi - \nabla\frac{\partial \chi}{\partial t}\right). \tag{12.9}$$

The last term here is nonzero for a moving vortex. Indeed,

$$\begin{aligned} -\frac{1}{2e}\left(\frac{\partial}{\partial t}\nabla\chi - \nabla\frac{\partial \chi}{\partial t}\right) &= \frac{1}{2e}\left[(\mathbf{v}_L \cdot \nabla)\nabla\chi - \nabla(\mathbf{v}_L \cdot \nabla)\chi\right] \\ &= -\frac{1}{2e}[\mathbf{v}_L \times \operatorname{curl}\nabla\chi] \\ &= -\frac{\Phi_0}{c}[\mathbf{v}_L \times \hat{\mathbf{z}}]\operatorname{sign}(e)\delta^{(2)}(\mathbf{r}). \end{aligned} \tag{12.10}$$

Here $\hat{\mathbf{z}}$ is the unit vector along the vortex axis in the positive direction of its circulation. The vortex circulation is determined by the sense of the phase increment, it coincides with the magnetic field direction for positive charge of carriers and is antiparallel to it for negative carriers: $\hat{\mathbf{z}} = \operatorname{sign}(e)\,\hat{\mathbf{H}}$, see Section 1.1.2. We now average eqn (12.9) over the vortex array. Since

$$\frac{\partial \mathbf{Q}}{\partial t} = -(\mathbf{v}_L \cdot \nabla)\mathbf{Q}_{st}$$

the average of this term vanishes because the gauge-invariant vector potential $\mathbf{Q}_{st}$ is a bounded oscillating function (periodic for a periodic vortex array). Its value at the boundary of a region occupied by each vortex, i.e., at the boundary of the vortex unit cell is zero, see eqn (1.40) on page 11. The average of $\nabla\Phi$ vanishes, too, since Φ is also a bounded oscillating function periodic for a regular vortex lattice. It turns to zero at the boundary of the vortex unit cell because of the Josephson relation, eqn (11.49). The only term which survives is the last term in eqn (12.9). Equation (12.10) gives

$$\langle \mathbf{E} \rangle_L = \frac{1}{c}\mathbf{B} \times \mathbf{v}_L \tag{12.11}$$

which is the Faraday's law.

Now we can calculate W. It is more instructive, however, to consider another useful relation which follows from the TDGL equations. It is expression for the force which acts on a moving vortex.

12.4 Force balance

Let us shift our vortex by a small arbitrary constant vector $\mathbf{d}$ and calculate the variation of the free energy caused by such a displacement. During the derivation we neglect surface contributions. We have

$$\begin{aligned}
\delta\mathcal{F}_{\mathrm{tot}} &= \int \left(\frac{\delta \mathcal{F}_{sn}}{\delta \Delta}(\mathbf{d}\cdot\nabla)\Delta + \frac{\delta \mathcal{F}_{sn}}{\delta \Delta^*}(\mathbf{d}\cdot\nabla)\Delta^* \right. \\
&\quad \left. + \frac{\delta \mathcal{F}_{sn}}{\delta \mathbf{A}}(\mathbf{d}\cdot\nabla)\mathbf{A} + \frac{\mathbf{H}}{4\pi}\mathrm{curl}\,[(\mathbf{d}\cdot\nabla)\mathbf{A}] \right) dV \\
&= \int \left[\frac{\delta \mathcal{F}_{sn}}{\delta \Delta}\mathbf{d}\left(\nabla - \frac{2ie}{c}\mathbf{A}\right)\Delta + \frac{\delta \mathcal{F}_{sn}}{\delta \Delta^*}\mathbf{d}\left(\nabla + \frac{2ie}{c}\mathbf{A}\right)\Delta^* \right. \\
&\quad + \frac{2ie}{c}(\mathbf{d}\cdot\mathbf{A})\left(\Delta\frac{\delta \mathcal{F}_{sn}}{\delta \Delta} - \Delta^*\frac{\delta \mathcal{F}_{sn}}{\delta \Delta^*}\right) \\
&\quad \left. + \frac{1}{c}(\mathbf{j} - \mathbf{j}_s)(\mathbf{d}\cdot\nabla)\mathbf{A} \right] dV \\
&= \mathbf{d}\int \left[\frac{\delta \mathcal{F}_{sn}}{\delta \Delta}\left(\nabla - \frac{2ie}{c}\mathbf{A}\right)\Delta + c.c. + \frac{1}{c}[\mathbf{j}_n \times \mathbf{H}] \right] dV \qquad (12.12)
\end{aligned}$$

where we use eqn (1.76) and the charge neutrality $\mathrm{div}\mathbf{j} = 0$.

Variation of the free energy determines the force which acts on a moving vortex from the environment

$$\delta\mathcal{F}_{\mathrm{tot}} = -\int \mathbf{d}\cdot\mathbf{F}_{\mathrm{env}}\, dV. \qquad (12.13)$$

It should be balanced by external Lorentz force due to the transport current. To include this force we add the free energy due to the transport current which is supplied by the external source. Its variation due to the vortex displacement is

$$\begin{aligned}
\delta\mathcal{F}_{\mathrm{ext}} &= -\frac{1}{c}\int \mathbf{j}_{\mathrm{tr}}\,(\mathbf{d}\cdot\nabla)\,\mathbf{A}\; dV \\
&= \int \left(\frac{1}{c}\mathbf{j}_{\mathrm{tr}}[\mathbf{d}\times\mathbf{H}] - \frac{1}{c}\mathrm{div}\,[\mathbf{j}_{\mathrm{tr}}(\mathbf{d}\cdot\mathbf{A})] + \frac{1}{c}(\mathbf{d}\cdot\mathbf{A})\mathrm{div}\,\mathbf{j}_{\mathrm{tr}} \right) dV.
\end{aligned}$$

On the next step, we omit the surface term and use that $\mathrm{div}\mathbf{j}_{\mathrm{tr}} = 0$. We find

$$\delta\mathcal{F}_{\mathrm{ext}} = -\mathbf{d}\int \frac{1}{c}[\mathbf{j}_{\mathrm{tr}} \times \mathbf{H}] dV. \qquad (12.14)$$

Conservation of $\mathcal{F}_{\mathrm{tot}}$ plus the external energy implies its translation invariance

$$\delta\mathcal{F}_{\rm tot} + \delta\mathcal{F}_{\rm ext} = 0.$$

It gives the balance of forces acting on a moving vortex of a unit length

$$\mathbf{F}_L - \int_{S_0} \left[\frac{\delta\mathcal{F}_{sn}}{\delta\Delta} \left(\nabla - \frac{2ie}{c}\mathbf{A} \right) \Delta + c.c. + \frac{1}{c}[\mathbf{j}_n \times \mathbf{H}] \right] dS = 0 \tag{12.15}$$

where $\mathbf{F}_L$ is the Lorentz force eqn (12.1). We use the fact that $\mathbf{d}$ is an arbitrary vector. In other words,

$$\begin{aligned} \frac{\Phi_0}{c}[\mathbf{j}_{\rm tr} \times \hat{\mathbf{H}}] = \int_{S_0} \Bigg(-\Gamma \Bigg[& \left(\frac{\partial \Delta^*}{\partial t} - 2ie\varphi\Delta^* \right) \left(\nabla - \frac{2ie}{c}\mathbf{A} \right) \Delta \\ & + \left(\frac{\partial \Delta}{\partial t} + 2ie\varphi\Delta \right) \left(\nabla + \frac{2ie}{c}\mathbf{A} \right) \Delta^* \Bigg] \\ & + \frac{1}{c}[\mathbf{j}_n \times \mathbf{H}] \Bigg) \, dS. \end{aligned} \tag{12.16}$$

Here the integration is over the unit cell of the vortex array.

To summarize, in eqns (12.12) or (12.16) the Lorentz force is balanced by the force produced by the interaction with the environment. The latter is proportional to the vortex velocity since the time-derivative, electric field and the scalar potential are all proportional to $\mathbf{v}_L$.

Note that the dissipation function integrated over the unit cell can be again obtained if we multiply eqn (12.16) with the vortex velocity $\mathbf{v}_L$:

$$\begin{aligned} \frac{\Phi_0}{c}\left(\mathbf{j}_{\rm tr} \cdot [\hat{\mathbf{H}} \times \mathbf{v}_L] \right) &= \int_{S_0} \Bigg[(\mathbf{v}_L \nabla\Delta - 2ie\varphi\Delta) \frac{\delta\mathcal{F}_{sn}}{\delta\Delta} + c.c. \\ & \quad - \left(\frac{1}{c}(\mathbf{v}_L\mathbf{A}) - \varphi \right) \operatorname{div}\mathbf{j}_n + \frac{1}{c}\left(\mathbf{j}_n[\mathbf{H} \times \mathbf{v}_L] \right) \Bigg] \, dS \\ &= \int_{S_0} \Bigg\{ - \left[\left(\frac{\partial \Delta}{\partial t} + 2ie\varphi\Delta \right) \frac{\delta\mathcal{F}_{sn}}{\delta\Delta} + c.c. \right] \\ & \quad + \mathbf{j}_n \left(\frac{1}{c}\nabla(\mathbf{v}_L\mathbf{A}) - \nabla\varphi \right) + \frac{1}{c}\left(\mathbf{j}_n[\mathbf{H} \times \mathbf{v}_L] \right) \Bigg\} \, dS \\ &= \int_{S_0} W \, dS. \end{aligned}$$

The left-hand side of this equation can be expressed through the effective flux flow conductivity. We write

$$\frac{\Phi_0}{c}\left(\mathbf{j}_{\rm tr} \cdot [\hat{\mathbf{H}} \times \mathbf{v}_L] \right) = \mathbf{j}_{\rm tr}\langle \mathbf{E} \rangle_L S_0 = \sigma_f \langle \mathbf{E} \rangle_L^2 S_0$$

since $S_0 = \Phi_0/B$. The right-hand side is also proportional to $\langle \mathbf{E} \rangle_L^2$. The coefficient of proportionality determines σ_f.

12.5 Flux flow

12.5.1 *Single vortex: Low fields*

Consider first a single vortex provided $\kappa \gg 1$ and $H \ll H_{c2}$. In this case we can neglect the magnetic field and the vector potential compared to the gradient of the order parameter since

$$(2e/c)A \sim (2e/c)H\xi \ll (2e/c)H_{c2}\xi \sim 1/\xi .$$

We have from eqn (12.15)

$$\begin{aligned}\frac{\Phi_0}{c}[\mathbf{j}_{tr} \times \hat{\mathbf{H}}] &= -\Gamma \int_{S_0} \left[\left(\frac{\partial \Delta^*}{\partial t} - 2ie\varphi\Delta^*\right)\nabla\Delta + c.c.\right] dS \\ &= 2\Gamma \int_{S_0} \left[(\mathbf{v}_L \cdot \nabla|\Delta|)\,\nabla|\Delta| - 2e|\Delta|^2\Phi\,\nabla\chi\right] dS\ . \end{aligned} \qquad (12.17)$$

The order parameter magnitude $|\Delta| = \Delta_\infty f(\rho)$ for a static vortex satisfies eqn (1.48). To calculate the integrals in eqn (12.17) we also need to know the gauge invariant potential Φ. It satisfies eqn (11.48) whence

$$\nabla^2\Phi - l_E^{-2} f^2 \Phi = 0$$

since $\eta\left(|\Delta|/\Delta_\infty\right) = f^2$ for the usual TDGL model, eqn (1.80). The potential φ should be finite. Therefore, we require

$$\Phi \to -\frac{\hbar}{2e}\left(\mathbf{v}_L \cdot \nabla\right)\chi \qquad (12.18)$$

for $\rho \to 0$. We put $\rho = \xi\tilde{\rho}$ and

$$\Phi = -\frac{v_{L\phi}\mu}{2e\xi} \qquad (12.19)$$

where ϕ is the azimuthal angle in the cylindrical frame (ρ, ϕ, z). The function $\mu(\tilde{\rho})$ satisfies the equation

$$\left(\frac{d^2\mu}{d\tilde{\rho}^2} + \frac{1}{\tilde{\rho}}\frac{d\mu}{d\tilde{\rho}} - \frac{\mu}{\tilde{\rho}^2}\right) - uf^2\mu = 0 \qquad (12.20)$$

where $u = \xi^2/l_E^2$, see eqn (1.81). The condition eqn (12.18) requires $\mu \to 1/\tilde{\rho}$ for $\tilde{\rho} \to 0$. We see that a moving vortex induces a dipole-like scalar potential proportional to the vortex velocity.

The two terms under the integral in eqn (12.17) represent two different mechanisms of dissipation during the vortex motion exactly as the two corresponding terms in the dissipation function eqn (12.4). The first term is the Tinkham's mechanism (Tinkham 1964) of relaxation of the order parameter: Due to the motion of a vortex, the order parameter at a given point varies in time which produces a relaxation accompanied by dissipation. The second is the so-called

Bardeen and Stephen (1965) contribution. It accounts for normal currents flowing through the vortex core.

Equation (12.17) gives

$$\frac{\Phi_0}{c}\left([\mathbf{j}_{\rm tr}\times\hat{\mathbf{H}}]\cdot\mathbf{d}\right) = 2\Gamma\int\left[(v_{L\rho}d_\rho)\left(\frac{d|\Delta|}{d\rho}\right)^2 + (v_{L\phi}d_\phi)\frac{|\Delta|^2\mu}{\rho}\right]dS$$

$$= 2\pi\Gamma(\mathbf{v}_L\cdot\mathbf{d})\int\left[\left(\frac{d|\Delta|}{d\tilde{\rho}}\right)^2 + \frac{|\Delta|^2\mu}{\tilde{\rho}}\right]\tilde{\rho}\,d\tilde{\rho}. \quad (12.21)$$

Here we have restored the arbitrary vector $\mathbf{d}$ that appears in eqns (12.12) and (12.14) to make the calculations more convenient. The transport current is

$$\mathbf{j}_{\rm tr} = \frac{2\pi c\Gamma\Delta_\infty^2 a}{\Phi_0 B}[\mathbf{B}\times\mathbf{v}_L]. \quad (12.22)$$

Here

$$a = \int\left[\left(\frac{df}{d\tilde{\rho}}\right)^2 + \frac{f^2\mu}{\tilde{\rho}}\right]\tilde{\rho}\,d\tilde{\rho}. \quad (12.23)$$

Expressing $\mathbf{v}_L$ through $\langle\mathbf{E}\rangle_L$ we obtain the flux flow conductivity

$$\sigma_f = \frac{2\pi c^2\Gamma\Delta_\infty^2 a}{B\Phi_0}.$$

With the microscopic values of Γ and Δ_∞ taken for dirty superconductors, we find

$$\sigma_f = \frac{ua}{2}\sigma_n\frac{H_{c2}}{B}. \quad (12.24)$$

As already mentioned, the first term in eqn (12.23) is due to the dissipation caused by relaxation of the order parameter after passage of a vortex, while the second one comes from the normal-current dissipation in the vortex core. In the TDGL model with $u \sim 1$, both mechanisms give comparable contributions to σ_f. The numerical calculations using the vortex order parameter obtained by solving the GL equation give for the first term (see the review by Gor'kov and Kopnin (1975) and references therein)

$$a_1 = \int\left(\frac{df}{d\tilde{\rho}}\right)^2\tilde{\rho}\,d\tilde{\rho} \approx 0.274$$

which is independent of u. The second term, however, does depend on u. For example, for $u = 5.79$ we have

$$a_2 = \int f^2\mu\,d\tilde{\rho} \approx 0.228.$$

To calculate a_2 we need first to solve eqn (12.20). In this case, the total coefficient is $a \approx 0.502$. The flux flow conductivity becomes

$$\sigma_f \approx 1.45\sigma_n \frac{H_{c2}}{B}. \tag{12.25}$$

It is close to the expression $\sigma_f = \sigma_n H_{c2}/B$ known as the Bardeen–Stephen model (Bardeen and Stephen 1965).This simple prediction implies that the flux flow resistivity is just a normal-state resistivity times the fraction of space occupied by vortex cores.

12.5.2 *Dense lattice: High fields*

Consider now high fields, $H \to H_{c2}$. We put $\mathbf{A} = (0, Hx, 0)$, and choose the static potential in the form $\varphi = -E_x x$, where the electric field $E_x = -(v_y/c)H$ is homogeneous in the leading approximation. Recall that the static order parameter in the vicinity of the upper critical field, see Section 1.1.2, eqn (1.36):

$$\Delta = \sum_n C_n e^{iqny} \exp\left[-\frac{1}{2\xi^2}\left(x - \frac{cqn}{2eH}\right)^2\right]. \tag{12.26}$$

Equation (12.16) gives

$$\frac{1}{c}[\mathbf{j}_{\rm tr} \times \mathbf{B}]\mathbf{d} = -\Gamma\left\langle\left[\mathbf{d}\left(\nabla - \frac{2ie}{c}\mathbf{A}\right)\Delta\right]\left(\frac{\partial\Delta^*}{\partial t} - 2ie\varphi\Delta^*\right) + c.c.\right\rangle_L$$
$$+\frac{1}{c}\left\langle[\mathbf{j}_n \times \mathbf{H}]\right\rangle_L \mathbf{d}$$

where $\mathbf{d}$ is again an arbitrary vector. We have

$$\frac{1}{c}[\mathbf{j}_{\rm tr} \times \mathbf{B}]\mathbf{d} = \Gamma\left\langle\left[d_x\frac{\partial\Delta}{\partial x} + d_y\left(\frac{\partial\Delta}{\partial y} - \frac{2ieHx}{c}\Delta\right)\right]\right.$$
$$\left.\times\, v_y\left[\frac{\partial\Delta^*}{\partial y} + \frac{2ieHx}{c}\Delta^*\right] + c.c.\right\rangle_L$$
$$+\frac{1}{c}[\mathbf{j}_n \times \mathbf{B}]\mathbf{d}. \tag{12.27}$$

We need to calculate the averages

$$\left\langle\frac{\partial\Delta}{\partial x}\left(\frac{\partial\Delta^*}{\partial y} + \frac{2ieHx}{c}\Delta^*\right) + c.c.\right\rangle_L \tag{12.28}$$

and

$$\left\langle\left(\frac{\partial\Delta}{\partial y} - \frac{2ieHx}{c}\Delta\right)\left(\frac{\partial\Delta^*}{\partial y} + \frac{2ieHx}{c}\Delta^*\right) + c.c.\right\rangle_L. \tag{12.29}$$

We make use of the identities

$$\frac{\partial\Delta}{\partial x} = \mathrm{sign}\,(e)\left(-i\frac{\partial\Delta}{\partial y} - \frac{2eHx}{c}\Delta\right),$$

$$\frac{\partial \Delta^*}{\partial x} = \operatorname{sign}(e)\left(i\frac{\partial \Delta^*}{\partial y} - \frac{2eHx}{c}\Delta^*\right), \tag{12.30}$$

derived for the order parameter from eqn (12.26). The sign of the electronic charge appears because the upper critical field contains the modulus of the electronic charge,

$$H_{c2} = \frac{\Phi_0}{2\pi\xi^2} = \frac{c}{2\,|e|\,\xi^2}.$$

For the average in eqn (12.28) we obtain

$$\left\langle \frac{\partial \Delta}{\partial x}\left(\frac{\partial \Delta^*}{\partial y} + \frac{2ieHx}{c}\Delta^*\right) + c.c.\right\rangle_L = \left\langle -i\frac{\partial \Delta}{\partial x}\frac{\partial \Delta^*}{\partial x} + c.c.\right\rangle_L = 0.$$

The second average is

$$\left\langle\left(\frac{\partial \Delta}{\partial y} - \frac{2ieHx}{c}\Delta\right)\left(\frac{\partial \Delta^*}{\partial y} + \frac{2ieHx}{c}\Delta^*\right)\right\rangle_L = \left\langle \frac{\partial \Delta}{\partial x}\frac{\partial \Delta^*}{\partial x}\right\rangle_L.$$

Therefore

$$\begin{aligned}
&\left\langle\left(\frac{\partial \Delta}{\partial y} - \frac{2ieHx}{c}\Delta\right)\left(\frac{\partial \Delta^*}{\partial y} + \frac{2ieHx}{c}\Delta^*\right)\right\rangle_L \\
&= -\frac{1}{2}\left\langle \Delta\left[{}^*\left(\frac{\partial}{\partial y} - \frac{2ieHx}{c}\right)^2\Delta + \frac{\partial^2 \Delta}{\partial x^2}\right]\right\rangle_L \\
&= \frac{1}{2\xi^2}\left\langle |\Delta|^2\right\rangle_L.
\end{aligned} \tag{12.31}$$

We obtain from eqn (12.27)

$$\frac{1}{c}[\mathbf{j}_{\text{tr}} \times \mathbf{B}] = \mathbf{v}_L\left(\Gamma\frac{\left\langle|\Delta|^2\right\rangle_L}{\xi^2} + \frac{\sigma_n B^2}{c^2}\right). \tag{12.32}$$

Finally

$$\mathbf{j}_{\text{tr}} = \frac{1}{c}[\mathbf{B} \times \mathbf{v}_L]\left(\sigma_n + 4e^2\Gamma\xi^2\left\langle|\Delta|^2\right\rangle_L\right) = \sigma_f \mathbf{E} \tag{12.33}$$

where the flux flow conductivity is

$$\sigma_f = \sigma_n + 4e^2\Gamma\xi^2\left\langle|\Delta|^2\right\rangle_L. \tag{12.34}$$

The average magnitude of the order parameter is from eqn (1.43)

$$\left\langle|\Delta|^2\right\rangle_L = \frac{2\kappa^2\Delta_\infty^2}{(2\kappa^2 - 1)\,\beta_A}\left(1 - \frac{B}{H_{c2}}\right).$$

Therefore

$$\sigma_f = \sigma_n + \frac{4e^2\Gamma\xi^2\Delta_\infty^2}{\beta_A[1 - 1/(2\kappa^2)]}\left(1 - \frac{B}{H_{c2}}\right)$$

$$= \sigma_n \left[1 + \frac{u}{2} \frac{(1 - B/H_{c2})}{\beta_A [1 - (1/2\kappa^2)]} \right]. \tag{12.35}$$

Equations (12.25) and (12.35) determine the flux flow conductivity in the limits of low and high magnetic fields. The effective resistivity $\rho_f = \sigma_f^{-1}$ is linear in the magnetic field in both limits. However, the slope at low fields is smaller than that for $H \to H_{c2}$. Indeed,

$$\frac{H_{c2}}{\rho_n} \frac{\partial \rho_f}{\partial B} = \begin{cases} 2/ua, & H \ll H_{c2}, \\ u/2\beta_A, & H \to H_{c2}. \end{cases}$$

For practical values of $u \geq 5.79$, one has $2/ua < u/2\beta_A$.

It is interesting to note that the linearized TDGL equation has an analytical solution describing the moving vortex lattice (Schmid 1966). Indeed, the TDGL equation

$$\tau_\Delta \left(\frac{\partial \Delta}{\partial t} + 2ie\varphi \Delta \right) = \xi^2 \left(\nabla - \frac{2ie}{c} \mathbf{A} \right)^2 \Delta + \Delta \tag{12.36}$$

with $\mathbf{A} = (0,\, Hx,\, 0)$ and $\phi = -E_x x$ has the solution within the first-order terms in E

$$\Delta = \sum_n C_n \exp \left[iqn \left(y + \frac{cE_x t}{H} \right) \right] \times \exp \left[-\frac{1}{2\xi^2} \left(x - \frac{qnc}{2eH} \right)^2 + ie\tau_\Delta E_x \left(x - \frac{qnc}{2eH} \right) \right]. \tag{12.37}$$

It corresponds to a vortex lattice moving with the velocity $v_y = -cE_x/H$.

12.5.3 *Direction of the vortex motion*

From eqns (12.22) and (12.33) we observe that vortices move at a right angle to the transport current, and generate the electric field parallel to it. The Lorentz force from the transport current is balanced by purely dissipative friction force:

$$\frac{\Phi_0}{c} [\mathbf{j}_{\rm tr} \times \hat{\mathbf{H}}] + \eta \mathbf{v}_L = 0$$

where the vortex viscosity is expressed through σ_f:

$$\eta = \frac{B\Phi_0 \sigma_f}{c^2}. \tag{12.38}$$

There is no Hall effect and no gyroscopic force, i.e., the force which would be perpendicular to the vortex velocity: the transverse force $\mathbf{F}_\perp$ in Fig. 12.1 is absent. This is the result of a fully dissipative dynamics of the TDGL model. We discuss what can be the origin of a gyroscopic force and of the Hall effect in the TDGL model later in Section 12.9.

12.6 Anisotropic superconductors

Almost all high temperature superconductors are highly anisotropic, and some of them have even a well-developed layered structure. The time-dependent Ginzburg–Landau model can also be used to calculate the flux flow conductivity in such superconductors. One should note, however, that the TDGL scheme works reasonably well only for dirty superconductors and in a very limited range of parameters. Therefore, the results of this section do not directly apply to real high temperature superconductors which are, in fact, clean materials. Nevertheless, it is instructive to see what predictions can be expected within the TDGL model for anisotropic superconductors. In this section we consider uniaxial anisotropic superconductors. Some aspects of the vortex dynamics in layered systems will be discussed in the following section.

Applying the general TDGL scheme described on page 19 to the anisotropic Ginzburg–Landau equation (7.21) we obtain the corresponding order parameter equation in the form

$$-\Gamma\left(\frac{\partial\Delta}{\partial t}+2ie\varphi\Delta\right)=\alpha\Delta+\beta\left|\Delta\right|^{2}\Delta-\gamma_{ab}\left(\nabla-\frac{2ie}{c}\mathbf{A}\right)^{2}\Delta$$
$$-\gamma_{c}\left(\frac{\partial}{\partial z}-\frac{2ie}{c}A_{z}\right)^{2}\Delta.$$

The total current in the ab plane is

$$\mathbf{j}=\sigma_{n}^{(ab)}\mathbf{E}-2ie\gamma_{ab}\left[\Delta^{*}\left(\nabla-\frac{2ie}{c}\mathbf{A}\right)\Delta-\Delta\left(\nabla+\frac{2ie}{c}\mathbf{A}\right)\Delta^{*}\right], \tag{12.39}$$

while the total current along c has the form

$$j_{z}=\sigma_{n}^{(c)}E_{z}-2ie\gamma_{c}\left[\Delta^{*}\left(\frac{\partial}{\partial z}-\frac{2ie}{c}A_{z}\right)\Delta-\Delta\left(\frac{\partial}{\partial z}+\frac{2ie}{c}A_{z}\right)\Delta^{*}\right] \tag{12.40}$$

where the z axis is directed along the crystallographic c axis.

We follow the approach developed by Ivlev and Kopnin (1991). Assume that the normal-state conductivity satisfies the condition

$$\frac{\sigma_{n}^{(ab)}}{\sigma_{n}^{(c)}}=\frac{\gamma_{ab}}{\gamma_{c}}=\frac{\xi_{ab}^{2}}{\xi_{c}^{2}}=\frac{m_{c}}{m_{ab}}. \tag{12.41}$$

since both the gradient terms in the Ginzburg–Landau equation and the conductivity tensor are proportional to the same factor

$$\frac{\partial E_{n}}{\partial p_{i}}\frac{\partial E_{n}}{\partial p_{j}}$$

averaged over the Fermi surface with the inverse impurity scattering cross section, see Section 9.3. This is true for dirty superconductors. In clean systems, eqn (12.41) is expected to hold only for isotropic scattering by impurities.

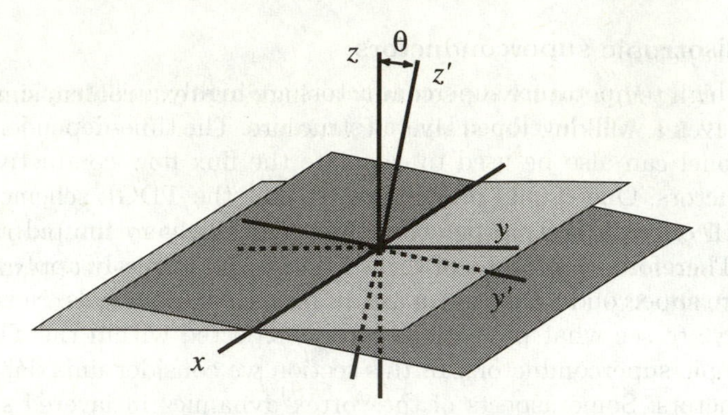

FIG. 12.2. The coordinate frame (x, y, z) is associated with the crystallographic axes; the frame (x', y', z') has the z' along the magnetic field.

Consider the orientation of the magnetic field at an angle θ to the crystallographic c-axis. We chose the coordinate frame (x', y', z') with the $x' = x$-axis perpendicular to the plane comprising the c axis and the magnetic field, see Fig. 12.2. The y'-axis makes the angle θ with the ab plane. The order parameter and the magnetic field depend only on x' and y'. The order parameter equation takes the form

$$-\Gamma\left(\frac{\partial\Delta}{\partial t} + 2ie\varphi\Delta\right) = -\left|\alpha\right|\Delta + \beta\left|\Delta\right|^2\Delta - \gamma_{ab}\left(\frac{\partial}{\partial x'} - \frac{2ie}{c}A_{x'}\right)^2\Delta$$
$$-\gamma\left(\theta\right)\left(\frac{\partial}{\partial y'} - \frac{2ie}{c}A_{y'}\right)^2\Delta, \qquad (12.42)$$

where we define

$$\gamma\left(\theta\right) = \gamma_{ab}\cos^2\theta + \gamma_c\sin^2\theta.$$

The current along the x-axis is given by eqn (12.39) while

$$j_{y'} = \sigma_n\left(\theta\right)E_{y'}$$
$$-2ie\gamma\left(\theta\right)\left[\Delta^*\left(\frac{\partial}{\partial y'} - \frac{2ie}{c}A_{y'}\right)\Delta - \Delta\left(\frac{\partial}{\partial y'} + \frac{2ie}{c}A_{y'}\right)\Delta^*\right]. \qquad (12.43)$$

We define also

$$\sigma_n\left(\theta\right) = \sigma_n^{(ab)}\cos^2\theta + \sigma_n^{(c)}\sin^2\theta = \sigma_n^{(ab)}\gamma(\theta)/\gamma_{ab}.$$

Comparison of eqn (12.42) with eqn (7.33) shows, in particular, that the upper critical magnetic field for this configuration is (Kats 1969, 1970, Lawrence and Doniach 1971)

$$H_{c2}\left(\theta\right) = \frac{\Phi_0}{2\pi\xi_{ab}\xi\left(\theta\right)}$$

where $\xi^2\left(\theta\right)/\xi_{ab}^2 = \gamma\left(\theta\right)/\gamma_{ab}$.

12.6.1 *Low fields*

For $H_{c1} \ll H \ll H_{c2}$, we have essentially a single-vortex problem similar to that in Section 12.5.1. Making the scaling of variables

$$x' = \xi_{ab}\,\tilde{x},\; y' = \xi(\theta)\,\tilde{y},$$

and $\mathbf{A} = (H/H_{c2})\,\tilde{\mathbf{A}}$ one can see that the static single-vortex problem reduces to an isotropic situation of Section 1.1.2. A static vortex determined by eqns (12.42), (12.39), and (12.43) has an axial symmetry in the space

$$\tilde{\rho} = \left(\frac{x'}{\xi_{ab}}, \frac{y'}{\xi(\theta)}\right)$$

such that

$$\Delta = \Delta_\infty f(\tilde{\rho}) \exp\left(i\tilde{\phi}\right)$$

where $\tilde{\rho}$ and $\tilde{\phi}$ are the radius and the azimuthal angle in the coordinate frame $(\tilde{x}, \tilde{y})$. The function $f(\tilde{\rho})$ satisfies eqn (7.39) on page 139.

To calculate the flux flow conductivity we can use eqn (12.17) which is also valid for anisotropic superconductors. To evaluate the integrals in the r.h.s. of eqn (12.17) we need to find the gauge-invariant potential Φ. Let us put

$$\Phi = \frac{1}{2e}\left[\frac{v_{Lx}\sin\tilde{\phi}}{\xi_{ab}} - \frac{v_{Ly}\cos\tilde{\phi}}{\xi(\theta)}\right]\mu(\tilde{\rho}).$$

The anisotropy is scaled out with this transformation due to eqn (12.41). The function μ satisfies eqn (12.20) where now

$$u = \frac{8e^2\Gamma\gamma_{ab}}{\beta\sigma_n^{(ab)}}.$$

We find from eqn (12.17)

$$\frac{\Phi_0}{c}\left([\mathbf{j}_{\rm tr} \times \hat{\mathbf{H}}] \cdot \mathbf{d}\right) = 2\pi\Gamma\Delta_\infty^2 \xi_{ab}\xi(\theta)\, a\left(\frac{v_{Lx}d_x}{\xi_{ab}^2} + \frac{v_{Ly}d_y}{\xi^2(\theta)}\right)$$

where a is determined by eqn (12.23).

Expressing the vortex velocity in terms of the average electric field

$$\mathbf{v}_L = (cE_{y'}/B,\; -cE_{x'}/B) \tag{12.44}$$

we obtain the flux flow conductivity components for currents along the x-axis and in the (x', y') plane:

$$\frac{\sigma_f^{(xx)}}{\sigma_n^{(ab)}} = \frac{\sigma_f^{(yy)}(\theta)}{\sigma_n(\theta)} = \frac{ua}{2}\frac{H_{c2}(\theta)}{B}. \tag{12.45}$$

This important result shows that, for the given orientation of the magnetic field, the ratio of the flux flow conductivity to the normal-state conductivity in the same direction in the plane perpendicular to the applied magnetic field scales as the upper critical field.

12.6.2 *High fields*

For $H \to H_{c2}(\theta)$ we calculate the flux flow conductivity using a solution of the linearized time-dependent equation as was done on page 242. We assume $\varphi = -E_{x'}x' - E_{y'}y'$, and $A_{y'} = Hx'$, $A_{x'} = 0$. Within the first-order terms in E we find instead of eqn (12.37)

$$\Delta = \sum_n C_n \exp\left[i\left(qn + 2eE_{y'}t\right)\left(y' + cE_{x'}t/H\right)\right]$$
$$\times \exp\left[-\frac{1}{2\xi_{ab}}\left(x - \frac{cE_{x'}t}{H} - \frac{cqn}{2eH}\right)^2\right]$$
$$\times \exp\left[\frac{\Gamma\xi_{ab}}{\gamma_{ab}}\left(ieE_{x'}\xi_{ab} - eE_{y'}\xi\left(\theta\right)\right)\left(x - \frac{cqn}{2eH}\right)\right].$$

The solution describes the vortex lattice moving with the velocity eqn (12.44). The coefficients C_n satisfy the periodicity conditions for an Abrikosov lattice and are normalized according to the nonlinear eqn (12.42). Calculating the components of current we find (keep in mind that we use $\kappa \gg 1$) the same scaling as for low fields,

$$\frac{\sigma_f^{(xx)}}{\sigma_n^{(ab)}} = \frac{\sigma_f^{(yy)}\left(\theta\right)}{\sigma_n\left(\theta\right)} = 1 + \frac{u}{2}\frac{H_{c2}\left(\theta\right) - B}{\beta_A H_{c2}\left(\theta\right)}.$$

For a general orientation of the transport current in the (x', y') plane, the electric field is not parallel to the current, of course. However, it does not give rise to the Hall effect because the direction of an electric field does not change if the magnetic field is reversed, $\mathbf{E}(\mathbf{H}) = \mathbf{E}(-\mathbf{H})$: both the vortex circulation and the vortex velocity change their signs keeping $\mathbf{E}$ unchanged.

12.7 Flux flow in layered superconductors

Flux flow in layered superconductors is a very rich and complicated phenomenon. Moreover, it has not been fully investigated so far. The fundamental difference with respect to homogeneous isotropic and anisotropic superconductors is a strong interaction between vortices and the crystal structure itself known as intrinsic pinning. This introduces new features to the vortex dynamics. Motion of vortices in presence of pinning (including intrinsic pinning) is associated with deformations of the vortex lattice and should be considered together with elastic properties of vortex arrays. The problem of vortex pinning is a special issue in the vortex physics. The reader can find some relevant references in the review by Blatter *et al.* (1994). In this section we consider two particular examples when the vortex deformations are either not important or can be easily taken into account. The first example is motion of vortices in a highly layered superconductor in a magnetic field inclined with respect to the layers. In this case a vortex makes sharp bends before it crosses superconducting layers and penetrates through them in a form of two-dimensional pancake vortices. The second example treats straight vortices aligned parallel to the layers when they

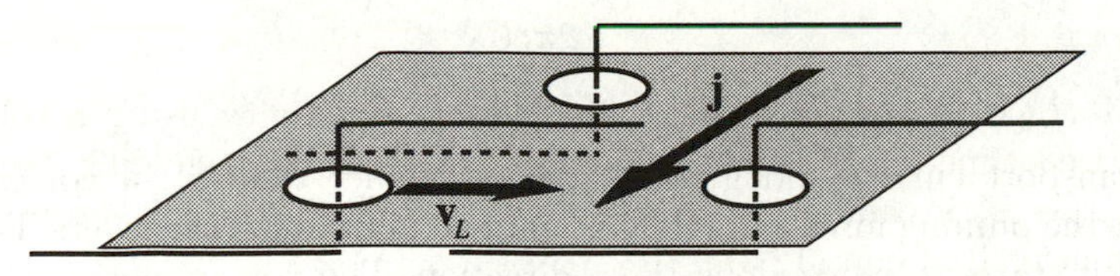

FIG. 12.3. Motion of vortices in layered superconductors. The shadowed plane is the conducting layer parallel to the crystal *ab* plane. Only pancake vortices move and produce dissipation in a presence of transport current **j**

move in the direction perpendicular to the layers such that they experience an intrinsic pinning as described in Section 7.4.2.

12.7.1 *Motion of pancake vortices*

Consider a highly layered superconductor in a magnetic field which is inclined to the conducting layers at some angle θ. We can describe a magnetic field penetration in terms of two systems of vortices (Kes *et al.* 1990, Ivlev *et al.* 1990). The component parallel to the layers B_{ab} produces Josephson vortices whose cores fit in between the layers so that superconductivity on the layers is not essentially affected. This is equivalent to the assumption of a very large value of the in-plane upper critical field $H_{c2}(\pi/2)$. These vortices are strongly pinned by an interaction with the crystal structure. The magnetic field component B_c penetrates by forming pancake vortices having normal cores on the layers and screening currents flowing in the layers. Combining these two vortex systems, one can say that, for a tilted magnetic field, Josephson vortices (produced by the component B_{ab}) cross the layers by making kinks in places where they meet pancake vortices (produced by B_c), see Fig. 12.3. If a transport current flows along the layers, the kinks, i.e., the pancake vortices, will move along the planes thus producing dissipation. Since the density of pancake vortices is proportional to $B_c/H_{c2}(0)$, the corresponding flux flow conductivity is

$$\frac{\sigma_f^{(xx)}}{\sigma_n^{(ab)}} = \frac{ua}{2}\frac{H_{c2}(0)}{B\,|\cos\theta|}$$

according to eqn (12.45). The flux flow conductivity depends only on the magnetic field component along the c axis.

12.7.2 *Intrinsic pinning*

Consider the situation when the magnetic field and vortices are aligned parallel to the layers. The transport current flows in the plane of layers such that the Lorentz force eqn (12.1) is perpendicular to the layers. We choose the z-axis to be along the crystallographic c-direction. The vortex displacement δz along the z-axis is opposed by the intrinsic pinning force. In this section, we consider the simplest case of a weakly layered structure. For this limit, the pinning force is given by eqn (7.44) on page 140. Balancing the Lorentz force and the pinning force we find the maximum depinning current

$$j_p = \frac{2\pi c U_0}{\phi_0 s}. \tag{12.46}$$

If the transport current along the x-axis is smaller than j_p, a vortex cannot overcome the pinning force and remains trapped in between the layers. For higher currents, it starts to move along the z direction. The vortex motion is opposed also by friction such that the full force balance is

$$\eta^{(xx)} \frac{\partial \delta z}{\partial t} = \frac{\Phi_0}{c} \left[j - j_p \sin\left(\frac{2\pi \delta z}{s} \right) \right]. \tag{12.47}$$

The vortex viscosity introduced in eqn (12.38)

$$\eta^{(xx)} = \frac{B\Phi_0 \sigma_f^{(xx)}}{c^2} = \frac{2\pi \Gamma \Delta_0^2 \xi_{ab}}{\xi_c}$$

is expressed through the corresponding flux flow conductivity $\sigma_f^{(xx)}$. It is the conductivity in eqn (12.45) for the magnetic field orientation $\theta = \pi/2$.

A vortex which moves according to eqn (12.47) needs a time

$$t_0 = \frac{\eta^{(xx)} cs}{\Phi_0} \sqrt{j^2 - j_p^2}$$

to cover the distance s. The averaged velocity is $v_z = s/t_0$ which gives rise to the electric field in the x-direction $E_x = Bv_z/c$ such that

$$E_x = \sigma_f^{(xx)} \sqrt{j^2 - j_p^2}.$$

The I–V curve becomes nonlinear in presence of pinning. A finite voltage appears when $j > j_c$. For high currents, $j \gg j_c$, the linear dependence is restored.

12.8 Flux flow within a generalized TDGL theory

12.8.1 *Dirty superconductors*

In this section we consider a situation where the superconductor cannot be described by a simple TDGL model but is still within the range of parameters such that the generalized TDGL scheme derived in Section 11.2 is applicable. As we know, it requires slow gradients and time derivatives such that

$$Dk^2, \omega \ll \tau_{\mathrm{ph}}^{-1}.$$

Since, in presence of vortices, the gradients are $k \sim \xi^{-1}(T)$, the condition of applicability is $\Delta^2/T \ll \tau_{\mathrm{ph}}^{-1}$ or $T_c - T \ll \tau_{\mathrm{ph}}^{-1}$. Note that, depending on temperature, the order parameter can be either outside of, $\Delta \gg \tau_{\mathrm{ph}}^{-1}$, or inside, $\Delta \ll \tau_{\mathrm{ph}}^{-1}$, the gapless region.

We start with eqn (12.15) and transform it as follows

$$\frac{\Phi_0}{c}[\mathbf{j}_{\rm tr} \times \hat{\mathbf{H}}] = \int_{S_0} \left[e^{i\chi} \nabla |\Delta| \frac{\delta \mathcal{F}_{sn}}{\delta \Delta} + e^{-i\chi} \nabla |\Delta| \frac{\delta \mathcal{F}_{sn}}{\delta \Delta^*} - \frac{2ie}{\hbar c} \mathbf{Q} \left(\Delta \frac{\delta \mathcal{F}_{sn}}{\delta \Delta} - \Delta^* \frac{\delta \mathcal{F}_{sn}}{\delta \Delta^*} \right) + \frac{1}{c} [\mathbf{j}_n \times \mathbf{H}] \right] dS$$

$$= \int_{S_0} \left[\nabla |\Delta| \frac{\delta \mathcal{F}_{sn}}{\delta |\Delta|} + \frac{1}{c} \mathbf{Q} \mathrm{div} \mathbf{j}_s + \frac{1}{c} [\mathbf{j}_n \times \mathbf{H}] \right] dS. \quad (12.48)$$

Using eqns (11.25) and (11.26) we can write the force balance equation (12.48) as

$$\frac{\Phi_0}{c}[\mathbf{j}_{\rm tr} \times \hat{\mathbf{H}}] = \nu(0) \int_{S_0} \left[-\left(\nabla |\Delta| \frac{\partial |\Delta|}{\partial t} \right) \frac{\pi}{4T} \sqrt{1 + 4|\Delta|^2 \tau_{\rm ph}^2} + \frac{\pi e^2}{cT} \frac{\Phi |\Delta|^2}{\sqrt{1 + 4|\Delta|^2 \tau_{\rm ph}^2}} \mathbf{Q} \right] dS + \int_{S_0} [\mathbf{j}_n \times \mathbf{H}] \, dS. \quad (12.49)$$

In this section, we only consider the limit of low fields when vortices are well separated. One can neglect the magnetic field in eqn (12.49) and put

$$\mathbf{Q} = -\frac{c}{2e} \nabla \chi.$$

Using eqn (11.27) we find for the function μ defined through eqn (12.19)

$$\frac{d^2\mu}{d\rho^2} + \frac{1}{\rho} \frac{d\mu}{d\rho} - \frac{\mu}{\rho^2} - \frac{\pi |\Delta|^2 \mu}{2DT \sqrt{1 + 4|\Delta|^2 \tau_{\rm ph}^2}} = 0.$$

This equation differs from eqn (12.20) by a more complicated Δ-dependence of the last term. The boundary condition is $\mu \to 1/\rho$ for $\rho \to 0$. Replacing $\partial |\Delta| / \partial t$ with $-(\mathbf{v}_L \cdot \nabla) |\Delta| = -v_{L\rho}(\partial |\Delta| / \partial \rho)$ in eqn (12.49) we obtain

$$\frac{\Phi_0}{c}[\mathbf{j}_{\rm tr} \times \hat{\mathbf{H}}] = \mathbf{v}_L \frac{\pi^2 \nu(0)}{4T} \int_0^{r_0} \left[\left(\frac{\partial |\Delta|}{\partial \rho} \right)^2 \sqrt{1 + 4|\Delta|^2 \tau_{\rm ph}^2} + \frac{\mu}{\xi \rho} \frac{|\Delta|^2}{\sqrt{1 + 4|\Delta|^2 \tau_{\rm ph}^2}} \right] \rho d\rho. \quad (12.50)$$

The integral is extended over the radius r_0 of the unit cell of the vortex lattice.

Equation (12.50) determines the flux flow conductivity. Using $\mathbf{v}_L = c\left[\mathbf{E} \times \hat{\mathbf{H}}\right] / B$ we obtain

$$\sigma_f = \frac{c^2 \nu(0) \pi^2 \Delta_\infty^2 F(q)}{4T \Phi_0 B} \quad (12.51)$$

where $q = 2\Delta_\infty \tau_{\rm ph}$, and Δ_∞ is the order parameter magnitude far from the vortex. The function $F(q) = F_1(q) + F_2(q)$ is defined by two equations,

$$F_1(q) = \int_0^{r_0} \left(\frac{\partial f}{\partial \rho}\right)^2 \sqrt{1 + q^2 f^2}\, \rho d\rho, \tag{12.52}$$

$$F_2(q) = \int_0^{r_0} \frac{\mu}{\xi \rho} \frac{f^2}{\sqrt{1 + q^2 f^2}}\, \rho d\rho, \tag{12.53}$$

where, as before, $f = |\Delta| / \Delta_\infty$. Using the dirty-limit value

$$\xi^2 = \frac{\pi^3 T_c D}{7\zeta(3)\Delta_\infty^2}$$

from eqns (1.9) and (1.15), we find

$$\sigma_f = \beta \sigma_n \frac{H_{c2}}{B} \tag{12.54}$$

where

$$\beta = \frac{uF(q)}{2}$$

and $u = \pi^4 / 14\zeta(3) \approx 5.79$.

In the gapless regime, $q \ll 1$, the functions $F_1(q)$ and $F_2(q)$ are of the same order so that $F(q) \to a$, and the result of the TDGL model eqn (12.24) with $\beta = ua/2$ is reproduced. A more realistic case, when the order parameter is outside the gapless region, corresponds to $q \gg 1$. For $q \gg 1$ when $T_c \tau_{\rm ph} \sqrt{1 - T/T_c} \gg 1$, the electric field penetration length is large,

$$l_E = \sqrt{(q/u)}\, \xi.$$

The function μ under the integral in eqn (12.53) is $\mu \approx 1/\rho$ within a region larger than the vortex core. The term $F_2(q)$ diverges logarithmically and is cut off at distances $\rho \sim R_0$ such that

$$R_0 = \min\{l_E, r_0\} \tag{12.55}$$

where $r_0 \sim \xi(T)\sqrt{H_{c2}/B}$ is the intervortex distance,

$$F_2(q) = \frac{1}{q\xi} \ln \frac{R_0}{\xi}. \tag{12.56}$$

However, the prefactor in eqn (12.56) is small. The largest contribution comes thus from $F_1(q)$ in eqn (12.52) which gives

$$F(q) \approx F_1(q) = q \int_0^\infty f \left(\frac{\partial f}{\partial \rho}\right)^2 \rho d\rho.$$

The integral is of the order of unity. The flux flow conductivity is thus much higher than what is predicted by the Bardeen and Stephen model. Indeed,

$\beta \sim \sqrt{1-T/T_c}(T_c\tau_{\rm ph}) \gg 1$. It is determined by a relaxation of the order parameter (Tinkham's mechanism) via slow inelastic electron–phonon interactions while the contribution from normal currents in the vortex core is smaller by a factor on order q^{-2}. The factor β decreases proportionally to $\sqrt{1-T/T_c}$ when T approaches T_c.

12.8.2 *d-wave superconductors*

The same approach can be applied to calculate the flux flow conductivity in a d-wave superconductor (Kopnin 1998 a) described by eqns (11.46) and (11.47). We know that d-wave superconductors should be clean, $T_c\tau_{\rm imp} \gg 1$. However, the applicability of eqns (11.46) and (11.47) requires that the temperatures are close to T_c such that $\ell \ll \xi(T)$ which is equivalent to $1-T/T_c \ll (\tau_{\rm imp}T_c)^2$. We obtain from eqn (12.48)

$$\begin{aligned}\frac{\Phi_0}{c}[\mathbf{j}_{\rm tr}\times\hat{\mathbf{H}}] = \frac{\pi\nu(0)}{8T_c}\int_{S_0}\Big[2\left\langle\nabla\left|\Delta_{\mathbf{p}}\right|(\mathbf{v}_L\cdot\nabla)\left|\Delta_{\mathbf{p}}\right|\right\rangle_{\mathbf{p}} \\ +\frac{8e^2}{c}\mathbf{Q}\Phi\left\langle\left|\Delta_{\mathbf{p}}\right|^2\right\rangle_{\mathbf{p}} - U\nabla\left\langle\left|\Delta_{\mathbf{p}}\right|^2\right\rangle_{\mathbf{p}}\Big]\,dS \\ +\frac{\sigma_n}{c}\int_{S_0}[\mathbf{E}\times\mathbf{H}]\,dS. \end{aligned} \quad (12.57)$$

We keep the subscript $\mathbf{p}$ to distinguish the momentum average from the average $\langle\ldots\rangle_L$ taken over the vortex array.

The function U is found from eqn (11.41) under the condition that it does not increase with distance. Assuming that the diffusion is faster than the inelastic relaxation we find

$$\nabla U = \mathbf{v}_L\frac{\tau_{\rm imp}}{D}\left[\left\langle\left|\Delta_{\mathbf{p}}\right|^2\right\rangle_{\mathbf{p}} - \left\langle\left\langle\left|\Delta_{\mathbf{p}}\right|^2\right\rangle_{\mathbf{p}}\right\rangle_L\right]. \quad (12.58)$$

In the low field limit $H \ll H_{c2}$ we have $\mathbf{Q} = -(c/2e)\nabla\chi$. Moreover, the scalar potential is $\Phi = (1/2e)\partial\chi/\partial t$ because $l_E \gg \xi$. For a vortex lattice with a symmetry not lower than tetragonal,

$$\int_{S_0}\left\langle\nabla_i\left|\Delta_{\mathbf{p}}\right|\nabla_k\left|\Delta_{\mathbf{p}}\right|\right\rangle_p dS = \frac{1}{2}\delta_{ik}\int_{S_0}\left\langle(\nabla\left|\Delta_{\mathbf{p}}\right|)^2\right\rangle_{\mathbf{p}} dS,$$

$$\int_{S_0}(\nabla_i\chi\nabla_k\chi)\left\langle\left|\Delta_{\mathbf{p}}\right|^2\right\rangle_p dS = \frac{1}{2}\delta_{ik}\int_{S_0}(\nabla\chi)^2\left\langle\left|\Delta_{\mathbf{p}}\right|^2\right\rangle_{\mathbf{p}} dS.$$

As a result, the flux flow conductivity becomes

$$\sigma_f = \sigma_n + \frac{\pi^2\nu(0)c^2\left\langle\left|\Delta_{\mathbf{p}}\right|_\infty^2\right\rangle_{\mathbf{p}} F_d}{4T_c\Phi_0 B}. \quad (12.59)$$

Equation (12.59) looks exactly as eqn (12.51) with the difference that the factor F_d is now

$$F_d = \left(2\pi\left\langle\left|\Delta_{\mathbf{p}}\right|_\infty^2\right\rangle_{\mathbf{p}}\right)^{-1}\int_{S_0}\left[\left\langle(\nabla\left|\Delta_{\mathbf{p}}\right|)^2\right\rangle_{\mathbf{p}} + (\nabla\chi)^2\left\langle\left|\Delta_{\mathbf{p}}\right|^2\right\rangle_{\mathbf{p}}\right.$$

$$+\frac{\tau_{\mathrm{imp}}}{D}\left[\left\langle|\Delta_{\mathbf{p}}|^{2}\right\rangle_{\mathbf{p}}-\left\langle\left\langle|\Delta_{\mathbf{p}}|^{2}\right\rangle_{\mathbf{p}}\right\rangle_{L}\right]^{2}\Bigg] dS \tag{12.60}$$

instead of eqns (12.52), (12.53). The largest contribution comes from the second term under the integral which is logarithmically diverging at distances $r \sim R_0$ where R_0 is determined by eqn (12.55). Therefore, $F_d \approx \ln[R_0/\xi(T)]$. We had already such logarithmic divergence of F due to a long relaxation length l_E in eqn (12.56) of the previous section. However, it was not important for the flux flow conductivity of dirty superconductors. On the contrary, it dominates in the present example. Writing eqn (12.51) in the form of eqn (12.54) we find that $\beta \sim F/T_c\tau_{\mathrm{imp}} \ll 1$ because $T_c\tau_{\mathrm{imp}} \gg 1$ for a clean superconductor.

12.8.3 *Discussion: Flux flow conductivity*

We can now summarize the results obtained for the flux flow conductivity within the TDGL model. The model identifies two principal sources for dissipation associated with a moving vortex. The first is due to variations of the order parameter with time as the vortex passes through a superconductor. The simple TDGL model describes the process of the order parameter relaxation in terms of a characteristic relaxation time τ_Δ which is of the order of $(T_c - T)^{-1}$. The generalized TDGL scheme treats this process in a more detailed manner. We encounter a concept of nonequilibrium excitations which relax through their interactions with phonons. Deviation from equilibrium is created by order parameter variations which cause changes in the energy spectrum and thus produce redistribution of excitations away from equilibrium. The order parameter returns to its undisturbed value only together with the relaxing excitations. Its characteristic relaxation time depends now on the electron–phonon mean free time τ_{ph}:

$$\tau_\Delta \sim \tau_{\mathrm{ph}}\left(1-\frac{T}{T_c}\right)^{-1/2}$$

and is much slower than $(T_c - T)^{-1}$ when the parameters are outside the gapless region. This results in a larger dissipation and larger conductivity as compared to the simple TDGL model.

The second mechanism is associated with currents which flow through vortex cores. Near a vortex core, the supercurrent converges into a normal current according to eqn (11.26) through the current conservation (the charge neutrality)

$$\mathrm{div}\mathbf{j}_s + \mathrm{div}\mathbf{j}_n = 0.$$

It is the normal current which dissipates energy. The conversion rate is determined by the r.h.s. of eqns (1.79) or (11.26), or by the r.h.s. of eqn (11.44) in the d-wave case. For a gapless regime, the conversion is fast. In our examples, the gapless regime in s-wave superconductors is associated with magnetic impurities or with an electron–phonon interaction very near the critical temperature. Here the dissipation caused by the normal currents is high. In a gapless situation for a d-wave superconductor near T_c, it gives even the largest contribution to the flux flow conductivity. However, the conversion rate is slow in superconductors which are far from the gapless regime, see eqn (12.56) in the limit $\Delta_\infty \tau_{\rm ph} \gg 1$.

The results obtained confirm our general expectation that the vortex dynamics is governed by kinetics of excitations driven out of equilibrium by a moving vortex. In the following chapter we consider an example when the kinetics of excitations creates even larger dissipation than that predicted by the generalized TDGL approach.

What the generalized TDGL model can not describe completely is the Hall effect associated with the transverse gyroscopic force on a moving vortex. To conclude the chapter on the TDGL model, we discuss how one should modify the TDGL scheme to incorporate at least some possible mechanisms responsible for the vortex Hall effect.

12.9 Flux flow Hall effect

We start with a very brief summary on the Hall effect in the normal state. A more detailed analysis of the Hall effect in normal metals can be found, for example, in the book by Abrikosov (1998). In the presence of a magnetic field, the current in a normal metal has generally both longitudinal and transverse components with respect to the electric field

$$\mathbf{j}_n = \sigma_n \mathbf{E} + \sigma_n^H [\mathbf{E} \times \hat{\mathbf{H}}] \tag{12.61}$$

where σ_n^H is the normal-state Hall conductivity. The Hall angle is defined as

$$\tan \Theta_H = \sigma_n^H / \sigma_n .$$

The Hall conductivity has a simple form for a metal with electronic spectrum $E_n = \mathbf{p}^2/2m^*$. For low magnetic fields,

$$\sigma_n^H = \sigma_n \omega_c \tau \tag{12.62}$$

where

$$\omega_c = \frac{eH}{m^* c}$$

is the cyclotron frequency. The low field limit implies $\omega_c \tau \ll 1$.

Let us estimate the parameter $\omega_c \tau$ for magnetic fields less than H_{c2}. For dirty superconductors

$$\omega_c \tau \sim \frac{\tau}{2m^* \xi^2} \frac{H}{H_{c2}} \sim \frac{T_c}{E_F} \left(1 - \frac{T}{T_c}\right) \frac{H}{H_{c2}} . \tag{12.63}$$

It is very small because, normally, $T_c/E_F \ll 1$. In this limit, the Hall conductivity is much smaller than the Ohmic component and the Hall angle is also small. We thus expect that the Hall effect in dirty superconductors cannot be large. We shall see nevertheless, that it exhibits interesting features specific only for the superconducting state. The most interesting property is that the Hall angle can change its sign after transition into a superconducting state, this is known as the Hall effect anomaly. The Hall anomaly has been observed in many experiments starting with the conventional niobium and vanadium type II superconductors (Niessen *et al.* 1967, Noto *et al.* 1976) and later re-discovered also in high temperature superconductors (see, for example, Hagen *et al.* 1993). This section describes the aspects of the flux flow Hall effect that can be identified within the TDGL theory.

12.9.1 *Modified TDGL equations*

As we have seen, the usual TDGL theory gives zero Hall effect. This is a result of pure dissipative nature of the TDGL equations. To incorporate nondissipative forces, we have to modify our TDGL equations (Kopnin *et al.* 1992, 1993, Dorsey 1992). This can be done in two ways. First, we write the normal current in the form of eqn (12.61). Second, we allow for an imaginary part of the relaxation constant Γ

$$\Gamma = \Gamma' + i\Gamma''. \tag{12.64}$$

The origin of an imaginary part of Γ is as follows (Aronov and Rapoport 1992). We know that a moving vortex induces a scalar potential proportional to the vortex velocity. The induced potential adds to the chemical potential of the superconductor: $\mu = \mu_0 - e\varphi$. On the other hand, the critical temperature depends on the chemical potential, therefore, the coefficient α in the GL equation becomes

$$\alpha = \alpha_0 - \frac{\partial \alpha}{\partial \mu} e\varphi = -\nu \left(1 - \frac{T}{T_c}\right) + \frac{\nu}{T_c}\frac{\partial T_c}{\partial \mu} e\varphi. \tag{12.65}$$

Since the critical temperature in the BCS theory is $T_c = \Omega_{\mathrm{BCS}} \exp(-1/\lambda)$ where Ω_{BCS} is the cut-off energy of pairing interaction, and $\lambda = |g|\nu$ is the coupling constant, we have

$$\frac{\partial T_c}{\partial \mu} = \frac{T_c}{\nu\lambda}\frac{\partial \nu}{\partial \mu}$$

Therefore, correction to α becomes

$$\delta\alpha = \frac{1}{\lambda}\frac{\partial \nu}{\partial \mu} e\varphi.$$

To preserve the gauge invariance, we should write the correction in the form

$$\delta\alpha\Delta = -i\frac{1}{2\lambda}\frac{\partial \nu}{\partial \mu}\left(\frac{\partial \Delta}{\partial t} + 2ie\varphi\Delta\right) \tag{12.66}$$

which results in an imaginary part of the coefficient in front of the order-parameter time-derivative

$$\Gamma'' = -\frac{1}{2\lambda}\frac{\partial \nu}{\partial \mu}. \tag{12.67}$$

The imaginary part Γ'' is small

$$\zeta = -\frac{\Gamma''}{\Gamma'} \sim \frac{1}{\lambda}\frac{T_c}{E_F}; \tag{12.68}$$

however, it has the correct order of magnitude to account for a Hall effect in dirty superconductors. For example, for usual superconductors, $\zeta \sim 10^{-2} \div 10^{-3}$ and $\zeta \sim 10^{-1} \div 10^{-2}$ for HTSC.

The imaginary part of Γ results in a modification of the equation for Φ. It takes the form

$$\begin{aligned} \nabla^2\Phi - \frac{8e^2\Gamma'|\Delta|^2}{\sigma_n}\Phi &= -\frac{1}{c}\mathrm{div}\,\frac{\partial \mathbf{Q}}{\partial t} + \frac{2e\Gamma''}{\sigma_n}\frac{\partial|\Delta|^2}{\partial t} \\ &\quad -\frac{4\pi(\sigma_n^H/H)}{\sigma_n^2 c}\left[\mathbf{Ej} + \frac{\partial}{\partial t}\left(\frac{H^2}{8\pi}\right)\right]. \end{aligned} \tag{12.69}$$

The last line is obtained under the assumption that $\sigma_n^H \propto H$.

12.9.2 *Hall effect: Low fields*

We calculate the Ohmic and Hall components of conductivity in the same way as in the previous section. The force balance is obtained from eqn (12.15):

$$\begin{aligned} \frac{\Phi_0}{c}[\mathbf{j}_{\mathrm{tr}} \times \hat{\mathbf{H}}] \cdot \mathbf{d} &= -\int_{S_0}\left[(\mathbf{d}\nabla\Delta)\,\Gamma\left(\frac{\partial \Delta^*}{\partial t} - 2ie\phi\Delta^*\right) + c.c.\right] dS \\ &= 2\Gamma'\int_{S_0}\left[(\mathbf{v}_L\nabla|\Delta|)\,(\mathbf{d}\nabla|\Delta|) - 2e|\Delta|^2(\mathbf{d}\nabla\chi)\Phi\right] dS \\ &\quad +2\Gamma''\int_{S_0}\left[|\Delta|\,(\mathbf{v}_L\nabla|\Delta|)\,[\mathbf{d}\nabla\chi] + 2e\Phi|\Delta|(\mathbf{d}\nabla|\Delta|)\right] dS. \end{aligned}$$

We put $\Phi = \Phi_1 + \Phi_2$ where, as in eqn (12.19),

$$\Phi_1 = -\frac{v_{L\phi}\mu_1}{2e\xi}, \tag{12.70}$$

while

$$\Phi_2 = -\zeta\frac{v_{L\rho}\mu_2}{2e\xi}. \tag{12.71}$$

Equations for $\mu_1(\rho)$ and $\mu_2(\rho)$ follow from eqn (12.69) where one can neglect the second line, since $\sigma_n^H/\sigma_n \sim (H/H_{c2})\zeta \ll \zeta$ for $H \ll H_{c2}$. The function μ_1 satisfies equation (12.20) while μ_2 is to be found from

$$\frac{d^2\mu_2}{d\rho^2} + \frac{1}{\rho}\frac{d\mu_2}{d\rho} - \frac{\mu_2}{\rho^2} - \frac{f^2\mu_2}{l_E^2} = -\frac{\xi}{2l_E^2}\frac{df^2}{d\rho} \tag{12.72}$$

with the condition $\mu_2 \to 0$ for both $\rho \to 0$ and $\rho \to \infty$.

The transport current becomes

$$\mathbf{j}_{\mathrm{tr}} = \frac{2\pi c\Gamma'\Delta_\infty^2}{\Phi_0}\left(a[\hat{\mathbf{H}} \times \mathbf{v}_L] + b\zeta\,\mathrm{sign}\,(e)\mathbf{v}_L\right). \tag{12.73}$$

The sign of the electron charge in the second term appears because the sense of the phase circulation is positive for positive charge of carriers when the direction of the magnetic field coincides with the vortex circulation, and it is negative for a negative charge of carriers. Expressing the vortex velocity through the average electric field we find

$$\mathbf{j}_{\mathrm{tr}} = \sigma_{\mathrm{O}}\mathbf{E} + \sigma_{\mathrm{H}}[\mathbf{E}\times\hat{\mathbf{H}}]$$

where the longitudinal (Ohmic) flux flow conductivity σ_{O} is given by eqn (12.24) with the constant a defined by eqn (12.23). The Hall conductivity is

$$\sigma_{\mathrm{H}} = \mathrm{sign}\,(e)\sigma_n\frac{ub\zeta}{2}\left(\frac{H_{c2}}{B}\right). \tag{12.74}$$

The constant b is

$$b = \int_0^\infty \left[\frac{1}{2}(\xi+\rho\mu_1)\frac{df^2}{d\rho} - f^2\mu_2\right]\frac{d\rho}{\xi}. \tag{12.75}$$

It is $b = 0.27$ for $u = 5.79$.

12.9.3 *High fields*

Equation (12.15) gives

$$\frac{1}{c}[(\mathbf{j}_{\mathrm{tr}} - \mathbf{j}_n)\times\mathbf{B}] = -\left\langle \Gamma^*\left[\mathbf{d}\left(\nabla - \frac{2ie}{\hbar c}\mathbf{A}\right)\Delta\right]\left[\frac{\partial\Delta^*}{\partial t} - \frac{2ie\phi}{\hbar}\Delta^*\right] + c.c.\right\rangle_L .$$

Using eqns (12.30) and eqn (12.31) we find

$$\begin{aligned}\mathbf{j}_{\mathrm{tr}} &= \mathbf{j}_n + 2|e|\Gamma'\left\langle|\Delta|^2\right\rangle_L\left([\hat{\mathbf{H}}\times\mathbf{v}_L] + \mathrm{sign}\,(e)\,\zeta\mathbf{v}_L\right)\\ &= \left(\sigma_n + 4e^2\xi^2\Gamma'\langle|\Delta|^2\rangle_L\right)\mathbf{E}\\ &\quad + \left(\sigma_n^H + 4e^2\xi^2\Gamma'\zeta\mathrm{sign}\,(e)\langle|\Delta|^2\rangle_L\right)[\mathbf{E}\times\hat{\mathbf{H}}].\end{aligned} \tag{12.76}$$

The longitudinal (Ohmic) conductivity is the same as in eqn (12.34), while the Hall conductivity has the form

$$\begin{aligned}\sigma_{\mathrm{H}} &= \sigma_n^H + 4e^2\xi^2\Gamma'\zeta\mathrm{sign}\,(e)\langle|\Delta|^2\rangle\\ &= \sigma_n^H + \mathrm{sign}\,(e)\sigma_n\frac{\zeta u}{2}\frac{(1-B/H_{c2})}{\beta_A[1-(1/2\kappa^2)]}.\end{aligned} \tag{12.77}$$

12.9.4 *Discussion: Hall effect*

The TDGL model gives, of course, an oversimplified description of the Hall effect. This model predicts a Hall angle $\Theta_H \sim \zeta \sim \lambda^{-1}(T_c/E_F)$. We shall see later that, in clean superconductors with $\ell \gg \xi\,(T)$, the mechanisms of the Hall effect are more complicated. The resulting Hall conductivity appears to be much larger than what we obtain from our simple TDGL model except for a relatively short mean free path $\ell \sim \xi\,(T)$. However, the contribution which we consider here is also present in clean superconductors. Indeed, eqn (12.65) is quite general and does not depend on the quasiparticle mean free path. To make a bridge between

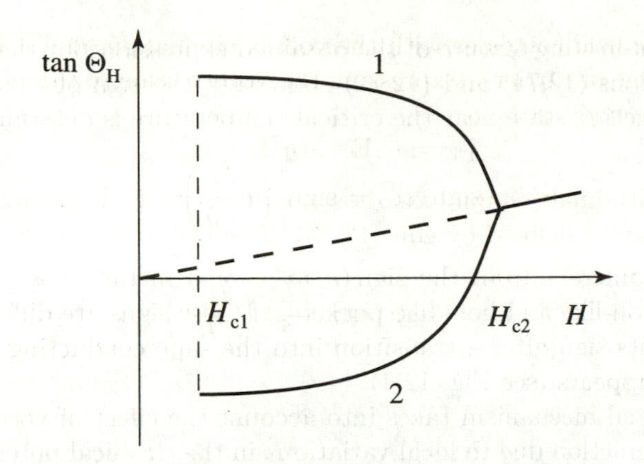

FIG. 12.4. The Hall angle as a function of magnetic field. The sign of the Hall angle is not changed after a transition into the superconducting state if the signs of ω_c and of ζ are the same (curve 1). The sign reverses if ω_c and of ζ have different signs (curve 2). The thick dashed line is an extrapolation of the normal-state Hall angle into the superconducting region.

the general situation and the TDGL model we note that, in superconductors which are far from the gapless conditions like the one considered in Section 12.8, the electric field penetration length l_E is usually long compared to $\xi(T)$. In this case, eqn (12.72) gives $\mu_2 = 0$ while eqn (12.20) results in $\mu_1 = \xi/\rho$. In this limit, we obtain $b = 1$ and

$$\sigma_{\mathrm{H}} = \mathrm{sign}\,(e)\sigma_n \frac{u\zeta}{2}\left(\frac{H_{c2}}{B}\right).$$

Using $\Gamma'' = -(1/2\lambda)(d\nu/d\epsilon)$ and $\Gamma' = \pi\nu/8T_c$ with the expression for H_{c2} and the clean-limit value

$$\xi^2(T) = \pi^3 T_c D/7\zeta(3)\,\Delta_\infty^2,$$

we find

$$\sigma_{\mathrm{H}} = \delta N_\Delta ec/B. \tag{12.78}$$

We denote here

$$\delta N_\Delta = \left(\frac{1}{\lambda}\frac{d\nu}{d\mu}\right)\Delta_\infty^2 \tag{12.79}$$

the "virtual" variation in the electron density caused by the change in the electronic spectrum after transition into the superconducting state. Equation (12.78) was obtained by Feigel'man *et al.* (1995) and by van Otterlo *et al.* (1995). Of course, there is no real density change in a superconductor because of the charge neutrality: all the variations are compensated by the corresponding variations in the chemical potential. Equation (12.78) appears to be exact from the point of view of the microscopic theory (Kopnin 1996) in the limit $T \to T_c$, see Section 14.6.2.

Note one interesting feature of the obtained expressions for the Hall conductivities. Equations (12.74) and (12.77) show that the sign of the Hall effect in the superconducting state near the critical temperature is determined by

$$\operatorname{sign}(e\zeta) = \operatorname{sign}\left(e\frac{\partial \nu}{\partial \mu}\right)$$

which can be different from the sign (e/m^*) for a metal whose Fermi surface has both electron-like and hole-like pockets. If these signs are different, the Hall angle changes its sign after a transition into the superconducting state and the Hall anomaly appears (see Fig. 12.4).

The suggested mechanism takes into account the effect of vortex motion on the pairing interaction due to local variations in the chemical potential of superconducting electrons $\mu_s = (1/2)\,\partial\chi/\partial t$. This mechanism is of a thermodynamic nature and is not related directly to the kinetics of nonequilibrium excitations. We shall see later that interaction of vortices with excitations in clean superconductors can produce much larger flux flow Hall effect: excitations play a prominent role again.

13

VORTEX DYNAMICS IN DIRTY SUPERCONDUCTORS

The force exerted on a vortex from the environment is derived microscopically. The kinetic equation is solved for the distribution function of excitations driven out of equilibrium by the moving vortex; the flux flow conductivity in a dirty superconductor is calculated. The vortex viscosity appears to be much larger than what is predicted by the TDGL model.

13.1 Microscopic derivation of the force on moving vortices

The TDGL scheme can only be applied in very special situations under rather restricting conditions of gapless superconductivity. This chapter treats the vortex dynamics in a more general case when nonstationary processes in a superconductor go beyond the TDGL model. We shall see that kinetics of nonequilibrium excitations makes the nonstationary behavior of a superconductor more complicated and diverse.

The complexity of the problem gives rise to the expectation that it would not be easy to calculate anything beyond the simple TDGL equations. Fortunately, this is not exactly the case, especially when we are interested in a linear response of a vortex array to applied perturbations. The major simplification is that the force on the moving vortex can be expressed through the characteristics of a static vortex and through the solutions to the kinetic equations which only contain the order parameter and magnetic field for a steady vortex array. The force thus does not contain distortions of the order parameter and of the magnetic field caused by the moving vortex. Restricting ourselves to the linear approximation in the vortex velocity we start with the derivation of the general expression for the force which acts on a moving vortex from the environment.

13.1.1 *Variation of the thermodynamic potential*

Let us consider a variation of the thermodynamic potential eqn (3.71) derived in Section 3.3. We have in the real frequency representation

$$\int \delta\Omega_{\,tot} d^3r = -\nu\,(0)\,\mathrm{Tr}\int \left[\delta\check{H}(\mathbf{r})\check{g}_\epsilon^{(\mathrm{st})}\,(\mathbf{p},\mathbf{r})\right]\frac{d\Omega_p}{4\pi}\,\frac{d\epsilon}{4}\,d^3r$$
$$+\int \frac{\delta\left(\Delta_{\mathbf{p}}(\mathbf{r})\Delta^*_{\mathbf{p}}(\mathbf{r})\right)}{|g|}\frac{d\Omega_{\mathbf{p}}}{4\pi}\,d^3r + \frac{1}{c}\int \mathbf{j}\cdot\delta\mathbf{A}\,d^3r. \qquad (13.1)$$

Here

$$\Omega_{tot} = \Omega_{sn} + \frac{H^2}{8\pi}.$$

The Hamiltonian is

$$\check{H} = \begin{pmatrix} -(e/c)\mathbf{v}_F \cdot \mathbf{A} & -\Delta_{\mathbf{p}} \\ \Delta^*_{\mathbf{p}} & (e/c)\mathbf{v}_F \cdot \mathbf{A} \end{pmatrix}$$

in the quasiclassical approximation. The variation eqn (13.1) is taken at a constant chemical potential.

When the system is in a nonstationary state, the concept of thermodynamic potential can be used only if deviations from the equilibrium are small. The total Green function describing the nonstationary state in frequency representation is

$$\check{g}^K_{\epsilon_+,\epsilon_-} = \check{g}^{(st)}_\epsilon \, 2\pi\delta(\omega) + \check{g}^{(nst)}_{\epsilon_+,\epsilon_-}, \tag{13.2}$$

where $\epsilon_\pm = \epsilon \pm \omega/2$. It is the stationary part $\check{g}^{(st)}$ which enters the variation of the thermodynamic potential in eqn (13.1). We substitute it with the difference $\check{g}^K_{\epsilon_+,\epsilon_-} - \check{g}^{(nst)}_{\epsilon_+,\epsilon_-}$, and take into account that the order parameter and the current in the nonstationary state satisfy equations (9.17) and (9.18). We obtain

$$\int \delta\left(\Omega_{sn} + \frac{H^2}{8\pi}\right) d^3r = \nu(0)\,\mathrm{Tr}\int \left[\delta\check{H}(\mathbf{r})\check{g}^{(nst)}_\epsilon(\mathbf{p},\mathbf{r},t)\right] \frac{d\Omega_p}{4\pi}\frac{d\epsilon}{4}\, d^3r. \tag{13.3}$$

The terms with $\check{g}^K$ disappear due to eqns (9.17) and (9.18) (we recall that $\lambda = |g|\nu(0)$).

Equation (13.3) determines the variation of the thermodynamic potential with respect to Δ and $\mathbf{A}$. For example,

$$\frac{\delta\Omega_{sn}}{\delta\Delta^*_{\mathbf{p}}} = \nu(0)\int V(\mathbf{p},\mathbf{p}')\, f^{(nst)}_\epsilon(\mathbf{p}'_F,\mathbf{k};t)\,\frac{d\epsilon}{4}\frac{d\Omega_{p'}}{4\pi}, \tag{13.4}$$

and

$$\frac{\delta}{\delta\mathbf{A}}\left(\Omega_{sn} + \frac{H^2}{8\pi}\right) = \frac{1}{c}\mathbf{j}^{(nst)} \tag{13.5}$$

where

$$\mathbf{j}^{(nst)} = -e\nu(0)\,\mathrm{Tr}\int \mathbf{v}_F\check{\tau}_3\check{g}^{(nst)}_\epsilon(\mathbf{p},\mathbf{r},t)\,\frac{d\Omega_p}{4\pi}\frac{d\epsilon}{4}. \tag{13.6}$$

Equations (13.4) and (13.5) are microscopic counterparts of the corresponding TDGL equations. The right-hand sides of these equations are proportional to deviations from equilibrium; they vanish in a stationary state.

13.1.2 *Force on vortices*

Consider a variation of the total thermodynamic potential in the form

$$\delta\Omega_{tot} = \int \left[\frac{\delta\Omega_{tot}}{\delta\Delta}(\mathbf{d}\cdot\nabla)\Delta + c.c. + \frac{\delta\Omega_{tot}}{\delta\mathbf{A}}(\mathbf{d}\cdot\nabla)\mathbf{A} + \frac{\delta\Omega_{tot}}{\delta\mu}(\mathbf{d}\cdot\nabla)\mu\right] dV \tag{13.7}$$

where $\mathbf{d}$ is an arbitrary constant vector of an infinitesimal translation. The last term can be written as

$$-\int N\nabla\mu\, dV = -\int \nabla\left(N\mu_1\right) dV + \int \mu_1 \nabla N\, dV$$

where $\mu_1 = \mu - E_F$ is the deviation of the chemical potential from its value in the normal state. Equation (13.7) takes the form

$$\delta\mathcal{F}_{tot} = \int \left[\frac{\delta\Omega_{tot}}{\delta\Delta}\left(\mathbf{d}\cdot\nabla\right)\Delta + c.c. + \frac{\delta\Omega_{tot}}{\delta\mathbf{A}}\left(\mathbf{d}\cdot\nabla\right)\mathbf{A}\right] dV \tag{13.8}$$

where

$$\mathcal{F} = \Omega + \int N\mu_1\, dV$$

is the free energy which is a function of N instead of μ. The gradient of the particle density is omitted because the density is constant due to the charge neutrality. Equation (13.8) follows from the general property of thermodynamic functions: Variation of the free energy $\mathcal{F}$ for constant N is equal to the variation of Ω for a constant μ. The gradient of free energy density for a constant number of particles gives the density of force acting on the vortex from environment.

The r.h.s. of eqn (13.8) can be transformed in exactly the same way as we did it earlier for the derivation of the force balance within the TDGL model in Section 12.4. The force from environment per unit vortex length becomes

$$\mathbf{F}_{\rm env} = -\int_{S_0}\left(\left[\left(\nabla - \frac{2ie}{c}\mathbf{A}\right)\Delta\right]\frac{\delta\Omega_{sn}}{\delta\Delta} + \left[\left(\nabla + \frac{2ie}{c}\mathbf{A}\right)\Delta^*\right]\frac{\delta\Omega_{sn}}{\delta\Delta^*}\right.$$
$$\left. + \frac{1}{c}\left[\mathbf{j}^{(\rm nst)} \times \mathbf{H}\right]\right) d^2r. \tag{13.9}$$

The relation

$$2\frac{\delta\Omega_{sn}}{\delta\chi} \equiv i\left(\Delta\frac{\delta\Omega_{sn}}{\delta\Delta} - \Delta^*\frac{\delta\Omega_{sn}}{\delta\Delta^*}\right) = -\frac{1}{2e}{\rm div}\left(\mathbf{j} - \mathbf{j}^{(\rm nst)}\right)$$

used during the derivation follows from eqn (13.5) and the requirement of the gauge invariance according to which the vector potential and the gradient of the order parameter phase always come in the combination $\mathbf{A} - (c/2e)\nabla\chi$.

Using eqns (13.4) and (13.6) we can finally present eqn (13.9) in the form (Larkin and Ovchinnikov 1986)

$$\mathbf{F}_{\rm env} = -\int_{S_0} d^2r \int \frac{d\epsilon}{4} \int \frac{dS_F}{(2\pi)^3 v_F} {\rm Tr}\left(\check{g}^{(\rm nst)}\, \hat{\nabla}\check{H}\right), \tag{13.10}$$

where

$$\hat{\nabla}\check{H} = \begin{pmatrix} (e/c)\left[\mathbf{H}\times\mathbf{v}_F\right] & -\left(\nabla - 2ie\mathbf{A}/c\right)\Delta \\ \left(\nabla + 2ie\mathbf{A}/c\right)\Delta^* & -(e/c)\left[\mathbf{H}\times\mathbf{v}_F\right] \end{pmatrix}. \tag{13.11}$$

Equation (13.10) is written in the form which does not assume a spherical Fermi surface.

In presence of a transport current, the force $\mathbf{F}_{\rm env}$ should be balanced by the Lorentz force. The corresponding term eqn (12.14) can be incorporated into the free energy exactly in the same way as we did it in Section 12.4. The force balance becomes

$$\frac{\Phi_0}{c}[\mathbf{j}_{\rm tr}\times\hat{\mathbf{H}}]+\mathbf{F}_{\rm env}=0. \tag{13.12}$$

Consider now the nonstationary part of the Green function. We write the Keldysh function eqn (10.62) in frequency representation

$$\begin{aligned}\check{g}^K_{\epsilon_+,\epsilon_-}&=\left(\check{g}^R_{\epsilon_+,\epsilon_-}-\check{g}^A_{\epsilon_+,\epsilon_-}\right)f^{(0)}(\epsilon)-\frac{df^{(0)}}{d\epsilon}\frac{\hat{\omega}}{2}\left(\check{g}^R_{\epsilon_+,\epsilon_-}+\check{g}^A_{\epsilon_+,\epsilon_-}\right)\\&+\left(\check{g}^R_{\epsilon_+,\epsilon_-}-\check{g}^A_{\epsilon_+,\epsilon_-}\right)f_1(\epsilon)+\left(\check{g}^R_{\epsilon_+,\epsilon_-}\check{\tau}_3-\check{\tau}_3\check{g}^A_{\epsilon_+,\epsilon_-}\right)f_2(\epsilon)\end{aligned}$$

where

$$\hat{\omega}\check{g}=\begin{pmatrix}\omega g_\epsilon(\omega) & (\omega-2e\varphi)\,f_\epsilon(\omega)\\ -(\omega+2e\varphi)\,f^\dagger_\epsilon(\omega) & \omega\bar{g}_\epsilon(\omega)\end{pmatrix}. \tag{13.13}$$

As in eqn (11.21) we define the nonstationary function as

$$\check{g}^{(\rm nst)}_{\epsilon_+,\epsilon_-}=\check{g}^K_{\epsilon_+,\epsilon_-}-\left(\check{g}^R_\epsilon-\check{g}^A_\epsilon\right)f^{(0)}(\epsilon)\,2\pi\delta(\omega).$$

Using the particle–hole symmetry of the Green functions $\check{g}^{R(A)}_{\epsilon_+,\epsilon_-}$ of the type of eqn (10.68) with respect to the transformation $\epsilon\to-\epsilon$ and $\mathbf{p}\to-\mathbf{p}$ one can show that

$$\nu(0)\int dr\int\frac{d\epsilon}{4}\int\frac{d\Omega_p}{4\pi}f^{(0)}(\epsilon)\,{\rm Tr}\left[\left(\check{g}^R_{\epsilon_+,\epsilon_-}-\check{g}^A_{\epsilon_+,\epsilon_-}\right)\hat{\nabla}\check{H}\right] \tag{13.14}$$

is even in ω. Therefore, the first derivatives of Δ and $\mathbf{A}$ in time drop out of eqn (13.14). As a result, the nonequilibrium part of the total Green function in eqn (13.10) can be written as

$$\begin{aligned}\check{g}^{(\rm nst)}_{\epsilon_+,\epsilon_-}&=-\frac{df^{(0)}}{d\epsilon}\frac{\hat{\omega}}{2}\left(\check{g}^R_{\epsilon_+,\epsilon_-}+\check{g}^A_{\epsilon_+,\epsilon_-}\right)\\&+\left(\check{g}^R_{\epsilon_+,\epsilon_-}-\check{g}^A_{\epsilon_+,\epsilon_-}\right)f_1(\epsilon)+\left(\check{g}^R_{\epsilon_+,\epsilon_-}\check{\tau}_3-\check{\tau}_3\check{g}^A_{\epsilon_+,\epsilon_-}\right)f_2(\epsilon)\end{aligned}$$

within the first-order terms in ω. Performing the Fourier transformation back to the center-of-mass time, we get

$$\begin{aligned}\check{g}^{(\rm nst)}&=-\frac{i}{2}\frac{\hat{\partial}}{\partial t}\left(\check{g}^R_\epsilon+\check{g}^A_\epsilon\right)\frac{df^{(0)}}{d\epsilon}\\&+\left(\check{g}^R_\epsilon-\check{g}^A_\epsilon\right)f_1(\epsilon)+\left(\check{g}^R_\epsilon\check{\tau}_3-\check{\tau}_3\check{g}^A_\epsilon\right)f_2(\epsilon)\end{aligned} \tag{13.15}$$

where the gauge-invariant time-derivative is defined by eqn (10.32).

Equations (13.10) and (13.15) make the basis for calculations of the force on a vortex within the microscopic theory. We shall use them for both dirty and clean

superconductors in the following sections. We observe that the force is expressed through the Green functions $\check{g}_\epsilon^R$ and $\check{g}_\epsilon^R$, as well as through the distribution functions f_1 and f_2. For the Green functions we can take their values for a steady vortex array because f_1 and f_2 are themselves proportional to deviation from equilibrium. In turn, f_1 and f_2 are solutions of kinetic equations which again contain only the steady-state Green functions. The problem thus reduces to two major steps. First, we need to find the Green functions for a stationary vortex. Second, we solve the resulting kinetic equations and find the distribution function of nonequilibrium excitations. The force can then be calculated performing the necessary operations in eqn (13.10).

13.2 Diffusion controlled flux flow

In this section we consider dirty s-wave superconductors such that

$$\tau_{\text{imp}}\Delta(T) \ll 1 \tag{13.16}$$

for temperatures close to T_c. Note that the gradients are not assumed to be small any more in the sense of eqn (11.9). On the contrary, we concentrate on the situation when the relaxation by diffusion is faster than the inelastic relaxation:

$$D\xi^{-2}(T) \gg \tau_{\text{ph}}^{-1}. \tag{13.17}$$

We recall also that the inelastic relaxation rate is much smaller than the critical temperature, $T_c \gg \tau_{\text{ph}}^{-1}$. The temperature range when the condition eqn (13.17) is fulfilled corresponds to

$$T_c \gg T_c - T \gg \tau_{\text{ph}}^{-1}.$$

This condition is much less restrictive than what we have considered in the previous chapter where, on the contrary, $D\xi^{-2}$ was assumed smaller than τ_{ph} and thus temperatures $T_c - T \ll \tau_{\text{ph}}^{-1}$ were required. The present conditions are more "natural" and can easily be realized in experiments.

Due to eqn (13.16) the stationary Green functions for dirty superconductors are determined by Usadel equations (5.97). For temperatures close to T_c, the diffusion terms in the Usadel equations can be neglected despite the fact that the gradients are of the order of $\xi^{-1}(T)$. Indeed,

$$D\xi^{-2}(T) \sim \left(1 - \frac{T}{T_c}\right)\Delta \ll \Delta$$

if $\tau_{\text{imp}}T_c \ll 1$, or

$$D\xi^{-2}(T) \sim \tau_{\text{imp}}\Delta^2 \ll \Delta$$

when $\tau_{\text{imp}}T_c \gg 1$ but still $\tau_{\text{imp}}\Delta(T) \ll 1$. The diffusion terms are thus small compared to Δ under conditions of eqn (13.16). We also neglect the small inelastic pair-breaking. As a result, we obtain the adiabatic expressions (10.104) and (10.105) for the regular functions $\langle g_-\rangle$ and $\langle g_+\rangle$.

Under these conditions, we can use kinetic equations (10.107) and (10.108) derived for adiabatic Green functions. Neglecting the inelastic scattering we obtain

$$D\nabla^2 f_1 = \frac{1}{2}\left[\langle f_- \rangle \frac{\hat{\partial}\Delta^*}{\partial t} + \left\langle f_-^\dagger \right\rangle \frac{\hat{\partial}\Delta}{\partial t}\right] \frac{\partial f^{(0)}}{\partial \epsilon}. \tag{13.18}$$

The boundary conditions for eqn (13.18) are formulated as follows. For $\epsilon^2 > |\Delta|^2_{\max}$, the quasiparticle Green functions extend over long distances from the vortex cores. For a periodic vortex array, the distribution function f_1 is translationally invariant with the period of the vortex array. Thus it should not increase at large distances from the vortex. This requires that the average gradient of f_1 vanishes

$$S_0^{-1} \int_{S_0} \nabla f_1 d^2 r = 0. \tag{13.19}$$

However, for particles with $\epsilon^2 < |\Delta|^2_{\max}$, large distances are not accessible. The boundary conditions are imposed at the surface determined by the condition

$$|\epsilon| = |\Delta(\mathbf{r})| \tag{13.20}$$

beyond which the particle cannot penetrate. The conditions can be obtained from eqns (10.106) and (13.18). Integrating eqn (13.18) across the surface in eqn (13.20) one finds that the derivative $\nabla_n f_1$ along its normal is continuous at this surface. At the same time, eqn (10.106) implies that, at the side of the surface where $|\epsilon| < |\Delta(\mathbf{r})|$, the gradient vanishes because $g_- = 0$. Therefore, one should also have

$$\nabla_n f_1|_{|\epsilon|=|\Delta(\mathbf{r})|} = 0 \tag{13.21}$$

at the other side of the surface defined by eqn (13.20), where ∇_n is the derivative along the normal to the surface. Equations (13.19) and (13.21) are the required boundary conditions to eqn (13.18) (Larkin and Ovchinnikov 1986).

The expression for the force eqn (13.10) contains the nonstationary part of the current eqn (13.6) proportional to the energy integral of $g_- \mathbf{f}_1$. According to eqn (10.108), we have

$$\mathbf{j}^{(\mathrm{nst})} = \sigma_n \int \left[\mathbf{E} \langle g_- \rangle^2 \frac{\partial f^{(0)}}{\partial \epsilon} + e^{-1} \langle g_- \rangle \nabla \langle g_- f_2 \rangle\right] \frac{d\epsilon}{2}.$$

The integral diverges logarithmically at $\epsilon \to \Delta$ due to a square-root singularity in $\langle g_- \rangle$. This divergence is cut off at $\epsilon - \Delta \sim D\xi^{-2}$. However, the prefactor at the logarithm is proportional to Δ/T; it is thus small near the critical temperature. The main contribution, however, comes from the region of energies $\epsilon \sim T$. For such energies, $\langle g_- \rangle = 1$. At the same time, the function f_2 is proportional to Δ/T (compare with eqn (11.19)) and can be neglected. As a result, we again have

$$\mathbf{j}^{(\mathrm{nst})} = \mathbf{j}_n = \sigma_n \mathbf{E}$$

According to eqn (13.18), the distribution function is, by the order of magnitude,

$$f_1 \sim \frac{\xi^2}{D} \frac{df^{(0)}}{d\epsilon} \frac{v_L}{\xi},$$

and gives a much larger contribution to the nonstationary Green function than the first term in the r.h.s. of eqn (13.15). Therefore, neglecting f_2 we have

$$\check{g}^{(\mathrm{nst})} = 2\check{g}_- f_1(\epsilon).$$

Equation (13.10) becomes

$$\begin{aligned} \mathbf{F}_{\mathrm{env}} &= -\frac{\nu(0)}{2} \int_{S_0} d^2r \int_{-\infty}^{\infty} d\epsilon\, f_1 \left(\hat{\nabla}\Delta^* \langle f_- \rangle + \hat{\nabla}\Delta \left\langle f_-^\dagger \right\rangle \right) \\ &\quad -\frac{1}{c} \int_{S_0} [\mathbf{j}^{(\mathrm{nst})} \times \mathbf{H}] d^2r \\ &= \nu(0) \int_{S_0} d^2r \int_{-\infty}^{\infty} d\epsilon\, f_1 \nabla \left[\sqrt{\epsilon^2 - |\Delta|^2} \right] \Theta\left(\epsilon^2 - |\Delta|^2\right) \\ &\quad -\frac{1}{c} \int_{S_0} [\mathbf{j}^{(\mathrm{nst})} \times \mathbf{H}] d^2r. \end{aligned} \tag{13.22}$$

The next step is to solve the kinetic equation (13.18) for f_1. Consider first the limit of low magnetic fields. For well-separated vortices, the order parameter magnitude depends only on the distance from the vortex axis. Equation (13.18) for $\epsilon^2 > |\Delta|^2$ takes the form

$$D\nabla^2 f_1 = -\frac{\partial}{\partial t} \sqrt{\epsilon^2 - |\Delta|^2}^2 \frac{df^{(0)}}{d\epsilon} = \left(v_{L\rho} \frac{d}{d\rho} \right) \sqrt{\epsilon^2 - |\Delta|^2}^2 \frac{df^{(0)}}{d\epsilon}.$$

We put

$$f_1 = D^{-1} v_{L\rho} w(\rho) \frac{df^{(0)}}{d\epsilon}$$

and obtain for $w(\rho)$

$$\left(\frac{d^2}{d\rho^2} + \frac{1}{\rho}\frac{d}{d\rho} - \frac{1}{\rho^2} \right) w = \frac{d}{d\rho} \sqrt{\epsilon^2 - |\Delta|^2}$$

The solution regular at $\rho \to 0$ is

$$w = \rho^{-1} \int_0^{\rho} \left(\sqrt{\epsilon^2 - |\Delta|^2} - C \right) \rho' \, d\rho'.$$

The constant should be found from the boundary conditions. Denote $\mathbf{d}$ a constant arbitrary vector. We have

$$(\mathbf{d} \cdot \nabla) f_1 = d_\rho \frac{df_1}{d\rho} + d_\phi \frac{1}{\rho} \frac{df_1}{d\phi}$$

$$
\begin{aligned}
&= D^{-1}\frac{df^{(0)}}{d\epsilon}\left(d_\phi v_{L\phi} - d_\rho v_{L\rho}\right) \\
&\times\rho^{-2}\int_0^\rho \left(\sqrt{\epsilon^2 - |\Delta|^2} - C\right)\rho'\, d\rho' \\
&+D^{-1}\frac{df^{(0)}}{d\epsilon} d_\rho v_{L\rho}\left(\sqrt{\epsilon^2 - |\Delta|^2} - C\right). \qquad (13.23)
\end{aligned}
$$

For $\epsilon^2 > |\Delta|^2_{\max}$, the boundary condition has the form of eqn (13.19). We average eqn (13.23) over the angles ϕ. Since

$$
\left(d_\phi v_{L\phi} - d_\rho v_{L\rho}\right) = -\left[\left(d_x v_x - d_y v_y\right)\cos(2\phi) + \left(d_x v_y + d_y v_x\right)\sin(2\phi)\right],
$$

the first term in the r.h.s. of eqn (13.23) vanishes. The second term gives

$$
\int (\mathbf{d}\cdot\nabla) f_1\, d\phi = \frac{\pi\,(\mathbf{d}\cdot\mathbf{v}_L)}{D}\frac{df^{(0)}}{d\epsilon}\left(\sqrt{\epsilon^2 - |\Delta|^2} - C\right).
$$

Integrating now over the radial distance, we have from eqn (13.19)

$$
\int_0^{r_0}\left(\sqrt{\epsilon^2 - |\Delta|^2} - C\right)\rho\, d\rho = 0
$$

where r_0 is the radius of the vortex lattice unit cell. Thus

$$
C = \left\langle \sqrt{\epsilon^2 - |\Delta|^2}\right\rangle_L
$$

where the average is taken over the vortex unit cell. The function $w(\rho)$ vanishes for $\rho = r_0$.

For $\epsilon^2 < |\Delta|^2_{\max}$ we use the boundary condition eqns (13.21). Equation (13.23) taken for the derivative along the ρ direction with $d_\phi = 0$ gives

$$
C = -\frac{2\pi}{S_\epsilon^2}\int_0^{\rho_\epsilon}\sqrt{\epsilon^2 - |\Delta|^2}\rho'\, d\rho' \equiv -\left\langle\sqrt{\epsilon^2 - |\Delta|^2}\right\rangle_{S_\epsilon}
$$

where $S_\epsilon = \pi\rho_\epsilon^2$ and ρ_ϵ is determined by $|\epsilon| = |\Delta(\rho_\epsilon)|$.

Now we calculate the force on a vortex. The balance of the Lorentz force and the force in eqn (13.22) reads

$$
\begin{aligned}
\mathbf{F}_L = &-\nu(0)\int_{-\infty}^{\infty} d\epsilon\int_{S_0} d^2r\, f_1\nabla\left[\sqrt{\epsilon^2 - |\Delta|^2}\right]\Theta\left(\epsilon^2 - |\Delta|^2\right) \\
&+\frac{1}{c}\int_{S_0}\left[\mathbf{j}^{(\mathrm{nst})}\times\mathbf{H}\right] d^2r.
\end{aligned}
$$

Integrating the first term in the r.h.s. by parts we find that it is equal to

$$
\nu(0)\int_{|\epsilon|>|\Delta|_{\max}} d\epsilon\int_{S_0} d^2r\,\left(\sqrt{\epsilon^2 - |\Delta|^2} - C\right)\nabla f_1
$$

$$+\nu(0)\int_{|\epsilon|<|\Delta|_{\max}} d\epsilon \int_{S_0} d^2r\, \sqrt{\epsilon^2-|\Delta|^2}\Theta\left(\epsilon^2-|\Delta|^2\right)\nabla f_1.$$

The surface term vanishes because $f_1 = 0$ at $\rho = r_0$ for $|\epsilon| > |\Delta|_{\max}$, and $\sqrt{\epsilon^2-|\Delta|^2} = 0$ at $\rho = \rho_\epsilon$ for $|\epsilon| < |\Delta|_{\max}$. The energy integral is determined by $\epsilon \sim |\Delta|$ and the spatial integral converges at distances of the order of $\xi(T)$. The equation of the force balance becomes

$$\mathbf{F}_L = \frac{\nu(0)\,\alpha(H)}{2TD}\left\langle|\Delta|^2\right\rangle_L^{3/2}\xi^2\mathbf{v}_L + \frac{1}{c}\int_{S_0}[\mathbf{j}^{(\mathrm{nst})}\times\mathbf{H}]d^2r$$

where we denote

$$\alpha(H) = \int_{\tilde\epsilon>\tilde\Delta_{\max}} d\tilde\epsilon\int_{S_0} d^2\tilde r\left[\sqrt{\tilde\epsilon^2-\tilde\Delta^2}-\left\langle\sqrt{\tilde\epsilon^2-\tilde\Delta^2}\right\rangle_L\right]^2$$
$$+\int_{\tilde\epsilon<\tilde\Delta_{\max}} d\tilde\epsilon\int_{S_\epsilon} d^2\tilde r\,\left[\left(\tilde\epsilon^2-\tilde\Delta^2\right)+\left\langle\sqrt{\tilde\epsilon^2-\tilde\Delta^2}\right\rangle_{S_\epsilon}^2\right].$$

We put here

$$\epsilon = \tilde\epsilon\sqrt{\left\langle|\Delta|^2\right\rangle_L},\ |\Delta| = \tilde\Delta\sqrt{\left\langle|\Delta|^2\right\rangle_L},\ \rho = \tilde\rho\,\xi(T).$$

As a result,

$$\mathbf{j}-B^{-1}\left\langle\mathbf{j}^{(\mathrm{nst})}H\right\rangle = \mathbf{v}_L\frac{\sigma_n H_{c2}\alpha(H)\,\Lambda^2}{2\pi c}\left(\frac{\pi^3}{7\zeta(3)}\right)^2\frac{T}{\Delta_\infty}\left\langle\frac{|\Delta|^2}{\Delta_\infty^2}\right\rangle_L^{3/2} \tag{13.24}$$

and, as usually,

$$\Delta_\infty^2 = \frac{8\pi^2T_c\,(T_c-T)}{7\zeta(3)}.$$

The coherence length is defined by eqn (6.26)

$$\xi^2(T) = \frac{\pi^3T\Lambda D}{7\zeta(3)\,\Delta_\infty^2}.$$

For high magnetic fields when vortices are closely packed, it is not easy to solve eqn (13.18) because the vortex lattice does not have an axial symmetry any more. To simplify calculations, we adopt the approximation of the round unit cell even for high vortex density. Within this approximation, we obtain the same solution as before even for high magnetic fields. Therefore, we shall use eqn (13.24) as an approximation for the whole range of magnetic fields.

For well separated vortices at low fields one can neglect the normal current in the l.h.s. of eqn (13.24). Indeed, it is of the order of $\sigma_n E$ while the term in the r.h.s. has an order of $(H_{c2}/B)\,\sigma_n E$. For fields close to the upper critical field

H_{c2}, one has $\mathbf{j}^{(\mathrm{nst})} = \sigma_n \mathbf{E}$. Putting $\mathbf{v}_L = c\left[\mathbf{E}\times\hat{\mathbf{H}}\right]/B$ one thus can write an interpolation expression

$$\mathbf{j} - \sigma_n \mathbf{E} = \sigma_n \frac{H_{c2}}{B}\frac{\alpha(H)\Lambda^2}{2\pi}\left(\frac{\pi^3}{7\zeta(3)}\right)^2 \frac{T_c}{\Delta_\infty}\left\langle\frac{|\Delta|^2}{\Delta_\infty^2}\right\rangle_L^{3/2}\mathbf{E}.$$

The flux flow conductivity becomes

$$\frac{\sigma_f}{\sigma_n} = 1 + \frac{\alpha(H)\Lambda^2}{2\pi}\left[\frac{\pi^3}{7\zeta(3)}\right]^2 \frac{T_c}{\Delta_\infty}\frac{H_{c2}}{B}\left\langle\frac{|\Delta|^2}{\Delta_\infty^2}\right\rangle_L^{3/2}. \tag{13.25}$$

This result was obtained by Gor'kov and Kopnin (1973 a) and by Larkin and Ovchinnikov (1986). The function $\alpha(H)$ is of the order of unity. To calculate it for well separated vortices at low fields we observe that $\left\langle|\Delta|^2\right\rangle_L = \Delta_\infty^2$ and $\tilde{\Delta}_{\max} = 1$; moreover

$$\left\langle\sqrt{\tilde{\epsilon}^2 - \tilde{\Delta}^2}\right\rangle_L = \sqrt{\tilde{\epsilon}^2 - 1}.$$

On the other hand, for high fields

$$\left\langle|\Delta|^2\right\rangle_L = \frac{2\kappa^2\Delta_\infty^2}{(2\kappa^2 - 1)\beta_A}\left(1 - \frac{B}{H_{c2}}\right).$$

The high-field flux flow conductivity becomes

$$\frac{\sigma_f}{\sigma_n} - 1 = \frac{\alpha(H)\Lambda^2}{\pi}\left[\frac{\pi^3}{7\zeta(3)}\right]^2 \frac{T_c}{\Delta_\infty}\left(\frac{1 - B/H_{c2}}{[1 - (1/2\kappa^2)]\beta_A}\right)^{3/2}.$$

As a function of the magnetic field, this expression is valid until

$$1 - \frac{H}{H_{c2}} \gg \left(1 - \frac{T}{T_c}\right).$$

For higher fields, the superconductor becomes essentially gapless, and the regular TDGL contribution [the first term in the r.h.s. of eqn (13.15)] dominates.

Equation (13.25) can be written in a form similar to eqn (12.54)

$$\frac{\sigma_f}{\sigma_n} - 1 = \beta\frac{H_{c2}}{B} \tag{13.26}$$

where β is a function of temperature, magnetic field, and also of the electron mean free path. The numerical values for β are listed in the review by Larkin and Ovchinnikov (1986). In the dirty limit $T_c\tau \ll 1$, one has for $H \ll H_{c2}$

$$\beta = 4.04\left(1 - \frac{T}{T_c}\right)^{-1/2},$$

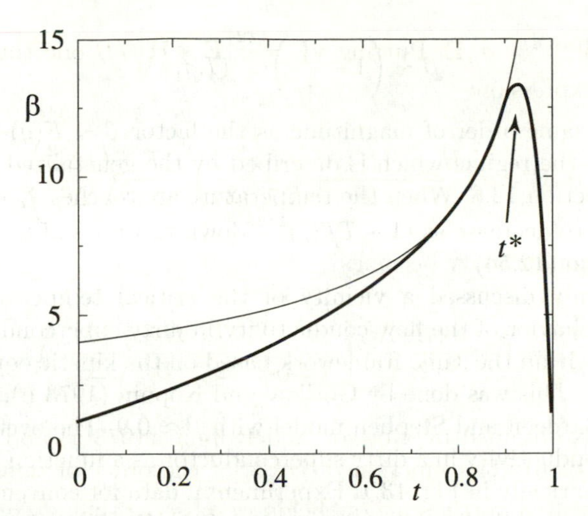

FIG. 13.1. The factor β as a function of temperature, $t = T/T_c$. The solid line is an interpolation between: (i) $\beta \approx 0.9$ at zero temperature, (ii) $\beta = 4.04(1 - T/T_c)^{-1/2}$ in the region close to T_c (thin line), and (iii) the TDGL result $\beta = 1.45$ for $T = T_c$. The temperature $t^* = T^*/T_c$ where β reaches its maximum is such that $1 - T^*/T_c \sim (T_c\tau_{\rm ph})^{-1}$.

and

$$\beta = 0.31\left(1 - \frac{T}{T_c}\right)^{-1/2}\left(\frac{1 - B/H_{c2}}{1 - (1/2\kappa^2)}\right)^{3/2}$$

for $H \to H_{c2}$. The factor $(1 - T/T_c)^{-1/2}$ follows form eqn (13.25) due to the term T_c/Δ_∞ in the right-hand side.

13.2.1 *Discussion*

We note first of all that the flux flow conductivity in dirty superconductors appears to be much larger than what is predicted by the simple TDGL model or by the Bardeen–Stephen model. Indeed, for low inductions the factor β in eqn (13.26) is $\beta = 1$ within the Bardeen–Stephen model. The microscopic theory, however, predicts a large β near T_c which grows as

$$\beta \sim \left(1 - \frac{T}{T_c}\right)^{-1/2}$$

when T approaches T_c. This is the consequence of a decrease in the diffusion relaxation rate when the size of the vortex core increases near T_c with an increasing coherence length. When temperature is close enough to T_c the diffusion relaxation rate becomes comparable with the inelastic relaxation rate when $D\xi^{-2}(T) \sim \tau_{\rm ph}^{-1}$. This happens for temperature $(1 - T/T_c) \sim (T_c\tau_{\rm ph})^{-1}$ at the border of applicability of eqn (13.25). At this point, the increase in β saturates at

$$\beta \sim \left(1 - \frac{T}{T_c}\right)^{1/2} (T_c \tau_{\rm ph})$$

which has the same order of magnitude as the factor $\beta \sim F(q)$ in eqn (12.54). This is exactly the regime which is described by the generalized TDGL scheme discussed in Section 12.8. When the temperature approaches T_c even closer, the factor β starts to decrease as $(1 - T/T_c)^{1/2}$ down to values of the order of unity according to eqn (12.54).

We have only discussed a vicinity of the critical temperature. The low-temperature behavior of the flow conductivity in dirty superconductors can also be considered within the same framework based on the kinetic equations derived in Section 10.5. This was done by Gor'kov and Kopnin (1973 b). The result approaches the Bardeen and Stephen model with $\beta \approx 0.9$. The overall behavior of the flux flow conductivity in a dirty superconductor as a function of temperature is shown schematically in Fig. 13.1. Experimental data for conventional dirty superconductors generally lie above the results of the Bardeen–Stephen model; on the other hand, they are below the predictions of the exact microscopic theory (see the reviews by Gor'kov and Kopnin 1975 and by Larkin and Ovchinnikov 1986). The reason is that the parameter $T_c\tau_{\rm ph}$ is not that large in practice, thus the increase in β is limited to lower values.

14

VORTEX DYNAMICS IN CLEAN SUPERCONDUCTORS

We discuss first the general features of the vortex dynamics in clean superconductors using the Boltzmann kinetic equation. We establish the conditions when the dissipative dynamics of vortices transforms into a Hamiltonian one. We emphasize the crucial importance of excitations localized in vortex cores. Next we use the quasiclassical Green function technique and the kinetic equations of the previous chapters to calculate the longitudinal and Hall components of the flux flow conductivity for s-wave superconductors. We discuss the forces on a vortex as functions of temperature and of purity of the superconductor.

14.1 Introduction

This chapter is devoted to the vortex dynamics in clean superconductors. Clean systems offer more intriguing physics than dirty superconductors considered in the previous section. For example, one of the fundamental problems can be formulated as follows: Speaking of clean superconductors one can, in particular, think of such a system where no relaxation processes are available, i.e., the mean free path of excitations is infinite. In this case, vortices should move without dissipation since there is no mechanism to absorb the energy. The vortex velocity should then be parallel to the transport current, which makes the electric field perpendicular to the current; the dissipation thus vanishes $\mathbf{j} \cdot \mathbf{E} = 0$. This contrasts with what we know from the previous chapters on dirty superconductors: the vortex motion is dissipative, and each moving vortex experiences a large friction; it generates an electric field parallel to the transport current which produces energy dissipation. Clearly, a crossover should occur from dissipative to nondissipative vortex motion as the quasiparticle mean free path increases. The question is what is the condition which controls the crossover?

This question is of a fundamental importance for our understanding of the dynamics of superconductors and, in a more broad sense, for the understanding of dynamic properties of quantum condensed matter in general. To illustrate the problem one can consider a simple example as follows. One can argue that a time-dependent nondissipative superconducting state, similarly to any other quantum state, can be described by a Hamiltonian dynamics based on a time-dependent Schrödinger equation. Such a description has been suggested for a weakly interacting Bose gas by Pitaevskii (1961) and Gross (1961, 1963); it is widely used also for superfluid helium II. The Gross–Pitaevskii equation is essentially a nonlinear Schrödinger equation, it has the imaginary factor $i\hbar$ in front of the time-derivative of the condensate wave function $\partial\Psi/\partial t$. On the other hand, the time-dependent Ginzburg–Landau model which is a particular case of a more general Model F dynamics (Hohenberg and Halperin 1977) is believed to describe a relaxation dynamics of superconductors near the transition temperature. In contrast to the Gross–Pitaevskii equation, it has the time-derivative $\partial\Psi/\partial t$ with a real factor in front of it. The question which we are interested in can be formulated as follows: What is the condition when the imaginary prefactor transforms into a real one?

It seems that there is no universal answer to this simple question in general. However, the problem of crossover from nondissipative to dissipative behavior of a condensed matter state can be solved for the particular example of superconducting vortex dynamics. We have already seen in Section 12.9 that a relaxation constant in the time-dependent Ginzburg–Landau model has in fact a small imaginary part which results in appearance of a small transverse component of the electric field with respect to the current. We shall see later that the transverse component of the electric field increases at the expense of the longitudinal component as the mean free path of excitations grows. The crossover condition, however, does not coincide simply with the condition which divides superconductors between dirty and clean ones. The criterion rather involves the spectrum of excitations localized in the vortex cores; the distance between their levels takes the part of the energy gap. The condition for a nondissipative vortex motion requires that the relaxation rate of localized excitations is smaller than the distance between the levels. This implies a much longer mean free path of excitations than the condition for a superconductor to be just in a clean limit.

14.1.1 *Boltzmann kinetic equation approach*

A vortex moving in a clean superconductor experiences both friction and transverse forces. The transverse force comes from several different sources including the hydrodynamic Magnus force, the force produced by excitations scattered from the vortex, and the force associated with the momentum flow from the heat bath to the vortex through the localized excitations. The whole rich and exciting physics involved in the vortex dynamics can be successfully described by the general formalism developed in Chapter 10. In this chapter we concentrate on isolated vortices such that their cores do not overlap, i.e., on the region of magnetic fields $H \ll H_{c2}$. It is this limit when the specifics of core states is more pronounced.

Let us first discuss a general picture using a simple approach based on the Boltzmann kinetic equation. For simplicity, we consider an s-wave superconductor. We remind that the profile of the order parameter $\Delta(\mathbf{r})$ near the vortex core produces a potential well where localized states with a discrete spectrum exist. The spectrum of excitations in the vortex core has been calculated in Section 6.4 (see Fig. 6.2). The localized states correspond to energies $|\epsilon| < \Delta_\infty$. The spectrum has the so-called anomalous branch with $n = 0$ whose energy varies from $-\Delta_\infty$ to $+\Delta_\infty$ as the particle impact parameter b changes from $-\infty$ to $+\infty$ and crosses $\epsilon = 0$, being an odd function of b. For low $\epsilon \ll \Delta_\infty$, the anomalous branch is $\epsilon_0 = -\omega_0 \mu$ where $\mu = -bp_\perp$ is the angular momentum, and $p_\perp$ is the momentum in the plane perpendicular to the vortex axis (Caroli *et al.* 1964). In an s-wave superconductor with an axisymmetric vortex, the angular momentum μ is quantized and so is the spectrum, ω_0 being the distance between the discrete levels in the vortex core. The spectrum may also have branches with $n \neq 0$ which are separated from the one with $n = 0$ by energies of the order of Δ. The states $n \neq 0$ do not cross zero of energy as functions of b; they approach the same energy $+\Delta_\infty$ or $-\Delta_\infty$ for both $b \to \pm\infty$. Numerical calculations (Gygi and Schlüter 1991) show that these states are practically absorbed by the continuous spectrum. However we include the states with $n \neq 0$ into consideration for generality. We denote the separation between the levels with neighboring angular momenta through

$$\omega_n = p_\perp^{-1} \frac{\partial \epsilon_n}{\partial b} = -\frac{\partial \epsilon_n}{\partial \mu}. \tag{14.1}$$

The interlevel spacing $\omega_n(b)$ is an even function of b for $n = 0$; the spacing ω_n decreases with increasing n.

We choose the direction of the z-axis in such a way that the vortex has a positive circulation. The z-axis is thus parallel the magnetic field for positive charge of carriers, and it is antiparallel to it for negative charge: $\hat{\mathbf{z}} = \hat{\mathbf{H}}\,\mathrm{sign}\,(e)$. Since the particle velocity $\mathbf{v}_\perp$ makes an angle α with the x-axis, the cylindrical coordinates of the position point (ρ, ϕ) are connected with the impact parameter and the coordinate along the trajectory through $\rho^2 = b^2 + s^2$ where

$$b = \rho \sin(\phi - \alpha)\ ;\ s = \rho \cos(\phi - \alpha). \tag{14.2}$$

The coordinates are shown in Fig. 6.1.

The first step is as follows. We assume that the quasiclassical spectrum $\epsilon_n\,(b)$ of a particle plays the role of its effective Hamiltonian. We can thus invoke the Boltzmann equation in the canonical form

$$\frac{\partial f}{\partial t} + \frac{\partial f}{\partial \alpha} \cdot \frac{\partial \epsilon_n}{\partial \mu} - \frac{\partial \epsilon_n}{\partial \alpha} \cdot \frac{\partial f}{\partial \mu} = \left(\frac{\partial f}{\partial t} \right)_{\rm coll}, \tag{14.3}$$

to describe the quasiparticle distribution (Stone 1996). We shall derive eqn (14.3) from our set of generalized kinetic equations in the following chapter.

Second, we assume that the force acting on a vortex from the environment is exerted via excitations localized in the vortex core. According to this picture, the force on a moving vortex can be written as the momentum transfer from the localized excitations to the vortex

$$\mathbf{F}_{\rm env} = \frac{1}{2} \sum_n \int \frac{dp_z}{2\pi} \frac{d\alpha d\mu}{2\pi} \frac{\partial \mathbf{p}_n}{\partial t} f_1 = -\frac{1}{2} \sum_n \int \frac{dp_z}{2\pi} \frac{d\alpha d\mu}{2\pi} \frac{\partial \epsilon_n}{\partial b} [\hat{\mathbf{z}} \times \hat{\mathbf{v}}_\perp] f_1. \tag{14.4}$$

Here we make use of the Hamilton equation

$$\frac{\partial \mathbf{p}_n}{\partial t} = -\nabla \epsilon_n = -\frac{\partial \epsilon_n}{\partial b} [\hat{\mathbf{z}} \times \hat{\mathbf{v}}_\perp]. \tag{14.5}$$

The second equality follows from the fact that, in the coordinate frame of Fig. 6.1, the energy only depends on the particle impact parameter b. The normalization in eqn (14.4) is chosen such that the sum over the two spin states enters as $(1/2)\sum_s$. We derive eqn (14.4) microscopically in Section 14.5.

We shall later demonstrate that eqn (14.4) indeed gives the full force on a vortex if the vortex density is small enough. The measure of the vortex density is set by the ratio of the Larmor radius $r_H = v_F/\omega_c$ of particles in the magnetic field $B = \Phi_0 n_L$ to their mean free path, ℓ. Here $\omega_c = eB/mc$ is the cyclotron frequency and n_L is the density of vortex lines. When the magnetic field is small (and so is the vortex density) such that $r_H \gg \ell$, i.e., $\omega_c \tau \ll 1$, the excitations with energies above the gap Δ_∞ are in equilibrium with the heat bath, $f_1 = 0$. According to eqn (14.4) they do not produce a force and thus do not influence considerably the vortex motion.

For a moving vortex, the force from the environment is balanced by the Lorentz force:

$$\mathbf{F}_\mathrm{L} = \frac{\Phi_0}{c}[\mathbf{j}_\mathrm{tr} \times \hat{\mathbf{z}}]\mathrm{sign}\,(e), \tag{14.6}$$

with the flux quantum $\Phi_0 = \pi c/|e|$. This form of the Lorentz force is the same as in eqn (12.1) but the vortex circulation axis $\hat{\mathbf{z}}$ is used. The force balance equation

$$\mathbf{F}_\mathrm{L} + \mathbf{F}_\mathrm{env} = 0, \tag{14.7}$$

determines the transport current in terms of the vortex velocity and thus allows to find the flux flow conductivity tensor.

14.1.2 *Forces in s-wave superconductors*

We consider a parabolic spectrum of quasiparticles in the normal state and thus a spherical Fermi surface. We shall simplify the collision integral in eqn (14.3) using the relaxation-time approximation

$$\left(\frac{\partial f}{\partial t}\right)_\mathrm{coll} = -\frac{f_1}{\tau_n}. \tag{14.8}$$

Equation (14.3) is easy to solve. Let us take the distribution function in the form

$$f_1 = -\frac{\partial f^{(0)}}{\partial \epsilon}\left[\gamma_\mathrm{O}([\mathbf{v}_L \times \mathbf{p}_\perp]\cdot\hat{\mathbf{z}}) + \gamma_\mathrm{H}(\mathbf{v}_L \cdot \mathbf{p}_\perp)\right] \tag{14.9}$$

where the factors $\gamma_\mathrm{O,H}$ are to be found. For an axi-symmetric s-wave vortex the energies ϵ_n do not depend on α and the term $\partial\epsilon_n/\partial\alpha$ vanishes. In the term with the time-derivative, the energy ϵ_n contains a time dependence through $\mu(t) = [(\mathbf{r} - \mathbf{v}_L t) \times \mathbf{p}]\cdot\hat{\mathbf{z}}$ such that

$$\frac{\partial f^{(0)}}{\partial t} = \frac{\partial f^{(0)}}{\partial \epsilon}\frac{\partial \epsilon_n}{\partial \mu}\frac{\partial \mu}{\partial t} = \frac{\partial f^{(0)}}{\partial \epsilon}\frac{\partial \epsilon_n}{\partial \mu}\left([\mathbf{p}_\perp \times \mathbf{v}_L]\cdot\hat{\mathbf{z}}\right). \tag{14.10}$$

With the Ansatz (14.9), the force eqn (14.4) splits into two terms $\mathbf{F}_\mathrm{env} = \mathbf{F}_\parallel + \mathbf{F}_\perp$, with the friction $\mathbf{F}_\parallel$ and transverse $\mathbf{F}_\perp$ forces given by

$$\mathbf{F}_\parallel = -\pi N\left\langle\!\!\left\langle \sum_n \int \omega_n \gamma_\mathrm{O}\frac{\partial f^{(0)}}{\partial \epsilon}\frac{d\mu}{2}\right\rangle\!\!\right\rangle \mathbf{v}_L, \tag{14.11}$$

$$\mathbf{F}_{\perp} = \pi N \left\langle\!\!\left\langle \sum_n \int \omega_n \gamma_{\rm H} \, \frac{\partial f^{(0)}}{\partial \epsilon} \, \frac{d\mu}{2} \right\rangle\!\!\right\rangle [\hat{\mathbf{z}} \times \mathbf{v}_L] \,, \tag{14.12}$$

where N is the electron density, $\langle\!\langle \ldots \rangle\!\rangle$ is the average over the Fermi surface with the weight $\pi p_{\perp}^2$,

$$\langle\!\langle \ldots \rangle\!\rangle = V_F^{-1} \int \pi p_{\perp}^2 dp_z \, (\ldots) \,, \tag{14.13}$$

and V_F is the volume encompassed by the Fermi surface. For an isotropic Fermi surface,

$$\langle\!\langle \ldots \rangle\!\rangle = \frac{3}{4} \int \sin^3 \theta \, d\theta \, (\ldots) \,.$$

where $p_z = p_F \cos \theta$.

The Boltzmann equation (14.3) gives the factors $\gamma_{\rm O,H}$ in the form

$$\gamma_{\rm O} = \frac{\omega_n \tau_n}{\omega_n^2 \tau_n^2 + 1}, \qquad \gamma_{\rm H} = \frac{\omega_n^2 \tau_n^2}{\omega_n^2 \tau_n^2 + 1}. \tag{14.14}$$

The *longitudinal force* $\mathbf{F}_{\parallel}$ defines the friction coefficient in the vortex equation of motion and determines the Ohmic component of the conductivity $\sigma_{\rm O}$. Expressing the vortex velocity $\mathbf{v}_L$ through the average electric field $\mathbf{E}$, as $\mathbf{v}_L = c\,[\mathbf{E} \times \hat{\mathbf{z}}]/B \, {\rm sign}(e)$, we find

$$\sigma_{\rm O} = \frac{N|e|c}{B} \left\langle\!\!\left\langle \sum_n \int \frac{\omega_n \tau_n}{\omega_n^2 \tau_n^2 + 1} \, \frac{\partial f^{(0)}}{\partial \epsilon} \, \frac{d\epsilon}{2} \right\rangle\!\!\right\rangle . \tag{14.15}$$

The *transverse force* determines the Hall conductivity

$$\sigma_{\rm H} = \frac{Nec}{B} \left\langle\!\!\left\langle \sum_n \int \frac{\omega_n^2 \tau_n^2}{\omega_n^2 \tau_n^2 + 1} \, \frac{\partial f^{(0)}}{\partial \epsilon} \, \frac{d\epsilon}{2} \right\rangle\!\!\right\rangle . \tag{14.16}$$

We emphasize that the force $\mathbf{F}_{\rm env}$ is defined as the response of the whole environment to the vortex displacement. It is therefore the *total* force acting on the vortex from the ambient system, including all partial forces such as the longitudinal friction force and the nondissipative transverse force. The transverse force, in turn, includes various parts which can be identified historically (Sonin 1987) as the Iordanskii force (Iordanskii 1964), the spectral flow force (Volovik 1986, Stone and Gaitan 1987, Kopnin *et al.* 1995), and the Magnus force. We shall discuss this later in more detail in Section 14.6.3.

The main conclusion is that the Ohmic and Hall conductivities depend on the purity of the sample through the parameter $\omega_0 \tau$ (recall that $\omega_n \lesssim \omega_0$ and decreases rapidly with n). One can distinguish two regimes: moderately clean $\omega_0 \tau \ll 1$ and superclean $\omega_0 \tau \gg 1$. Note that the moderately clean regime still requires that the superconductor is clean in the usual sense $\Delta_\infty \tau \gg 1$.

In the moderately clean limit where $\omega_0 \tau \ll 1$, the conductivity roughly follows the Bardeen and Stephen (1965) expression at low temperatures though it exhibits an extra temperature-dependent factor Δ_∞/T_c on approaching T_c (Kopnin and Lopatin 1995),

$$\sigma_{\rm O} \sim \sigma_n \frac{H_{c2}}{H}\frac{\Delta_\infty}{T_c}.$$

The factor Δ_∞/T_c appears because the number of delocalized quasiparticles contributing to the vortex dynamics decreases near T_c. Note that the flux flow conductivity of a dirty superconductor has just the inverse extra factor T_c/Δ_∞ as compared to the Bardeen–Stephen model (see Section 13.2.1). We discuss this later in Section 14.6.2. The Hall conductivity and the Hall angle are small $\sigma_{\rm H} \sim (\omega_0\tau)\,\sigma_{\rm O}$ and $\tan\Theta_H \sim (\omega_0\tau)$, respectively.

In the superclean limit, on the contrary, the Ohmic conductivity is small. The vortex dynamics becomes nondissipative. The friction force vanishes while the transverse force eqn (14.12) takes the form

$$\mathbf{F}_\perp = \pi N \tanh\frac{\Delta_\infty}{2T}\,[\hat{\mathbf{z}} \times \mathbf{v}_L].$$

We have accounted for the fact that

$$\int \omega_0 \frac{\partial f^{(0)}}{\partial \epsilon}\frac{d\mu}{2} = \tanh\frac{\Delta_\infty}{2T}$$

since $\epsilon_0(\mu)$ varies from $-\Delta_\infty$ to $+\Delta_\infty$. Similar integrals with ω_n vanish for $n \neq 0$ because $\epsilon_n(\mu)$ returns to the same energy $+\Delta_\infty$ (or to the same $-\Delta_\infty$) for both $\mu \to \pm\infty$. The force balance equation tell us that the transport current is coupled to the vortex velocity by

$$\mathbf{j}_{\rm tr} = eN\tanh\frac{\Delta_\infty}{2T}\mathbf{v}_L.$$

It shows that $N\tanh(\Delta_\infty/2T)$ electrons move together with the vortex. The factor $\tanh(\Delta_\infty/2T)$ accounts for the fact that delocalized excitations are at rest with respect to the heat bath according to our assumption made on page 274. At low temperatures, delocalized excitations are absent, and the current is proportional to the density of all the electrons. We shall discuss this in more detail in Section 14.6.3. The corresponding Hall conductivity is

$$\sigma_H = \frac{Nec}{B}\tanh\left(\frac{\Delta_\infty}{2T}\right).$$

We conclude that it is the parameter $\omega_0\tau$ which determines the crossover from dissipative to nondissipative vortex dynamics. We stress again that these results apply for the low-field limit, $\omega_c\tau \ll 1$. When $\omega_c\tau \sim 1$, delocalized excitations also contribute to the force $\mathbf{F}_{\rm env}$. We consider this in more detail later using the microscopic approach which allows us to get a more comprehensive picture of this phenomenon.

14.2 Spectral representation for the Green functions

We turn now to the microscopic description of the vortex dynamics in clean superconductors. We consider first some general properties of Green functions of clean superconductors which will be used later. For clean systems, the quasiparticle spectrum is well defined. According to the general properties of the Green functions, the combination $G^R - G^A$ is a sum of a spectral weight multiplied by δ -functions at energies corresponding to the quasiparticle spectrum. We show that the same holds also for quasiclassical Green functions taken on a discrete spectrum. We consider states in a vortex core for definiteness.

Let us expand the Green function $\check{G}_\epsilon^{R(A)}$ in the eigenfunctions of a quasiclassical particle with the momentum $p \sim p_F \gg \xi^{-1}$ moving along a definite trajectory. In case of a linear vortex, we specify the trajectory by the direction of particle velocity $\mathbf{v}_F$ and by the impact parameter b with respect to the vortex axis. Consider the Bogoliubov–de Gennes wave function $\mathcal{U}(s, b, z)$

$$\mathcal{U} = \begin{pmatrix} u \\ -v \end{pmatrix}, \quad \mathcal{U}^\dagger = (u^* \ v^*)$$

at a point with coordinates s and b in the coordinate frame (s, b) rotated by an angle α with respect to the (x, y) frame as shown in Fig. 6.1 and with z-coordinate along the vortex axis. This satisfies the equation

$$\left[-\frac{1}{2m^*}\left[\left(\frac{\partial}{\partial s}\right)^2 + \left(\frac{\partial}{\partial b}\right)^2\right] - E_F + \check{H}\right]\mathcal{U} = \epsilon\,\check{\tau}_3\mathcal{U}.$$

For a two-dimensional problem, we ignore the trivial z dependence for brevity. The Hamiltonian is

$$\check{H} = \begin{pmatrix} -(e/c)\,\mathbf{v}_F \cdot \mathbf{A} & -\Delta \\ \Delta^* & (e/c)\,\mathbf{v}_F \cdot \mathbf{A} \end{pmatrix}.$$

We take a fixed point b_0 in the vicinity of b and put $b = b_0 + b_1$ where $p_F^{-1} \ll b_1 \ll \xi$. Next, we perform a Fourier transformation in the coordinate b_1. As a result we obtain a wave function $\mathcal{U}(b_0, p_b, s)$ which depends on the wave vector p_b and on the coordinate s; in addition, it also has a parametric dependence on b_0. The obtained wave function satisfies the equation

$$\left[\frac{1}{2m^*}\left[-\left(\frac{\partial}{\partial s}\right)^2 + p_b^2\right] - E_F + \check{H}(b_0, s)\right]\mathcal{U} = \epsilon\,\check{\tau}_3\mathcal{U}.$$

Here we neglect a small correction b_1 in the Hamiltonian $\check{H}$. We now take the limit of small p_b and fix it within the interval $\xi^{-1} \ll p_b \ll \sqrt{p_F/\xi}$. This means that the particle momentum becomes parallel to the s-axis. In other words, the

s-axis is now directed along the particle trajectory while b_0 is thus the impact parameter. The equation takes the form

$$\left[\frac{1}{2m^*}\left[-\left(\frac{\partial}{\partial s}\right)^2\right] - E_F + \check{H}(b,s)\right]\mathcal{U} = \epsilon_n(b)\,\check{\tau}_3\mathcal{U}. \tag{14.17}$$

The term $p_b^2/2m$ can be neglected as compared to $\check{H} \sim \Delta$. We also omit the index 0 at b. The wave function depends now on the distance along the trajectory s and on the parameters b and α where α is the angle between the particle momentum $\mathbf{p}_\perp$ and the x axis of the coordinate frame. They satisfy the orthogonality conditions eqns (3.59) and (3.60) which now read

$$\sum_n \check{\tau}_3 \mathcal{U}_n(s_1)\mathcal{U}_n^\dagger(s_2) = \check{1}\delta(s_1 - s_2), \tag{14.18}$$

$$\int ds\, \mathcal{U}_n^\dagger(s)\,\check{\tau}_3\mathcal{U}_{n'}(s) = \delta_{n,n'}. \tag{14.19}$$

We also denote the energy spectrum as $\epsilon_n(b)$; it is a function of the impact parameter b and also on the momentum along the z axis p_z as well as (possibly) on α . In addition, it can depend on the principal quantum number n and on other quantum numbers (for example, particle–hole isotopic spin, etc.) relevant for the particular problem. The impact parameter b is a good quantum number for a quasiclassical particle which moves in a potential Δ whose magnitude is much smaller than the kinetic energy E_F. Under these conditions, the trajectory is a straight line with a constant b even if the problem does not have an axial symmetry.

We can now expand the Green function of the particle in terms of these wave functions:

$$\begin{aligned}\check{G}_\epsilon^{R(A)}(\mathbf{v}, b;\; s_1, s_2) &= \begin{pmatrix} G^{R(A)} & F^{R(A)} \\ -F^{\dagger R(A)} & \bar{G}^{R(A)} \end{pmatrix} \\ &= -\sum_n \frac{\mathcal{U}_n(\mathbf{v}, b; s_1)\mathcal{U}_n^\dagger(\mathbf{v}, b; s_2)}{\epsilon - \epsilon_n(b) \pm i\delta}.\end{aligned} \tag{14.20}$$

The Green function matrix obeys the equation

$$\left[E_n\left(-i\frac{\partial}{\partial s}\right) - E_F - \epsilon\check{\tau}_3 + \check{H}\right]\check{G}_\epsilon^{R(A)} = \check{1}\,\delta(s_1 - s_2) \tag{14.21}$$

where $E_n(\mathbf{p})$ is the normal-state electronic spectrum.

Next, we determine the relation between the Green function $\check{G}_\epsilon^{R(A)}$ and its quasiclassical counterpart $\check{g}^{R(A)}$. We represent the Green function (both retarded and advanced) as

$$\check{G}_\epsilon = e^{ip_\perp(s_1 - s_2)}[\check{a}_+\Theta(s_1 - s_2) - \check{a}_-\Theta(s_2 - s_1)] \tag{14.22}$$

where $\check{a}_\pm(s)$ are slow functions of $s = (s_1 + s_2)/2$; they vary over distances of the order of ξ as compared to atomic-scale variations of the exponents. From eqn

(14.21) we find $\check{a}_+ + \check{a}_- = i/v_\perp$ where $v_\perp$ is the particle velocity in the plane perpendicular to the vortex axis. It is easy to check that the functions $\check{a}_\pm$ satisfy the same Eilenberger equations as the quasiclassical Green functions $\check{g}^{R(A)}$. To do this we insert eqn (14.22) into eqns (14.21) and expand in small gradients of $a_\pm$. Using the boundary conditions at large distances, we obtain (Gor'kov and Kopnin 1973 a)

$$\check{a}_\pm = \frac{i}{2v_\perp}\left(\check{1} \pm \check{g}\right).$$

Thus,

$$\check{G}_\epsilon^{R(A)}(s_1 \to s_2) = \frac{i}{2v_\perp}\left[\check{g}^{R(A)} + \check{1}\,\mathrm{sign}\,(s_1 - s_2)\right].$$

This relation was obtained by Kopnin and Lopatin (1995). Finally

$$\check{G}_\epsilon^{R}(s_1 \to s_2) - \check{G}_\epsilon^{A}(s_1 \to s_2) = \frac{i}{2v_\perp}\left(\check{g}^R - \check{g}^A\right). \tag{14.23}$$

Comparing eqns (14.20) and (14.23), one can write

$$\check{g}_-(\mathbf{p}, \mathbf{r}) = \sum_n \check{g}_n\left(p_z, \alpha, b; s\right)\delta\left[\epsilon - \epsilon_n\left(p_z, \alpha, b\right)\right] \tag{14.24}$$

where $\check{g}_n$ are the spectral weights for the quasiclassical Green function.

14.3 Useful identities

Here we concentrate on the Green functions in presence of vortices and derive several identities (Kopnin and Lopatin 1995, Blatter *et al.* 1999) which will be used in what follows. Consider the energies $|\epsilon| < \Delta_\infty$. We find from eqn (14.19)

$$\begin{aligned}
\int \mathrm{Tr}\ \left[\check{\tau}_3 \check{g}_-\right] ds &= -iv_\perp \int \mathrm{Tr}\ \left[\check{\tau}_3\left(\check{G}_\epsilon^R(s,s) - \check{G}_\epsilon^A(s,s)\right)\right] ds \\
&= 2\pi v_\perp \sum_n \delta(\epsilon - \epsilon_n) \int \mathrm{Tr}\left[\mathcal{U}_n^\dagger\left(s\right)\check{\tau}_3 \mathcal{U}_n\left(s\right)\right] ds \\
&= 2\pi v_\perp \sum_n \delta(\epsilon - \epsilon_n).
\end{aligned} \tag{14.25}$$

We turn to derivation of another identity. Consider the matrix $\hat{\nabla}\check{H}$ defined by eqn (13.11) and calculate

$$\begin{aligned}
&\mathrm{Tr}\ \left[\check{g}_-(\mathbf{d}\cdot\hat{\nabla}\check{H})\right] \\
&= f_-\mathbf{d}\cdot\left(\nabla + \frac{2ie}{c}\mathbf{A}\right)\Delta^* + f_-^\dagger\mathbf{d}\cdot\left(\nabla - \frac{2ie}{c}\mathbf{A}\right)\Delta + \frac{2e}{c}g_-\mathbf{d}\cdot\left[\mathbf{H}\times\mathbf{v}_F\right] \\
&= \left(f_-\mathbf{d}\cdot\nabla\Delta^* + f_-^\dagger\mathbf{d}\cdot\nabla\Delta\right) - \frac{2e}{c}g_-\mathbf{d}\cdot\nabla\left(\mathbf{v}_F\cdot\mathbf{A}\right) \\
&\quad + \frac{2ie}{c}\mathbf{d}\cdot\mathbf{A}\left(f_-\Delta^* - f_-^\dagger\Delta\right) + \frac{2e}{c}g_-\left(\mathbf{v}_F\cdot\nabla\right)\mathbf{d}\cdot\mathbf{A}
\end{aligned}$$

$$= \mathrm{Tr}\left[\check{g}_-(\mathbf{d}\cdot\nabla\check{H})\right] + \frac{2e}{c}(\mathbf{v}_F\cdot\nabla)(\mathbf{d}\cdot\mathbf{A}g_-) \tag{14.26}$$

where $\mathbf{d}$ is an arbitrary constant vector. In the fourth line we use the Eilenberger equation

$$\mathbf{v}_F\cdot\nabla g^{R(A)} - i\left(f^{R(A)}\Delta^* - f^{\dagger R(A)}\Delta\right) = 0 \tag{14.27}$$

neglecting the small collision integral. Since $g^R - g^A$ vanishes for large distances from the vortex if $\epsilon < \Delta_\infty$ we obtain

$$\int \mathrm{Tr}\left[(\mathbf{d}\cdot\hat{\nabla}\check{H})\check{g}_-\right] ds = \int \mathrm{Tr}\left[(\mathbf{d}\cdot\nabla\check{H})\check{g}_-\right] ds. \tag{14.28}$$

We use eqn (14.23) again and find that the r.h.s. of eqn (14.28) is

$$\begin{aligned}
-iv_\perp &\int \mathrm{Tr}\ \left[(\mathbf{d}\cdot\nabla\check{H})\left(\check{G}^R_\epsilon(s,s) - \check{G}^A_\epsilon(s,s)\right)\right] ds \\
&= 2\pi v_\perp \sum_n \delta(\epsilon-\epsilon_n)\int \mathrm{Tr}\ \left[\mathcal{U}_n^\dagger\left(d_s\frac{\partial\check{H}}{\partial s} + d_b\frac{\partial\check{H}}{\partial b}\right)\mathcal{U}_n\right] ds \\
&= 2\pi v_\perp d_b \sum_n \delta(\epsilon-\epsilon_n)\int \mathrm{Tr}\ \left[\mathcal{U}_n^\dagger\frac{\partial\check{H}}{\partial b}\mathcal{U}_n\right] ds \\
&= 2\pi\left([\mathbf{v}_\perp\times\mathbf{d}]\cdot\hat{\mathbf{z}}\right)\sum_n \frac{\partial\epsilon_n}{\partial b}\delta(\epsilon-\epsilon_n).
\end{aligned}$$

This proves the identity

$$\int \mathrm{Tr}\ \left[(\hat{\nabla}\check{H})\hat{g}_-\right] ds = 2\pi[\hat{\mathbf{z}}\times\mathbf{v}_\perp]\sum_n \frac{\partial\epsilon_n(b)}{\partial b}\delta\left[\epsilon-\epsilon_n(b)\right]. \tag{14.29}$$

It is interesting to note that the integral

$$\begin{aligned}
&\int \mathrm{Tr}\ \left[\left(\hat{\nabla}\check{H}\right)\hat{g}_-\right] d^2r \\
&\quad = 2\pi[\hat{\mathbf{z}}\times\mathbf{v}_\perp]\sum_n\int_{-\infty}^{\infty}\frac{\partial\epsilon_n(b)}{\partial b}\delta\left[\epsilon-\epsilon_n(b)\right] db \\
&\quad = 2\pi[\hat{\mathbf{z}}\times\mathbf{v}_\perp]\sum_n \mathrm{sign}\left(\frac{\partial\epsilon_n(b)}{\partial b}\right)
\end{aligned} \tag{14.30}$$

gives $2\pi[\hat{\mathbf{z}}\times\mathbf{v}_\perp]N(\epsilon)$ where an integer $N(\epsilon)$ is the algebraic sum of numbers of spectrum branches which cross the energy ϵ as functions of b. Here $+1$ is assigned to the branch which crosses ϵ upwards and -1 is for the branch going in the opposite direction. Therefore, the branch which starts at $b=-\infty$ from an energy above ϵ, crosses ϵ an even number of times and then returns at $b=\infty$ to an energy above ϵ, does not contribute to N. For $|\epsilon| < \Delta_\infty$ it is the anomalous branch with $n=0$ only which crosses any energy once, and, therefore, $N=1$.

Application of eqn (14.30) to energies $|\epsilon| > \Delta_\infty$ needs a discussion. First, an integral over the remote surface may appear in course of transformations involved in the derivation of eqn (14.28). We will see how to deal with the surface terms later. Second, for energies above the gap, the energy spectrum is continuous. One can, however, make it quasi-discrete by placing the system in a large box. Now we can count the energy branches which cross an energy ϵ once. This counting can be performed for impact parameters larger than the core radius. This implies that the number N does not depend on the actual structure of the vortex core: to calculate N one can use any convenient model for the core. The surface contributions are determined by large distances and are also independent of the actual core structure.

The simplest model is an "artificial" vortex with a core size much larger than ξ. In such a vortex the order parameter magnitude varies on distances much longer than ξ so that the Green functions can be found by expanding the Eilenberger equations (5.84), (5.85) in powers of small gradients. We get for $|\epsilon| > |\Delta(\rho)|$:

$$f_- = \left(f - \frac{\partial g}{\partial |\Delta|} \frac{e}{c} \mathbf{v}_F (\mathbf{A} - \frac{c}{2e} \nabla \chi) - \frac{i}{2|\Delta|} \mathbf{v}_F \nabla g \right) e^{i\chi}, \tag{14.31}$$

$$g_- = g + \frac{\partial g}{\partial \epsilon} \frac{e}{c} \mathbf{v}_F (\mathbf{A} - \frac{c}{2e} \nabla \chi), \tag{14.32}$$

and $f_- = (f_-^\dagger)^*$. For $|\epsilon| < |\Delta(\rho)|$ the functions are $f_- = f_-^\dagger = g_- = 0$. Here the "adiabatic" Green functions are

$$g = \frac{\epsilon}{\sqrt{\epsilon^2 - |\Delta|^2}} \Theta\left(\epsilon^2 - |\Delta|^2\right), \; f = \frac{|\Delta|}{\sqrt{\epsilon^2 - |\Delta|^2}} \Theta\left(\epsilon^2 - |\Delta|^2\right). \tag{14.33}$$

We now find

$$\begin{aligned} &\frac{1}{2} \int_{S_0} \mathrm{Tr}\left[(\hat{\nabla} \check{H})(\hat{g}^R - \hat{g}^R) \right] d^2 r \\ &= - \int_{S_0} \left[\mathbf{v}_\perp \times \left(\mathrm{curl}\,[(\frac{2e}{c}\mathbf{A} - \nabla\chi) g] + g\, \mathrm{curl}\, \nabla \chi \right) - 2 f \nabla |\Delta| \right] d^2 r \\ &= [\hat{\mathbf{z}} \times \mathbf{v}_\perp] \left(g(\infty) \oint d\mathbf{r}\, (\frac{2e}{c}\mathbf{A} - \nabla\chi) + 2\pi g(0) \right). \end{aligned}$$

The last term in the second line vanishes being the full derivative of the quantity $\sqrt{\epsilon^2 - |\Delta|^2}$ because $|\Delta|$ is periodic in the vortex lattice. The contour integral taken along the boundary of the unit cell represents the surface term mentioned earlier (note that for large distances, $g = g_-$). It vanishes because the magnetic flux through the unit cell is just the flux quantum Φ_0. Since $|\Delta(0)| = 0$, the function $g(0) = 1$, and we obtain

$$\frac{1}{2} \int_{S_0} \mathrm{Tr}\left[(\hat{\nabla} \check{H})(\hat{g}^R - \hat{g}^R) \right] d^2 r = 2\pi [\hat{\mathbf{z}} \times \mathbf{v}_\perp]. \tag{14.34}$$

The right-hand side of eqn (14.34) corresponds to the number $N = 1$ of the branches crossing ϵ once. Note that the same model of the vortex core would result in $N = 1$ for energies $|\epsilon| < \Delta_\infty$, as well, in accordance with the statement that this topological property of the spectrum does not depend on the core structure.

14.4 Distribution function

In this section we concentrate on s-wave superconductors. The case of d-wave superconductors will be discussed later. The distribution function is determined by the kinetic equations (10.41) and (10.56) which are derived in Section 15.5. We write down these equations again

$$\left(e\mathbf{v}_F \cdot \mathbf{E} g_- + \frac{1}{2}\left[f_- \frac{\hat{\partial}\Delta^*}{\partial t} + f_-^\dagger \frac{\hat{\partial}\Delta}{\partial t}\right]\right)\frac{\partial f^{(0)}}{\partial \epsilon} + \mathbf{v}_F \cdot \nabla(f_2 g_-)$$
$$-\left(\frac{1}{2}\left[(\hat{\nabla}\Delta)f_-^\dagger + (\hat{\nabla}\Delta^*)f_-\right] - \frac{e}{c}[\mathbf{v}_F \times \mathbf{H}]g_-\right)\frac{\partial f_1}{\partial \mathbf{p}} = J_1, \tag{14.35}$$

and

$$\mathbf{v}_F \nabla f_1 = 0. \tag{14.36}$$

In the first equation, we omitted $\partial f_1/\partial t$ for a steady vortex motion as well as $\partial\Delta/\partial\mathbf{p}$ which vanishes for an s-wave superconductor. We already know from Section 15.5 that $f_2 \ll f_1$. Equations (14.35) and (14.36) are the basis for our derivations.

14.4.1 *Localized excitations*

Consider first excitations with $|\epsilon| < \Delta_\infty$ localized in the vortex core. The constant function f_1 can be found by integrating eqn (14.35) along the trajectory:

$$-\int\left(\frac{1}{2}\left[(\hat{\nabla}\Delta)f_-^\dagger + (\hat{\nabla}\Delta^*)f_-\right] - \frac{e}{c}[\mathbf{v}_F \times \mathbf{H}]g_-\right)\frac{\partial f_1}{\partial \mathbf{p}}\, ds$$
$$+\frac{\partial f^{(0)}}{\partial \epsilon}\int\left(e\mathbf{v}_F\mathbf{E} g_- + \frac{1}{2}\left[f_- \frac{\hat{\partial}\Delta^*}{\partial t} + f_-^\dagger \frac{\hat{\partial}\Delta}{\partial t}\right]\right) ds$$
$$= \int J_1^{(1)}\{f_1\}\, ds. \tag{14.37}$$

In the collision integral J, only the term with f_1 is important. The term with $g_- f_2$ vanishes at large distances from the vortex core.

For a moving vortex,

$$\frac{\partial\Delta}{\partial t} = -(\mathbf{v}_L \cdot \nabla)\,\Delta, \quad \frac{\partial\mathbf{A}}{\partial t} = -(\mathbf{v}_L \cdot \nabla)\,\mathbf{A}$$

up to the leading terms in the vortex velocity $\mathbf{v}_L$. The expression

$$e\mathbf{v}_F\mathbf{E} g_- + \frac{1}{2}\left[f_- \frac{\hat{\partial}\Delta^*}{\partial t} + f_-^\dagger \frac{\hat{\partial}\Delta}{\partial t}\right]$$

in the second line in the l.h.s. of eqn (14.37) can be written as

$$g_- \frac{e}{c} (\mathbf{v}_L \nabla) (\mathbf{v}_F \mathbf{A}) - \frac{\mathbf{v}_L}{2} \left[f_- \nabla \Delta^* + f_-^\dagger \nabla \Delta \right]$$
$$-e g_- \mathbf{v}_F \nabla \varphi - ie\varphi \left(f_- \Delta^* - f_-^\dagger \Delta \right)$$

and transformed into

$$-\frac{e}{c} \mathbf{v}_L \left[\mathbf{H} \times \mathbf{v}_F \right] g_- - \frac{\mathbf{v}_L}{2} \left[f_- \hat{\nabla} \Delta^* + f_-^\dagger \hat{\nabla} \Delta \right]$$
$$+ \frac{ie}{c} (\mathbf{v}_L \cdot \mathbf{A}) \left(f_- \Delta^* - f_-^\dagger \Delta \right) + g_- \frac{e}{c} (\mathbf{v}_F \nabla) (\mathbf{v}_L \mathbf{A}) - e \mathbf{v}_F \nabla (\varphi g_-)$$
$$= -\frac{\mathbf{v}_L}{2} \mathrm{Tr} \left[\hat{\nabla} \check{H} \check{g}_- \right] - \mathbf{v}_F \nabla \left[g_- \left(e\varphi - \frac{e}{c} \mathbf{v}_L \mathbf{A} \right) \right]. \tag{14.38}$$

Here we twice used the Eilenberger equation (14.27).

The integral in the first line of eqn (14.37) is

$$\frac{1}{2} \int \mathrm{Tr} \, (\hat{g}_- \hat{\nabla} \check{H}) \, \frac{\partial f_1}{\partial \mathbf{p}} \, ds, \tag{14.39}$$

The integration along the trajectory in eqn (14.37) can now be performed employing the identity of eqn (14.29) and the fact that g_- vanishes for $s \to \pm\infty$. We find

$$\left[\pi ([\mathbf{v}_\perp \times \mathbf{v}_L] \hat{\mathbf{z}}) \frac{\partial f^{(0)}}{\partial \epsilon} \frac{\partial \epsilon_n}{\partial b} + \pi \left(\left[\mathbf{v}_\perp \times \frac{\partial f_1}{\partial \mathbf{p}} \right] \hat{\mathbf{z}} \right) \frac{\partial \epsilon_n}{\partial b} \right] \delta(\epsilon - \epsilon_n)$$
$$+ \int_{-\infty}^{\infty} J_1 \, ds = 0.$$

We use that $\partial f_1 / \partial \mathbf{p}$ is constant along the trajectory. Consider this in more detail. The distribution function f_1 depends on the variables p_z, α, b, where p_z is the momentum along the z-axis. The momentum derivative $\partial f_1 / \partial \mathbf{p}$ can be presented as

$$\frac{\partial f_1}{\partial \mathbf{p}} = \pm \frac{\partial f_1}{\partial p_\perp} \frac{\mathbf{v}_\perp}{v_\perp} \pm \frac{1}{p_\perp} \left(\frac{\partial f_1}{\partial \alpha} - s \frac{\partial f_1}{\partial b} \right) \frac{\hat{\mathbf{z}} \times \mathbf{v}_\perp}{v_\perp}. \tag{14.40}$$

The plus sign is for quasi-electrons while the minus sign is for quasi-holes, since the directions of $\mathbf{v}_\perp$ and $\mathbf{p}_\perp$ are either the same or opposite for these two types of quasiparticles, respectively. The terms $\partial f_1 / \partial p_\perp$ and $\partial f_1 / \partial \alpha$ in eqn (14.40) are indeed constant along the trajectory but the term $s \, \partial f_1 / \partial b$ is not. However, one can show that this term vanishes identically for an axisymmetric vortex in s-wave superconductors. We shall prove this later in Section 15.3 in a more general form.

The collision integral can be also presented as a spectral sum

$$J_1 = \sum_n J_n \delta(\epsilon - \epsilon_n) \tag{14.41}$$

because it contains terms proportional to either g_- or f_-, $f_-^\dagger$. The kinetic equation becomes

$$\pi([\mathbf{v}_\perp \times \mathbf{v}_L]\hat{\mathbf{z}})\frac{\partial f^{(0)}}{\partial \epsilon}\frac{\partial \epsilon_n}{\partial b} + \pi\left(\left[\mathbf{v}_\perp \times \frac{\partial f_1}{\partial \mathbf{p}}\right]\hat{\mathbf{z}}\right)\frac{\partial \epsilon_n}{\partial b} + \int_{-\infty}^{\infty} J_n\, ds = 0 \quad (14.42)$$

where all terms are taken at $\epsilon = \epsilon_n$. The term $\partial f_1/\partial \mathbf{p}$ in eqn (14.42) should be now understood as

$$\left(\left[\mathbf{v}_\perp \times \frac{\partial f_1}{\partial \mathbf{p}}\right]\hat{\mathbf{z}}\right) = \pm\frac{v_\perp}{p_\perp}\frac{\partial f_1}{\partial \alpha}. \quad (14.43)$$

The component in eqn (14.40) parallel to $\mathbf{v}_\perp$ drops out of eqn (14.42).

14.4.1.1 *Low temperatures* The integro-differential equation (14.42) can hardly be solved analytically in a general case. Some simplifications arise when temperatures are much below T_c. Only the level with $n = 0$ is excited at such temperatures. The corresponding solution was first obtained by Kopnin and Kravtsov (1976 a). For simplicity, we restrict ourselves to a spherical Fermi surface.

We assume that the scattering by impurities is the primary source of relaxation. This is the most natural assumption for practically all superconducting materials. Consider the impurity collision integral for energies $\epsilon \ll \Delta_\infty$. The spectral weight for $n = 0$ in the expansion eqn (14.41) of the collision integral can be found from eqn (10.74). It has the form

$$\begin{aligned} J_0 = -\frac{1}{\tau}&\left[(f_1\langle g_-\rangle - \langle f_1 g_-\rangle)\, g_0\right. \\ &\left.-\frac{1}{2}\left(\left(f_1\left\langle f_-^\dagger\right\rangle - \left\langle f_1 f_-^\dagger\right\rangle\right) f_0 + (f_1\langle f_-\rangle - \langle f_1 f_-\rangle)\, f_0^\dagger\right)\right]. \end{aligned}$$

Recall that $\langle\cdots\rangle$ is the usual average over the Fermi surface. The Green functions for low energies were found in Section 6.4. According to eqn (6.83) the spectral weights are

$$g_0 = \frac{\pi v_\perp e^{-K}}{2C},\; f_0 = g_0 e^{i\phi}\frac{is}{\rho}.$$

The bound state energy is given by eqn (6.84).

$$\epsilon_0 = C^{-1} b\int_0^\infty (\Delta_0/\rho)\, e^{-K} d\rho \quad (14.44)$$

where

$$K(\rho) = \frac{2}{v_\perp}\int_0^\rho \Delta_0 d\rho,\; C = \int_0^\infty e^{-K} d\rho$$

and we neglect a small ω_c. Due to the δ-function in $\check{g}_-$, the integration over $d\alpha$ selects a narrow region of angles near $\alpha = \phi$ and $\alpha = \pi + \phi$ such that the impact parameter is small $b \sim \epsilon/\Delta_\infty$ and the distance is either $s \approx \rho$ or

$s \approx -\rho$, respectively. Calculating the Fermi-surface averages we will perform the integration over $d\alpha$ using the rule

$$\int \frac{d\alpha}{2\pi} F(s;b) = \int \frac{db}{2\pi |s|} \left[F(s=\rho;b) + F(s=-\rho;b) \right].$$

As a result

$$\int \frac{d\alpha}{2\pi} g_0 \delta(\epsilon - \epsilon_0) = \frac{g_0}{\pi\rho(\partial\epsilon_0/\partial b)}; \int \frac{d\alpha}{2\pi} f_0 \delta(\epsilon - \epsilon_0) = 0.$$

Let us take the distribution function in the form of eqn (14.9) and consider the collision integral. Performing the averaging, we have

$$\int \frac{d\alpha}{2\pi} f_1 g_0 \delta(\epsilon - \epsilon_0) = 0$$

and

$$\int \frac{d\alpha}{2\pi} f_1 f_0 \delta(\epsilon - \epsilon_0) = -\frac{ip_\perp g_0 e^{i\phi}}{\pi\rho(\partial\epsilon_0/\partial b)} \frac{\partial f^{(0)}}{\partial \epsilon} \left[\gamma_O v_{L\phi} + \gamma_H v_{L\rho} \right].$$

Here we omit contributions proportional to small b. Using

$$p_\perp v_{L\rho} = \left[\frac{s}{\rho} (\mathbf{v}_L \cdot \mathbf{p}_\perp) + \frac{b}{\rho} ([\mathbf{v}_L \times \mathbf{p}_\perp] \cdot \mathbf{z}) \right],$$

$$p_\perp v_{L\phi} = \left[-\frac{b}{\rho} (\mathbf{v}_L \cdot \mathbf{p}_\perp) + \frac{s}{\rho} ([\mathbf{v}_L \times \mathbf{p}_\perp] \cdot \mathbf{z}) \right],$$

we obtain for the collision integral

$$\begin{aligned} \int J_0 ds &= -\frac{1}{\tau_{\mathrm{imp}}} \int_{-\infty}^{\infty} \left[f_1 \langle g_- \rangle g_0 + \frac{1}{2} \left(\left\langle f_1 f_-^\dagger \right\rangle f_0 + \langle f_1 f_- \rangle f_0^\dagger \right) \right] ds \\ &= \frac{\pi v_\perp}{\tau_{\mathrm{imp}}} \frac{\partial f^{(0)}}{\partial \epsilon} \left[\hat{T}\{\gamma_H\} (\mathbf{v}_L \cdot \mathbf{p}_\perp) + \hat{T}\{\gamma_O\} ([\mathbf{v}_L \times \mathbf{p}_\perp] \mathbf{z}) \right]. \end{aligned} \tag{14.45}$$

where $\hat{T}\{\gamma_{\mathrm{H,O}}\}$ denotes the operator

$$\hat{T}\{\gamma_{\mathrm{H,O}}\} = \frac{1}{2C} \left[\left\langle \frac{v_\perp}{C(\partial\epsilon_0/\partial b)} \right\rangle \gamma_{\mathrm{H,O}} + \frac{1}{p_\perp} \left\langle \frac{p_\perp v_\perp \gamma_{\mathrm{H,O}}}{C(\partial\epsilon_0/\partial b)} \right\rangle \right] \ln\left(\frac{\Delta}{\epsilon}\right).$$

The integral $\int J_0 ds$ is logarithmic; the divergence is cut off at $\rho \sim (\epsilon/\Delta)\xi$.

To solve the kinetic equation we insert this into eqn (14.42) and find the expressions for γ_O and γ_H in terms of the vortex velocity. These expressions also contain the averages

$$\left\langle \frac{p_\perp v_\perp \gamma_{\mathrm{H,O}}}{C(\partial\epsilon_0/\partial b)} \right\rangle.$$

Next, we average γ_O and γ_H with the weight $p_\perp v_\perp / C(\partial\epsilon_0/\partial b)$, solve the obtained equation for these averages, and exclude them from expressions for γ_O and γ_H. This completes the calculation of γ_O and γ_H.

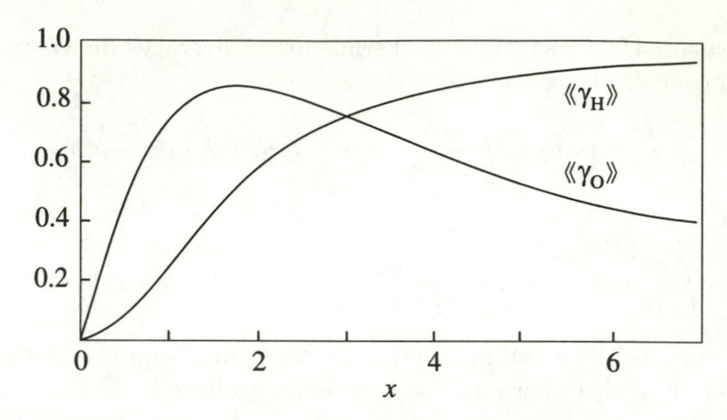

FIG. 14.1. The factors $\langle\langle\gamma_{\rm O}\rangle\rangle$ and $\langle\langle\gamma_{\rm H}\rangle\rangle$ as functions of the purity parameter x.

This general scheme of calculations can be carried out explicitly if we make use of the so-called Kramer–Pesch effect: the compression of the vortex core at low temperatures. According to Kramer and Pesch (1974) the size of the vortex core decreases as $\xi_1 \sim (T/\Delta)\,\xi_0$ for $T \ll \Delta$. In this case one has from eqn (14.44)

$$C\,(\partial\epsilon_0/\partial b) = \Delta_\infty \ln\left(\Delta_\infty/T\right).$$

Equation (14.42) yields a set of equations

$$\begin{aligned}\left(\sin\theta\,\langle\sin\theta\rangle\,\gamma_{\rm O} + \left\langle\sin^2\theta\gamma_{\rm O}\right\rangle\right) &= \pi x\,(1-\gamma_{\rm H})\,/4,\\ \left(\sin\theta\,\langle\sin\theta\rangle\,\gamma_{\rm H} + \left\langle\sin^2\theta\gamma_{\rm H}\right\rangle\right) &= \pi x\gamma_{\rm O}/4,\end{aligned} \tag{14.46}$$

where we put $v_\perp = v_F\sin\theta$ and

$$x = \frac{4\Delta_\infty^2\tau_{\rm imp}\ln\left(\Delta_\infty/T\right)}{\pi E_F}. \tag{14.47}$$

One finds from equations (14.46)

$$\left\langle\sin^2\theta\gamma_{\rm H}\right\rangle = \frac{z/2}{(1+y)^2 + (2z/\pi x)^2} \tag{14.48}$$

$$\left\langle\sin^2\theta\gamma_{\rm O}\right\rangle = \frac{\pi x}{4}\left[1 - \frac{1+y}{(1+y)^2+(2z/\pi x)^2}\right] \tag{14.49}$$

where

$$y\,(x) = \frac{4}{\pi}\left\langle\frac{\sin^3\theta}{\sin^2\theta + x^2}\right\rangle,\; z\,(x) = 2x^2\left\langle\frac{\sin^2\theta}{\sin^2\theta + x^2}\right\rangle.$$

We shall see later that these are the values

$$\langle\langle\gamma_{\rm O,H}\rangle\rangle \equiv \frac{3}{2}\left\langle\sin^2\theta\gamma_{\rm O,H}\right\rangle$$

which are needed to calculate the corresponding components of the flux flow conductivity. The function $y(x) = 1$ for $x \to 0$ and $y(x) \to 0$ for $x \to \infty$ while $z(x) = 2x^2$ for $x \to 0$ and $z(x) = 4/3$ for $x \to \infty$. Correspondingly,

$$\langle\langle\gamma_{\rm H}\rangle\rangle = \begin{cases} 3x^2/8, & x \to 0, \\ 1, & x \to \infty, \end{cases} \quad \langle\langle\gamma_{\rm O}\rangle\rangle = \begin{cases} 3\pi x/16, & x \to 0, \\ 4.41/x, & x \to \infty. \end{cases}$$

The factors $\langle\langle\gamma_{\rm O,H}\rangle\rangle$ are plotted in Fig. 14.1 as functions of the purity parameter x.

14.4.1.2 *Relaxation-time approximation, arbitrary temperatures* The solution of the kinetic equations obtained for low temperatures by Kopnin and Kravtsov (1976 a) can be generalized to arbitrary temperatures (Kopnin and Lopatin 1995). To do this we need to simplify the collision integral. Inspection of eqn (14.45) shows that the collision integral takes the form

$$\int_{-\infty}^{\infty} J_0\, ds = -\frac{\pi v_\perp}{\tau_0} f_1$$

if one neglects the p_z dependence of $\gamma_{\rm O}$ and $\gamma_{\rm H}$. Here $\tau_0 \sim \tau$ is some effective relaxation time for the low-energy level. To generalize our consideration for temperatures not much below T_c we adopt a τ-approximation for the collision integral. We assume that

$$\int_{-\infty}^{\infty} J_n\, ds = -\frac{\pi v_\perp}{\tau_n} f_1 \tag{14.50}$$

for all states where $\tau_n(b) \sim \tau_{\rm imp}$ is the effective scattering time. Within the relaxation-time approximation the relaxation rate τ_n^{-1} can also include electron–phonon and electron–electron collisions. The electron–phonon scattering can be important for very low concentration of impurities when $\ell_{\rm imp} \gg \ell_{\rm ph}$.

We also write the distribution function in a more general form

$$f_1 = -\frac{\partial f^{(0)}}{\partial \epsilon}\left[\frac{p_\perp \gamma_{\rm O}}{v_\perp}([\mathbf{v}_L \times \mathbf{v}_\perp]\hat{\mathbf{z}}) \pm \frac{p_\perp \gamma_H}{v_\perp}(\mathbf{v}_L \cdot \mathbf{v}_\perp)\right] \tag{14.51}$$

where we take into consideration a possible existence of both particle-like (the upper sign) and hole-like (lower sign) carriers in the normal state. The factors $\gamma_{\rm O,H}$ are found from eqn (14.42) with help of eqn (14.50). They are determined by eqn (14.14) and coincide with the result obtained earlier within the Boltzmann kinetic equation. Equation (14.14) reproduces reasonably well the most important feature of the exact expression (14.48) and (14.49) for the lowest level $n = 0$: As a function of $\omega_0\tau$, the parameter $\gamma_{\rm O}$ first increases linearly, reaches the maximum value of the order unity for $\omega_0\tau \sim 1$ and then decreases as $(\omega_0\tau)^{-1}$ for $\omega_0\tau \gg 1$. The parameter $\gamma_{\rm H}$ grows monotonously first as $(\omega_0\tau)^2$ and then saturates at $\gamma_{\rm H} = 1$ for $\omega_0\tau \gg 1$, see Fig. 14.1. The use of τ-approximation allows us to find the distribution function for levels with higher energies, as well. We see that the general trend of the parameters $\gamma_{\rm O}$ and $\gamma_{\rm H}$ as functions of $\omega_n\tau$ is the same as for the low-energy level.

14.4.2 *Delocalized excitations*

One would expect that, if the density of moving vortices is small enough, the deviation of the distribution function from equilibrium for the excitations with energies $|\,\epsilon\,| > \Delta_\infty$ should vanish at large distances, where the excitations are in equilibrium with the crystal lattice or sample boundaries. However, for a finite vortex density, when the distance between vortices is shorter than the magnetic field penetration depth, one has to take into account also the motion of electrons in the magnetic field with the induction $B \approx H$, where H is the applied field. Their behavior is determined by the relation between the Larmor radius $r_H = v_F/\omega_c \sim (E_F/\Delta)\,(H_{c2}/H)\xi$ and the mean free path ℓ.

The trajectory of a delocalized quasiparticle crosses many vortex unit cells at various distances from vortex axes, i.e., at different impact parameters. Since the distribution function f_1 is constant along the trajectory it should thus be independent of the impact parameter as well. Let us integrate eqn (14.35) over the unit cell of the vortex lattice. Using eqn (14.38) we obtain

$$\begin{aligned} -\frac{1}{2}\frac{\partial f^{(0)}}{\partial \epsilon}\int_{S_0} \mathrm{Tr}\left[\mathbf{v}_L\cdot\hat{\nabla}\check{H}\check{g}_-\right] d^2r - \frac{1}{2}\frac{\partial f_1}{\partial \mathbf{p}}\int_{S_0} \mathrm{Tr}\left[\hat{\nabla}\check{H}\check{g}_-\right] d^2r \\ + \int_{S_0} \mathbf{v}_F\cdot\nabla\left[g_- f_2 - g_-\left(e\varphi - \frac{e}{c}\mathbf{v}_L\cdot\mathbf{A}\right)\frac{\partial f^{(0)}}{\partial\epsilon}\right] d^2r \\ = \int_{S_0} J_1 d^2r. \end{aligned} \tag{14.52}$$

The second line gives zero because the functions under the gradient do not increase with distance; the average gradient should thus vanish. For example, one can see that

$$e\varphi - \frac{e}{c}\mathbf{v}_L\cdot\mathbf{A} = e\Phi - \frac{e}{c}\mathbf{v}_L\cdot\mathbf{Q}$$

where $\mathbf{Q}$ and Φ are the gauge-invariant potentials eqns (1.46) and (1.75) periodic in a vortex lattice. Using the identity eqn (14.34) we find

$$\pi\left([\mathbf{z}\times\mathbf{v}_\perp]\,\mathbf{v}_L\right)\frac{\partial f^{(0)}}{\partial\epsilon} + \pi\left([\mathbf{z}\times\mathbf{v}_\perp]\frac{\partial f_1}{\partial \mathbf{p}}\right) + \int_{S_0} J_1 d^2r = 0. \tag{14.53}$$

The integral of the last term is determined by large distances from the vortex core where the Green functions g_- and f_- are determined by eqn (14.33). Equation (10.74) results in the collision integral

$$J_1 = -\frac{1}{\tau_{\mathrm{imp}}}(f_1 - \langle f_1\rangle).$$

Therefore,

$$\int_{S_0} J_1 d^2r = -\frac{f_1}{\tau_{\mathrm{imp}}}\frac{\Phi_0}{B}.$$

Inserting this into eqn (14.53) we find the factors $\gamma_{\mathrm{O,H}}$ in eqn (14.51)

$$\gamma'_{\rm O} = \frac{\omega_c\tau}{\omega_c^2\tau^2+1};\ \gamma'_{\rm H} = \frac{\omega_c^2\tau^2}{\omega_c^2\tau^2+1}, \tag{14.54}$$

where we replace $\tau_{\rm imp}$ with τ for brevity. The primes refer to the fact that these expressions apply for $|\epsilon| > \Delta_\infty$. The cyclotron frequency is

$$\omega_c = H|e|v_\perp/cp_\perp. \tag{14.55}$$

The modulus of charge appears due to our choice of the z-axis.

Comparing eqns (14.14) and (14.54) one concludes that localized particles in the vortex core move like a charge in a magnetic field corresponding to the cyclotron frequency $\omega_c \sim \omega_0$. By the order of magnitude, $\omega_n \sim \Delta_0^2/E_F \sim eH_{c2}/mc$. One can say that the behavior of localized excitations in the moving vortex is similar to that of a charge in a magnetic field of the order of H_{c2}. This contrasts to the Bardeen and Stephen (1965) assumption of the effective field being the external field H.

For high fields, $\omega_c\tau \gg 1$, we obtain for the distribution function in eqn (14.51)

$$f_1 = -(\mathbf{p}\cdot\mathbf{v}_L)\frac{\partial f^{(0)}}{\partial\epsilon}. \tag{14.56}$$

This equation suggests the total distribution function in the form $f = f^{(0)}(\epsilon - \mathbf{p}\cdot\mathbf{v}_L)$ showing that the excitations move together with the vortex. Indeed, delocalized excitations cannot relax at the heat bath, since they cannot escape to distances longer than $r_H \ll \ell$. On the contrary, for low fields such that $\omega_c\tau \ll 1$ and $r_H \gg \ell$, we have $f_1 \ll pv_L(\partial f^{(0)}/\partial\epsilon)$. In this case the delocalized excitations are almost in equilibrium with the heat bath and do not participate in the vortex motion.

One can look at this from a slightly different point of view. Consider the mean free path of delocalized excitations with respect to their collisions with vortices. If the vortex cross section is σ_v, the mean free path is $\ell_v = 1/\sigma_v n_L$ where $n_L = B/\Phi_0$ is the density of vortices. We shall see later in Section 14.6.3 that the vortex cross section is $\sigma_v \sim p_F^{-1}$ [compare with eqn (14.100)] so that

$$\ell_v \sim p_F/n_L \sim v_F/\omega_c, \tag{14.57}$$

i.e., $\ell \sim r_H$. In the limit $\omega_c\tau \gg 1$, the vortex mean free path ℓ_v becomes shorter than the impurity mean free path $\ell_{\rm imp} = v_F\tau_{\rm imp}$ so that the delocalized excitations scatter on vortices more frequently than on impurities and thus come to equilibrium with moving vortices. A similar consideration also applies to localized excitations: In the limit $\omega_0\tau_{\rm imp} \gg 1$ interaction with a vortex is more effective than relaxation on impurities, the excitations thus relax to equilibrium with the moving vortex.

The case of low fields corresponds to electrically neutral superfluids, where $r_H = \infty$. At the first glance, it is simply because ω_c vanishes together with the charge of carriers. However, this is not completely correct. In fact, to estimate a deviation from equilibrium of delocalized excitations in this case one has again to compare the mean free path of excitations with their mean free path with respect to scattering by vortices. Keeping in mind that the vortex density is $n_L = 2\Omega/\kappa$ where Ω is an angular velocity of a rotating container and $\kappa = \pi/m$ is the circulation quantum, eqn (14.57) gives $\ell_v \sim v_F/\Omega$. We observe that the cyclotron frequency is replaced with the rotation velocity in a full compliance with the Larmor theorem. The ratio of the particle–particle mean free path ℓ to the vortex mean free path is $\ell/\ell_v \sim \Omega\tau$. With the practical rotation velocity Ω of a few radians per second one always has ℓ_v exceedingly larger than ℓ. Delocalized excitations are thus at rest in the container frame.

Consider this in more detail. The kinetic equations (10.41) and (10.44) at large distances from the vortex give

$$\mathbf{v}_\perp \cdot \nabla(g_- f_2) = -\frac{1}{\tau_{\text{imp}}}(f_1 - \langle f_1 \rangle), \tag{14.58}$$

$$\mathbf{v}_\perp \cdot \nabla f_1 = -\frac{g_-}{\tau_{\text{imp}}}(f_2 - \langle f_2 \rangle). \tag{14.59}$$

We keep the collision integral in the second equation because the distances which we are interested in are of the order and larger than ℓ, thus eqn (14.36) is not sufficient, see page 198 . The functions f_1 and f_2 are independent of the distance along the trajectory for $\xi \ll \rho \ll \ell$. These constants are coupled to each other by the conditions $f_1 = g_- f_2$ for $s \to +\infty$ and $f_1 = -g_- f_2$ for $s \to -\infty$ which follow from the fact that eqns (14.58, 14.59) should not have exponentially increasing solutions. They are the boundary conditions to be imposed at distances s larger than the core size of the vortex, but shorter than ℓ. For delocalized excitations with ϵ above the gap at infinity, these boundary conditions give $f_1 \sim f_2$. As a result both f_1 and f_2 are small. Indeed, since $f_1 \sim f_2$, one can neglect both the term with $\partial f_1/\partial \mathbf{p}$ and the collision integral in eqn (14.35) compared to the term containing f_2. The distribution functions f_1 and f_2 which come from eqn (14.35) are then, by the order of magnitude, $(v_L \Delta / v_F)(\partial f_0/\partial \epsilon)$. Hence the distribution function f_1 for $\epsilon > \Delta_\infty$ is by the factor of $\Delta / (E_F \min\{\omega_0 \tau, 1\})$ smaller than its magnitude for $\epsilon < \Delta_\infty$ and should be considered as zero. This agrees with eqn (14.54) in the limit $\omega_c \to 0$.

14.5 Flux flow conductivity

The force from environment on a vortex is determined by the equation (13.10). Remind that the nonequilibrium Green function is, according to eqn (13.15)

$$\hat{g}^{(\text{nst})} = -\frac{i}{2}\frac{\hat{\partial}(\hat{g}^R + \hat{g}^A)}{\partial t}\frac{\partial f^{(0)}}{\partial \epsilon} + (\hat{g}^R - \hat{g}^A) f_1 + (\hat{g}^R \hat{\sigma}_z - \hat{\sigma}_z \hat{g}^A) f_2. \tag{14.60}$$

The main contribution to eqn (14.60) comes from the part containing f_1. The force from environment eqn (13.10) becomes

$$\mathbf{F}_{\text{env}} = -\int d^2 r \int \frac{d\epsilon}{4} \int \frac{dS_F}{(2\pi)^3 v_F} \text{Tr}\left[(\hat{g}^R - \hat{g}^A)\, \hat{\nabla} \check{H} \right] f_1. \tag{14.61}$$

Consider first localized excitations in the vortex core with $|\epsilon| < \Delta_\infty$. Using the identity of eqn (14.29) we obtain the contribution from the localized excitations to eqn (14.61)

$$\mathbf{F}_{\rm env}^{(\rm loc)} = -\pi \int db \int d\epsilon \int \frac{dS_F}{(2\pi)^3 v_F} f_1 [\hat{\mathbf{z}} \times \mathbf{v}_\perp] \sum_n \frac{\partial \epsilon_n}{\partial b} \delta(\epsilon - \epsilon_n). \tag{14.62}$$

The integration over db and ds is extended from $-\infty$ to $+\infty$ since the integral converges at distances of the order of ξ which is much shorter than the unit cell size. Equation (14.62) coincides exactly with eqn (14.4) obtained phenomenologically using the concept of semiclassical particles localized in the core.

The Fermi-surface area element $dS_F = dp_\perp \, dp'$ is constructed out of two tangential momentum increments: $d\mathbf{p}_\perp$ with the magnitude $dp_\perp = p_\perp d\alpha$ in the $(x\,y)$ plane, and $d\mathbf{p}'$ perpendicular to both $d\mathbf{p}_\perp$ and $\mathbf{v}_F$. One has $(v_\perp/v_F)dp' = dp_z$. Therefore, for a Fermi surface isotropic in the $(a\,b)$ plane,

$$\int \frac{v_\perp p_\perp dS_F}{(2\pi)^3 v_F} = \int \frac{2S(p_z)\, dp_z}{(2\pi)^3} \frac{d\alpha}{2\pi}, \tag{14.63}$$

where $S(p_z) = \pi p_\perp^2$ is the area of the cross section of the Fermi surface by the plane $p_z = const$. The integral in eqn (14.63) gives

$$\frac{2V_F}{(2\pi)^3} = N$$

where V_F is the volume within the Fermi surface and N is the density of electrons (or holes). Equation (14.62) reduces to eqn (14.4) for a spherical Fermi surface. Performing the average over $d\alpha$ in eqn (14.62) we get with help of eqn (14.51)

$$\mathbf{F}_{\rm env}^{(\rm loc)} = \pi \sum_n \int \frac{2S(p_z)\, dp_z}{(2\pi)^3} \int \frac{db}{2} \frac{\partial \epsilon_n}{\partial b} \frac{\partial f^{(0)}(\epsilon_n)}{\partial \epsilon} \times [-\gamma_{\rm O}(\epsilon_n)\mathbf{v}_L \pm \gamma_{\rm H}(\epsilon_n)[\hat{\mathbf{z}} \times \mathbf{v}_L]] \tag{14.64}$$

which coincides with eqns (14.11, 14.12).

To evaluate the part of the force in eqn (14.61) for excitations with energies $|\epsilon| > \Delta_\infty$ we use the identity eqn (14.34) derived earlier. We obtain

$$\mathbf{F}_{\rm env}^{(\rm del)} = \pi \int \frac{2S(p_z)\, dp_z}{(2\pi)^3} \int_{|\epsilon| > \Delta_\infty} \frac{d\epsilon}{2} \frac{\partial f^{(0)}(\epsilon)}{\partial \epsilon} \times [-\gamma_{\rm O}' \mathbf{v}_L \pm \gamma_{\rm H}' [\hat{\mathbf{z}} \times \mathbf{v}_L]] \,. \tag{14.65}$$

Balancing the force from environment by the Lorentz force eqn (14.6) according to eqn (14.7) determines the transport current in terms of the electric field

$$\mathbf{j}_{\rm tr} = \sigma_{\rm O}\mathbf{E} + \sigma_{\rm H}\, [\mathbf{E} \times \mathbf{B}]/B.$$

The Ohmic and Hall components of conductivity are $\sigma_{\rm O,H} = \sigma^{\rm (loc)}_{\rm O,H} + \sigma^{\rm (del)}_{\rm O,H}$. The contribution from localized states is

$$\sigma^{\rm (loc)}_{\rm O} = \frac{2|e|c}{(2\pi)^3 B}\sum_n \left(\int_e + \int_h\right) S(p_z)\, dp_z \int \frac{db}{2}\frac{\partial \epsilon_n}{\partial b}\frac{df^{(0)}(\epsilon_n)}{d\epsilon}\gamma_{\rm O}(\epsilon_n), \quad (14.66)$$

$$\sigma^{\rm (loc)}_{\rm H} = \frac{2ec}{(2\pi)^3 B}\sum_n \left(\int_e - \int_h\right) S(p_z)\, dp_z \int \frac{db}{2}\frac{\partial \epsilon_n}{\partial b}\frac{df^{(0)}(\epsilon_n)}{d\epsilon}\gamma_{\rm H}(\epsilon_n). \quad (14.67)$$

The integrals over the Fermi surface are shown for the electron-like and hole-like parts separately. Equations (14.66) and (14.67) coincide exactly with eqns (14.15) and (14.16) obtained by the Boltzmann scheme.

The delocalized states give

$$\sigma^{\rm (del)}_{\rm O} = \frac{2|e|c}{(2\pi)^3 B}\left[\int_e S(p_z)\gamma'_{\rm O}\, dp_z + \int_h S(p_z)\gamma'_{\rm O}\, dp_z\right]\left[1 - \tanh\frac{\Delta_\infty}{2T}\right], \quad (14.68)$$

$$\sigma^{\rm (del)}_{\rm H} = \frac{2ec}{(2\pi)^3 B}\left[\int_e S(p_z)\gamma'_{\rm H}\, dp_z - \int_h S(p_z)\gamma'_{\rm H}\, dp_z\right]\left[1 - \tanh\frac{\Delta_\infty}{2T}\right]. \quad (14.69)$$

We use $\gamma'_{\rm O,H}$ as independent of ϵ.

14.6 Discussion

14.6.1 *Conductivity: Low temperatures*

The total conductivities are sums of the contributions from both localized and delocalized excitations. However, the contribution from delocalized excitations is small compared to $\sigma^{\rm (loc)}$ in most of the cases since the ratio $\omega_c/\omega_0 \sim H/H_{c2}$ is small in our limit. The exceptions are: (i) the case of very pure samples such that $\omega_c\tau > 1$ even for fields $H \ll H_{c2}$, (ii) the immediate vicinity of T_c.

We start the discussion with the exact solution of the kinetic equation available for low temperatures (14.48), (14.49). Delocalized excitations do not contribute. Equations (14.67), (14.66) give

$$\sigma_{\rm O} = \frac{(N_e + N_h)\,|e|\,c}{B}\langle\langle\gamma_{\rm O}\rangle\rangle\,, \quad \sigma_{\rm H} = \frac{(N_e - N_h)\,ec}{B}\langle\langle\gamma_{\rm H}\rangle\rangle \quad (14.70)$$

where x is determined by eqn (14.47). The electron (hole) densities are $N_{e,h} = 2V_{e,h}/(2\pi)^3$, where $V_{e,h}$ are the volumes confined within the particle- and hole-like parts of the Fermi surface, respectively. The factors $\langle\langle\gamma_{\rm O}\rangle\rangle$ and $\langle\langle\gamma_{\rm H}\rangle\rangle$ are shown in Fig. 14.1 as functions of x. The Hall angle defined through $\tan\Theta_{\rm H} = \sigma_{\rm H}/\sigma_{\rm O}$ is plotted in Fig. 14.2.

For a moderately clean limit, $x \ll 1$, the dissipative (Ohmic) component of the conductivity tensor can be written in a form similar to the Bardeen–Stephen model. The upper critical field for clean superconductors at low temperatures has been calculated by Gor'kov (1959 b)

$$H_{c2} = \left(\frac{e^2_{\rm ln}\gamma}{2}\right)\frac{c\Delta^2_\infty}{|e|v_F^2} \approx 1.04\frac{\Phi_0}{2\pi\xi_0^2}.$$

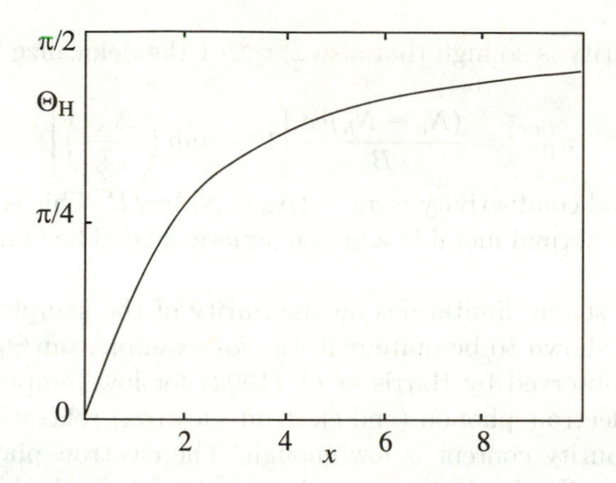

FIG. 14.2. The Hall angle as a function of the purity parameter x.

Here $e_{\ln} \approx 2.718$ is the base of natural logarithm (we denote it $e_{\ln}$ to distinguish from the electronic charge e), $\gamma \approx 1.78$ is the Euler constant, and $\xi_0 = v_F/2\pi T_c$ is the zero temperature coherence length. Equation (3.87) was used to write the upper critical field in terms of ξ_0. We find with help of this expression

$$\sigma_{\rm O} = \frac{3\pi x\,(N_e + N_h)\,|e|\,c}{16B} \approx 0.23\sigma_n \frac{H_{c2}}{B} \ln \frac{\Delta_\infty}{T}. \tag{14.71}$$

Equation (14.71) was first obtained by Larkin and Ovchinnikov (1976). It has an extra logarithmic factor as compared to the Bardeen–Stephen model. The Hall component is small, $\sigma_{\rm H}/\sigma_{\rm O} \sim x$.

In the superclean limit $x \gg 1$ the factor $\langle\langle \gamma_{\rm H} \rangle\rangle = 1$ so that the Hall conductivity is large

$$\sigma_{\rm H} = (N_e - N_h)\,ec/B$$

while the dissipative component vanishes as x^{-1}. The Hall angle is close to $\pi/2$.

14.6.2 *Conductivity: Arbitrary temperatures*

14.6.2.1 *Superclean limit* The superclean limit is reached when $\omega_0\tau \gg 1$. The parameter $\omega_c\tau$ can be either small or large. In both cases the Hall angle is close to $\pi/2$. Consider first the contribution from localized excitations. In this case $\gamma_{\rm H} = 1$, and $\gamma_{\rm O}$ is small. The Hall conductivity from the localized excitations becomes

$$\sigma_{\rm H}^{(\rm loc)} = \frac{2ec}{(2\pi)^3 B}\left(\int_e - \int_h\right) S(p_z)\,dp_z \int \frac{db}{2}\frac{\partial\epsilon_0}{\partial b}\,\frac{\partial f^{(0)}(\epsilon_0)}{\partial\epsilon}. \tag{14.72}$$

Since $\epsilon_0(b)$ runs from $-\Delta_\infty$ to $+\Delta_\infty$ as b varies from $-\infty$ to $+\infty$ (see Fig. 6.2), we obtain

$$\sigma_{\rm H}^{(\rm loc)} = \frac{(N_e - N_h)ec}{B} \tanh\left(\frac{\Delta_\infty}{2T}\right). \tag{14.73}$$

The contribution from the terms with $n \neq 0$ vanishes (see the discussion on page 276).

When the purity is so high that also $\omega_c\tau \gg 1$ the delocalized excitations eqn (14.69) give

$$\sigma_{\rm H}^{(\rm loc)} = \frac{(N_e - N_h)ec}{B}\left[1 - \tanh\left(\frac{\Delta_\infty}{2T}\right)\right]$$

such that the total conductivity is $\sigma_{\rm H} = (N_e - N_h)ec/B$. This is exactly the Hall conductivity of a normal metal in a high magnetic field. The Ohmic conductivity is small.

Despite very strong limitations on the purity of the sample, the superclean regime has been shown to be quite realistic: for example, $\tan\Theta_{\rm H}$ of the order of unity has been observed by Harris *et al.* (1994) for low temperatures. In principle, inelastic electron–phonon (and electron–electron) relaxation can also play a role if the impurity content is low enough. The electron–phonon mechanism can become more effective in the superclean regime since, for high-T_c superconductors, the electron–phonon relaxation rate is quite substantial at T_c, though it decreases rapidly for lower temperatures. It is possible that the increase in the Hall angle towards the superclean limit observed by Harris *et al.* (1994) is due to the increase in $\tau_{\rm ph}$ with lowering of the temperature.

14.6.2.2 *Moderately clean limit* This limit corresponds to $\omega_0\tau \ll 1$, and is the most common situation in clean superconductors. Moreover, this regime is always reached as T approaches T_c even for a very pure sample: for temperatures close to T_c, the parameter $\omega_0\tau_0$ decreases with Δ_∞ since $\partial\epsilon_0/\partial b \sim \Delta_\infty^2/v_F$. In the moderately clean limit the conductivities are mainly determined by the localized excitations because $\omega_c\tau \ll \omega_0\tau \ll 1$. One has in the moderately clean limit $\gamma_{\rm H} = (\omega_n\tau_n)^2$, $\gamma_{\rm O} = \omega_n\tau_n$. The Hall conductivity becomes

$$\sigma_{\rm H} = \frac{ec}{(2\pi)^2 B}\left(\int_e - \int_h\right) dp_z \sum_n \int_0^\infty db \, \frac{\partial f^{(0)}(\epsilon_n)}{\partial\epsilon}\, \tau_n^2 \left(\frac{\partial\epsilon_n}{\partial b}\right)^3 . \qquad (14.74)$$

The Ohmic conductivity is

$$\sigma_{\rm O} = \frac{|e|c}{(2\pi)^2 B}\left(\int_e + \int_h\right) p_\perp \, dp_z \sum_n \int_0^\infty db \, \frac{\partial f^{(0)}(\epsilon_n)}{\partial\epsilon}\, \tau_n \left(\frac{\partial\epsilon_n}{\partial b}\right)^2 . \qquad (14.75)$$

The largest contribution in eqns (14.74) and (14.75) comes from the term $n = 0$ since ω_n is considerably smaller than ω_0 according to numerical results by Gygi and Schlüter (1991).
The Hall angle is small: $\tan\Theta_{\rm H} \sim \omega_0\tau$.

For low temperatures, $T \ll T_c$,

$$\sigma_{\rm H} = \frac{ec}{(2\pi)^2 B}\left(\int_e - \int_h\right) dp_z \, \tau_0^2 \left(\frac{\partial\epsilon_0}{\partial b}\right)^2 , \qquad (14.76)$$

where the derivative of the energy is taken at $b = 0$. The Ohmic conductivity is

$$\sigma_{\rm O} = \frac{|e|c}{(2\pi)^2 B}\left(\int_e + \int_h\right) p_\perp \, dp_z \, \tau_0 \frac{\partial\epsilon_0}{\partial b} \qquad (14.77)$$

since only the level with $n = 0$ is important for $T \ll T_c$. These expressions are generalizations of the results obtained earlier for low temperatures by the exact solution of the kinetic equation assuming the Kramer–Pesch contraction of the vortex core. With $\tau_0 \sim \tau$ and $\partial\epsilon_0/\partial b \sim (\Delta_\infty/\xi_0)\ln(\Delta_\infty/T)$, eqns (14.76) and (14.77) agree with eqns (14.70) in the limit $x \ll 1$.

For temperatures close to T_c, the Ohmic conductivity eqn (14.75) for $B \ll H_{c2}$ is (Kopnin and Lopatin 1995)

$$\sigma_{\mathrm{O}} \sim \sigma_n \frac{H_{c2}}{B}\frac{\Delta_\infty}{T_c}. \tag{14.78}$$

Equation (14.78) implies a temperature dependence of the conductivity $\sigma_{\mathrm{O}} \propto (1 - T/T_c)^{3/2}$. The extra factor Δ_∞/T_c in eqn (14.78) as compared to the Bardeen–Stephen model has indeed been observed experimentally by Kambe *et al.* (1999). The Hall conductivity near T_c is

$$\sigma_{\mathrm{H}}^{(\mathrm{loc})} \sim \frac{Nec}{B}\left(\frac{T_c^2\tau}{E_F}\right)^2\left(\frac{\Delta_\infty}{T_c}\right)^5. \tag{14.79}$$

It is proportional to $(1 - T/T_c)^{5/2}$. This temperature dependence has been confirmed experimentally by Kim *et al.* (1996).

The temperature dependence of the flux-flow conductivities in clean superconductors differs from that obtained using the TDGL model (see Section 12.5). This is not unexpected because the TDGL model does not work for clean superconductors. Moreover, eqn (14.78) differs also from the prediction of the microscopic theory for dirty superconductors near T_c (see Section 13.2). However, the microscopic results for dirty superconductors transform into expression (14.78) for clean superconductors at the border where these two limiting cases meet. Equation (13.25) of Section 13.2 can be written as

$$\sigma_{\mathrm{O}} \sim \sigma_n \frac{H_{c2}}{B}\frac{T_c}{\Delta_\infty}\Lambda^2. \tag{14.80}$$

Equation (14.80) is valid as long as $\Delta(T)\,\tau \ll 1$. To match it with eqn (14.78) we need to take eqn (14.80) in the regime when $T_c\tau \gg 1$. For $T_c\tau \gg 1$ the function $\Lambda^2(T_c\tau) \sim (T_c\tau)^{-2}$ thus eqn (14.80) becomes

$$\sigma_{\mathrm{O}} \sim \sigma_n \frac{H_{c2}}{B}\frac{\Delta_\infty}{T_c}(\Delta_\infty\tau)^{-2}.$$

We can see that this expression transforms into eqn (14.78) at the limit of its applicability $\Delta_\infty(T)\,\tau \sim 1$. Equation (14.78) also transforms into eqn (12.59) obtained in Section 12.8.2 for a clean d-wave superconductor if we put $\Delta_\infty(T)\tau_{\mathrm{imp}} \sim 1$ in eqn (14.78).

The reason for the different dependences of σ_{O} is that, for clean superconductors with $\ell \gg \xi(T)$, the main contribution to conductivity comes from localized excitations only whose number decreases as $T \to T_c$. On the contrary, in dirty superconductors with $\ell \ll \xi$, all the excitations contribute to the conductivities.

Deviation of the distribution function from equilibrium increases as T approaches T_c because the diffusion relaxation becomes slow due to an increase in the vortex core size.

We can make an interesting observation concerning the sign of the Hall effect. Comparing eqn (14.76) with the Hall conductivity in the normal state (Abrikosov 1998):

$$\sigma_{\rm H}^{(\rm del)} = \frac{e^3 B \tau^2}{(2\pi)^2 c^2} \left(\int_e v_\perp^2 \, dp_z - \int_h v_\perp^2 \, dp_z \right), \tag{14.81}$$

we see that the sign of the normal-state Hall conductivity can differ from that in the superconducting state. The reason is that the sign is determined by the result of subtraction of electron and hole contributions which are given by integrals over the Fermi surface taken of the functions with different momentum dependences. It is the cyclotron frequency in the normal state, while it is the interlevel spacing in the superconducting state. For example, the hole contribution in the normal state can be larger than the electron contribution, however, electrons can give more than holes in the superconducting state. The sign of the Hall effect in this case will change after the transition from the normal to the superconducting state. The sign reversal depends on the shape of the Fermi surface; it is absent for a simple parabolic spectrum $E_n = p^2/2m^*$ when the flux-flow and the normal-state Hall conductivities have the same sign as was assumed in the earlier models by Bardeen and Stephen (1965) and by Nozières and Vinen (1966). The possibility for the Hall angle anomaly, i.e., for the sign reversal of the Hall angle exists also within the modified TDGL model discussed in Section 12.9.1. It was attributed to the energy dependent density of states at the Fermi surface. In clean superconductors, the origin of the Hall angle sign reversal is also associated with a complicated Fermi surface in the normal state. However, the exact reason for the anomaly is slightly different: In clean superconductors, the energy spectrum of localized excitations in the vortex core plays the most important role. Being dependent on the shape and on the topology of the Fermi surface, the Hall effect anomaly can appear or disappear as the chemical potential E_F varies for different doping levels of the superconducting material. This behavior has been indeed observed in many experiments (see, for example, Nagaoka *et al.* (1998)).

To conclude the discussion we note that the origin of the Hall effect in clean superconductors differs from the mechanism discussed in Section 12.9. In clean superconductors, it is the dynamics of nonequilibrium excitations which determines the conductivity. It gives a much larger contribution to the Hall conductivity than what we have obtained in Section 12.9. Indeed, the Hall angle in a moderately clean limit is $\tan \Theta_H \sim \omega_0 \tau$ while it only is $\tan \Theta_H \sim \lambda^{-1}(T_c/E_F)$ for the mechanism associated with the variation of the pairing interaction due to the chemical potential changes induced by a moving vortex. The latter effect does not exist within the quasiclassical approximation. It appears due to violation of the particle–hole symmetry of eqn (13.14); the symmetry breaks down when the density of states assumes an energy dependence beyond the quasiclassical approximation. As a result, eqn (13.14) acquires a small term linear in the time-derivative of the order parameter.

We outline briefly the origin of this effect. As we noticed already, it is proportional to the inverse strength of the pairing interaction λ. The $1/\lambda$ contribution to the force on a vortex comes from the regular part of the nonstationary Green function in the expression of the type of eqn (13.14). To calculate it we need to return to the full non-quasiclassical Green functions $\check{G}^{R(A)}$. Equation (13.14) becomes

$$\int d^2r \int \frac{d\epsilon}{4\pi i} \int \frac{d^3p}{(2\pi)^3} f^{(0)}(\epsilon) \operatorname{Tr}\left[\left(\check{G}^R_{\epsilon_+,\epsilon_-} - \check{G}^A_{\epsilon_+,\epsilon_-}\right) \hat{\nabla} \check{H}\right]. \tag{14.82}$$

We now expand eqn (14.82) in powers of ω up to the first-order terms. The zero-order term represents the static contribution and drops out of the nonequilibrium part of the Green function. The first-order term vanishes within the quasiclassical approximation due to the particle–hole symmetry. As we shall see in a moment, the non-quasiclassical correction in eqn (14.82) from the first-order term is logarithmically divergent due to integration over $d\epsilon$. This divergence is to be cut off at the BCS frequency $\Omega_{\rm BCS}$. Since the characteristic frequencies are large, one can look for the Green function by expanding the Gor'kov equation (8.20) in powers of the small ratio Δ/ϵ up to the leading term. We obtain

$$F^{R(A)}_{corr} = -\frac{\hat{\omega}\Delta_\omega}{2} \frac{\partial}{\partial \xi_{\mathbf{p}}} [G^{R(A)}_\epsilon \bar{G}^{R(A)}_\epsilon], \tag{14.83}$$

$F^{\dagger R(A)}_{corr}$ is obtained by substituting $\hat{\omega}\Delta_\omega$ with $-\hat{\omega}\Delta^*_\omega$. The operator $\hat{\omega}$ is defined by eqn (13.13). It is because of the full derivative with respect to $d\xi_{\mathbf{p}}$ that the quasiclassical contribution to eqn (14.82) vanishes. However, the contribution becomes nonzero if one takes into account an energy dependence of the normal-state density of states in the momentum–space volume element

$$\frac{d^3p}{(2\pi)^3} = \nu(\xi_{\mathbf{p}}) d\xi_{\mathbf{p}} \frac{d\Omega_{\mathbf{p}}}{4\pi}.$$

After an integration over $d\xi_{\mathbf{p}}$ we obtain an expression which has an asymptotics $1/\epsilon$ for large ϵ. It diverges logarithmically and represents a correction to the pairing interaction. Cutting the integral at $\epsilon \sim \Omega_{\mathbf{BCS}}$ we find the correction to the force in the form

$$\mathbf{F}^{(corr)}_{\rm env} = \frac{i}{2}\left(\frac{1}{\lambda}\frac{d\nu}{d\xi_{\mathbf{p}}}\right) \int d^2r \left[\hat{\nabla}\Delta \frac{\hat{\partial}\Delta^*}{\partial t} - \hat{\nabla}\Delta^* \frac{\hat{\partial}\Delta}{\partial t}\right]. \tag{14.84}$$

Here $\hat{\partial}/\partial t = \partial/\partial t \pm 2ie\varphi$ with the upper sign for Δ and the lower sign for Δ^*.

We put $\partial/\partial t = -\mathbf{v}_L \cdot \nabla$ for a moving vortex and calculate the integral in eqn (14.84). We neglect the vector potential for an extreme type II superconductor with a large Ginzburg–Landau parameter. The scalar potential is proportional to the vortex velocity $\mathbf{v}_L$. As in eqn (12.19), the potential φ contains $v_{L\phi}$, where

(ρ, ϕ, z) are the coordinates in the cylindrical frame. Putting $e\varphi = v_{L\phi}\psi$ we observe that the correction to the force is perpendicular to the vortex velocity,

$$\mathbf{F}_{\perp}^{(corr)} = -\pi \delta N_{\Delta} w [\mathbf{v}_L \times \hat{\mathbf{z}}], \tag{14.85}$$

where

$$w = 1 - \Delta_{\infty}^{-2} \int_0^{\infty} \psi \frac{d|\Delta|^2}{d\rho} \rho d\rho.$$

and δN_{Δ} is defined by eqn (12.79). Note that $\partial\nu/\partial\xi_{\mathbf{p}} = \partial\nu/\partial E_n = \partial\nu/\partial\mu$ in the normal state. Due to the extra transverse force eqn (14.85), the Hall conductivity acquires a correction, $\sigma_{\mathrm{H}} \to \sigma_{\mathrm{H}} + \sigma_{\mathrm{H}}^{(corr)}$, where σ_{H} is determined by eqns (14.67) and (14.69) while

$$\sigma_{\mathrm{H}}^{(corr)} = \delta N_{\Delta} wec/B. \tag{14.86}$$

Equation (14.84) is independent of the impurity scattering because the characteristic energies are large. It is only the potential φ which depends on the real kinetics of a superconductor. Therefore, the function w in eqn (14.85) may vary depending on the parameters of the superconductor. It has been shown by Kopnin (1996) that $w \to 1$ near the critical temperature when $\psi \to 0$. In this limit, eqn (14.86) coincides with eqn (12.78) obtained within the modified TDGL theory and with the result of van Otterlo *et al.* (1995).

We emphasize once again that, in the clean limit $\Delta\tau \gg 1$, nonequilibrium excitations give a much larger contribution to the Hall effect. Indeed, $\sigma_{\mathrm{H}} \sim (Nec/B)(\omega_0\tau)^2$ in the moderately clean case while $\sigma_{\mathrm{H}}^{(corr)} \sim \lambda^{-1}(Nec/B)(\Delta/E_F)^2$. It is only in a close vicinity of T_c such that $\Delta_{\infty}^2\tau \sim T_c/\lambda$ when the variations of the pairing interaction become important.

14.6.3 *Forces*

Let us now discuss the forces acting on a moving vortex. The friction force is determined by the longitudinal components of eqns (14.64) and (14.65):

$$\mathbf{F}_{\parallel} = -\pi N \left[\sum_n \left\langle\!\!\left\langle \int \frac{db}{2} \frac{\partial \epsilon_n}{\partial b} \frac{\partial f^{(0)}(\epsilon_n)}{\partial \epsilon} \gamma_{\mathrm{O}} \right\rangle\!\!\right\rangle + \left(1 - \tanh \frac{\Delta_{\infty}}{2T}\right) \gamma_{\mathrm{O}}' \right] \mathbf{v}_L. \tag{14.87}$$

It is proportional to the mean free path of excitations in the moderately clean regime and vanishes in the superclean limit. The transverse force deserves a more detailed discussion because it has been a matter of controversy for a long time since first calculated for vortices in helium II by Lifshitz and Pitaevskii (1957) and then by Iordanskii (1964). Consider for simplicity a spherical Fermi surface. The full transverse force from eqns (14.64) and (14.65) is

$$\mathbf{F}_{\perp} = \pi N \left[\sum_n \left\langle\!\!\left\langle \int \frac{db}{2} \frac{\partial \epsilon_n}{\partial b} \frac{\partial f^{(0)}(\epsilon_n)}{\partial \epsilon} \gamma_{\mathrm{H}} \right\rangle\!\!\right\rangle + \left(1 - \tanh \frac{\Delta_{\infty}}{2T}\right) \gamma_{\mathrm{H}}' \right] [\hat{\mathbf{z}} \times \mathbf{v}_L]. \tag{14.88}$$

In the superclean limit, when both $\omega_c\tau$ and $\omega_0\tau$ are much larger than unity, the factors $\gamma_H = \gamma'_H = 1$, and the transverse force becomes (see page 276)

$$\mathbf{F}_\perp = \pi N[\hat{\mathbf{z}} \times \mathbf{v}_L]. \tag{14.89}$$

The balance eqn (14.7) against the Lorentz force gives the transport current in the form

$$\mathbf{j}_{tr} = Ne\mathbf{v}_L. \tag{14.90}$$

This equation is consistent with the Helmholtz theorem of conservation of circulation in an ideal fluid: vortices move together with the flow. On the contrary, the transverse force disappears in the limit $\omega_0\tau, \omega_c\tau \to 0$. This is the overall behavior of the transverse force as a function of the quasiparticle mean free path.

From the historical point of view it is interesting to identify several contribution to the transverse force and consider them one by one. We can present the full transverse force in the form

$$\mathbf{F}_\perp = -\pi N_s[\mathbf{v}_L \times \hat{\mathbf{z}}] - \pi N_n[\mathbf{v}_L \times \hat{\mathbf{z}}] + \mathbf{F}_{sf} \tag{14.91}$$

where N_n is the density of normal quasiparticles at large distances from the vortex core

$$N_n = N \int_{\epsilon > \Delta_\infty} \frac{\epsilon}{\sqrt{\epsilon^2 - \Delta_\infty^2}} \frac{\partial f^{(0)}}{\partial \epsilon}\, d\epsilon, \tag{14.92}$$

while superconducting density is $N_s = N - N_n$. The force

$$\mathbf{F}_{sf} = \pi N \sum_n \left\langle\!\!\left\langle \int \frac{d\mu}{2} \frac{\partial f^{(0)}}{\partial \epsilon} \frac{\omega_n}{\omega_n^2 \tau_n^2 + 1} \right\rangle\!\!\right\rangle [\mathbf{v}_L \times \hat{\mathbf{z}}]$$
$$+\pi N \left(1 - \tanh \frac{\Delta_\infty}{2T}\right) \frac{1}{\omega_c^2 \tau^2 + 1} [\mathbf{v}_L \times \hat{\mathbf{z}}] \tag{14.93}$$

is called the spectral flow force (Kopnin *et al.* 1995).

Let us consider now the transport current that enters the force balance equation $\mathbf{F}_L + \mathbf{F}_\perp + \mathbf{F}_\parallel = 0$ through the Lorentz force. In presence of an electric field, the transport current is not entirely due to a supercurrent, a part of it being carried by delocalized quasiparticles. Far from the vortex core, the quasiparticle (normal) current is

$$\mathbf{j}^{(qp)} = -2\nu(0)\, e \int_{|\epsilon| > \Delta_\infty} \mathbf{v}_\perp g_- f_1 \frac{d\epsilon}{2} \frac{d\Omega_{\mathbf{p}}}{4\pi}. \tag{14.94}$$

Using eqns (14.9) and (14.54) we find

$$\mathbf{j}^{(qp)} = N_n e \gamma'_H \mathbf{v}_L + N_n e \gamma'_O \left[\mathbf{z} \times \mathbf{v}_L\right]. \tag{14.95}$$

Writing the transport current as $\mathbf{j}_{tr} = N_s e \mathbf{v}_s + \mathbf{j}^{(qp)}$ we get the force balance eqn (14.4) in the form

$$\mathbf{F}_{\mathrm{M}} + \mathbf{F}_{\mathrm{L}}^{(\mathrm{qp})} + \mathbf{F}_{\mathrm{I}} + \mathbf{F}_{\mathrm{sf}} + \mathbf{F}_{\parallel} = 0. \tag{14.96}$$

Here

$$\mathbf{F}_{\mathrm{M}} = \pi N_s [(\mathbf{v}_s - \mathbf{v}_L) \times \hat{\mathbf{z}}]$$

is the Magnus force;

$$\mathbf{F}_{\mathrm{L}}^{(\mathrm{qp})} = \frac{\Phi_0}{c} [\mathbf{j}^{(\mathrm{qp})} \times \hat{\mathbf{z}}] \mathrm{sign}\,(e) \tag{14.97}$$

is the Lorentz force from the quasiparticle current eqn (14.95). The force

$$\mathbf{F}_{\mathrm{I}} = \pi N_n \left[(\mathbf{v}_n - \mathbf{v}_L) \times \hat{\mathbf{z}} \right]$$

is called the Iordanskii force (Iordanskii 1964). The Iordanskii force is the counterpart of the Magnus force for normal excitations. We have included the normal velocity $\mathbf{v}_n$ to make this similarity more transparent. In our consideration, the normal velocity is always zero $\mathbf{v}_n = 0$ since the equilibrium corresponds to excitations at rest in the reference frame associated with the crystal lattice.

The spectral flow force $\mathbf{F}_{\mathrm{sf}}$ is mostly due to the momentum flow from the Fermi sea of normal excitations to the moving vortex via the gapless spectral branch (the term $n = 0$ in the sum) going through the vortex core from negative to positive energies (Volovik 1986, Stone and Gaitan 1987). Due to the time-dependent angular momentum of the excitations $\mu(t) = [(\mathbf{r} - \mathbf{v}_L t) \times \mathbf{p}] \cdot \hat{\mathbf{z}}$, there appears a flow (with the velocity $\partial\mu/\partial t$) of spectral levels characterized by the angular momentum μ. Each particle on a level carries a momentum p_F. The momentum transfer to the vortex is effective if the quasiparticle relaxation occurs quickly: the factor $(\omega_0^2 \tau_0^2 + 1)^{-1}$ in the first line of eqn (14.93) accounts for the relaxation on localized levels: the relaxation and hence the momentum transfer is complete for $\omega_0 \tau \ll 1$, and vanishes in the opposite limit. The first term in eqn (14.93) thus describes a disorder-mediated momentum flow along the anomalous chiral branch $\epsilon_0\,(\mu)$ for energies below the gap. The second line in eqn (14.93) accounts for the spectral flow for energies above the gap. The corresponding factor which takes into account the relaxation rate is $(\omega_c^2 \tau^2 + 1)^{-1}$.

The spectral flow force vanishes in the limit $\tau \to \infty$ when both $\omega_0 \tau \gg 1$ and $\omega_c \tau \gg 1$ such that the transverse force is given by eqn (14.89). The quasiparticle current is $\mathbf{j}^{(\mathrm{qp})} = N_n e \mathbf{v}_L$ so that the force from the quasiparticle current compensates the Iordanskii force. The force balance eqn (14.96) reduces to $\mathbf{F}_{\mathrm{M}} = 0$. The vortex thus moves with the superfluid velocity $\mathbf{v}_L = \mathbf{v}_s$. As follows from eqn (14.56), quasiparticles also have a velocity $\mathbf{v}_L$ so that all the particles move together which amounts to the total current as in eqn (14.90).

On the contrary, the spectral flow force has its maximum value for moderately clean limit, $\omega_0 \tau \ll 1$. Equation (14.93) gives in this limit

$$\mathbf{F}_{\mathrm{sf}} = \pi N [\mathbf{v}_L \times \hat{\mathbf{z}}].$$

This completely compensates the first two terms in eqn (14.91), i.e, the Iordanskii force and the part of the Magnus force that contains the vortex velocity; the transverse force vanishes. The quasiparticle current vanishes even faster because $\omega_c \ll \omega_0$. The Lorentz force is balanced only by a friction force $\mathbf{F}_{\parallel}$. As a result, the dissipative dynamics is restored.

The low-field limit when $\omega_c\tau \ll 1$ is most practical for superconductors. Moreover, this regime is realized in electrically neutral superfluids such as ^{3}He. As we already know, the low field limit $\omega_c\tau \ll 1$ corresponds to a situation when the delocalized excitations are at rest with the heat bath. Since delocalized excitations are in equilibrium the quasiparticle current vanishes. The force balance becomes

$$\mathbf{F}_{\mathrm{M}} + \mathbf{F}_{\mathrm{I}} + \mathbf{F}_{\mathrm{sf}} + \mathbf{F}_{\parallel} = 0 \tag{14.98}$$

where the spectral flow force is

$$\begin{aligned} \mathbf{F}_{\mathrm{sf}} = \pi N \sum_n \left\langle\!\!\left\langle \int \frac{d\mu}{2} \frac{\partial f^{(0)}}{\partial \epsilon} \frac{\omega_n}{\omega_n^2 \tau_n^2 + 1} \right\rangle\!\!\right\rangle [\mathbf{v}_L \times \hat{\mathbf{z}}] \\ + \pi N \left(1 - \tanh \frac{\Delta_\infty}{2T}\right) [\mathbf{v}_L \times \hat{\mathbf{z}}]. \end{aligned} \tag{14.99}$$

It is interesting to note that the spectral flow force from above gap states in this case is related to the anomalous contribution to the transverse vortex cross section for scattering of delocalized quasiparticles. The transverse vortex cross section was calculated by Kopnin and Kravtsov (1976 b) and by Galperin and Sonin (1976):

$$\sigma_\perp = \frac{\pi}{p_\perp} \left[\frac{\epsilon}{\sqrt{\epsilon^2 - \Delta_\infty^2}} - 1 \right]. \tag{14.100}$$

Inserting eqn (14.100) into the expression for the force exerted on the vortex by scattered excitations,

$$\mathbf{F}_\perp^{(\mathrm{sc})} = \int_{|\epsilon| > \Delta_0} \frac{d\epsilon}{4} \frac{\partial f^{(0)}}{\partial \epsilon} \int \frac{dp_z}{2\pi} p_\perp^3 \sigma_\perp [\hat{\mathbf{z}} \times \mathbf{v}_L], \tag{14.101}$$

we recover the last two terms in eqn (14.91) that are due to the normal excitations, namely the term $-\pi N_n [\mathbf{v}_L \times \hat{\mathbf{z}}]$ and the last term in the spectral flow force eqn (14.99).

The first term in eqn (14.100) corresponds to the cross section of a vortex in a Bose superfluid (Sonin 1987)

$$\sigma'_\perp = 2\pi / m_B v_g \tag{14.102}$$

where v_g is the group velocity defined by eqn (1.88) which, in our case, is $v_g = v_F\sqrt{\epsilon^2 - \Delta_\infty^2}/\epsilon$, and m_B is the mass of a Bosonic atom. Note that, in our case, $m_B = 2m$. The corresponding part of the transverse cross section can be easily obtained from the semi-classical description. Indeed, the Doppler energy due to the vortex velocity is $\mathbf{p}\cdot\mathbf{v}_s = \mathbf{p}\cdot\nabla\chi/m_B$. Its contribution to the quasiparticle action is

$$A = -\frac{1}{m_B}\int \mathbf{p}\cdot\nabla\chi\, dt = -\frac{p}{m_B v_g}\int \frac{\partial\chi}{\partial s}\, ds = -\frac{p}{m_B v_g}\delta\chi.$$

Here $s = v_g t$ is the coordinate along the particle trajectory as in Fig. 6.1, and $\delta\chi$ is the variation of the order parameter phase along the trajectory. The change in the transverse momentum of the particle is $\delta p_\perp = \partial A/\partial b$ hence the transverse cross section becomes

$$\sigma'_\perp = \int \frac{\delta p_\perp}{p}\, db = \frac{A_+ - A_-}{p}$$

where $A_\pm$ is the action along the trajectory passing on the left (right) side of the vortex. Since $\delta\chi_+ - \delta\chi_- = -2\pi$ we recover eqn (14.102). With this expression for the cross section, eqn (14.101) gives the Iordanskii force $\mathbf{F}_\mathrm{I} = -\pi N_n[\mathbf{v}_L \times \hat{\mathbf{z}}]$.

The second term in eqn (14.100) originates from the fact that here, as distinct from the situation in a Bose superfluid, the phase of the single-particle wave function changes by π upon encircling the vortex, while it is the order parameter phase which changes by 2π. It is this singularity, produced by the vortex in the single-particle wave function, which results in the anomalous contribution to the cross section in eqn (14.100). Inserted into eqn (14.101), it exactly reproduces the second term in eqn (14.93) taken for $\omega_c\tau \ll 1$. We see that the spectral flow force is related to a single-particle anomaly associated with the vortex.

15

BOLTZMANN KINETIC EQUATION

We derive the canonical Boltzmann kinetic equation for two particular examples. First case is a superconductor with homogeneous in space order-parameter magnitude and current. The second example treats the excitations in the vortex core. The Boltzmann equation is applied to calculate the vortex mass and to study the vortex dynamics in a d-wave superconductor.

15.1 Canonical equations

The microscopic nonstationary theory of superconductivity based on the Green function technique is a powerful method to describe the vortex dynamics both in dirty and in clean superconductors, as well as in other superfluid Fermi systems. However, a practical disadvantage of the method is its mathematical complexity, which tends to hide the physical picture of the phenomenon. For a clean system, where the excitation spectrum is well defined, an alternative way to deal with dynamical processes is expected to be based on the quasiclassical Boltzmann kinetic equation. For normal metals, the equivalence of the quasiclassical Green function approach to the Boltzmann equation has been demonstrated by Keldysh (1964). In Chapter 10 we have derived the set of kinetic equations which are suitable for many nonstationary problems in the theory of superconductivity. In the present chapter we consider a few examples for which the generalized kinetic equations can be reduced to the canonical Boltzmann form

$$\frac{\partial f}{\partial t}+\frac{\partial f}{\partial \mathbf{q}}\cdot\frac{\partial \epsilon_n}{\partial \mathbf{p}}-\frac{\partial \epsilon_n}{\partial \mathbf{q}}\cdot\frac{\partial f}{\partial \mathbf{p}}=\left(\frac{\partial f}{\partial t}\right)_{\rm coll}, \tag{15.1}$$

where $(\partial f/\partial t)_{\rm coll}$ is the collision integral and $\epsilon_n(\mathbf{q},\mathbf{p})$ is the quasiclassical excitation spectrum, characterized by the canonically conjugated generalized "coordinate" $\mathbf{q}$ and "momentum" $\mathbf{p}$ of the excitation (n denotes the set of other quantum numbers). The first example is the case of constant in space order parameter and supercurrent. We shall see that the Boltzmann equation is in this case completely similar to its normal-metal counterpart. Other examples refer to the dynamics of excitations in the presence of moving vortices.

15.2 Uniform order parameter

One can easily derive the canonical form of the Boltzmann equation for a situation where the order parameter magnitude, supercurrent, and external fields only slowly vary in space and in time (Betbeder-Matibet and Nozières 1969, Aronov *et al.* 1981, 1986). We have already discussed this equation in Section 1.2.3. Let us define the energy of excitations as in eqn (1.54)

$$\epsilon_{\mathbf{p}} = \tilde{\epsilon}_{\mathbf{p}} + \mathbf{p}_s \cdot \mathbf{v}_F \tag{15.2}$$

where

$$\tilde{\epsilon}_{\mathbf{p}} = \sqrt{\tilde{\xi}_{\mathbf{p}}^2 + |\Delta|^2},\ \tilde{\xi}_{\mathbf{p}} = \xi_{\mathbf{p}} + e\Phi, \tag{15.3}$$

and

$$\mathbf{p}_s = \frac{1}{2}\left(\nabla\chi - \frac{2e}{c}\mathbf{A}\right).$$

According to the definition eqn (15.2), the excitation energy is always positive. This coincides with the Landau picture of the Fermi liquid. We define the distribution function for the excitations with the energy spectrum eqn (15.2) through our generalized distribution functions $f_1(\epsilon, \mathbf{p})$ and $f_2'(\epsilon, \mathbf{p})$ as

$$n_{\mathbf{p}} = \frac{1}{2}\left[1 - \left(\mathrm{sign}\left(\tilde{\xi}_{\mathbf{p}}\right)\left[f^{(0)}(\epsilon_{\mathbf{p}}) + f_1(\epsilon_{\mathbf{p}}, \mathbf{p})\right] + \frac{\tilde{\epsilon}_{\mathbf{p}}}{\left|\tilde{\xi}_{\mathbf{p}}\right|} f_2'(\epsilon_{\mathbf{p}}, \mathbf{p})\right)\right]. \tag{15.4}$$

In equilibrium, it is the Fermi function

$$n_0 = \left(1 + \exp\left[\mathrm{sign}(\xi_{\mathbf{p}})\frac{\epsilon_{\mathbf{p}}}{T}\right]\right)^{-1}.$$

We also note that for a constant $|\Delta|$ and $\mathbf{p}_s$, the Green functions are

$$g_- = \frac{\epsilon - \mathbf{p}_s \cdot \mathbf{v}_F}{\sqrt{(\epsilon - \mathbf{p}_s \cdot \mathbf{v}_F)^2 - |\Delta|^2}}\Theta\left[(\epsilon - \mathbf{p}_s \cdot \mathbf{v}_F)^2 - |\Delta|^2\right],$$

$$f_- = \frac{\Delta}{\sqrt{(\epsilon - \mathbf{p}_s \cdot \mathbf{v}_F)^2 - |\Delta|^2}}\Theta\left[(\epsilon - \mathbf{p}_s \cdot \mathbf{v}_F)^2 - |\Delta|^2\right].$$

The radicals are defined as analytical functions of $x = \epsilon - \mathbf{p}_s \cdot \mathbf{v}_F$ on the complex plane of x with the cut between $-|\Delta|$ and $|\Delta|$ (see Fig. 5.1); they change sign when x goes from positive $x > |\Delta|$ to negative $x < -|\Delta|$. If ϵ is replaced with a positive $\epsilon_{\mathbf{p}}$, the Green functions become

$$g_- = \frac{\tilde{\epsilon}_{\mathbf{p}}}{\left|\tilde{\xi}_{\mathbf{p}}\right|},\ f_- = \frac{\Delta}{\left|\tilde{\xi}_{\mathbf{p}}\right|}.$$

One can check after some algebra that in the linear approximation with respect to the deviation from equilibrium, the two equations (10.52) and (10.53) for the functions $f_1(\epsilon, \mathbf{p})$ and $f_2(\epsilon, \mathbf{p})$ possessing a definite parity in $(\epsilon, \mathbf{p})$ can be written as one equation

$$\frac{\partial n_{\mathbf{p}}}{\partial t} + \mathbf{v}_g \cdot \nabla n_{\mathbf{p}} + \mathbf{f} \frac{\partial n_{\mathbf{p}}}{\partial \mathbf{p}} = I\{n_{\mathbf{p}}\} \tag{15.5}$$

for the function $n_{\mathbf{p}}$ which has no definite parity in $\tilde{\xi}_{\mathbf{p}}$. Equations (10.52) and (10.53) can be recovered from eqn (15.5) by separating the even and odd parts in $\tilde{\xi}_{\mathbf{p}}$. The group velocity $\mathbf{v}_g$ and the elementary force $\mathbf{f}$ including the Lorentz force are defined in eqn (1.88). Equation (15.5) coincides with eqn (1.87) used in the beginning of this book.

The collision integral in eqn (15.5) has the form

$$I\{n_{\mathbf{p}}\} = -\frac{1}{2}\left(\frac{\tilde{\xi}_{\mathbf{p}}}{\tilde{\epsilon}_{\mathbf{p}}} J_1\{f_1\} + \frac{\tilde{\xi}_{\mathbf{p}}^2}{\tilde{\epsilon}_{\mathbf{p}}^2} J_2\{f_2\}\right). \tag{15.6}$$

For example, the impurity collision integral becomes

$$\begin{aligned} I^{(\mathrm{imp})}\{n_{\mathbf{p}}\} = -\frac{1}{\nu(0)\,\tau_{\mathrm{imp}}} \int & (u_{\mathbf{p}} u_{\mathbf{p}'} - v_{\mathbf{p}} v_{\mathbf{p}'})^2 \\ & \times (n_{\mathbf{p}} - n_{\mathbf{p}'})\,\delta(\epsilon_{\mathbf{p}} - \epsilon_{\mathbf{p}'}) \frac{d^3p'}{(2\pi)^3}. \end{aligned} \tag{15.7}$$

The Bogoliubov–de Gennes coherence factors are as in eqn (1.56)

$$u_{\mathbf{p}}^2 = \frac{1}{2}\left(1 + \frac{\tilde{\xi}_{\mathbf{p}}}{\tilde{\epsilon}_{\mathbf{p}}}\right), \quad v_{\mathbf{p}}^2 = \frac{1}{2}\left(1 - \frac{\tilde{\xi}_{\mathbf{p}}}{\tilde{\epsilon}_{\mathbf{p}}}\right).$$

During the transformation we use the relation

$$\frac{\partial \tilde{\epsilon}_{\mathbf{p}}}{\partial \tilde{\xi}_{\mathbf{p}}} = \frac{\partial \epsilon_{\mathbf{p}}}{\partial \xi_{\mathbf{p}}} = \frac{\tilde{\xi}_{\mathbf{p}}}{\tilde{\epsilon}_{\mathbf{p}}}.$$

Equation (15.5) has the form of the canonical Boltzmann equation. However, its region of applicability is limited to the situation where the order parameter and the supercurrent are almost uniform in space. One would think that the required scale of spatial variations should be larger than ξ. However, this condition does not guarantee that the obtained results are correct. The point is that the spatial variations also result in distortions of the quasiparticle spectrum. These distortions are not included into eqns (15.2), (15.3) though they can be important for some of the responses. We consider now one such example when the distortion of the spectrum provided by eqns (14.31) and (14.32) is important for calculating the flux flow conductivity.

15.2.1 *Boltzmann equation in presence of vortices*

A delocalized particle moves mostly far from the vortex core where the order parameter is constant and the superfluid velocity potential $\mathbf{p}\mathbf{v}_s$ is small compared to Δ. One could thus expect that the kinetic equation for delocalized excitations has its conventional form eqn (15.5) derived above. However, this is not exactly the case. To see this let us put the elementary Lorentz force in eqn (15.5) to be

$$\mathbf{f} = \frac{\partial \mathbf{p}}{\partial t} = \frac{e}{c}\mathbf{v}_g \times \mathbf{H} = \frac{\omega_c}{g}\left[\mathbf{p}_F \times \hat{\mathbf{z}}\right]. \tag{15.8}$$

This expression contains the group velocity

$$\mathbf{v}_g = \frac{\partial \epsilon_p}{\partial \mathbf{p}} = \frac{\mathbf{v}_F}{g},\; g = \frac{\epsilon}{\sqrt{\epsilon^2 - \Delta_\infty^2}} \tag{15.9}$$

in contrast to the previous definition eqn (1.88) that simply has $\mathbf{v}_F$. The driving term is also written in a slightly different way. We put

$$\frac{\partial n_{\mathbf{p}}}{\partial t} = \frac{\partial n_0}{\partial \epsilon}\frac{\partial \epsilon}{\partial t}$$

where the time-derivative of energy

$$\frac{\partial \epsilon}{\partial t} = e\mathbf{v}_g \cdot \mathbf{E} = \frac{e}{c}\mathbf{v}_g \cdot \left[\mathbf{H} \times \mathbf{v}_L\right] = \frac{\omega_c}{g}\mathbf{p}_\perp \cdot \left[\hat{\mathbf{z}} \times \mathbf{v}_L\right]. \tag{15.10}$$

now also contains the group velocity instead of $\mathbf{v}_F$ as in eqn (1.90).

The kinetic equation for the function f_1 becomes

$$\frac{\partial f_1}{\partial t} + \frac{\partial f^{(0)}}{\partial \epsilon}\frac{\omega_c}{g}\mathbf{p}_\perp \cdot \left[\hat{\mathbf{z}} \times \mathbf{v}_L\right] + \frac{\partial f_1}{\partial \mathbf{p}} \cdot \frac{\omega_c}{g}\left[\mathbf{p}_\perp \times \hat{\mathbf{z}}\right] = \left(\frac{\partial f}{\partial t}\right)_{\text{coll}}. \tag{15.11}$$

The spatial derivative of the distribution function vanishes since f_1 is constant in space.

For the energy spectrum of eqns (15.2, 15.3) the collision integral eqn (15.6) reduces to

$$\left(\frac{\partial f}{\partial t}\right)_{\text{coll}} = -\frac{1}{g\tau}f_1.$$

Finally, the kinetic equation (15.11) for a steady distribution takes the form

$$\left(\mathbf{v}_L \cdot \left[\hat{\mathbf{z}} \times \mathbf{p}_\perp\right]\right)\frac{\partial f^{(0)}}{\partial \epsilon} + \left(\left[\hat{\mathbf{z}} \times \mathbf{p}_\perp\right] \cdot \frac{\partial f_1}{\partial \mathbf{p}_\perp}\right) = \frac{1}{\omega_c \tau}f_1 \tag{15.12}$$

which coincides with the microscopic equation (14.53). Its solution provides the distribution function in the form of eqn (14.51) with the factors $\gamma'_{\mathrm{O,H}}$ determined by eqn (14.54).

It is thus eqn (15.11) that should be used for quasiparticles traveling through the vortex array rather than eqn (15.5) derived assuming a constant order parameter and uniform supercurrents. The correct equation contains the group velocity rather than $\mathbf{v}_F$ in the elementary force eqn (15.9) and in the energy gain eqn (15.10). Looking back at the derivation of eqn (14.53) we observe that if we had omitted the gradient terms from $\hat{\nabla}\check{H}$ in eqn (14.52) we would only have $(e/c)[\mathbf{v_F} \times \mathbf{H}]g_-$ left in $\hat{\nabla}\check{H}$ [see eqn (13.11)] and would thus obtain an extra factor g_- both in eqn (15.10) and in front of the elementary Lorentz force. The compensating contribution comes from the gradient expansion eqns (14.31, 14.32) through the identity eqn (14.34).

15.3 Quasiparticles in the vortex core

Consider another example of application of the Boltzmann equation to the vortex dynamics in clean superconductors. We study kinetic of excitations localized in the vortex cores (Stone 1996, Kopnin and Volovik 1997). We start with an observation that eqn (14.42) which we derived from kinetic equations (14.35) and (14.36) already looks almost like the canonical equation (15.1). In this section we present a microscopic verification of (15.1) for a general case (Blatter *et al.* 1999) when Δ can depend on the quasiparticle momentum as it is the case, for example, in *d*-wave superconductors. We demonstrate that our kinetic equations for the generalized distribution function derived from the quasiclassical Green function formalism can be further transformed into the simple and physically transparent canonical equation (15.1). We restrict ourselves to the particular example of vortex dynamics; the calculation can be generalized to include the dynamics of other topological defects in superfluid Fermi systems.

In Section 14.5, we have derived microscopically the force acting on a moving vortex, eqn (14.62). This equation shows that, within the quasiclassical approximation, the force $\mathbf{F}_{\rm env}$ can be represented as the momentum transfer from the heat bath via the localized quasiparticle excitations to the vortex [compare with eqn (14.4)],

$$\mathbf{F}_{\rm env} = -\sum_n \int \frac{d^d q\, d^d p}{(2\pi)^d} f(\mathbf{q}, \mathbf{p}) \frac{\partial \mathbf{p}_n}{\partial t}, \tag{15.13}$$

where $\partial \mathbf{p}_n/\partial t = -\partial \epsilon_n(\mathbf{q}, \mathbf{p})/\partial \mathbf{q}$ and d is the dimensionality of the problem ($d = 1$ in case of vortices). Our analysis thus provides a microscopic verification of the phenomenological approach to the vortex dynamics based on the concept of semiclassical particles obeying the Boltzmann kinetic equation where the quasiclassical spectrum ϵ_n of excitations in the vortex core plays the role of the Hamiltonian, and the force acting on the vortex results from the elementary force $\partial \mathbf{p}_n/\partial t$ the quasiparticles exert on the vortex core. Later in this chapter, we consider two more examples of implementation of this scheme. In Section 15.4 we calculate the vortex mass. Another example is the vortex dynamics in *d*-wave superconductors to be discussed in Section 15.5.

15.3.1 *Transformation into the Boltzmann equation*

Let us write the kinetic equations (10.41) and (10.56) again. We have

$$\left(e\mathbf{v}_F \cdot \mathbf{E} g_- + \frac{1}{2}\left[f_- \frac{\partial \Delta^*}{\partial t} + f_-^\dagger \frac{\partial \Delta}{\partial t}\right]\right) \frac{\partial f^{(0)}}{\partial \epsilon} + \mathbf{v}_F \cdot \nabla (f_2 g_-) + g_- \frac{\partial f_1}{\partial t}$$
$$- \left(\frac{1}{2}\left[(\hat{\nabla}\Delta) f_-^\dagger + (\hat{\nabla}\Delta^*) f_- \right] - \frac{e}{c}[\mathbf{v}_F \times \mathbf{H}] g_- \right) \frac{\partial f_1}{\partial \mathbf{p}}$$
$$+ \frac{1}{2}\left(\frac{\partial \Delta}{\partial \mathbf{p}} f_-^\dagger + \frac{\partial \Delta^*}{\partial \mathbf{p}} f_- \right) \nabla f_1 = J_1. \tag{15.14}$$

and

$$g_- \mathbf{v}_F \cdot \nabla f_1 = 0 \tag{15.15}$$

We start with a few preparatory steps before transforming the kinetic equation (15.14) into the Boltzmann equation. We again use the coordinate frame associated with the vortex as shown in Fig. 6.1. Recall that the z-axis is chosen parallel to the axis of the vortex, with the positive direction along the vortex circulation $\hat{\mathbf{z}} = \mathrm{sign}\,(e)\,\hat{\mathbf{H}}$. The distance along the quasiparticle trajectory as well as the impact parameter are $s = \rho\cos(\phi - \alpha)$ and $b = \rho \sin(\phi - \alpha)$, respectively, where ρ and ϕ are the radial distance and the azimuthal angle in the cylindrical frame, and α is the angle between $\mathbf{v}_\perp$ and the x-axis. In this representation, the quasiclassical Green function is specified by the momentum projection on the vortex axis $p_z = p_F \cos\theta$, the momentum direction α in the plane perpendicular to the vortex axis, and the impact parameter b, which is related to the angular momentum $\mu = -bp_\perp$, with $p_\perp$ the momentum projection on the plane perpendicular to the vortex axis. For simplicity, we consider a spherical Fermi surface here. Up to corrections of order $(p_F\xi)^{-1}$ we can assume straight quasiparticle trajectories and thus the angular momentum μ is a conserved quantity even for a non-axisymmetric vortex.

Next, we transform the operators in eqn (15.14). The momentum derivative $\partial\mathbf{p}$ in eqn (15.14) is taken at a constant position vector $\mathbf{r} = (\rho, \phi)$ with respect to variations in the momentum direction, with the magnitude of the momentum being fixed at the Fermi surface. The planar projection can be written as

$$\left(\frac{\partial}{\partial \mathbf{p}_\perp} \right)_{\mathbf{r}} = \hat{\mathbf{v}}_\perp \frac{\partial}{\partial p_\perp} + \frac{[\hat{\mathbf{z}} \times \hat{\mathbf{v}}_\perp]}{p_\perp} \left(\frac{\partial}{\partial \alpha} \right)_{\mathbf{r}},$$

with $\hat{\mathbf{v}}_\perp$ the unit vector in the direction of $\mathbf{v}_\perp$. Changing to variables s and b, the derivative with respect to α becomes

$$\left(\frac{\partial}{\partial \alpha} \right)_{\mathbf{r}} = b\frac{\partial}{\partial s} - s\frac{\partial}{\partial b} + \frac{\partial}{\partial \alpha}$$

and the spatial gradient in the (s, b) frame is

$$\nabla = \hat{\mathbf{v}}_\perp \frac{\partial}{\partial s} + [\hat{\mathbf{z}} \times \hat{\mathbf{v}}_\perp] \frac{\partial}{\partial b}. \tag{15.16}$$

In the presence of a vortex, the order parameter has the form $\Delta_{\mathbf{p}}(\rho,\phi) = \Delta_{p_\perp}(\alpha,s,b)e^{i\alpha}$. Here $\Delta_{p_\perp}(\alpha,s,b)$ is the order parameter expressed in the coordinate frame (s,b), where the azimuthal angle is measured from the momentum direction. For a non-axisymmetric vortex and/or in a d-wave superconductor, $\Delta_{p_\perp}(\alpha,s,b)$ can have an explicit dependence on the angular coordinate α.

To work with various terms in the kinetic equation (15.14) we use transformations very similar to those we made in Section 14.2 to derive identities for the quasiclassical Green functions. Keeping in mind that f_1 is independent of s (see eqn (15.15)), we rewrite the terms in the second and third lines of eqn (15.14) in the form

$$\left(f_-\nabla\Delta_{\mathbf{p}}^* + f_-^\dagger\nabla\Delta_{\mathbf{p}}\right)\cdot\frac{\partial f_1}{\partial\mathbf{p}} = \frac{1}{p_\perp}\left(f_-\frac{\partial\Delta^*}{\partial b} + f_-^\dagger\frac{\partial\Delta}{\partial b}\right)\left(\frac{\partial f_1}{\partial\alpha} - s\frac{\partial f_1}{\partial b}\right) + \left(f_-\frac{\partial\Delta^*}{\partial s} + f_-^\dagger\frac{\partial\Delta}{\partial s}\right)\frac{\partial f_1}{\partial p_\perp}, \tag{15.17}$$

and

$$\left(f_-\frac{\partial\Delta_{\mathbf{p}}^*}{\partial\mathbf{p}} + f_-^\dagger\frac{\partial\Delta_{\mathbf{p}}}{\partial\mathbf{p}}\right)\cdot\nabla f_1 = \frac{1}{p_\perp}\left[\left(f_-\frac{\partial\Delta^*}{\partial\alpha} + f_-^\dagger\frac{\partial\Delta}{\partial\alpha}\right) - s\left(f_-\frac{\partial\Delta^*}{\partial b} + f_-^\dagger\frac{\partial\Delta}{\partial b}\right)\right]\frac{\partial f_1}{\partial b} + \frac{1}{p_\perp}\left[\left(f_-\frac{\partial\Delta^*}{\partial s} + f_-^\dagger\frac{\partial\Delta}{\partial s}\right)b\right]\frac{\partial f_1}{\partial b}. \tag{15.18}$$

Moreover, since $\mathbf{A}$ does not depend on the momentum direction, we can subtract the zero term

$$\frac{e}{c}v_{\perp i}\left(\frac{\partial A_i}{\partial\mathbf{p}}\right)_{\mathbf{r}}\cdot\nabla f_1 = 0$$

from the third line of eqn (15.14). Next, we integrate (15.14) along the quasiclassical trajectory, using $\partial f_1/\partial s = 0$ and one of the Eilenberger equations,

$$i\left(\mathbf{v}_F\cdot\nabla\right)g_- = \Delta f_-^\dagger - \Delta^* f_-. \tag{15.19}$$

After some algebra the second and third lines in the l.h.s. of eqn (15.14) take the form

$$\int\left[\frac{e}{c}\left[\mathbf{v}_F\times\mathbf{H}\right]g_- - \frac{1}{2}\left(f_-\hat{\nabla}\Delta_{\mathbf{p}}^* + f_-^\dagger\hat{\nabla}\Delta_{\mathbf{p}}\right)\right]\frac{\partial f_1}{\partial\mathbf{p}}ds + \int\left[-\frac{e}{c}v_{\perp i}\frac{\partial A_i}{\partial\mathbf{p}}g_-\nabla f_1 + \frac{1}{2}\left(f_-\frac{\partial\Delta_{\mathbf{p}}^*}{\partial\mathbf{p}} + f_-^\dagger\frac{\partial\Delta_{\mathbf{p}}}{\partial\mathbf{p}}\right)\right]\nabla f_1 ds$$
$$= -\frac{1}{p_\perp}\int\left[\frac{1}{2}\left(f_-\frac{\partial\Delta^*}{\partial b} + f_-^\dagger\frac{\partial\Delta}{\partial b}\right) - \frac{e}{c}v_\perp\frac{\partial A_s}{\partial b}g_-\right]\frac{\partial f_1}{\partial\alpha}ds + \frac{1}{p_\perp}\int\left[\frac{1}{2}\left(f_-\frac{\partial\Delta^*}{\partial\alpha} + f_-^\dagger\frac{\partial\Delta}{\partial\alpha}\right) - \frac{e}{c}v_\perp\frac{\partial A_s}{\partial\alpha}g_-\right]\frac{\partial f_1}{\partial b}ds. \tag{15.20}$$

The term $\partial A_s/\partial\alpha$ accounts for the explicit α-dependence picked up by the vector potential when expressed in the coordinate frame (s,b). The s-derivatives present in eqns (15.17) and (15.18) disappear in eqn (15.20), as can be seen from the identity (14.29). In particular, taking the $\mathbf{v}_\perp$ projection of eqn (14.29) we find,

$$\int \mathrm{Tr}\left[\frac{\partial \check{H}}{\partial s}\check{g}_-\right] ds = 0.$$

It is this relation which serves to eliminate the s-derivatives from eqn (15.20).

The first term in the r.h.s. of eqn (15.20) has the form

$$-\frac{1}{2p_\perp}\frac{\partial f_1}{\partial \alpha}\int \mathrm{Tr}\left[\check{g}_-\frac{\partial \check{H}}{\partial b}\right] ds$$

which can also be transformed with the help of eqn (14.29). The second term has the form

$$\frac{1}{2p_\perp}\frac{\partial f_1}{\partial b}\int \mathrm{Tr}\left[\check{g}_-\frac{\partial \check{H}}{\partial \alpha}\right] ds.$$

It can be transformed using the relation

$$\int ds\,\mathrm{Tr}\left[\frac{\partial \mathcal{H}}{\partial \alpha}\check{g}_-\right] = 2\pi v_\perp \sum_n \delta(\epsilon-\epsilon_n)\frac{\partial \epsilon_n}{\partial \alpha} \tag{15.21}$$

which can easily be derived in exactly the same way as eqn (14.29). As a result, the r.h.s. of eqn (15.20) takes the simple form

$$-\frac{1}{p_\perp}\sum_n \left[\frac{\partial \epsilon_n}{\partial b}\frac{\partial f_1}{\partial \alpha} - \frac{\partial \epsilon_n}{\partial \alpha}\frac{\partial f_1}{\partial b}\right]\pi v_\perp \delta\left(\epsilon-\epsilon_n\right). \tag{15.22}$$

Next, we concentrate on the first line of eqn (15.14). The driving term $\propto \partial f^{(0)}/\partial \epsilon$ is the source of the nonequilibrium state as produced by time variations of the order parameter together with the electric field. According to eqn (14.38), the first term of the kinetic equation (15.14) can be written in the form

$$-\frac{1}{2}\frac{\partial f^{(0)}}{\partial \epsilon}\int \mathrm{Tr}\left[(\mathbf{v}_L\cdot\nabla)\,\check{H}g_-\right] ds,$$

which then can be further transformed using eqn (14.29). The second term in the first line of eqn (15.14) vanishes after integration over ds. The next term can be transformed with help of the identity (14.25). We finally obtain

$$\begin{aligned}&([\hat{\mathbf{v}}_\perp \times \mathbf{v}_L]\cdot\hat{\mathbf{z}})\,\frac{\partial f^{(0)}}{\partial \epsilon}\frac{\partial \epsilon_n}{\partial b} + \frac{1}{p_\perp}\frac{\partial f_1}{\partial \alpha}\frac{\partial \epsilon_n}{\partial b}\\ &-\frac{1}{p_\perp}\frac{\partial \epsilon_n}{\partial \alpha}\frac{\partial f_1}{\partial b} - \frac{\partial f_1}{\partial t} + \frac{1}{\pi v_\perp}\int_{-\infty}^{\infty} J_n\,ds = 0.\end{aligned} \tag{15.23}$$

A similar equation (14.42) has been already derived in Section 14.4 for s-wave superconductors. Note that the term with $\partial f_1/\partial b$ is absent in eqn (14.42) because $\partial \epsilon_n/\partial \alpha = 0$ for an axisymmetric vortex in an s-wave superconductor.

Equation (15.23) is nothing but the Boltzmann equation (14.3) used in the beginning of this chapter. Indeed, consider various terms in eqn (15.23) and compare them with eqn (14.3). Since

$$\frac{\partial \epsilon_n}{\partial b} = -p_\perp \frac{\partial \epsilon_n}{\partial \mu}$$

we have from eqn (15.23)

$$([\mathbf{p}_\perp \times \mathbf{v}_L]\cdot\hat{\mathbf{z}})\frac{\partial f^{(0)}}{\partial \epsilon}\frac{\partial \epsilon_n}{\partial \mu} + \frac{\partial f_1}{\partial t} + \frac{\partial f_1}{\partial \alpha}\frac{\partial \epsilon_n}{\partial \mu} - \frac{\partial \epsilon_n}{\partial \alpha}\frac{\partial f_1}{\partial \mu} = \left(\frac{\partial f}{\partial t}\right)_{\text{coll}}, \tag{15.24}$$

with $\epsilon_n = \epsilon_n(\mu, \alpha; p_z)$ and $\mu = -bp_\perp$. We denoted the collision integral as

$$\left(\frac{\partial f}{\partial t}\right)_{\text{coll}} = \frac{1}{\pi v_\perp}\int_{-\infty}^{\infty} J_n \, ds. \tag{15.25}$$

The time-derivative term in eqn (14.3) is obtained from eqn (15.24) if we assume that the energy ϵ_n contains a time dependence through the impact parameter $b(t) = -\mu(t)p_\perp^{-1}$, where $\mu(t) = [(\mathbf{r} - \mathbf{v}_L t)\times \mathbf{p}]\cdot\hat{\mathbf{z}}$:

$$\frac{\partial f^{(0)}}{\partial t} = \frac{\partial f^{(0)}}{\partial \epsilon}\frac{\partial \epsilon_n}{\partial \mu}\frac{\partial \mu}{\partial t} = \frac{\partial f^{(0)}}{\partial \epsilon}\frac{\partial \epsilon_n}{\partial \mu}([\mathbf{p}_\perp \times \mathbf{v}_L]\cdot\hat{\mathbf{z}}), \tag{15.26}$$

and combine it with $\partial f_1/\partial t$ in the total distribution function $f = f^{(0)}(\epsilon_n) + f_1$. Equation (14.3) can be rewritten into the generic form (15.1) with the canonical variables $q = \alpha$ and $p = \mu$.

To summarize the results of this section, we have started with the exact microscopic description of the nonstationary processes in terms of the Green function technique. Using the quasiclassical approximation, we have been able to reduce the problem of finding the nonequilibrium state of the superconductor with a moving vortex to the problem of solving the canonical Boltzmann equation for the distribution of nonequilibrium excitations localized in the vortex core. The only information needed to find the distribution function is the energy spectrum of the equilibrium excitations in the vortex core. The knowledge of the energy spectrum is also sufficient to calculate the force acting on the moving vortex. The full problem of the vortex dynamics thus reduces to several much easier and more transparent steps which are: finding the equilibrium energy spectrum of the excitations in the vortex core, solving the Boltzmann equation, and, finally, calculating the momentum transfer from the localized excitations.

15.4 Vortex mass

The vortex mass in superfluids and superconductors has been a long-standing problem in vortex physics and remains to be an issue of controversies. There are different approaches to its definition. In early works on this subject, the vortex mass was determined through an increase in the free energy of a superconductor calculated as an expansion of retarded and advanced Green functions up to the second order terms in slow time-derivatives of the order parameter. The quasiparticle distribution was assumed to be essentially in equilibrium. First used by Suhl (1965) (see also Duan and Leggett 1992) this approach yields the mass of the order of *one quasiparticle mass* (electron, in the case of a superconductor) per atomic layer. Another approach consists in calculating the energy $E^2/8\pi$ of electric field which is proportional to the square of the vortex velocity. This gives rise to the so-called electromagnetic mass (Coffey and Hao 1991) which, in good metals, is of the same order of magnitude (see Sonin *et al.* 1998 for a review).

A serious disadvantage of the above definitions of the vortex mass is that they do not take into account the kinetics of excitations disturbed by a moving vortex. We shall see that the inertia of excitations contributes much more to the vortex mass than what the old calculations predict. The kinetic equation approach described here is able to incorporate this effect. To implement this method we find the force necessary to support an unsteady vortex motion. Identifying then the contribution to the force proportional to the vortex acceleration, one defines the vortex mass as a coefficient of proportionality. This method was first applied for vortices in superclean superconductors by Kopnin (1978) and then was used by other authors (see for example, Kopnin and Salomaa 1991, Šimánek 1995). The resulting mass is of the order of the *total mass of all electrons* within the area occupied by the vortex. We will refer to this mass as to the *dynamic mass*. The dynamic mass originates from the inertia of excitations and can also be calculated as the momentum carried by localized excitations (Volovik 1997).

In the present section we describe how one can apply our kinetic equation approach for calculating the dynamic vortex mass in a general case of a finite relaxation time of nonequilibrium excitations produced by the moving vortex. Following Kopnin and Vinokur (1998) we use the Boltzmann kinetic equation to derive the equation for the vortex dynamics which contains the inertia term together with all the forces acting on a moving vortex. We shall see that dynamic mass displays a nontrivial feature: it is a tensor whose components depend on the quasiparticle mean free time. In s-wave superconductors, this tensor is diagonal in the superclean limit. The diagonal mass decreases rapidly as a function of the mean free time, and the off-diagonal components dominate in the moderately clean regime. Our results agree with the previous work (Kopnin 1978, Kopnin and Salomaa 1991, Volovik 1997) in the limit $\tau \to \infty$.

15.4.1 *Equation of vortex dynamics*

To introduce the vortex momentum we consider a non-steady motion of a vortex with a small acceleration. We start with localized excitations. Multiplying eqn (15.24) by $\mathbf{p}_\perp/2$ and summing up over all the quantum numbers, we obtain

$$\mathbf{F}_{\mathrm{env}}^{(\mathrm{loc})} = \mathbf{F}_{\mathrm{coll}}^{(\mathrm{loc})} - \frac{\partial \mathbf{P}^{(\mathrm{loc})}}{\partial t} - \pi N \tanh\left(\frac{\Delta_\infty}{2T}\right)[\mathbf{v}_L \times \hat{\mathbf{z}}] \tag{15.27}$$

where the l.h.s. of eqn (15.27) is the force from the environment on a moving vortex eqn (14.4) derived in Section 14.5. The first term in the r.h.s. of eqn (15.27) is the force exerted on the vortex by the heat bath via excitations localized in the vortex core:

$$\mathbf{F}^{(\mathrm{loc})}_{\mathrm{coll}} = -\frac{1}{2}\sum_n \int \mathbf{p}_\perp \left(\frac{\partial f}{\partial t}\right)_{\mathrm{coll}} \frac{dp_z}{2\pi}\frac{d\alpha\, d\mu}{2\pi}. \tag{15.28}$$

The second term in the r.h.s. of (15.27) is the change in the vortex momentum

$$\mathbf{P}^{(\mathrm{loc})} = -\frac{1}{2}\sum_n \int \mathbf{p}_\perp f_1 \frac{dp_z}{2\pi}\frac{d\alpha\, d\mu}{2\pi}. \tag{15.29}$$

To find the contribution from the states with $|\epsilon| > \Delta_\infty$ we multiplying eqn (15.11) by $(g/\omega_c)\mathbf{p}_\perp/2$ and integrate it over $(d\alpha/2\pi)\,(dp_z/2\pi)\,d\epsilon$. We obtain

$$\mathbf{F}^{(\mathrm{del})}_{\mathrm{env}} = \mathbf{F}^{(\mathrm{del})}_{\mathrm{coll}} - \frac{\partial \mathbf{P}^{(\mathrm{del})}}{\partial t} - \pi N \left[1 - \tanh\left(\frac{\Delta_\infty}{2T}\right)\right][\mathbf{v}_L \times \hat{\mathbf{z}}]. \tag{15.30}$$

Here $\mathbf{F}^{(\mathrm{del})}_{\mathrm{env}}$ is determined by eqn (14.65),

$$\mathbf{F}^{(\mathrm{del})}_{\mathrm{coll}} = -\int_{\epsilon>\Delta_\infty} \frac{g}{\omega_c}\mathbf{p}_\perp \left(\frac{\partial f}{\partial t}\right)_{\mathrm{coll}} \frac{dp_z}{2\pi}\frac{d\alpha}{2\pi}\, d\epsilon = \int_{\epsilon>\Delta_\infty} \mathbf{p}_\perp \frac{f_1}{\omega_c \tau}\frac{dp_z}{2\pi}\frac{d\alpha}{2\pi}\, d\epsilon$$

is the force from the heat bath, while the corresponding contribution to the vortex momentum is

$$\mathbf{P}^{(\mathrm{del})} = -\int_{\epsilon>\Delta_\infty} \mathbf{p}_\perp \frac{g}{\omega_c} f_1 \frac{dp_z}{2\pi}\frac{d\alpha}{2\pi}\, d\epsilon. \tag{15.31}$$

Note that the factor $\omega_c^{-1} = mS_0/\pi$ is proportional to the particle mass and to the area occupied by each vortex. The total momentum is $\mathbf{P} = \mathbf{P}^{(\mathrm{loc})} + \mathbf{P}^{(\mathrm{del})}$.

After a little algebra, the total force from the heat bath $\mathbf{F}_{\mathrm{coll}} = \mathbf{F}^{(\mathrm{loc})}_{\mathrm{coll}} + \mathbf{F}^{(\mathrm{del})}_{\mathrm{coll}}$ can be written in the form

$$\mathbf{F}_{\mathrm{coll}} = \mathbf{F}_{\mathrm{sf}} + \mathbf{F}_{\parallel}$$

where the friction and spectral flow forces are determined by eqns (14.87) and (14.93), respectively. The total force from environment eqns (15.27) and (15.30) takes the form

$$\mathbf{F}_{\mathrm{env}} = \mathbf{F}_{\mathrm{sf}} + \mathbf{F}_{\parallel} - \frac{\partial \mathbf{P}}{\partial t} - \pi N[\mathbf{v}_L \times \hat{\mathbf{z}}]. \tag{15.32}$$

This equation agrees with eqn (14.91) for a steady motion of vortices.

The equation of vortex dynamics is obtained by variation of the superconducting free energy plus the external field energy with respect to the vortex displacement. The variation of the superfluid free energy gives the force from the environment, $\mathbf{F}_{\mathrm{env}}$, while the variation of the external field energy produces the external Lorentz force. In the absence of pinning the total energy is translationally invariant. Therefore, the requirement of zero variation of the free energy again gives the condition $\mathbf{F}_{\mathrm{L}} + \mathbf{F}_{\mathrm{env}} = 0$ in the form of the force balance. Using our expression for $\mathbf{F}_{\rm env}$, the force balance can now be written in the form similar to eqn (14.96)

$$\mathbf{F}_{\rm M}+\mathbf{F}_{\rm L}^{(\rm qp)}+\mathbf{F}_{\rm I}+\mathbf{F}_{\rm sf}+\mathbf{F}_{\|}=\frac{\partial \mathbf{P}}{\partial t} \tag{15.33}$$

where the r.h.s. contains the time-derivative of the vortex momentum due to a non-steady vortex motion. For a steady vortex motion, eqn (15.33) reduces to eqn (14.96). The physical meaning of the eqn (15.33) is simple. The l.h.s. of this equation accounts for all the forces acting on a moving straight vortex line. The r.h.s. of eqn (15.33) comes from the inertia of excitations and is identified as the change in the vortex momentum. This definition of momentum is similar to that used by Stone (1996) and by Volovik (1997).

15.4.2 *Vortex momentum*

Having defined the vortex momentum, we calculate the vortex mass. The distribution function is given by eqn (14.9). The vortex momentum becomes $P_i = M_{ik}u_k$; it has both longitudinal and transverse components with respect to the vortex velocity. For a vortex with the symmetry not less than the four-fold, the effective mass tensor per unit length is $M_{ik} = M_{\|}\delta_{ik} - M_{\perp}e_{ikj}\hat{z}_j$ where $M_{\|} = M_{\|\,e} + M_{\|\,h}$ and $M_{\perp} = M_{\perp\,e} - M_{\perp\,h}$. Each component contains contributions from localized and delocalized states such that $M_{\|,\perp} = M_{\|,\perp}^{(\rm loc)} + M_{\|,\perp}^{(\rm del)}$ where

$$M_{\|\,e,h}^{(\rm loc)}=\frac{1}{4}\sum_n\int\frac{df^{(0)}}{d\epsilon}\,p_\perp^2\gamma_{\rm H}\,\frac{dp_z}{2\pi}\,\frac{d\mu\,d\alpha}{2\pi} \tag{15.34}$$

and

$$M_{\|\,e,h}^{(\rm del)}=\frac{1}{2}\int_{\epsilon>\Delta_\infty}\frac{g}{\omega_c}\frac{df^{(0)}}{d\epsilon}\,p_\perp^2\gamma_{\rm H}'\,\frac{dp_z}{2\pi}\,\frac{d\alpha}{2\pi}\,d\epsilon. \tag{15.35}$$

The same expression holds for $M_{\perp\,e,h}$ where $\gamma_{\rm H}$ is replaced with $\gamma_{\rm O}$. The e,h subscripts indicate the corresponding momentum integrations over the electron and hole parts of the Fermi surface, respectively.

If the vortex acceleration is slow, one can use expressions (14.14), (14.54) for a steadily moving vortex to calculate the vortex inertia. Consider first the contribution of the states with $|\epsilon| > \Delta_\infty$. Since $\gamma_{\rm H,O}'$ do not depend on energy and momentum, eqn (15.35) gives

$$M_{\|,\perp}^{(\rm del)}=\frac{\pi N_n}{\omega_c}\gamma_{\rm H,O}'=mN_nS_{\rm O}\gamma_{\rm H,O}'$$

Here N_n is the density of normal particles (or holes). The contribution of the delocalized states decreases as $\omega_c\tau$ gets smaller. In the limit of vanishing $\omega_c\tau$ the vortex mass is determined by localized excitations. The localized excitations dominate also at low temperatures $T \ll \Delta_\infty$. One has in this case

$$M_{\|}=\pi N\left\langle\!\!\left\langle\frac{\gamma_{\rm H}}{\omega_0}\right\rangle\!\!\right\rangle,\ M_{\perp}=\pi N\left\langle\!\!\left\langle\frac{\gamma_{\rm O}}{\omega_0}\right\rangle\!\!\right\rangle.$$

For $\omega_c\tau \gg 1$ the mass tensor for delocalized states is diagonal $M_{ik} = M_\parallel \delta_{ik}$ where $M_\parallel^{(\mathrm{del})} = mN_n S_0$; it is equal to the mass of *all* normal particles in the area occupied by the vortex. The mass tensor for localized states becomes diagonal in the superclean limit where $T_c^2\tau/E_F \gg 1$ with $M_\parallel^{(\mathrm{loc})} \sim \pi N \left\langle\!\left\langle \omega_0^{-1} \right\rangle\!\right\rangle \sim \pi\xi_0^2 mN$. This is the mass of all electrons in the area occupied by the vortex core. The mass decreases with τ. In the moderately clean regime $T_c^2\tau/E_F \ll 1$ where $\omega_n\tau \ll 1$, the diagonal component vanishes as τ^2, and the mass tensor is dominated by the off-diagonal part.

We should emphasize an important point that, in contrast to a conventional physical body, the mass of a vortex is not a constant quantity for a given system: it may depend on the frequency ω of the external drive. Indeed, for a nonzero ω we find from eqn (15.24)

$$\gamma_{\mathrm{H}} = \frac{\omega_n^2\tau_n^2}{\omega_n^2\tau_n^2 + (1 - i\omega\tau_n)^2};\ \gamma_{\mathrm{O}} = \frac{\omega_n\tau_n(1 - i\omega\tau_n)}{\omega_n^2\tau_n^2 + (1 - i\omega\tau_n)^2}.$$

As a result, all the dynamic characteristics of vortices including the conductivity and the effective mass acquire a frequency dispersion. In the limit $\tau \to \infty$, poles in $\gamma_{\mathrm{O,H}}$ appear at a frequency equal to the energy spacing between the quasiparticle states in the vortex core which gives rise to resonances in absorption of an external electromagnetic field (Kopnin 1978, 1998 b).

15.5 Vortex dynamics in d-wave superconductors

In this section we consider the vortex dynamics in d-wave superconductors at low temperatures. We specify later which temperatures can be considered as low. In d-wave superconductors, the vortex dynamics is expected to be more intricate due to a peculiar structure of the vortex-core states. The presence of gap nodes introduces the most important difference in the structure of core states compared to an s-wave superconductor. As was shown in Section 6.4, instead of a well-defined quasiclassical Caroli–de Gennes–Matricon interlevel spacing ω_0, the true quantum states in d-wave superconductors have a much smaller separation Ω_0 between quantum levels, $E_0 = -\Omega_0\left(m + \frac{1}{2}\right)$, which depends on the magnetic field. As a result, there appears another parameter $\Omega_0\tau$ which influences the vortex dynamics. We shall see that a new regime can be reached in superclean superconductors with longitudinal and transverse components of the conductivity tensor independent of the quasiparticle mean free time (Kopnin and Volovik 1997, Kopnin 1998 b). It is realized when the relaxation rate is smaller than the average distance between the quasiclassical energy levels $\langle\omega_0\rangle \sim T_c^2/E_F$ but larger than the separation between the true quantum-mechanical states Ω_0.

15.5.1 *Distribution function*

Here we employ the Boltzmann kinetic equation derived earlier in this chapter. We also use the spectrum of excitations in the vortex core calculated in Section 6.4. In d-wave superconductors, the gap at large distances from the vortex has the form $\Delta_\infty\left|\sin\left(2\alpha_0\right)\right|$. A particle is classically localized within the vortex core if it has an energy $|\epsilon| < \Delta_\infty$ and moves within the range of angles $\alpha > \alpha_0$ counted from one of the nodes. The angle α_0 is defined such that $|\epsilon| = \Delta_\infty |\sin(2\alpha_0)|$. For classically localized quasiparticles, the derivative $\partial\epsilon_n/\partial\mu = -\omega_n(\alpha)$ is the quasiclassical interlevel spacing. We shall consider low temperatures such that the level $n = 0$ is only important. Such temperatures correspond to $T \ll T_c\sqrt{H/H_{c2}}$. Keep in mind that according to eqn (6.100)

$$\omega_0(\alpha) = 8\tilde{\omega}_0\alpha^2 + \omega_c/2 \tag{15.36}$$

where $\tilde{\omega}_0 = (\Delta_\infty^2/v_\perp p_\perp)L$, and L denotes the logarithmic function of α.

The kinetic equation for the distribution function of localized excitations in a moving vortex has the form of eqn (15.24). We use a τ-approximation eqn (14.8) for the collision integral. We look for a solution in the form $f_1 = f_1[\epsilon_0(\alpha,\mu),\alpha]$ which does not contain an explicit dependence on μ. Equation (15.24) gives

$$\frac{\partial f_1}{\partial\alpha} - ([\mathbf{u}\times\mathbf{p}_\perp]\hat{\mathbf{z}})\frac{df^{(0)}}{d\epsilon} = U(\alpha)f_1 \tag{15.37}$$

where

$$U(\alpha) = \frac{1}{\omega_0(\alpha)\tau}. \tag{15.38}$$

We take the nonequilibrium distribution function again in the form of eqn (14.9). The factors γ_O and γ_H depend now on α. We assume a Fermi surface with not less than the tetragonal symmetry in the plane perpendicular to the vortex axis. For simplicity, we consider only the particle-like Fermi surface. Taking into account that, for the tetragonal symmetry, the responses do not depend on the direction of the vortex motion with respect to the crystal lattice, one finds two coupled first-order differential equations for $\gamma_O(\alpha)$ and $\gamma_H(\alpha)$:

$$\frac{\partial\gamma_O}{\partial\alpha} - \gamma_H - U(\alpha)\gamma_O + 1 = 0,$$
$$\frac{\partial\gamma_H}{\partial\alpha} + \gamma_O - U(\alpha)\gamma_H = 0. \tag{15.39}$$

Equations (15.39) can easily be solved. We have $\gamma_H(\alpha) = \operatorname{Re} W$; $\gamma_O(\alpha) = \operatorname{Im} W$ where

$$W = e^{[i\alpha+F(\alpha)]}\left(C - i\int_0^\alpha e^{-[i\alpha'+F(\alpha')]}\,d\alpha'\right)$$

with

$$F(\alpha) = \int_0^\alpha U(\alpha')\,d\alpha'.$$

In the moderately clean limit, $\tilde{\omega}_0\tau \ll 1$, the potential $U(\alpha)$ is always large, and we obtain the local solution as in an s-wave superconductor

$$\gamma_O(\alpha) = \omega_0(\alpha)\tau,\ \gamma_H(\alpha) = [\omega_0(\alpha)\tau]^2. \tag{15.40}$$

Consider temperatures $T \ll T_c\sqrt{H/H_{c2}}$ which correspond to the energy range $\epsilon \ll \Delta_\infty\sqrt{H/H_{c2}}$. For these temperatures, excitations are mostly localized in the vortex core as we discussed in Section 6.4.2. Let us find the distribution function in the region of angles α not specifically close to the gap nodes. In the superclean limit $\tilde{\omega}_0\tau \ll 1$, the potential $U(\alpha)$ is small almost everywhere except for vicinities of the gap nodes where it becomes large. For angles larger than $\alpha_1 \sim (\tilde{\omega}_0\tau)^{-1/2}$ from the nodes one can neglect the potential U. As a result, one has for $\delta\alpha < \alpha < \pi/2 - \delta\alpha$ where $\alpha_1 \ll \delta\alpha \ll 1$,

$$\begin{aligned} \gamma_{\mathrm{O}}(\alpha) &= A\cos\alpha + B\sin\alpha, \\ \gamma_{\mathrm{H}}(\alpha) &= 1 - A\sin\alpha + B\cos\alpha. \end{aligned} \tag{15.41}$$

The constants A and B and the overall behavior of the distribution function are to the highest extent determined by what happens in a close vicinity of the gap nodes. It is this region which is responsible for the whole build-up of nonequilibrium distribution. The constants can be found by matching with the solution in the vicinity of a node, $\alpha \ll 1$, where $\gamma_{\mathrm{O,H}}(\alpha) \propto e^{F(\alpha)}$. This provides the boundary condition

$$\gamma_{\mathrm{O,H}}(+\delta\alpha) = e^{2\lambda_0}\gamma_{\mathrm{O,H}}(-\delta\alpha)$$

across the node. Here $2\lambda_0 = F(+\delta\alpha) - F(-\delta\alpha)$. For energies $\epsilon \ll \Delta_\infty\sqrt{H/H_{c2}}$, the region of angles $\alpha < \alpha_0$ with delocalized trajectories is not important. The range of angles $\alpha \sim \alpha_1$ sets the width of the transition region near a gap node where the distribution function jumps from its value at $\alpha = -\delta\alpha$ to its value at $\alpha = +\delta\alpha$. We obtain using eqn (6.100)

$$\lambda_0 = \tau^{-1}\int_0^\infty \frac{d\alpha}{\omega_0(\alpha)} = \frac{\pi}{4\Omega_0\tau}. \tag{15.42}$$

The factor λ_0 contains exactly the integral which determines the true bound states in eqn (6.99). It converges and is determined by angles $\alpha \sim \sqrt{H/H_{c2}}$. The interlevel spacing Ω_0 is as in eqn (6.102)

$$\Omega_0 = \sqrt{\tilde{\omega}_0\omega_c}.$$

The solution for $\gamma_{\mathrm{O,H}}$ should be periodically continued to the rest of angles with the period $\pi/2$ since the response function has the same tetragonal symmetry as the underlying system: $\gamma(\pi/2 - \delta\alpha) = \gamma(-\delta\alpha)$. The behavior of the factors $\gamma_{\mathrm{O,H}}$ as functions of the angle α is illustrated schematically in Fig. 15.1.

Together with the above boundary condition, this gives

$$A = \frac{e^{\lambda_0}\sinh\lambda_0}{2\sinh^2\lambda_0 + 1};\ B = \frac{e^{-\lambda_0}\sinh\lambda_0}{2\sinh^2\lambda_0 + 1}. \tag{15.43}$$

We have after averaging over the azimuthal angle α

$$\langle\gamma_{\mathrm{H}}\rangle_\alpha = 1 - \frac{4}{\pi}\frac{\tanh^2\lambda_0}{1 + \tanh^2\lambda_0},\ \langle\gamma_{\mathrm{O}}\rangle_\alpha = \frac{4}{\pi}\frac{\tanh\lambda_0}{1 + \tanh^2\lambda_0}. \tag{15.44}$$

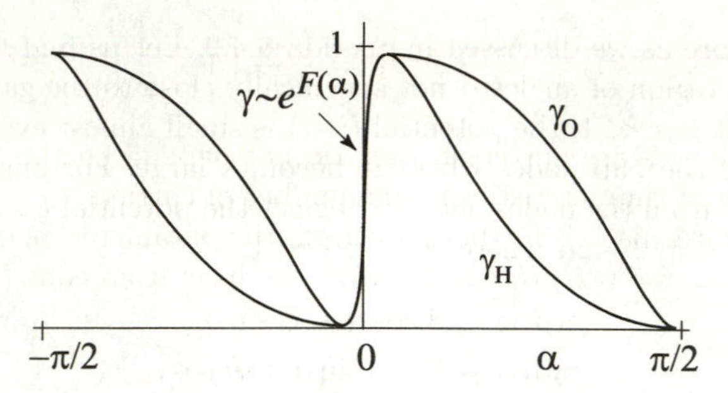

FIG. 15.1. The factors $\gamma_{\rm H,O}$ as functions of α. The limit $\lambda_0 \gg 1$ is shown when $\gamma_{\rm O} \approx \cos\alpha$ while $\gamma_{\rm H} \approx 1 - \sin\alpha$.

15.5.2 *Conductivity*

The force on a vortex from the environment where it moves is the momentum transferred from the excitations, eqn (14.4). Only the low energy levels with $n = 0$ are excited. The force takes the form

$$\begin{aligned}\mathbf{F}_{\rm env} &= \frac{1}{2}\int \frac{dp_z}{2\pi}\frac{d\mu\, d\alpha}{2\pi}[\hat{\mathbf{z}}\times\mathbf{p}_\perp]\frac{\partial\epsilon_0}{\partial\mu}f_1 \\ &= -[\hat{\mathbf{z}}\times\mathbf{v}_L]\int\frac{p_\perp^2\, dp_z}{8\pi}\left\langle\int\gamma_{\rm H}\frac{df^{(0)}}{d\epsilon}\frac{\partial\epsilon_0}{\partial\mu}\,d\mu\right\rangle_\alpha \\ &\quad +\mathbf{v}_L\int\frac{p_\perp^2\, dp_z}{8\pi}\left\langle\int\gamma_{\rm O}\frac{df^{(0)}}{d\epsilon}\frac{\partial\epsilon_0}{\partial\mu}\,d\mu\right\rangle_\alpha . \end{aligned} \tag{15.45}$$

Since $\gamma_{\rm O}$ and $\gamma_{\rm H}$ do not depend on μ, the integration over $d\mu$ in eqn (15.45) can be reduced to the integration over $d\epsilon$. This results in the components of the flux flow conductivity in the form

$$\sigma_{\rm O} = \frac{|e|c}{B}N\left\langle\!\left\langle\gamma_{\rm O}\right\rangle\!\right\rangle,\ \sigma_{\rm H} = \frac{ec}{B}N_e\left\langle\!\left\langle\gamma_{\rm H}\right\rangle\!\right\rangle . \tag{15.46}$$

where the average over the Fermi surface $\langle\!\langle\gamma_{\rm H,O}\rangle\!\rangle$ is defined by eqn (14.13).

In the moderately clean case, the factors $\gamma_{\rm O}$ and $\gamma_{\rm H}$ are given by eqn (15.40). The conductivities are

$$\begin{aligned}\sigma_{\rm O} &\sim \frac{Nec}{B}\frac{\Delta_\infty^2\tau}{E_F}\ln\left(\frac{T_c}{T}\right) \sim \sigma_n\frac{H_{c2}}{B}\ln\left(\frac{T_c}{T}\right); \\ \frac{\sigma_{\rm H}}{\sigma_{\rm O}} &\sim \frac{\Delta_\infty^2\tau}{E_F}\ln\left(\frac{T_c}{T}\right). \end{aligned} \tag{15.47}$$

This is similar to the results for an s-wave superconductor (see Section 14.6).

In the superclean limit $\tilde{\omega}_0\tau \gg 1$ the factors $\gamma_{\rm O}$ and $\gamma_{\rm H}$ for low temperatures $T \ll T_c$ are determined by eqn (15.44). The universal regime is reached when $\lambda_0 \gg 1$, i.e., when $T/T_c \ll \sqrt{H/H_{c2}} \ll 1/(\tilde{\omega}_0\tau)$. One has from eqn (15.44) $\gamma_{\rm H} = 1 - 2/\pi$, $\gamma_{\rm O} = 2/\pi$. The Ohmic and Hall conductivities become

$$\sigma_{\mathrm{O}} = \frac{N|e|c}{B}\frac{2}{\pi}; \sigma_{\mathrm{H}} = \frac{Nec}{B}\left(1 - \frac{2}{\pi}\right). \tag{15.48}$$

These conductivities are universal: they do not depend on the scattering time. The only material characteristics is the number of carriers.

If the mean free path is further increased, the parameter λ_0 decreases. In the limit $\lambda_0 \ll 1$, i.e., for $\sqrt{H/H_{c2}} \gg 1/(\tilde{\omega}_0\tau)$, we have from eqns (15.44, 15.46)

$$\sigma_{\mathrm{O}} \sim \frac{N|e|c}{B}\frac{1}{\Omega_0\tau}, \ \sigma_{\mathrm{H}} = \frac{Nec}{B}. \tag{15.49}$$

The longitudinal conductivity decreases and finally vanishes while the Hall conductivity reaches its nondissipative value.

The true interlevel distance Ω_0 introduces a new energy scale which does not exist for superconductors without gap nodes. For conventional superconductors, there is a single characteristic value, the minigap ω_0, which marks a crossover in the relaxation rate $1/\tau$ between two regimes: a dissipative viscous flux flow for $\omega_0\tau \ll 1$, and a dissipation-less vortex motion for $\omega_0\tau \gg 1$. For superconductors with gap nodes, there are two critical values: the average quasiclassical interlevel distance $\langle\omega_0\rangle \sim \tilde{\omega}_0$ and the true interlevel distance Ω_0 which is less than $\langle\omega_0\rangle$. Therefore one has three regimes: (i) moderately clean regime $\langle\omega_0\rangle\tau \ll 1$ with a high dissipation, (ii) intermediate universal regime $\Omega_0\tau \ll 1$ while $\langle\omega_0\rangle\tau \gg 1$ with a finite dissipation, where the quasiclassical levels are well resolved on average but the true minigap is smaller than the scattering rate; and (iii) the extreme superclean case $\Omega_0\tau \gg 1$ when the true levels are well separated and thus dissipation is small.

REFERENCES

Abrahams, E. and Tsuneto, T. (1966). *Phys. Rev.* **152**, 416.

Abrikosov, A. A. (1957). *Zh. Eksp. Teor. Fiz.* **32**, 1442 [*Sov. Phys. JETP* **5**, 1174 (1957)].

Abrikosov, A. A. (1998). *Fundamentals of The Theory of Metals.* North Holland, Amsterdam.

Abrikosov, A. A. and Gor'kov, L. P. (1958). *Zh. Eksp. Teor. Fiz.* **35**, 1158 [*Sov. Phys. JETP* **8**, 1090 (1959)].

Abrikosov, A. A. and Gor'kov, L. P. (1960). *Zh. Eksp. Teor. Fiz.* **39**, 1781 [*Sov. Phys. JETP* **12**, 1243 (1961)].

Abrikosov, A. A., Gor'kov, L. P., and Dzyaloshinskii, I. E. (1965). *Methods of Quantum Field Theory in Statistical Physics.* Pergamon Press.

Achucarro, A. and Vachaspati, T. (2000). *Phys. Rep.* **327**, 347.

Aronov, A. G., Galperin, Yu. M., Gurevich, V. L., and Kozub, V. I. (1981). *Adv. Phys.* **30**, 539.

Aronov, A. G., Galperin, Yu. M., Gurevich, V. L., and Kozub, V. I. (1986). In: *Nonequilibrium Superconductivity* (edited by D.N. Langenberg and A.I. Larkin). Elsevier Science Publishers, p. 325.

Aronov, A. G. and Rapoport, A. B. (1992). *Mod. Phys. Lett.* B **6**, 1093.

Artemenko, S. N. and Kruglov, A. N. (1990). *Phys. Lett.* A, **143**, 485.

Artemenko, S. N. and Volkov, A. F. (1975). *Phys. Lett.* A **55**, 113.

Artemenko, S. N. and Volkov, A. F. (1976). *Zh. Eksp. Teor. Fiz.* **70**, 1051 [*Sov. Phys. JETP* **43**, 548 (1976)].

Artemenko, S. N. and Volkov, A. F. (1979). *Usp. Fiz. Nauk* **128**, 3 [*Sov. Phys. Usp.* **22**, 295 (1980)].

Aslamazov, L. G. and Volkov, A. F. (1986). In: *Nonequilibrium Superconductivity* (edited by D.N. Langenberg and A.I. Larkin). Elsevier Science Publishers, p. 65.

Bardeen, J. and Stephen, M. J. (1965). *Phys. Rev.* **140** A, 1197.

Bardeen, J., Cooper, L. N., and Schrieffer, J. R. (1957). *Phys. Rev.* **108**, 1175.

Bardeen, J., Rickaizen, G., and Tewordt, L. (1959). *Phys. Rev.* **113**, 982.

Barone, A., Larkin, A. I., and Ovchinnikov, Yu. N. (1990). *Journal of Superconductivity* **3**, 155.

Betbeder-Matibet, O. and Nozières, P. (1969). *Ann. Phys.* **51**, 392.

Blatter, G., Feigel'man, M. V., Geshkenbein, V. B., Larkin, A. I., and Vinokur, V. M. (1994). *Rev. Mod. Phys.* **66**, 1125.

Blatter, G. Geshkenbein, V., and Kopnin, N. B. (1999). *Phys. Rev.* B **59**, 14 663.

Bogoliubov, N. N., Tolmachev, V. V., and Shirkov, D. V. (1958). *A New Method in the Theory of Superconductivity.* Acad. Sci. USSR [English translation: Consultants Bureau, Inc., New York (1959)].

Buchholtz, L. and Zwicknagl, G. (1981). *Phys. Rev.* B **23**, 5788.

Bulyzhenkov, I. E. and Ivlev, B. I. (1976 a). *Zh. Eksp. Teor. Fiz.* **70**, 1405 [*Sov. Phys. JETP* **43**, 731 (1976)].

Bulyzhenkov, I. E. and Ivlev, B. I. (1976 b). *Zh. Eksp. Teor. Fiz.* **71**, 1112 [*Sov. Phys. JETP* **44**, 613 (1976)].

Caroli, C., de Gennes, P. G., and Matricon, J. (1964). *Phys. Lett.* **9**, 307.

Clem, J. R. (1991). *Phys. Rev.* B **43**, 7837.

Coffey, M. and Hao, Z. (1991). *Phys. Rev.* B **44**, 5230.

de Gennes, P. G. (1966). *Superconductivity of Metals and Alloys.* W.A. Benjamin, Inc.

Dorsey, A.T. (1992). *Phys. Rev.* B **46**, 8376.

Duan, J.-M. and Leggett, A. J. (1992). *Phys. Rev. Lett.* **68**, 1216.

Eilenberger, G. (1968). *Z. Phys.* **214**, 195.

Eckern, U. and Schmid, A. (1981). *J. Low Temp. Phys.* **45**, 137.

Eliashberg, G. M. (1960). *Zh. Eksp. Teor. Fiz.* **38**, 966 [*Sov. Phys. JETP* **11**, 696 (1960)].

Eliashberg, G. E. (1968). *Zh. Eksp. Teor. Fiz.* **55**, 2443 [*Sov. Phys. JETP* **28**, 1298 (1969)].

Eliashberg, G. M. (1970). *Pis'ma Zh. Eksp. Teor. Fiz.* **11**, 186 [*JETP Lett.* **11**, 114 (1970)].

Eliashberg, G. M. (1971). *Zh. Eksp. Teor. Fiz.* **61**, 1254 [*Sov. Phys. JETP* **34**, 668 (1972)].

Eliashberg, G. M. and Ivlev, B. I. (1986). In: *Nonequilibrium Superconductivity* (edited by D.N. Langenberg and A.I. Larkin). Elsevier, p. 211.

Feigel'man, M. V., Geshkenbein, V. B., Larkin, A. I., and Vinokur, V. M. (1995). *Pis'ma Zh. Eksp. Teor. Fiz.* **62**, 811 [*JETP Lett.* **62**, 834 (1995)].

Galperin, Yu. M. and Sonin, E. B. (1976). *Fiz. Tverd. Tela* **18**, 3034 [*Sov. Phys. Solid State* **18**, 1768 (1976)].

Geilikman, B. T. and Kresin, V. Z. (1974). *Kinetic and Nonsteady-State Effects in Superconductors.* John Wiley and Sons, New York–Toronto.

Ginzburg, V. L. and Landau, L. D. (1950). *Zh. Eksp. Teor. Fiz.* **20**, 1064 ; in: Landau, L. D. *Collected Papers.* Pergamon Press, 1965.

Gor'kov, L. P. (1958). *Zh. Eksp. Teor. Fiz.* **34**, 735 [*Sov. Phys. JETP* **7**, 505 (1958)].

Gor'kov, L. P. (1959 a). *Zh. Eksp. Teor. Fiz.* **36**, 1918 [*Sov. Phys. JETP* **9**, 1364 (1959)].

Gor'kov, L. P. (1959 b). *Zh. Eksp. Teor. Fiz.* **37**, 833 [*Sov. Phys. JETP* **10**, 593 (1960)].

Gor'kov, L. P. (1959 c). *Zh. Eksp. Teor. Fiz.* **37**, 1407.

Gor'kov, L. P. and Eliashberg, G. M. (1968). *Zh. Eksp. Teor. Fiz.* **54**, 612 [*Sov. Phys. JETP* **27**, 328 (1968)].

Gor'kov, L. P. and Kalugin, P. (1985). *Pis'ma Zh. Eksp. Teor. Fiz.* **41**, 208 [*JETP Lett.* **41**, 253 (1985)].

Gor'kov, L. P. and Kopnin, N. B. (1973 a). *Zh. Eksp. Teor. Fiz.* **64**, 356 [*Sov. Phys. JETP* **37**, 183 (1973)].

Gor'kov, L. P. and Kopnin, N. B. (1973 b). *Zh. Eksp. Teor. Fiz.* **65**, 396 [*Sov. Phys. JETP* **38**, 195 (1974)].

Gor'kov, L. P. and Kopnin, N. B. (1975). *Usp. Fiz. Nauk* **116**, 413 [*Sov. Phys. Usp.* **18**, 496 (1976)].

Graf, D., Rainer, M. J., and Sauls, J. A. (1993). *Phys. Rev.* B **47**, 12 089.

Graf, M. J., Palumbo, M., Rainer, D., and Sauls, J. A. (1995). *Phys. Rev.* B **52**, 10 588.

Graf, M. J., Yip, S.-K., Sauls, J. A., and Rainer, D. (1996). *Phys. Rev.* B **53**, 15 147.

Gross, E. P. (1961). *Il Nouvo Cimento* **20**, 454.

Gross, E. P. (1963). *J. Math. Phys.* **4**, 195.

Gygi, F. and Schlüter, M. (1991). *Phys. Rev. B* **43**, 7609.

Hagen, S. J., Smith, A. W., Rajeswari, M., Peng, J. L., Li, Z. Y., Green, R. L., Mao, S. N., Xi, X. X., Bhattacharya, S., Li, Qi, and Lobb, C. J. (1993). *Phys. Rev.* B **47**, 1064.

Hansen, Brun E. (1968). *Phys. Lett.* **27** A, 576.

Harris, J. M., Yan, Y. F., Tsui, O. K. C., Matsuda, Y., and Ong, N. P. (1994). *Phys. Rev. Lett.* **73**, 1711.

Heeb, R., van Otterlo, A., Sigrist, M., and Blatter, G. (1996). *Phys. Rev.* B **54**, 9385.

Hohenberg, P. C. and Halperin, B. I. (1977). *Rev. Mod. Phys.* **49**, 435.

Iordanskii, S. V. (1964). *Ann. Phys.* **29**, 335.

Ivlev, B. I., Ovchinnikov, Yu. N., and Pokrovsky, V. L. (1990). *Europhys. Lett.* **13**, 187.

Ivlev, B. I. and Kopnin, N. B. (1989). *J. Low Temp. Phys.* **77**, 413.

Ivlev, B. I. and Kopnin, N. B. (1991). *Europhys. Lett.* **15**, 349.

Kambe, S., Huxley, A. D., Rodiere, P., and Flouquet, J. (1999). *Phys. Rev. Lett.* **83**, 1842.

Kats, E. I. (1969). *Zh. Eksp. Teor. Fiz.* **56**, 1675 [*Sov. Phys. JETP* **29**, 897 (1969)];

Kats, E. I. (1970). *Zh. Eksp. Teor. Fiz.* **58**, 1471 [*Sov. Phys. JETP* **31**, 787 (1970)].

Keldysh, L. V. (1964). *Zh. Eksp. Teor. Fiz.* **47**, 1515 [*Sov. Phys. JETP* **20**, 1018 (1965)].

Kes, P., Aarts, J., Vinokur, V. M., and van der Beek, C. J. (1990). *Phys. Rev. Lett.* **64**, 1063.

Kim, Y. B., Hempdtead, C. F., and Strnad, A. R. (1965). *Phys. Rev.* **139**, A 1163.

Kim, J.–T., Giapintzakis, J., and Ginsberg, D. M. (1996). *Phys. Rev.* B **53**, 5922.

Klemm, R. A., Luther, A., and Beasley, M. R. (1975). *Phys. Rev.* B **12**, 877.

Kopnin, N. B. (1978). *Pis'ma Zh. Eksp. Teor. Fiz.* **27**, 417 [*JETP Lett.* **27**, 390 (1978)].

Kopnin, N. B. (1987). *Zh. Eksp. Teor. Fiz.* **92**, 2105 [*Sov. Phys. JETP* **65**, 1187 (1987)].

Kopnin, N. B. (1994). *J. Low Temp. Phys.* **97**, 157.

Kopnin, N. B. (1996). *Phys. Rev.* B **54**, 9475.

Kopnin, N. B. (1998 a). *J. Low Temp. Phys.* **110**, 885.

Kopnin, N. B. (1998 b). *Phys. Rev.* **57**, 11 775.

Kopnin, N. B. and Kravtsov, V. E. (1976 a). *Pis'ma Zh. Eksp. Teor. Fiz.* **23**, 631 [*JETP Lett.* **23**, 578 (1976)].

Kopnin, N. B. and Kravtsov, V. E. (1976 b). *Zh. Eksp. Teor. Fiz.* **71**, 1644 [*Sov. Phys. JETP* **44**, 861 (1976)].

Kopnin, N. B. and Lopatin, A. V. (1995). *Phys. Rev. B* **51**, 15 291.

Kopnin, N. B. and Salomaa, M. M. (1991). *Phys. Rev. B* **44**, 9667.

Kopnin, N. B. and Vinokur, V. M. (1998). *Phys. Rev. Lett.* **81**, 3952.

Kopnin, N. B. and Volovik, G. E. (1997). *Phys. Rev. Lett.* **79**, 1377.

Kopnin, N. B., Ivlev, B. I., and Kalatsky, V. A. (1992). *Pis'ma Zh. Eksp. Teor. Fiz.* **55**, 717 [*JETP Lett.* **55**, 750 (1992)].

Kopnin, N. B., Ivlev, B. I., and Kalatsky, V. A. (1993). *J. Low Temp. Phys.* **90**, 1.

Kopnin, N. B., Volovik, G. E., and Parts, Ü. (1995). *Europhys. Lett.* **32**, 651.

Kramer, L. and Pesch, W. (1974). *Z. Phys.* **269**, 59.

Landau, L. D. and Lifshitz, E.M. (1959 a). *Quantum Mechanics* (3rd edn). Pergamon Press.

Landau, L. D. and Lifshitz, E. M. (1959 b). *Statistical Physics* (3rd edn). Pergamon Press.

Larkin, A. I. and Ovchinnikov, Yu. N. (1975). *Zh. Eksp. Teor. Fiz.* **68**, 1915 [*Sov. Phys. JETP* **41**, 960 (1976)].

Larkin, A. I. and Ovchinnikov, Yu. N. (1976). *Pis'ma Zh. Eksp. Teor. Fiz.* **23**, 210 [*JETP Lett.* **23** 187 (1976)].

Larkin, A. I. and Ovchinnikov, Yu. N. (1977). *Zh. Eksp. Teor. Fiz.* **73**, 299 [*Sov. Phys. JETP* **46**, 155 (1977)].

Larkin, A. I. and Ovchinnikov, Yu. N. (1986). In: *Nonequilibrium Superconductivity* (edited by D. N. Langenberg and A. I. Larkin) Elsevier Science Publishers, p. 493.

Larkin, A. I. and Ovchinnikov, Yu. N. (1995). *Phys. Rev.* B **51**, 5965.

Lawrence, W. E. and Doniach, S. (1971). In: *Proceedings of the Twelfth Iternational Conference on Low Temperature Physics* (edited by E. Kanda). Academic Press of Japan, Kyoto, p. 361.

Leggett, A. J. (1975). *Rev. Mod. Phys.* **47**, 331.

Lifshitz ,E. M. and Pitaevskii, L. P. (1957). *Zh. Eksp. Teor. Fiz.* **33**, 535 [*Sov. Phys. JETP* **6**, 418 (1957)].

Likharev, K. K. (1986). *Dynamics of Josephson Junctions and Circuits.* Gordon and Breach, New York.

Maki, K. (1969). *J. Low Temp. Phys.* **1**, 45.

Migdal, A. B. (1958). *Zh. Eksp. Teor. Fiz.* **34**, 1438.

Mineev, V. P. and Samokhin, K. V. (1999). *Introduction to Nonconventional Superconductivity.* Gordon and Breach, New York.

Nagaoka, T., Matsuda, Y., Obara, H., Sawa, A., Terashima, T., Chong, I., Takano, M., and Suzuki, M. (1998). *Phys. Rev. Lett.* **80**, 3594.

Niessen, A. K., Staas, F. A., and Weijsenfeld, C. H. (1967). *Phys. Lett.* **25 A**, 33.

Noto, K., Shinzawa, S., and Muto, Y. (1976). *Solid State Commun.* **18**, 1081.

Nozières, P. and Vinen, W. F. (1996). *Phil. Mag.* **14**, 667.

van Otterlo, A., Feigel'man, M., Geshkenbein, V., and Blatter, G. (1995). *Phys. Rev. Lett.* **75**, 3736.

Ovchinnikov, Yu. N. (1969). *Zh. Eksp. Teor. Fiz.* **56**, 1590 [*Sov. Phys. JETP* **29**, 853 (1969)].

Pitaevskii, L. P. (1961). *Sov. Phys. JETP* **13**, 451.

Rammer, J. and Smith, H. (1986). *Rev. Mod. Phys.* **58**, 323.

Saint-James, D., Sarma, G., and Thomas, E. J. (1969). *Type II Superconductivity.* Pergamon Press.

Schmid, A. (1966). *Phys. Kond. Materie* **5**, 302.

Schmid, A. and Schön, G. (1975). *J. Low Temp. Phys.* **20**, 207.

Serene, J. W. and Rainer, D. (1983). *Phys. Reports* **101**, 222.

Shelankov, A. L. (1985). *J. Low Temp. Phys.* **60**, 29.

Shellard, E. P. S. and Vilenkin, A. (1994). *Cosmic Strings and Other Topological Defects.* Cambridge University Press, Cambridge.

Šimánek, E. (1995). *J. Low Temp. Phys.* **100**, 1.

Sonin, E. B. (1987). *Rev. Mod. Phys.* **59**, 87.

Sonin, E. B., Geshkenbein, V. B., van Otterlo, A., and Blatter, G. (1998). *Phys. Rev.* B **57**, 575.

Stone, M. (1996). *Phys. Rev.* B **54**, 13 222.

Stone, M. and Gaitan, F. (1987). *Ann. Phys.* **178**, 89.

Suhl, H. *Phys. Rev. Lett.* **14**, 226 (1965).

Svidzinskii, A. V. (1982). *Spatially Inhomogeneous Problems in the Theory of Superconductivity.* Nauka, Moscow.

Tachiki, M. and Takahashi, S. (1989). *Solid State Commun.* **70**, 291.

Tinkham, M. (1964). *Phys. Rev. Lett.* **13**, 804.

Tinkham, M. (1972). *Phys. Rev.* B **6**, 1747.

Tinkham, M. (1996). *Introduction to Superconductivity.* (2nd edn) McGraw-Hill, New York.

Usadel, K. D. (1970). *Phys. Rev. Lett.* **25**, 507.

Volovik, G. E. (1986). *Pis'ma Zh. Eksp. Teor. Fiz.* **43**, 428 [*JETP Lett.* **43**, 551 (1986)].

Volovik, G. E. (1988). *J. Phys.* C **21**, L221.

Volovik, G. E. (1992). *Exotic Properties of Superfluid* ^3He. World Scientific.

Volovik, G. E. (1993). *Pis'ma Zh. Eksp. Teor. Fiz.* **58**, 457 [*JETP Lett.* **58**, 469 (1993)].

Volovik, G. E. (1997). *Pis'ma Zh. Eksp. Teor. Fiz.* **65**, 201 [*JETP Lett.* **65**, 217 (1997)].

Watts-Tobin, R. J., Krähenbühl, Y., and Kramer, L. (1981). *J. Low Temp. Phys.* **42**, 459.

Wölfle, P. and Vollhardt, D. (1990). *The Superfluid Phases of Helium 3.* Taylor and Francis, London, New York.

Zaitsev, A. V. (1984). *Zh. Eksp. Teor. Fiz.* **86**, 1742 [*Sov. Phys. JETP* **59**, 1015 (1984)].

INDEX